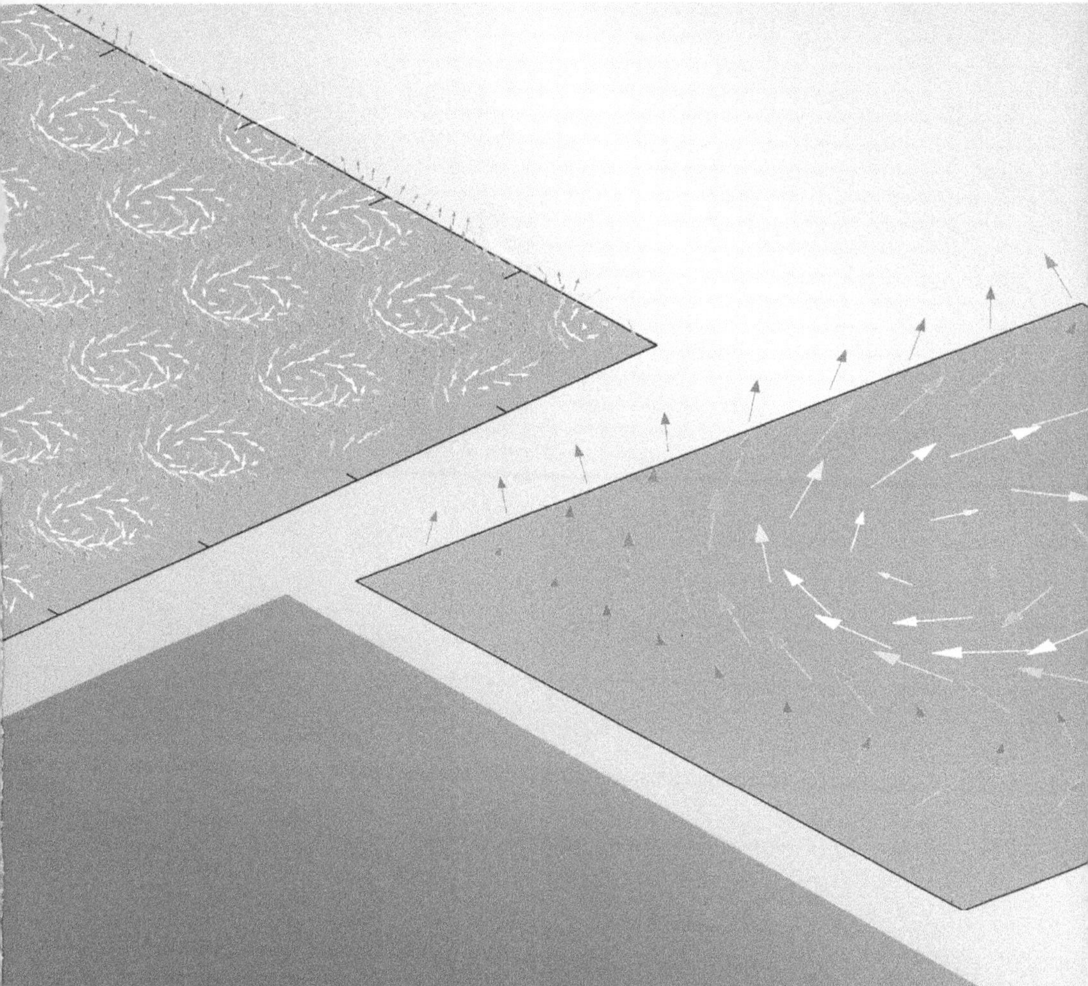

Physics of Magnetic Thin Films

Physics of Magnetic Thin Films

Theory and Simulation

Hung T. Diep

Jenny Stanford Publishing

Published by

Jenny Stanford Publishing Pte. Ltd.
Level 34, Centennial Tower
3 Temasek Avenue
Singapore 039190

Email: editorial@jennystanford.com
Web: www.jennystanford.com

British Library Cataloguing-in-Publication Data
A catalogue record for this book is available from the British Library.

Physics of Magnetic Thin Films: Theory and Simulation

ISBN 978-981-4877-42-8 (Hardcover)
ISBN 978-1-003-12110-7 (eBook)

Contents

Part II: Magnetism of Thin Films

Preface

This book is intended for graduate students and researchers who wish to understand theoretical mechanisms lying behind macroscopic properties of magnetic thin films. It provides a full description of basic theoretical methods and techniques of simulation to help readers in their research projects.

The idea of writing this book comes from the observation of what graduate students and research beginners need for understanding theoretically properties of magnetic thin films, and for mastering fundamental theoretical and numerical techniques in their research. Over the years, as a professor and thesis supervisor, I have seen that it is possible to master and to use various theoretical techniques by practical training on research subjects. The present book is written with this spirit in mind.

The first part of the book presents six chapters. Chapters 1 to 5 focus on the fundamental theory of bulk magnetic materials. Chapter 6 is devoted to the presentation of the Monte Carlo simulation methods. These chapters are part of lectures I have given during many years in a master program of physics.

The second part of the book contains 11 chapters (Chapters 7 to 17), all devoted to the main purpose of the book, namely "Physics of Magnetic Thin Films: Theory and Simulation." Each chapter is devoted to a subject, written as a research paper or a review with the presentation of the state-of-the-art literature on the subject and the motivation of the chapter. A detailed description of the techniques and the presentation of the results are then shown with discussion.

For the first five chapters on the basic theory of magnetism and Chapter 8, a number of exercises and problems are proposed at the end of each of these chapters for self-training. The detailed solutions are given in Part III.

Two appendixes, respectively, on fundamental elements of statistical physics and the second quantization method, are given at the end of the book to make it self-contained.

The works used for illustrations in Part II of this book are works from the author and his doctorate students over the years. I mention at the beginning of each chapter the main references of their contributions. They all know that I am very grateful to them for uncountable wonderful moments we have spent together, and for innumerable exciting and passionate discussions we have had.

Let me summarize the contents of the chapters.

Chapter 1 is devoted to the properties of the spin, object at the heart of magnetic materials. Several properties of systems of independent spins are shown. One can mention the paramagnetism and diamagnetism observed in some materials. Spin models such as Ising, XY, Heisenberg and Potts models are introduced. Different kinds of interactions between spins are described. In particular, the exchange interaction between Heisenberg spins is microscopically calculated. This model is widely used in the subsequent chapters of the book.

Chapter 2 shows the mean-field theory applied to a variety of bulk magnetic materials such as ferromagnets, antiferromagnets and ferrimagnets. Basic notions such as the transition temperature, the heat capacity, the magnetization and the susceptibility are calculated. The mean-field theory paves the way for more sophisticated theories presented afterward.

Chapter 3 is devoted to the spin-wave theory which shows how to calculate the spin-wave spectrum and the principal thermodynamic properties of magnetic materials including ferromagnets, antiferromagnets, ferrimagnets and helimagnets. Spin waves, or magnons when quantized, are collective excitations of a system of interacting spins. Spin waves dominate thermodynamic properties of magnetic materials at low temperatures.

Chapter 4 presents the Green's function method used in magnetism. Unlike the general Green's function theory in quantum many-body problems, its version used in magnetism is simple to manipulate. It gives correct results in a wide range of temperature. It can be used to treat any kind of spin ordering, from collinear ones in ferromagnets, antiferromagnets and ferrimagnets to non-collinear orderings in helimagnets and frustrated spin systems.

Chapter 5 is devoted to the theory of phase transitions and critical phenomena in magnetic materials. Basic notions and techniques such as the renormalization group, exact methods, transfer matrix, . . ., are introduced here with examples. Phase transitions in thin films are in all subjects presented in Part II. A familiarity with these techniques and their language is necessary to understand the remaining chapters.

Chapter 6 shows in details the Metropolis principle of Monte Carlo simulation and its implementation. Advanced techniques such as energy-histogram and multiple energy-histogram techniques, cluster-flipping methods and Wang–Landau flat energy-histogram algorithm are presented. This chapter is useful and necessary for understanding the works presented in Part II of the book.

Part II contains a number of my personal works carried out until this year on the subjects of the book. This choice of these materials was guided by the desire to show by an author the methods and techniques used to treat the subject of each chapter.

Chapter 7 is devoted to an introduction of the frustration and the presentation of a number of exactly solved models in two dimensions. These systems contain already most of the striking features of the frustration such as the high degeneracy of the ground state (GS), many phases in the GS phase diagram, the reentrance occurring near the boundaries of these phases, the disorder lines and the partial disorder. These phenomena are found in other non-solvable systems studied in the following chapters.

In Chapter 8, the spin-wave theory is introduced to calculate the spectrum in ferromagnetic and antiferromagnetic thin films where the loss of translational invariance is caused by the existence of a surface or an interface. Many examples for self-training are presented.

In Chapters 9 and 10, the effects of frustration in an antiferromagnetic of face-centered cubic lattice and the effects of a frustrated surface on the thermodynamic properties at finite temperatures are shown using the Heisenberg spin model and Monte Carlo simulations and the Green's function method.

Chapters 11 and 12 are devoted to the study of the spin waves in helimagnetic films in zero field and under an applied field. The surface spin configuration is calculated and the spin-wave spectrum is obtained analytically. The phase transition is studied using Monte Carlo simulation. Partial phase transitions are found in these films.

Chapter 13 shows the method to calculate the spin-wave spectrum and its effects on thermodynamic properties of a thin film with a Dzyaloshinskii–Moriya (DM) interaction. It is shown that the DM interaction affects strongly the acoustic spin waves at long wavelength.

Chapter 14 presents a study of a crystal of skyrmions generated in two dimensions using a Heisenberg Hamiltonian including the ferromagnetic interaction J, the Dzyaloshinskii–Moriya interaction D, and an applied magnetic field H. It is found that the relaxation of the skyrmion crystal is very slow and follows a stretched exponential law. The skyrmion crystal phase is shown to undergo a transition to the paramagnetic state at a finite temperature. Effects of the lattice elasticity on the skyrmion crystal is also presented.

Chapter 15 is devoted to a study of properties of a magneto-ferroelectric superlattice with a Dzyaloshinskii–Moriya interaction at the interface. It is shown that skyrmions are excited in the magnetic layers under an applied magnetic field, and they are stable up to a transition temperature. Effects of a frustration in the magnetic layer are shown to enhance the skyrmion creation.

Chapter 16 is devoted to a study of the critical exponents of magnetic thin films as a function of the film thickness. The film is studied using the ferromagnetic Ising model and the high-resolution multiple-histogram Monte Carlo technique. There is a systematic continuous deviation of the critical exponents from their 2D values. An explanation is given. For a strongly frustrated face-centered antiferromagnetic film, it is shown, using the Wang–Landau high-performance flat energy histogram, that the phase transition changes from the first order to the second order when the thickness decreases.

Chapter 17 shows how to calculate the spin resistivity in magnetically ordered materials by Monte Carlo simulations in various types of crystal: ferromagnetic, antiferromagnetic and frustrated spin systems. In simulations, various interactions are taken into account, in particular interaction between itinerant spins, interaction between lattice spins, and interaction between lattice spins and itinerant spins. A chemical potential as well as an electric field are included. To show the efficiency of the simulation method, the Monte Carlo spin resistivity as a function of temperature is compared with recent experimental data performed on semiconducting MnTe.

I would like to thank my colleagues and friends at the University of Cergy-Pontoise for their friendship and collaboration over the years. I wish to thank again my numerous former doctorate students for uncountable happy moments we have shared together in our search for the understanding of the world of physics and beyond.

Dr. Hung T. Diep
Professor of Physics
CY Cergy Paris University
Formerly University of Cergy-Pontoise, France
October 2020

Part I

Basic Theory of Magnetism

Chapter 1

Spin: Origin of Magnetism

1.1 Introduction

Magnetic properties of a material are due to spins of atoms, molecules or nucleus which compose the material. The spins can be independent of each other or they interact with each other leading to various collective phenomena. There are many spin models and various kinds of interaction. This chapter treats some properties of systems of independent spins and explains the origin of the exchange interaction in materials. Exchange interactions, or magnetic interactions, between spins of neighboring atoms can give rise to a magnetic ordering which is responsible for the principal low-temperature properties of the system. We will consider various spin models and several kinds of magnetic interaction and their consequences in the following chapters.

We consider an electron of charge $-e$ ($e > 0$), of mass m and of spin $\mathbf{S}$. The magnetic moment associated with the spin is written as

$$\mu_s = -g\mu_B\mathbf{S} \tag{1.1}$$

where $g = 2.0023$ is the Landé factor and μ_B the Bohr magneton defined by

$$\mu_B = \frac{e\hbar}{2m} \tag{1.2}$$

Physics of Magnetic Thin Films: Theory and Simulation
Hung T. Diep

ISBN 978-981-4877-42-8 (Hardcover), 978-1-003-12110-7 (eBook)
www.jennystanford.com

The negative sign of μ_s in (1.1) is due to the negative charge of the electron $(-e)$.

We define $\sigma = 2\mathbf{S}$. The components of σ are the well-known Pauli matrices

$$\sigma_x = \begin{pmatrix} 0 & 1 \\ 1 & 0 \end{pmatrix} \tag{1.3}$$

$$\sigma_y = \begin{pmatrix} 0 & -i \\ i & 0 \end{pmatrix} \tag{1.4}$$

$$\sigma_z = \begin{pmatrix} 1 & 0 \\ 0 & -1 \end{pmatrix} \tag{1.5}$$

We see that σ_z is diagonal. The eigenvalues of S_z are thus $-1/2$ and $1/2$. Using the above matrices, one can show that they obey the following commutation relations

$$\left[\sigma^+, \sigma^-\right] = 2\sigma_z \tag{1.6}$$

$$\left[\sigma_z, \sigma^\pm\right] = \pm 2\sigma^\pm \tag{1.7}$$

$$\left[\sigma_x, \sigma_y\right] = i\sigma_z + \text{relations by circular permutations of } x, y, z \tag{1.8}$$

where $\sigma^\pm = \sigma_x \pm i\sigma_y$.

In addition to the magnetic moment defined from its spin in (1.1), the electron has also an orbital moment due to its motion

$$\mu_l = -\frac{e}{2}(\mathbf{r} \wedge \mathbf{v}) = -\frac{e}{2m}(\mathbf{r} \wedge \mathbf{p}) = -\mu_B \mathbf{l} \tag{1.9}$$

where $\mathbf{r}$ and $\mathbf{v}$ are the position and the velocity of the electron. The kinetic orbital moment $\mathbf{l}$ is given by

$$\mathbf{l} = \frac{1}{\hbar}\mathbf{r} \wedge \mathbf{p} \tag{1.10}$$

This moment yields, as seen below, the so-called diamagnetic phenomenon observed under the application of a magnetic field.

The nuclear magnetic moment μ_I associated to the nuclear spin I is defined by a relation similar to (1.1) with μ_B replaced by $\mu_N = \frac{e\hbar}{2M}$ where M is the mass of the nucleus (proton or neutron) which is equal to $1836m$. The heavy mass of a nucleus makes the effect of μ_I very small with respect to that of μ_S. Therefore, when these two kinds of moment exist in the system, the effect of the nuclear moment is often neglected.

We consider now a system of independent spins with random orientations $\pm 1/2$, such as a free-electron gas. Since the spins are independent, their states are affected only by a variation of an external parameter such as the temperature and/or an applied magnetic field. In the absence of an applied magnetic field, the random distribution of spin magnetic moments results in a zero total moment. Under an applied magnetic field **B**, a number of spins will turn themselves into the field direction giving rise to a nonzero total moment **M**. The susceptibility χ defined as $\chi = \frac{dM}{dB}$ is then positive. The system is paramagnetic.

We consider the case of a system of atoms where each atom has initially zero magnetic moment. Under an applied field, the electrons of each atom may modify their states so as to create an induced moment to resist the field effect: The induced moment **M** is in the direction opposite to that of the field. This gives rise to a negative susceptibility. The system is diamagnetic.

We study the paramagnetism and the diamagnetism in the following.

1.2 Paramagnetism of a Free-Electron Gas

Under an applied magnetic field **H**, the energy of an electron of spin parallel to **H** decreases by an amount $\mu_B H$, and that of an electron of spin antiparallel to **H** increases by the same amount. This effect is called "Zeeman effect." We write

$$E_\uparrow = E - \mu_B H \tag{1.11}$$

$$E_\downarrow = E + \mu_B H \tag{1.12}$$

For a system of free electrons, the total magnetic moment induced by the field is

$$M = \mu_B (N_\uparrow - N_\downarrow) \tag{1.13}$$

where $N_\uparrow$ and $N_\downarrow$ are, respectively, the numbers of $\uparrow$ and $\downarrow$ spins.

Using the density of states for each type of spin given by Eq. (A.44) we have

$$M = \mu_B \int_0^\infty [\rho(E_\uparrow) f(E_\uparrow) - \rho(E_\downarrow) f(E_\downarrow)] dE \tag{1.14}$$

where $f(E_{\uparrow,\downarrow})$ is the Fermi–Dirac distribution function given by Eq. (A.37). In the case where H is small, we can use the approximation $\rho(E_\uparrow) \simeq \rho(E_\downarrow) \simeq \rho(E)$ because $\rho(E)$ is a smooth function of E and $\mu_B H$ is small. We can also replace $f(E_\uparrow)$ and $f(E_\downarrow)$ by their first-order Taylor expansion around E:

$$f(E_\uparrow) \simeq f(E) + \frac{\partial f}{\partial E}(E_\uparrow - E) \tag{1.15}$$

$$f(E_\downarrow) \simeq f(E) + \frac{\partial f}{\partial E}(E_\downarrow - E) \tag{1.16}$$

Replacing these into Eq. (1.14) we have

$$M = \mu_B^2 H \int_0^\infty \rho(E) \left(-\frac{\partial f}{\partial E}\right) dE \tag{1.17}$$

At low temperatures, $\frac{\partial f}{\partial E}$ is not zero only near E_F (see Problem 3 in Section 1.7). We have therefore

$$M \simeq 2\mu_B^2 H \rho(E_F) \int_0^\infty \left(-\frac{\partial f}{\partial E}\right) dE = 2\mu_B^2 H \rho(E_F)[f(0) - f(\infty)]$$

$$= 2\mu_B^2 H \rho(E_F) \tag{1.18}$$

The susceptibility is thus

$$\chi = \frac{dM}{dH} = 2\mu_B^2 \rho(E_F) = \mu_B^2 \rho_t(E_F) \tag{1.19}$$

where $\rho_t(E_F) = 2\rho(E_F)$ is the "total" density of states with the spin degeneracy [Eq. (A.45)].

Equation (1.19) is known as "Pauli paramagnetism" susceptibility which is independent of T. To calculate M at higher orders of T, we can use a low-T expansion shown in Problem 5 in Section 1.7. At high temperatures, χ is proportional to $1/T$ (Curie's law) as seen in that exercise. Experimental data for normal metals confirm Eq. (1.19), but strong variations of χ with temperature have been observed in some transition metals (Pd, Ti, ...) [186]. To explain these variations, it is necessary to take into account various interactions neglected in the free-electron gas model.

1.3 Paramagnetism of a System of Free Atoms

We consider a system of N independent atoms each of which has a total moment $\mathbf{J} = \mathbf{S} + \mathbf{L}$. The maximum modulus J of $\mathbf{J}$ is the sum of the amplitudes S and L: $J = S + L$. The magnetic moment of the i-th atom is

$$\mu_i = -g\mu_B \mathbf{J}_i \tag{1.20}$$

where the Landé factor is given by

$$g = \frac{3}{2} + \frac{S(S+1) - L(L+1)}{2J(J+1)} \tag{1.21}$$

Under an applied magnetic field **B** along the z direction, the Zeeman energy is

$$\mathcal{H} = -\sum_{i=1}^{N} \mu_i \cdot \mathbf{B} = \sum_{J_i^z} g\mu_B J_i^z B = \sum_{i=1}^{N} E_i \tag{1.22}$$

where $J_i^z = J, J-1, \cdots, -J$ ($2J+1$ values) and E_i is the energy of the i-th atom. The partition function of N atoms is written as the product of single-atom partition function z [see Eq. (A.9)]

$$Z = z^N \tag{1.23}$$

where

$$z = \sum_{J_i^z=-J}^{J} \mathrm{e}^{-\beta E_i} = \sum_{J_i^z=-J}^{J} \mathrm{e}^{-\beta g\mu_B B J_i^z} = \frac{\sinh[\beta g\mu_B B(J+1/2)]}{\sinh(\beta g\mu_B B/2)} \tag{1.24}$$

The last equality was obtained by using the formula for the geometric series of $2J+1$ terms, of ratio $\mathrm{e}^{-\beta g\mu_B B}$. For $J = 1/2$ one has

$$z = 2\cosh(\beta g\mu_B B/2) \tag{1.25}$$

The free energy F is (see Appendix A)

$$F = -k_B T \ln Z = -Nk_B T \ln z \tag{1.26}$$

The energy of the microscopic state l is

$$E_l = \sum_i E_i = -B\sum_i \mu_i^l = -BM_l \tag{1.27}$$

where μ_i^l is the magnetic moment of atom i and M_l the total magnetic moment, in the state l. The average magnetic moment is calculated by

$$\begin{aligned}\overline{M} &= \frac{1}{Z}\sum_l M_l \mathrm{e}^{-\beta E_l} = \frac{1}{Z}\sum_l M_l \mathrm{e}^{\beta B M_l} \\ &= \frac{1}{\beta}\frac{1}{Z}\frac{\partial}{\partial B}\sum_l \mathrm{e}^{\beta B M_l} = \frac{1}{\beta}\frac{1}{Z}\frac{\partial Z}{\partial B} = \frac{1}{\beta}\frac{\partial \ln Z}{\partial B} \\ &= -\frac{\partial F}{\partial B}\end{aligned} \tag{1.28}$$

Replacing Eq. (1.25) in Eq. (1.26) to obtain F, then using F in the last equality, one obtains for $J = 1/2$

$$\overline{M} = \frac{Ng\mu_B}{2}\tanh\left(\frac{g\mu_B B}{2k_B T}\right) \tag{1.29}$$

The magnetization $\overline{m}$, defined as magnetic moment per unit volume, is given by

$$\overline{m} = \frac{Ng\mu_B}{2V}\tanh\left(\frac{g\mu_B B}{2k_B T}\right) \tag{1.30}$$

At high T, $\tanh\left(\frac{g\mu_B B}{2k_B T}\right) \to \frac{g\mu_B B}{2k_B T}$, one has

$$\overline{m} = \frac{N}{V}\left(\frac{g\mu_B}{2}\right)^2 \frac{B}{k_B T} \tag{1.31}$$

The susceptibility is thus

$$\chi = \frac{\partial \overline{m}}{\partial B} = \frac{N}{V}\left(\frac{g\mu_B}{2}\right)^2 \frac{1}{k_B T} \tag{1.32}$$

This is the Curie's law, similar to that of a gas of free electrons at high T (see Problem 5 in Section 1.7).

At low T, one has $\tanh\left(\frac{g\mu_B B}{2k_B T}\right) \to 1$, so that the magnetization is maximum, i.e., $\overline{m} = \frac{Ng\mu_B}{2V}$ in the case $J = 1/2$. This result is different from that of Pauli paramagnetism of free electrons at low T given by Eq. (1.19).

The average energy of the system is (see Appendix A)

$$\begin{aligned}\overline{E} &= -\frac{\partial \ln Z}{\partial \beta} \\ &= -\frac{Ng\mu_B B}{2}\tanh\left(\frac{g\mu_B B}{2k_B T}\right) \qquad \text{for } J = 1/2 \\ &= -Ng\mu_B J\, B B_J\left(\frac{g\mu_B J\, B}{k_B T}\right) \qquad \text{if } J \neq 1/2 \end{aligned} \tag{1.33}$$

The paramagnetic heat capacity for $J = 1/2$ is thus

$$C_V = Nk_B\left(\frac{g\mu_B B}{2k_B T}\right)^2 \frac{1}{\cosh^2\left(\frac{g\mu_B B}{2k_B T}\right)} \tag{1.34}$$

Figure 1.1 shows C_V per atom versus $k_B T$ for $g\mu_B B = 1.5$. The peak separates two regions. In the low-T region the number of moments parallel to **B** is larger than that antiparallel to **B**, so that

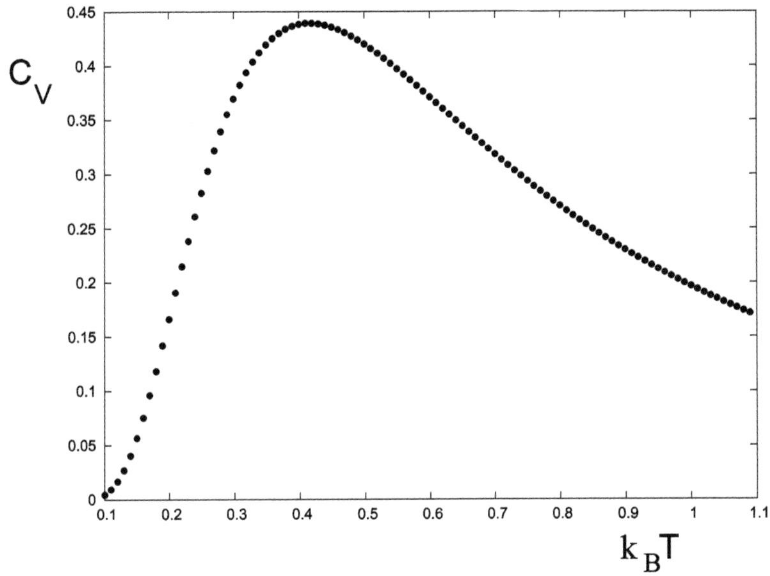

Figure 1.1 Specific heat C_V as a function of $k_B T$ for $g\mu_B B = 1.5$.

there is a non-zero magnetic moment. In the high-T region, they are equal, there is thus no induced magnetic moment. For a higher B, the low-T region is extended so that the peak of C_V moves to a higher temperature. Beyond the peak position, the magnetic moment induced by the magnetic field at low T is destroyed by the temperature.

1.4 Diamagnetism of Many-Electron Atoms

We consider the case where the valence orbital of an atom is completely occupied, i.e., the atom has no permanent magnetic moment from its electrons. As seen below, the applied magnetic field will result in a diamagnetic effect. However, when the valence orbital is only partially occupied, the applied magnetic field will give rise to both paramagnetism and diamagnetism. The paramagnetism is studied in the previous section. In the following, we study the diamagnetism.

We consider here an atom which has N_e electrons in its valence orbital. We suppose that the valence orbital is full: the orbital moment is $L = 0$ and the total spin is $S = 0$. The Hamiltonian of the valence electrons under the applied magnetic field $\mathbf{B}$ is written as

$$\mathcal{H} = \sum_{i=1}^{N_e} \left\{ \frac{1}{2m} [\mathbf{p}_i + e\mathbf{A}(\mathbf{r}_i)]^2 + 2\mu_B \mathbf{B} \cdot \mathbf{S}_i + U(\mathbf{r}_i) \right\} \quad (1.35)$$

where $\mathbf{A}(\mathbf{r}_i)$ is the vector potential associated with $\mathbf{B}$, namely $\text{rot}\mathbf{A}(\mathbf{r}_i) = \mathbf{B}$, and $U(\mathbf{r}_i)$ represents the interaction between electron i with the remaining electrons of the orbital. For $\mathbf{B} \parallel \mathbf{Oz}$, one can choose $\mathbf{A}(\mathbf{r}_i)$ as follows: $A_x = -yB/2$, $A_y = xB/2$, $A_z = 0$. From Eq. (1.35), one obtains

$$\mathcal{H} = \mathcal{H}_0 + \mu_B(L_z + 2S_z)B + \frac{e^2B^2}{8m} \sum_i (x_i^2 + y_i^2) \quad (1.36)$$

where $\mathcal{H}_0$ is the Hamiltonian in zero field, namely

$$\mathcal{H}_0 = \sum_i \left[\frac{p_i^2}{2m} + U(\mathbf{r}_i) \right] \quad (1.37)$$

L_z and S_z are the z components of the total orbital moment $\mathbf{L}$ and the total spin $\mathbf{S}$ defined by

$$\mathbf{L} = \frac{1}{\hbar} \sum_i \mathbf{r}_i \wedge \mathbf{p}_i \quad (1.38)$$

$$\mathbf{S} = \sum_i \mathbf{S}_i \quad (1.39)$$

The ground-state energy is

$$E_0(B) = E_0(B = 0) + \frac{e^2B^2}{8m} < 0| \sum_i (x_i^2 + y_i^2) |0 > \quad (1.40)$$

where $E_0(B = 0)$ is the energy in zero field. For a system of N free identical atoms, the total energy is equal to $NE_0(B)$. At $T = 0$, one has $F = E - TS = E$. The magnetization is

$$\overline{m} = \frac{\overline{M}}{V} = -\frac{1}{V}\frac{\partial F}{\partial B} = -\frac{Ne^2B}{4mV} < 0| \sum_i (x_i^2 + y_i^2) |0 > \quad (1.41)$$

The susceptibility is thus negative (diamagnetic) and given by

$$\chi = -\frac{Ne^2}{4mV} < 0| \sum_i (x_i^2 + y_i^2) |0 > \quad (1.42)$$

1.5 Magnetic Interactions in Solids

1.5.1 Exchange Interaction: Origin of Magnetism

In this paragraph, we show that the magnetic interaction between neighboring atoms leads to the Heisenberg spin model. We suppose that the reader has some knowledge of the Hartree–Fock approximation and is familiar with the second quantization method. If not, he/she can skip the following demonstration and go directly to Eq. (1.57).

We consider the Coulomb interaction between two electrons written in the second quantization (see Appendix B)

$$\hat{\mathcal{H}} = -\frac{1}{2}\sum_{\sigma;\sigma'}\iint \hat{\psi}^{+}_{\sigma}(\mathbf{r}_1)\hat{\psi}^{+}_{\sigma'}(\mathbf{r}_2)\frac{e^2}{|\mathbf{r}_1-\mathbf{r}_2|}\hat{\psi}_{\sigma}(\mathbf{r}_1)\hat{\psi}_{\sigma'}(\mathbf{r}_2)d\mathbf{r}_1 d\mathbf{r}_2 \tag{1.43}$$

where $\hat{\psi}_{\sigma}$ and $\hat{\psi}^{+}_{\sigma}$ are field operators defined by

$$\hat{\psi}_{\sigma}(\mathbf{r}) = \sum_{m,n} b_{mn\sigma}\varphi_{nm}(\mathbf{r}) \tag{1.44}$$

$$\hat{\psi}^{+}_{\sigma}(\mathbf{r}) = \sum_{m,n} b^{+}_{mn\sigma}\varphi^{+}_{nm}(\mathbf{r}) \tag{1.45}$$

where $\varphi_{nm}(\mathbf{r})$ and $\varphi^{+}_{nm}(\mathbf{r})$ are wave functions of orbital m at the site n of the crystal, b and b^{+} fermion annihilation and creation operators. The wave functions φ_{nm} constitute an orthogonal set. Equation (1.43) becomes

$$\hat{\mathcal{H}} = -\frac{1}{2}\sum\langle\varphi_{n_1m_1\sigma_1}(\mathbf{r}_1)\varphi_{n_2m_2\sigma_2}(\mathbf{r}_2)|\frac{e^2}{|\mathbf{r}_1-\mathbf{r}_2|}| \\ \times\varphi_{n_3m_3\sigma_3}(\mathbf{r}_1)\varphi_{n_4m_4\sigma_4}(\mathbf{r}_2)\rangle b^{+}_{n_1m_1\sigma_1}b^{+}_{n_2m_2\sigma_2}b_{n_3m_3\sigma_3}b_{n_4m_4\sigma_4} \tag{1.46}$$

where the sum runs over $(n_1, m_1, \sigma_1, \cdots, n_4, m_4, \sigma_4)$.

If $n_1 = n_2 = n_3 = n_4$, the interactions are between electrons of the same site. Equation (1.46) is the origin of Hund's atomic empirical rules. In addition, if $m_1 = m_2 = m_3 = m_4$, this equation is the Coulomb term in the Hubbard Hamiltonian which will be shown below.

For simplicity, we suppose one electron per site and one orbital per electron in the following.

If $n_1 = n_3$ and $n_2 = n_4$, the Coulomb term is given by

$$\hat{\mathcal{H}}_c = -\frac{1}{2}\sum_{n_1,n_2}\langle n_1 n_2|\frac{e^2}{|r_{12}|}|n_2 n_1\rangle \sum_{\sigma_1,\sigma_2} b^+_{n_1\sigma_1} b^+_{n_2\sigma_2} b_{n_1\sigma_1} b_{n_2\sigma_2} \tag{1.47}$$

If $n_1 = n_4$ and $n_2 = n_3$ (by consequence, $\sigma_1 = \sigma_2$), the exchange term becomes

$$\begin{aligned}\hat{\mathcal{H}}_{ex} &= -\frac{1}{2}\sum_{n_1,n_2,\sigma_1=\sigma_2}\langle n_1 n_2|\frac{e^2}{|r_{12}|}|n_2 n_1\rangle b^+_{n_1\sigma_1} b^+_{n_2\sigma_2} b_{n_1\sigma_1} b_{n_2\sigma_2} \\ &= -\frac{1}{2}\sum_{n_1,n_2,\sigma_1=\sigma_2}\langle n_1 n_2|\frac{e^2}{|r_{12}|}|n_2 n_1\rangle b^+_{n_1\sigma_1} b_{n_1\sigma_1} b^+_{n_2\sigma_2} b_{n_2\sigma_2} \\ &= -\frac{1}{2}\sum_{n_1,n_2} J_{n_1 n_2} \sum_{\sigma_1,\sigma_2,\sigma_1=\sigma_2} b^+_{n_1\sigma_1} b_{n_1\sigma_1} b^+_{n_2\sigma_2} b_{n_2\sigma_2}\end{aligned} \tag{1.48}$$

where

$$J_{n_1 n_2} \equiv \langle n_1 n_2|\frac{e^2}{|r_{12}|}|n_2 n_1\rangle \tag{1.49}$$

$$\begin{aligned}&\sum_{\sigma_1,\sigma_2,\sigma_1=\sigma_2} b^+_{n_1\sigma_1} b_{n_1\sigma_1} b^+_{n_2\sigma_2} b_{n_2\sigma_2} \\ &= b^+_{n_1\uparrow} b_{n_1\uparrow} b^+_{n_2\uparrow} b_{n_2\uparrow} + b^+_{n_1\downarrow} b_{n_1\downarrow} b^+_{n_2\downarrow} b_{n_2\downarrow} \\ &= \frac{1}{2}\left(b^+_{n_1\uparrow} b_{n_1\uparrow} + b^+_{n_1\downarrow} b_{n_1\downarrow}\right)\left(b^+_{n_2\uparrow} b_{n_2\uparrow} + b^+_{n_2\downarrow} b_{n_2\downarrow}\right) \\ &\quad +\frac{1}{2}\left(b^+_{n_1\uparrow} b_{n_1\uparrow} - b^+_{n_1\downarrow} b_{n_1\downarrow}\right)\left(b^+_{n_2\uparrow} b_{n_2\uparrow} - b^+_{n_2\downarrow} b_{n_2\downarrow}\right) \\ &\quad + b^+_{n_1\uparrow} b_{n_1\downarrow} b^+_{n_2\downarrow} b_{n_2\uparrow} + b^+_{n_1\downarrow} b_{n_1\uparrow} b^+_{n_2\uparrow} b_{n_2\downarrow}\end{aligned} \tag{1.50}$$

where the last two terms have been added. Note that these terms do not affect the result because their averages are zero in the diagonal representation: $\langle\hat{\psi}|b^+_{n_1\uparrow} b_{n_1\downarrow} b^+_{n_2\downarrow} b_{n_2\uparrow}|\hat{\psi}\rangle = \langle\hat{\psi}|b^+_{n_1\downarrow} b_{n_1\uparrow} b^+_{n_2\uparrow} b_{n_2\downarrow}|\hat{\psi}\rangle = 0$.

We define next the following spin operators

$$S^z_n = \frac{1}{2}(b^+_{n\uparrow} b_{n\uparrow} - b^+_{n\downarrow} b_{n\downarrow}) \tag{1.51}$$

$$S^+_n \equiv S^x_n + iS^y_n = b^+_{n\uparrow} b_{n\downarrow} \tag{1.52}$$

$$S^-_n \equiv S^x_n - iS^y_n = b^+_{n\downarrow} b_{n\uparrow} \tag{1.53}$$

As we suppose one electron per site, we have

$$\left(b^+_{n_1\uparrow}b_{n_1\uparrow} + b^+_{n_1\downarrow}b_{n_1\downarrow}\right) = 1$$
$$\left(b^+_{n_2\uparrow}b_{n_2\uparrow} + b^+_{n_2\downarrow}b_{n_2\downarrow}\right) = 1$$

The right-hand side of (1.50) becomes

$$\frac{1}{2} + 2\mathbf{S}_{n_1}\cdot\mathbf{S}_{n_2} \tag{1.54}$$

Using

$$\begin{aligned}\mathbf{S}_{n_1}.\mathbf{S}_{n_2} &= S^z_{n_1}S^z_{n_2} + S^y_{n_1}S^y_{n_2} + S^x_{n_1}S^x_{n_2}\\ &= S^z_{n_1}S^z_{n_2} + \frac{1}{2}\left(S^+_{n_1}S^-_{n_2} + S^-_{n_1}S^+_{n_2}\right)\end{aligned} \tag{1.55}$$

we rewrite (1.48) as

$$\hat{\mathcal{H}}_{ex} = -\frac{1}{2}\sum_{n_1,n_2} J_{n_1n_2}\left(\frac{1}{2} + 2\mathbf{S}_{n_1}\cdot\mathbf{S}_{n_2}\right) \tag{1.56}$$

where we added a factor $\frac{1}{2}$ to remove the double counting of each pair (n_1, n_2). We use now the notation $\sum_{(n_1,n_2)}$ instead of $\frac{1}{2}\sum_{n_1,n_2}$ where (n_1, n_2) indicates *the pair* (n_1n_2) counted only once. Finally, one has

$$\hat{\mathcal{H}}_{ex} = -\sum_{(n_1n_2)} J_{n_1n_2}\left(\frac{1}{2} + 2\mathbf{S}_{n_1}\cdot\mathbf{S}_{n_2}\right) \tag{1.57}$$

The first term does not depend on spins. The second term is the Heisenberg model which shall be used later throughout this book.

Hamiltonian (1.57) is thus the origin of ferromagnetism observed in ferromagnetic materials. If the interaction is a Coulomb interaction as we suppose here, then $J_{n_1n_2}$ is positive as shown below. We write

$$J_{n_1n_2} = \iint \varphi^*_{n_1}(\mathbf{r}_1)\varphi^*_{n_2}(\mathbf{r}_2)\frac{e^2}{|\mathbf{r}_1 - \mathbf{r}_2|}\varphi_{n_1}(\mathbf{r}_1)\varphi_{n_2}(\mathbf{r}_2)d\mathbf{r}_1d\mathbf{r}_2 \tag{1.58}$$

where the wave functions $\varphi_n(\mathbf{r})$ have been supposed to be orthogonal. They are Wannier wave functions constructed from linear combinations of Bloch wave functions. Making use of

$$\frac{e^2}{|\mathbf{r}_1 - \mathbf{r}_2|} = \frac{1}{\Omega}\sum_{\mathbf{k}}\frac{4\pi e^2}{k^2}e^{i\mathbf{k}\cdot(\mathbf{r}_1-\mathbf{r}_2)} \tag{1.59}$$

we obtain

$$
\begin{aligned}
J_{n_1n_2} &= \frac{1}{\Omega}\sum_k \frac{4\pi e^2}{k^2}\int \varphi^*_{n_1}(\mathbf{r}_1)\varphi_{n_2}(\mathbf{r}_1)e^{-i\mathbf{k}.\mathbf{r}_1}d\mathbf{r}_1 \\
&\times \int \varphi^*_{n_2}(\mathbf{r}_2)\varphi_{n_1}(\mathbf{r}_2)e^{i\mathbf{k}.\mathbf{r}_2}d\mathbf{r}_2 \\
&= \frac{1}{\Omega}\sum_{\mathbf{k}} \frac{4\pi e^2}{k^2} I^2
\end{aligned} \tag{1.60}
$$

where

$$
I = \int_\Omega \varphi^*_{n_1}(\mathbf{r}_1)\varphi_{n_2}(\mathbf{r}_1)e^{-i\mathbf{k}.\mathbf{r}_1}d\mathbf{r}_1 = \int_\Omega \varphi^*_{n_2}(\mathbf{r}_2)\varphi_{n_1}(\mathbf{r}_2)e^{-i\mathbf{k}.\mathbf{r}_2}d\mathbf{r}_2
$$

The above two integrals are identical because the indices and variables are dummy. $J_{n_1n_2}$ is thus positive. From (1.57), we see that if $\mathbf{S}_{n_1}$ and $\mathbf{S}_{n_2}$ are parallel, then the energy is lowest. This state of spin ordering is called "ferromagnetic."

1.5.2 Spin Models: Magnetic Materials

In magnetic materials, depending on the nature of the spins and the interaction between them, one can use several spin models as described below.

1.5.3 Heisenberg Model

The Heisenberg model for the interaction between two spins localized at the lattice sites i and j is given by

$$
-2J_{ij}\mathbf{S}_i \cdot \mathbf{S}_j \tag{1.61}
$$

where J_{ij} is the exchange integral resulting from the Coulomb interaction between two electrons of spins $\mathbf{S}_i$ and $\mathbf{S}_j$, localized at $\mathbf{r}_i$ and $\mathbf{r}_j$. We have demonstrated (1.61) in the previous section [see Eq. (1.57)]. In general, the value of J_{ij} depends on the distance between the spins and on the orientation of $\mathbf{r}_j - \mathbf{r}_i$ with respect to the crystalline axes.

In the quantum model, $\mathbf{S}_i$ is a quantum spin whose components obey the spin commutation relations. For example, in the case of spin one-half its components are the Pauli matrices (1.3)–(1.5). In the classical model, $\mathbf{S}_i$ is considered as a vector.

We consider hereafter the simplest case where the exchange interaction is limited between nearest neighbors and this interaction is identical and equal to J for all pairs of nearest neighbors. In this case, the Hamiltonian reads

$$\mathcal{H} = -2J \sum_{<i,j>} \mathbf{S}_i \cdot \mathbf{S}_j \tag{1.62}$$

where the sum is performed over all pairs of nearest neighbors. We see that if $J > 0$, $\mathcal{H}$ is minimum when all spins are parallel. This spin configuration corresponds to the ferromagnetic ground state.

In the case where $J <, 0$, we have to distinguish classical and quantum spin models:

(i) In the classical model, the spins are vectors. Except for lattices composed of equilateral triangular faces in their elementary cell such as the two-dimensional triangular lattice, the face-centered cubic lattice and the hexagonal-close-pack lattice, $\mathcal{H}$ is minimum in the other lattices when all nearest neighbors are antiparallel throughout the crystal. This spin configuration is called "classical antiferromagnetic ground state" or "Néel state".

(ii) In the quantum model, the spins can be decomposed into spin operators [see (1.51)–(1.53)]: $(S_i^z, S_i^+ \equiv S_i^x + iS_i^y, S_i^- \equiv S_i^x - iS_i^y)$. These spin operators obey the spin commutation relations

$$[S_l^+, S_m^-] = 2S_l^z \delta_{l,m} \quad \text{and} \quad [S_l^z, S_m^\pm] = \pm S_l^\pm \delta_{l,m} \tag{1.63}$$

It is known that the Néel state is not the quantum ground state for antiferromagnets. As will be shown in Chapter 3, at zero temperature the so-called zero-point quantum fluctuations make the spins contracted from its full length. The zero-point spin contraction is of the order of a few percents depending on the lattice structure. The real quantum ground state of antiferromagnets is not known, though we know that it is not far from the classical Néel state, except in low dimensions where strong quantum fluctuations can destroy the Néel order. Note that in the case of ferromagnets, the perfect parallel spin configuration is the real ground state in both classical and quantum spin models.

When the geometry of the lattice does not allow us to satisfy all the interactions between a spin with its neighbors, the system is said "frustrated." This happens when the elementary lattice cell is composed of equilateral triangles such as in the triangular lattice. There are many striking effects due to the frustration. The reader is referred to the book "Frustrated Spin Systems" for reviews [85]. We will outline some remarkable properties of frustrated systems in Section 5.7.3.

1.5.4 Ising, XY, Potts Models

Besides the Heisenberg model, there are three other spin models which are very popular in magnetism and in statistical physics:

1. Ising model: The Ising model is defined by

$$\mathcal{H} = -\sum_{(i,j)} J_{ij}\sigma_i\sigma_j \tag{1.64}$$

where σ_i (σ_j) is the spin at lattice site i (j) with two possible values $+1$ and -1. Such a spin is called "Ising spin." The Ising model can be used to study spin systems with a strong uniaxial anisotropy. It is also used as a simple model for phase transition investigations (see Chapter 5). Any system of interacting particles where each of them has two individual states can be mapped into a system of Ising spins. For example, in a binary alloy where each lattice site can be occupied by an A atom or a B atom, we can assign an up spin ($+1$) for an A atom and a down spin (-1) for a B atom. We can then study for example the structure of the alloy by studying the Ising model at various concentrations of A and B.

2. XY model: The XY model is defined by the interaction between two-component spins. It is sometimes called "model of plane rotators." The Hamiltonian is given by

$$\mathcal{H} = -\sum_{(i,j)} J_{ij}(S_{ix}S_{jx} + S_{iy}S_{jy}) \tag{1.65}$$

The XY model is used to study spin systems with a strong planar anisotropy called "easy-plane anisotropy" which exists in some magnetic materials. It is also used in statistical physics. A very interesting ferromagnetic XY model in two dimensions has been extensively studied because it gives rise to a very special phase transition known as the "Kosterlitz–Thouless" transition discovered in the 1970s [10, 58, 191, 380]. A discussion on this phase transition is given in Chapter 5.

3. Potts model: The q-state Potts model is defined by

$$\mathcal{H} = -\sum_{(i,j)} J_{ij}\delta_{\sigma_i,\sigma_j} \tag{1.66}$$

where σ_i (σ_j) is a parameter taking q values, $\sigma_i = 1, 2, \cdots, q$ for example, and $\delta_{\sigma_i,\sigma_j}$ denotes the Kronecker symbol. The sum is performed over pairs of nearest neighbors. If interaction $J_{ij} = J > 0$ for nearest neighbors (i, j), then in the ground state there is only one value of q: It is ferromagnetic. Note that if $q = 2$, the model is equivalent to the Ising model. We define the Potts order parameter Q by

$$Q = \frac{[q \max(Q_1, Q_2, \cdots, Q_q) - 1]}{q-1} \tag{1.67}$$

where Q_n is the spatial average defined by

$$Q_n = \frac{1}{N}\sum_{j=1}^{N} \delta_{\sigma_j,n} \tag{1.68}$$

where $n = 1, \ldots, q$, the sum runs over all lattice sites, and N is the total site number. From this definition we see that the ground state containing only one kind of spin has $Q = 1$, while in the disordered state q kinds of spin are equally present in the system, namely $Q_1 = Q_2 = \cdots = Q_q = 1/q$, so that $Q = 0$.

The q-state Potts model is used to study systems of interacting particles where each particle has q individual states. Exact methods to treat the Potts models in two dimensions are shown in a book by Baxter [25].

1.6 Conclusion

This chapter introduces the spin and shows some principal behaviors of a system of independent spins under the application of a magnetic field. We have examined three cases. The first case concerns free electrons at low temperatures. We have obtained the so-called Pauli paramagnetism where the susceptibility is a constant at the first-order approximation. The second case is the system of free atoms where each atom has a permanent magnetic moment. Under the application of a magnetic field, the susceptibility is positive (paramagnetic) and proportional to $1/T$ (Curie's law). The third case is a diamagnetic case: The reaction of the electrons in an atom to an applied magnetic field gives rise to a negative susceptibility. This phenomenon is called "atomic diamagnetism."

We have also demonstrated the Heisenberg model describing the magnetic interaction between two spins which leads to a magnetic ordering in solids. We have also presented several spin models such as the Ising, XY and Potts models. These models are used in the following chapters to study theoretically behaviors of bulk materials and thin films.

1.7 Problems

Problem 1. Orbital and spin moments of an electron:

Using the theory of angular momentum, calculate the orbital and spin moments of an electron. Determine the total magnetic moment.

Problem 2. Zeeman effect:

(a) Calculate the magnetic moment per atom for Fe, provided the saturated magnetization under an applied magnetic field equal to 1.7×10^6 A/m, the mass density of Fe $\rho = 7970$ kg/m^3 and the atomic mass of Fe $M = 56$.

(b) Calculate ΔE the separation of the energy levels due to the Zeeman effect on the atomic level corresponding to the wavelength $\lambda = 643.8$ nm of a cadmium atom. Calculate the variation of frequency $\Delta \nu$ of the initial level.
Numerical application: Calculate ΔE and $\Delta \nu$ for the following fields $\mu_B H = 0.5$, 1, and 2 Tesla.

Problem 3. Fermi–Dirac distribution for free-electron gas:

Electrons are fermions which obey Pauli's exclusion principle. Microscopic states follow the Fermi–Dirac statistics.

The Fermi–Dirac distribution is given by (see Appendix A)

$$f(E, T, \mu) = \frac{1}{e^{\beta(E-\mu)} + 1} \tag{1.69}$$

where μ is the chemical potential, $\beta = \frac{1}{k_B T}$, k_B the Boltzmann constant and T the temperature. The function $f(E, T, \mu)$ is the number of electrons of the microscopic state of energy E at temperature T.
Give the properties of $f(E, T, \mu)$ at $T = 0$. Plot $f(E, T, \mu)$ as a function of E for an arbitrary $\mu(> 0)$, at $T = 0$ and at low T.

Problem 4. Sommerfeld's expansion :
Consider the function

$$I = \int_0^\infty h(E) f(E) dE \tag{1.70}$$

where $h(E)$ is a function differentiable at any order with respect to E.
Show that $h(E)$ can be expanded in powers of T at low T as follows:

$$I = \int_0^\mu h(E) dE + \frac{\pi^2}{6}(k_B T)^2 h^{(1)}(E)|_{E=\mu} + \frac{7\pi^4}{360}(k_B T)^4 h^{(3)}(E)|_{E=\mu} + \cdots \tag{1.71}$$

where $h^{(n)}(E)|_{E=\mu}$ is the n-th derivative of $h(E)$ at $E = \mu$.

Problem 5. Pauli paramagnetism:
Calculate the susceptibility of a three-dimensional electron gas in an applied magnetic field **B**, at low and high temperatures. One supposes that B is small.

Problem 6. Paramagnetism of free atoms for arbitrary **J**:
Consider a gas of N atoms of moment **J** in a volume V. Show that the average of the total magnetic moment per volume unit of the gas is

$$\overline{m} = \frac{N g \mu_B J}{V} B_J\left(\frac{g \mu_B B J}{k_B T}\right) \tag{1.72}$$

where $B_J(x)$ is the Brillouin function given by

$$B_J(x) = \frac{2J+1}{2J} \coth\left(\frac{2J+1}{2J} x\right) - \frac{1}{2J} \coth\left(\frac{x}{2J}\right) \tag{1.73}$$

Show that at high temperature one has

$$\chi = \frac{N}{V}(g\mu_B)^2\frac{J(J+1)}{3}\frac{1}{k_B T} \tag{1.74}$$

Find the limit of $\overline{m}$ at $T = 0$.

Problem 7. Langevin's theory of diamagnetism:

Consider an electron in an atom. In the theory of diamagnetism by Langevin, the motion of the electron around the nucleus is equivalent to the motion of a magnetic moment **m** generated by a current i which circulates in a closed loop of surface A.

(a) Write a relation between i, **m** and A.

(b) Show that the magnetic moment of the electron is written as $m = evr/2$ where e is the charge of the electron, v its velocity and r its orbital radius.

(c) Show that an applied magnetic field **H**, perpendicular to the orbital plane, gives rise to the following variation of its magnetic moment $\Delta m = -\frac{\mu_B e^2 r^2 \mathbf{H}}{4m_e}$ (m_e: electron mass). Comment on the negative sign.

(d) What will be the result if **H** makes an angle θ with the surface normal ?

(e) Calculate the susceptibility of a material of mass density ρ made of atoms of Z electrons, of mass M.

Numerical application: $\rho = 2220$ kg/m^3, $e = 1.6 \times 10^{-19}$ C, $Z = 6$, $r = 0.7 \times 10^{-10}$ m.

Problem 8. Langevin's theory of paramagnetism:

Consider an atom of permanent magnetic moment **m** (atom having an odd number of electrons). Using the Maxwell–Boltzmann statistics, show that the magnetic moment resulting from the application of a magnetic field **H** on a material of N atoms per volume unit, in an arbitrary direction is given by

$$\mathbf{M} = N\mathbf{m}\mathcal{L}\left(\frac{\mu_B \mathbf{m}\cdot\mathbf{H}}{k_B T}\right)$$

where $\mathcal{L}(x) = \coth(x) - \frac{1}{x}$ (Langevin function). Calculate the susceptibility in the case of a weak field.

Chapter 2

Mean-Field Theory of Magnetic Materials

The mean-field theory, or molecular-field approximation, is considered as the first-order approximation to treat a system of interacting spins. This chapter shows how this theory is applied to ferromagnets, antiferromagnets and ferrimagnets. As a first-order approximation, its results give a quick look at the system's properties. We will discuss the validity of the mean-field theory below and show how to improve it in Chapter 5.

2.1 Mean-Field Theory of Ferromagnets

We consider the Heisenberg model for a ferromagnet with the following Hamiltonian

$$\mathcal{H} = -2\sum_{(i,j)} J_{ij}\mathbf{S}_i \cdot \mathbf{S}_j - g\mu_B \sum_i \mathbf{H}_0 \cdot \mathbf{S}_i \qquad (2.1)$$

where $\mathbf{H}_0$ is a magnetic field applied in the z direction, g the Landé factor and μ_B the Bohr magneton. The first sum is performed over spin pairs $(\mathbf{S}_i, \mathbf{S}_j)$ occupying lattice sites i and j. For simplicity, we suppose in the following only the interaction between nearest

Physics of Magnetic Thin Films: Theory and Simulation
Hung T. Diep

ISBN 978-981-4877-42-8 (Hardcover), 978-1-003-12110-7 (eBook)
www.jennystanford.com

neighbors is not zero. Note that this hypothesis is not a hypothesis of the mean-field theory because the mean-field theory can be applied to systems including far-neighbor interactions as seen in Problem 8.

We consider the spin at the site i. The interaction energy with its nearest neighbors and with the magnetic field are written as

$$\mathcal{H}_i = -2J \sum_{\vec{\rho}} \mathbf{S}_i \cdot \mathbf{S}_{i+\vec{\rho}} - g\mu_B H_0 S_i^z \tag{2.2}$$

where $\vec{\rho}$ are vectors connecting the site i to its nearest neighbors and J denotes the exchange integral between $\mathbf{S}_i$ with its nearest neighbors.

2.1.1 Mean-Field Equation

The only assumption of the mean-field theory is to suppose that all neighboring spins have the same average value, namely $< \mathbf{S}_{i+\vec{\rho}} >=< S^z >$ for all $(i + \vec{\rho})$. This value is to be computed in the following. We choose the z axis as the spin quantization axis. The average values of the x and y spin components are then zero since the spin precesses circularly around the z axis:

$$< S_{i+\vec{\rho}}^x >=< S_{i+\vec{\rho}}^y >= 0 \tag{2.3}$$

For the z component, we have for all neighbors

$$< S_{i+\vec{\rho}}^z >=< S^z > + < \Delta S^z > \tag{2.4}$$

where $< S^z >$ is the average value in the absence of the magnetic field, and $< \Delta S^z >$ is the variation of $< S^z >$ induced by the field. Equation (2.2) is rewritten as

$$\mathcal{H}_i \simeq -2CJ \left[S_i^z (< S^z > + < \Delta S^z >)\right] - g\mu_B H_0 S_i^z \tag{2.5}$$

where C is the coordination number (number of nearest neighbors). We can express $\mathcal{H}_i$ as

$$\mathcal{H}_i = -g\mu_B \overline{H} S_i^z \tag{2.6}$$

where

$$\overline{H} = \frac{2CJ\ [< S^z > + < \Delta S^z >]}{g\mu_B} + H_0 \tag{2.7}$$

$\overline{H}$ is called "molecular field" acting on the spin $\mathbf{S}_i$.

Let us suppose $H_0 = 0$ for the moment. In that case $< \Delta S^z >= 0$ in Eq. (2.7). We have

$$\overline{H} = \frac{2CJ < S^z >}{g\mu_B} \tag{2.8}$$

The average value $< S^z >$ is calculated using the canonical description (see Appendix A) as follows:

$$< S^z >= \frac{\sum_{S_i^z=-S}^{S} S_i^z \mathrm{e}^{-\beta\mathcal{H}_i}}{Z_i} \tag{2.9}$$

where $\beta = \frac{1}{k_B T}$ and Z_i the partition function defined by [see Eq. (A.9)]

$$\begin{aligned} Z_i &= \sum_{S_i^z=-S}^{S} \exp(\beta g\mu_B \overline{H} S_i^z) \\ &= \frac{\sinh[\beta g\mu_B \overline{H}(S + \frac{1}{2})]}{\sinh[\frac{1}{2}\beta g\mu_B \overline{H}]} \end{aligned} \tag{2.10}$$

where $S = |\mathbf{S}_i|$. We obtain

$$\begin{aligned} &\sum_{S_i^z=-S}^{S} S_i^z \mathrm{e}^{-\beta\mathcal{H}_i} = \frac{\partial}{\partial\alpha} \sum_{S_i^z=-S}^{S} \mathrm{e}^{\alpha S_i^z} \ (\alpha \equiv \beta g\mu_B \overline{H}_0) = \frac{\partial}{\partial\alpha} Z_i \\ &= \frac{(S + \frac{1}{2})\cosh(S + \frac{1}{2})\alpha \sinh\frac{\alpha}{2} - \frac{1}{2}\sinh(S + \frac{1}{2})\alpha \cosh\frac{\alpha}{2}}{\sinh^2\frac{\alpha}{2}} \end{aligned} \tag{2.11}$$

from which one gets

$$< S^z >= S B_S(x) \tag{2.12}$$

$B_S(x)$ is the Brillouin function defined by

$$B_S(x) = \frac{2S+1}{2S} \coth\frac{(2S+1)x}{2S} - \frac{1}{2S}\coth\frac{x}{2S} \tag{2.13}$$

where

$$x = x_0 = \beta g\mu_B S\overline{H} = \beta[2CJS < S^z >] \tag{2.14}$$

Equation (2.12) is called "mean-field equation." Since the argument x of $B_S(x)$ contains $< S^z >$, (2.12) is therefore an implicit equation of $< S^z >$ which depends on the temperature. In the case of spin one-half, $S = \frac{1}{2}$, the Brillouin function is

$$B_{\frac{1}{2}}(x) = \tanh x \tag{2.15}$$

In the case where $S \to \infty$, we have from Eq. (2.13)

$$B_\infty(x) = \coth x - \frac{1}{x} \equiv \text{Langevin function} \tag{2.16}$$

Now, suppose that H_0 is not zero but very weak. We use $\overline{H}$ of Eq. (2.7) with $< \Delta S^z >$ being very small. The mean-field equation is

$$< S^z > + < \Delta S^z > = S B_S(x) \tag{2.17}$$

where

$$x = \beta g \mu_B S \overline{H} = \beta[2CJ\, S(< S^z > + < \Delta S^z >) + g\mu_B S H_0] \tag{2.18}$$

Expanding the Brillouin function near $x = x_0 = \beta 2CJ\, S < S^z >$ and identifying the second terms of the two sides of (2.17), we have

$$< \Delta S^z > = S B_S'(x_0) \left[\frac{1}{k_b T} (g\mu_B S H_0 + 2CJ\, S < \Delta S^z >) \right] \tag{2.19}$$

where $B_S'(x_0)$ is the derivative of $B_S(x)$ with respect to x taken at x_0.

2.1.2 Mean-Field Critical Temperature

Let us study the mean-field equation with respect to T.

At high T, $\beta < S^z > \ll 1$, we obtain from (2.13)

$$B_S(x) \simeq \frac{S+1}{3S} x - \frac{[S^2 + (S+1)^2](S+1)}{90 S^3} x^3 + O(x^5) \tag{2.20}$$

Equation (2.12) becomes

$$\begin{aligned} &< S^z > \left[\frac{2CJ\, S(S+1)}{3k_B T} - 1 \right] \\ &= \frac{S(S+1)[S^2 + (S+1)^2]}{90} \left(\frac{2CJ}{k_B T} \right)^3 < S^z >^3 + O(x^5) \end{aligned} \tag{2.21}$$

This equation has a solution $< S^z > \neq 0$ only if

$$\left[\frac{2}{3} C S(S+1) - \frac{k_B T}{J} \right] > 0 \tag{2.22}$$

namely

$$T < \frac{2CJ\, S(S+1)}{3k_B} \equiv T_c \tag{2.23}$$

Once this condition is satisfied, $< S^z >$ is given by

$$< S^z >^2 \simeq \frac{10}{3} \frac{S^2(S+1)^2}{[S^2 + (S+1)^2]} \left[\frac{T_c - T}{T_c} \right] \tag{2.24}$$

T_c is called "critical temperature." When $T \geq T_c$, the solution of (2.21) is $< S^z >= 0$.

At low temperatures, $2CJS < S^z >$ is much larger than $k_B T$, the expansion of (2.13) gives

$$B_S(x) \simeq 1 - \frac{1}{S} e^{(-x/S)} + \cdots \quad (2.25)$$

which leads to

$$< S^z >= SB_S(x) \simeq S - e^{-2JCS/k_B T} + \cdots \quad (2.26)$$

If $T = 0$, we have $< S^z >= S$.

2.1.3 Graphical Solution

In general, we solve (2.12) by a graphical method: We look for the intersection of the two curves $y_1 = \frac{<S^z>}{S} = \frac{k_B T}{2CJS^2} x$ and $y_2 = B_S(x)$ which represent the two sides of (2.12). The first curve y_1 versus x is a straight line with a slope proportional to the temperature. For a given value of T, there are two symmetric intersections at $\pm M$ as shown in Fig. 2.1. It is obvious that if the slope of y_1 is larger than the slope of y_2 at $x = 0$, there is no intersection other than the one at $x = 0$. The solution is then $< S^z >= 0$. The slope of y_2 at $x = 0$ thus determines the critical temperature T_c, namely

$$\frac{k_B T_c}{2CJS^2} = B'_S(x = 0) \quad (2.27)$$

which is identical to T_c given by Eq. (2.23).

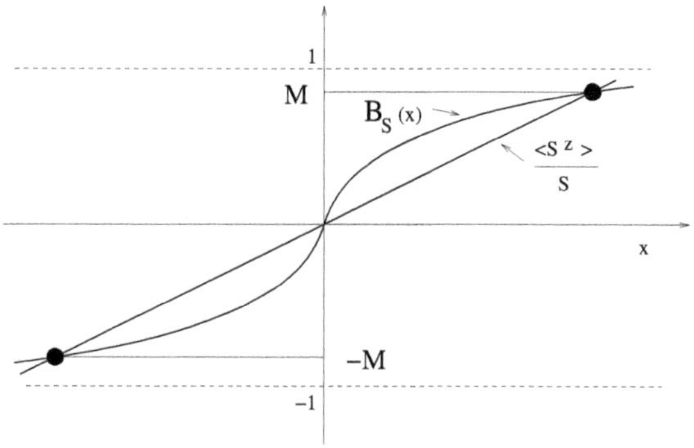

Figure 2.1 Graphical solutions of Eq. (2.12).

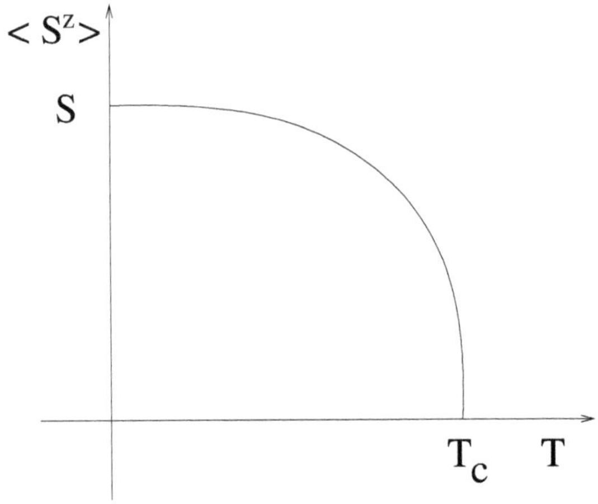

Figure 2.2 Thermal average $< S^z >$ versus T.

We display the positive solution of $< S^z >$ as a function of T in Fig. 2.2.

2.1.4 Specific Heat

The average energy of a spin when $H_0 = 0$ is calculated by [see Eq. (A.10)]

$$\overline{E}_i = -\frac{\partial \ln Z_i}{\partial \beta} \simeq -2CJ < S^z >^2 \tag{2.28}$$

The total ferromagnetic energy of the crystal is

$$\overline{E} = \frac{1}{2} N \overline{E}_i \tag{2.29}$$

where the factor $\frac{1}{2}$ is added in order to count each interaction just once. The specific heat is

$$C_V = \left(\frac{\partial \overline{E}}{\partial T}\right)_V = -2NCJ < S^z > \frac{\partial < S^z >}{\partial T} \tag{2.30}$$

At low temperatures, $< S^z > \cong S - e^{-2CJ\,S/k_B T}$ [see (2.26)], we have

$$C_V(T \simeq 0) \simeq 4Nk_B \left[\frac{CJ\,S}{k_B T}\right]^2 e^{-2CJ\,S/k_B T} \tag{2.31}$$

When $T \to 0$, we have $C_V \simeq 0$.

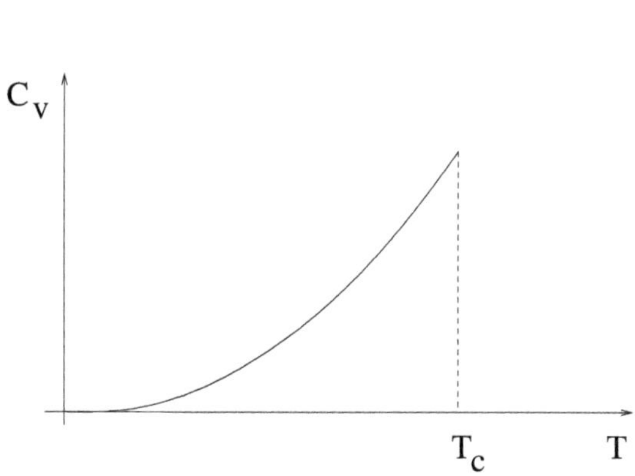

Figure 2.3 C_V calculated by the mean-field theory versus T.

For $T > T_c$, we have $\overline{E} = 0$; therefore $C_V = 0$. Let us calculate C_V when $T \to T_c^-$. We have from (2.24)

$$\lim_{T \to T_c^-} < S^z > \frac{\partial < S^z >}{\partial T} = -\frac{5}{2}\frac{k_B}{JC}\frac{S(S+1)}{[S^2 + (S+1)^2]} \tag{2.32}$$

so that

$$C_V(T \to T_c^-) = 5Nk_B\frac{S(S+1)}{S^2 + (S+1)^2} \tag{2.33}$$

The discontinuity of C_V at T_c is thus

$$\text{For } S = \frac{1}{2} \Rightarrow \Delta C_V = \frac{3}{2}Nk_B$$
$$\text{For } S = \infty \Rightarrow \Delta C_V = \frac{5}{2}Nk_B \tag{2.34}$$

This discontinuity is an artifact of the mean-field theory resulting from the fact that critical fluctuations near T_c have been neglected by replacing all spins by a uniform average. When fluctuations around the average values of spins are taken into account, C_V diverges at T_c when we approach T_c from both sides. Some more details on this point are given in Chapter 5. We show in Fig. 2.3 C_V calculated by the mean-field theory as a function of T.

2.1.5 Susceptibility

The susceptibility is defined by

$$\chi_{\parallel} = \left(\frac{\partial M}{\partial H_0}\right)_{H_0=0} = Ng\mu_B\left(\frac{\partial < S^z >}{\partial H_0}\right)_{H_0=0} \tag{2.35}$$

where N is the total number of spins ($M = Ng\mu_B < S^z >$ is the total magnetic moment). From Eq. (2.19), we have

$$< \Delta S^z >= \frac{SB'_S(x)\frac{g\mu_B S}{k_B T}H_0}{1 - SB'_S(x)\left(\frac{2CJS}{k_B T}\right)} \tag{2.36}$$

therefore,

$$\chi_{\parallel} = \frac{N(g\mu_B)^2 S^2 B'_S(x)}{k_B T - 2CJ\, S^2 B'_S(x)} \tag{2.37}$$

where $x = \frac{2CJS<S^z>}{k_B T}$.

When $T \geq T_c$, we have $< S^z >= 0$ and $B'_x(0) = \frac{S+1}{3S}$. We get

$$\chi_{\parallel}(T \geq T_c) = \frac{N(g\mu_B)^2 S(S+1)}{3k_B(T - T_c)} \qquad \text{Curie-Weiss law} \tag{2.38}$$

When $T \lesssim T_c$, we have $< S^z >\to 0$. Expanding $B'_S(x)$ with respect to $< S^z >$, we obtain

$$\chi_{\parallel}(T \lesssim T_c) = \frac{N(g\mu_B)^2 S(S+1)}{6k_B(T_c - T)} \tag{2.39}$$

It is noted that the coefficient in this case is twice smaller than that in (2.38). When $T \to 0$, $\chi_{\parallel} \to 0$ because $M \to$ constant. The inverse of the susceptibility is schematically shown as a function of T in Fig. 2.4.

In reality, a ferromagnetic crystal can have several ferromagnetic domains with spins pointing in different directions. This is due to the presence of defects, dislocations and imperfections during the formation of the crystal. The region between two magnetic domains is called "domain wall" in which the matching of two spin orientations is progressively realized. We show schematically magnetic domains and a domain-wall spin configuration in Fig. 2.5. The presence of domain walls makes it difficult to compare calculated and experimental susceptibilities.

2.1.6 Validity of Mean-Field Theory

The mean-field theory assumes that all spins have the same value, meaning that it neglects instantaneous fluctuations of each spin. Fluctuations favor disorder, so when taken into account, fluctuations cause a transition at a temperature lower than T_c given by (2.23).

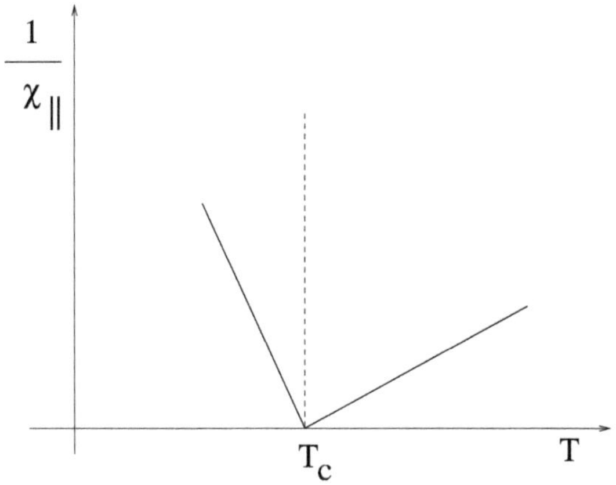

Figure 2.4 Inverse of the susceptibility obtained by mean-field theory versus T.

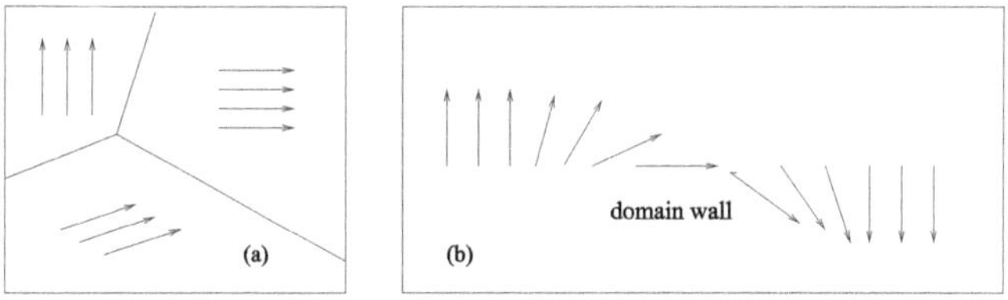

Figure 2.5 (a) Ferromagnetic domains in an imperfect crystal (b) Example of a spin structure in a domain wall.

Due to the approximation of uniform spins, the mean-field theory thus overestimates the critical temperature T_c. This point is studied in Section 5.3 with the Landau–Ginzburg theory.

Another artifact of the mean-field theory is that it results in a phase transition at a finite temperature in spin systems in any space dimension: T_c given by Eq. (2.23) is not zero even for one dimension ($C = 2$). This is not correct because we know that in dimensions $d = 1$ and $d = 2$, fluctuations are so strong that they destroy magnetic long-range order at any finite temperature in many systems. The mean-field theory, however, becomes exact for dimension $d > 4$ (see Chapter 5 for more details).

2.2 Antiferromagnetism in Mean-Field Theory

In Section 1.5.2, we have seen that depending on the sign of the exchange interaction a spin system can have an antiferromagnetic order at zero and low temperatures. We study here some properties of antiferromagnets by the mean-field theory.

We consider a system of Heisenberg spins interacting with each other via the Hamiltonian

$$\mathcal{H} = \sum_{(i,j)} J_{ij} \mathbf{S}_i \cdot \mathbf{S}_j - g\mu_B \sum_i \mathbf{H}_0 \cdot \mathbf{S}_i \tag{2.40}$$

where g and μ_B are the Landé factor and the Bohr magneton, respectively. $\mathbf{H}_0$ is a magnetic field applied along the z axis. To simplify the presentation, we suppose that the exchange interaction J_{ij} is limited to the nearest neighbors with $J_{ij} = J$. We have

$$\mathcal{H} = J \sum_{(i,j)} \mathbf{S}_i \cdot \mathbf{S}_j - g\mu_B \sum_i \mathbf{H}_0 \cdot \mathbf{S}_i \tag{2.41}$$

Note that we have defined the exchange terms in the Hamiltonian with a positive sign so that the antiferromagnetic interaction corresponds to $J > 0$. In zero applied field, the neighboring spins are antiparallel, except in geometrically frustrated systems (see Chapter 5). A few antiferromagnetic systems are displayed in Fig. 2.6.

2.2.1 Mean-Field Theory

In the case of non frustrated lattice, the antiferromagnetic ordering has two sublattices (see Fig. 2.6): sublattice of ↑ spins and sublattice

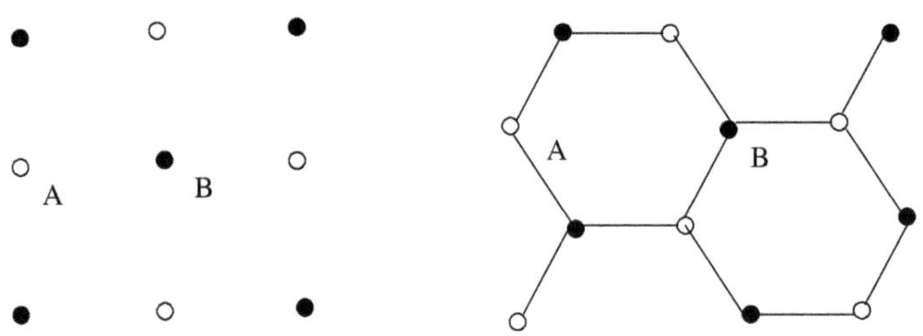

Figure 2.6 Antiferromagnetic ordering: Black and white circles denote ↑ and ↓ spins, respectively.

of ↓ spins, indicated hereafter by indices l and m, respectively. For simplicity, we treat the case of weak field $H_0 \ll J$ so that the antiferromagnetic ordering remains.

The mean-field theory is applied to an antiferromagnet as follows. We write the following mean-field energies of spins l and m:

$$H_l = CJ < S_-^z > S_l^z + [CJ < \Delta S_- > -g\mu_B H_0]\, S_l^z \quad (2.42)$$

$$H_m = CJ < S_+^z > S_m^z + [CJ < \Delta S_+ > -g\mu_B H_0]\, S_m^z \quad (2.43)$$

where C is the coordination number, $< S_l^z > = < S_+^z > + < \Delta S_+ >$ denotes the average value of S_l^z, and $< \Delta S_+ >$ the spin variation induced by the applied field. Using H_l, we calculate $< S_l^z >$ as follows:

$$\begin{aligned} < S_l^z > &= < S_+^z > + < \Delta S_+ > \\ &= \frac{\mathrm{Tr} S_l^z \mathrm{e}^{-\beta H_l}}{\mathrm{Tr e}^{-\beta H_l}} \\ &= S B_S(x) \end{aligned} \quad (2.44)$$

where $B_S(x)$ is the Brillouin function given by

$$B_S(x) = \frac{2S+1}{2S} \coth \frac{(2S+1)x}{2S} - \frac{1}{2S} \coth \frac{x}{2S} \quad (2.45)$$

with

$$x = \beta[-CJS(< S_-^z > + < \Delta S_- >) + g\mu_B S H_0] \quad (2.46)$$

For weak fields, we expand the function $B_S(x)$ around

$$x_0 = -\beta CJS < S_-^z > .$$

We then obtain

$$\begin{aligned} &< S_+^z > + < \Delta S_+ > \\ &\quad \simeq S B_S(-\beta CJS < S_-^z >) - \beta[CJS^2 < \Delta S_- > \\ &\quad -g\mu_B S^2 H_0] B_S'(x_0) \end{aligned}$$

therefore

$$< S_+^z > \simeq S B_S(-\beta CJS < S_-^z >) \quad (2.47)$$

$$< \Delta S_+ > \simeq -\beta[CJS^2 < \Delta S_- > -g\mu_B S^2 H_0] B_S'(x_0) \quad (2.48)$$

$B'_S(x_0)$ being the derivative of $B_S(x_0)$ with respect to x taken at x_0. In the same manner, we obtain for down sublattice spin $< S^z_m >$

$$< S^z_- > \simeq SB_S(-\beta CJS < S^z_+ >) \tag{2.49}$$

$$< \Delta S_- > \simeq -\beta[CJS^2 < \Delta S_+ > -g\mu_B S^2 H_0]B'_S(x_0^-) \tag{2.50}$$

with $x_0^- = -\beta CJS < S^z_+ >$.

If the two sublattices are symmetric, namely $< S^z_+ > = - < S^z_- > \equiv < S^z >$, then Eqs. (2.47) and (2.49) are equivalent because the Brillouin function is an odd function. We then have only one implicit equation for $< S^z >$ to solve

$$< S^z >= SB_S(\beta CJS < S^z >) \tag{2.51}$$

This mean-field equation for a sublattice spin is the same as that for ferromagnets, Eq. (2.12) [note that there is no factor 2 in Eq. (2.51) because we did not use the factor 2 for the exchange interaction in the Hamiltonian (2.40)]. We have thus the same result on the temperature dependence of $< S^z >$ and on the critical temperature. Therefore, the critical temperature for antiferromagnets, called "Néel temperature" and denoted by T_N, is given by

$$\frac{k_B T_N}{J} = \frac{CS(S+1)}{3} \tag{2.52}$$

We calculate $< \Delta S_\pm >$. Since H_0 induces a positive amount of the z component for both sublattices, and by symmetry, we have $< \Delta S_+ >=< \Delta S_- >\equiv< \Delta S >$. Note that $B'_S(x)$ is an even function of x; therefore from Eqs. (2.48) and (2.50), we have

$$< \Delta S >= -\beta[CJS^2 < \Delta S > -g\mu_B S^2 H_0]B'_S(\beta CJS < S^z >) \tag{2.53}$$

The susceptibility is given by

$$\chi_\parallel = \left(\frac{\partial M}{\partial H_0}\right)_{H_0=0} = \frac{Ng\mu_B < \Delta S >}{H_0}$$
$$= \frac{N(g\mu_B S)^2 B'_S(\beta CJS < S^z >)}{k_B T + CJS^2 B'_S(\beta CJS < S^z >)} \tag{2.54}$$

When $T \to 0$, $B'_S(\cdots)$ tends to 0 faster than T. We deduce that $\chi_\parallel = 0$. On the contrary, for $T \geq T_N$, $B'_S(\cdots) \simeq \frac{S+1}{3S}$, we get

$$\chi_\parallel = \frac{N(g\mu_B)^2 S(S+1)}{3k_B(T+T_N)} \tag{2.55}$$

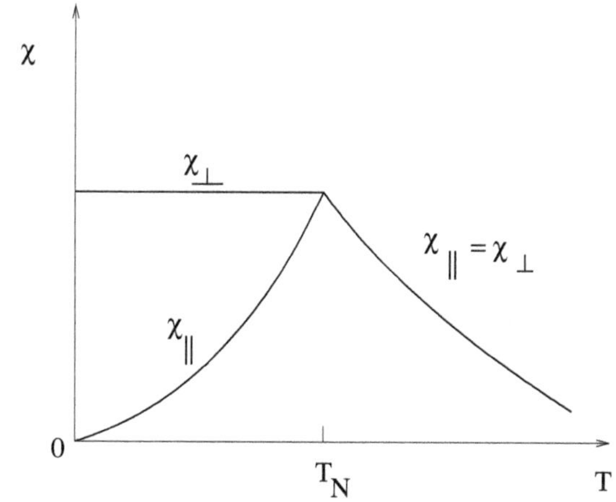

Figure 2.7 Susceptibility $\chi_{\|}$ and $\chi_{\perp}$ of an antiferromagnet versus T.

where we notice the + sign in front of T_N, in contrast to the ferromagnetic case. There is thus no divergence of the susceptibility at the phase transition for an antiferromagnet.

In the case where the applied field is also weak but perpendicular to the z axis, for example $\mathbf{H}_0 \parallel \mathbf{Ox}$, we modify (2.42) and (2.43) to obtain

$$\chi_{\perp}(T \geq T_N) = \frac{N(g\mu_B)^2 S(S+1)}{3k_B(T + T_N)} = \chi_{\|}(T \geq T_N) \tag{2.56}$$

and

$$\chi_{\perp}(T \leq T_N) = \frac{N(g\mu_B)^2}{4CJ} = \text{constant} \tag{2.57}$$

We show in Fig. 2.7 $\chi_{\|}$ and $\chi_{\perp}$ versus T.

In materials which have magnetic domains or in powdered systems, experimental susceptibility at $T \leq T_N$ is an average with spatial weight coefficients 1/3 and 2/3:

$$\chi(T \leq T_N) = \frac{1}{3}\chi_{\|} + \frac{2}{3}\chi_{\perp} \tag{2.58}$$

2.2.2 Spin Orientation in a Strong Applied Magnetic Field

The results shown above have been calculated with the assumption of weak field. When $\mathbf{H}_0$ is sufficiently strong, the results will be different as seen below.

We suppose that $\mathbf{H}_0$ is parallel to the z axis. The ↑ spins have their energy lowered by the Zeeman effect $-g\mu_B S_l^z H_0$ while the ↓ spins have their energy increased by $-g\mu_B S_m^z H_0 > 0 (S_m^z < 0)$. Contrary to the weak field case where the spins remain approximately antiparallel because of the dominant J, in the case of strong field the competition between the Zeeman effect and the exchange interaction determines the stable spin configuration as seen below.

We consider the general case where we add a uniaxial anisotropy term to the Hamiltonian (2.41) to fix the easy-magnetization axis. We suppose that $\mathbf{H}_0$ is applied in the direction which forms an angle ζ ($\zeta \in [0, \pi]$) with respect to the easy-magnetization axis. The competition between the Zeeman effect and J gives rise to a configuration of the two sublattices shown in Fig. 2.8 where θ is the angle of $\mathbf{H}_0$ with respect to the $+z$ axis.

The exchange energy is written as

$$E_e = \lambda M^2 \cos(\pi - 2\varphi) \tag{2.59}$$

where

$$\lambda = \frac{2CJ}{N(g\mu_B)^2} \tag{2.60}$$

with M being the magnetization modulus.

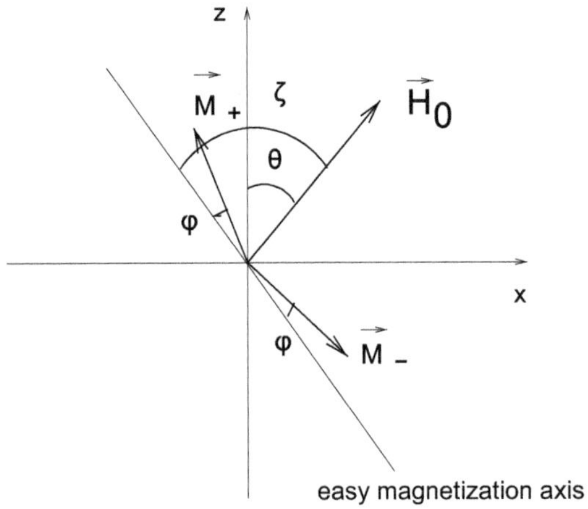

Figure 2.8 Spin orientation with respect to the direction of $\mathbf{H}_0$.

The anisotropy energy is written for sublattices $\mathbf{M}_+$ and $\mathbf{M}_-$ (see Fig. 2.8) as

$$E_a = \frac{K}{2}\left[\cos^2(\zeta - \theta - \varphi) + \cos^2(\zeta - \theta + \varphi)\right] \tag{2.61}$$

where K is the anisotropy constant.

The Zeeman energy is

$$E_Z = -H_0 M\left[\cos(\zeta - \varphi) + \cos(\pi - \zeta - \varphi)\right] \tag{2.62}$$

The energy induced by the variation of M under the applied field is

$$E_i = -\frac{1}{2}\chi_{\parallel}(H_0 \cos\zeta)^2 \tag{2.63}$$

because, by definition,

$$dM = \chi_{\parallel} dH \cos\zeta \tag{2.64}$$

$$\Rightarrow \quad E_i = -\int H \cos\zeta \, dM = -\chi_{\parallel}\int_0^{H_0} H\, dH \cos^2\zeta$$

$$= -\frac{1}{2}\chi_{\parallel}(H_0 \cos\zeta)^2 \tag{2.65}$$

The total energy is thus

$$E = E_e + E_a + E_Z + E_i \tag{2.66}$$

By minimizing E of (2.66) with respect to φ and ζ we obtain

$$0 = \frac{\partial E}{\partial \varphi}$$
$$= \cos\varphi\{4\lambda M^2 \sin\varphi + 2K\cos[2(\zeta - \theta)]\sin\varphi - 2H_0 M \sin\zeta\}$$

therefore,

$$\sin\varphi = \frac{H_0 M \sin\zeta}{2\lambda M^2 + K\cos[2(\zeta - \theta)]} \tag{2.67}$$

Since, by definition, $\chi_\perp = \frac{2M\sin\varphi}{H_0 \sin\zeta}$, we get

$$\chi_\perp = \frac{2M^2}{2\lambda M^2 + K\cos[2(\zeta - \theta)]} \tag{2.68}$$

Before minimizing (2.66) with respect to ζ, we consider the regime where $\lambda \gg K, H_0$. In this case $\varphi \simeq 0$ (see Fig. 2.8), so that

$\cos(\zeta - \varphi) \simeq \cos(\pi - \zeta - \varphi)$. Replacing this and (2.67)-(2.68) in (2.66), we arrive at

$$E = -\lambda M^2 - K\cos^2(\zeta - \varphi) - \frac{1}{2}\chi_{\parallel} H_0^2 \cos^2\zeta - \frac{1}{2}\chi_{\perp} H_0^2 \sin^2\zeta \quad (2.69)$$

The minimization with respect to ζ leads to

$$0 = \frac{\partial E}{\partial \zeta}$$

$$\tan(2\zeta) = \frac{\sin(2\theta)}{\cos(2\theta) - \frac{\chi_{\perp} - \chi_{\parallel}}{2K} H_0^2} \quad (2.70)$$

We examine a particular case where $\theta = 0$ ($\mathbf{H}_0 \parallel \mathbf{Oz}$). The solutions are

- $\zeta = 0$ if $H_0 < H_c$,
- $\zeta \simeq \pi/2$ if $H_0 > H_c$ where

$$H_c(\text{critical field}) = \sqrt{\frac{2K}{\chi_{\perp} - \chi_{\parallel}}} \quad (2.71)$$

The spin configurations corresponding to these two solutions are displayed in Fig. 2.9. The transition between these phases when $H_0 = H_c$ is called "spin-flop transition": the spins are approximately perpendicular to $\mathbf{H}_0$ for $H_0 > H_c$. H_c est called "critical field."

This result has been obtained with the hypothesis $\lambda \gg K, H_0$. In the case where H_0 is larger than H_c and larger than the local exchange field acting on a spin, all spins will turn into the direction of $\mathbf{H}_0$.

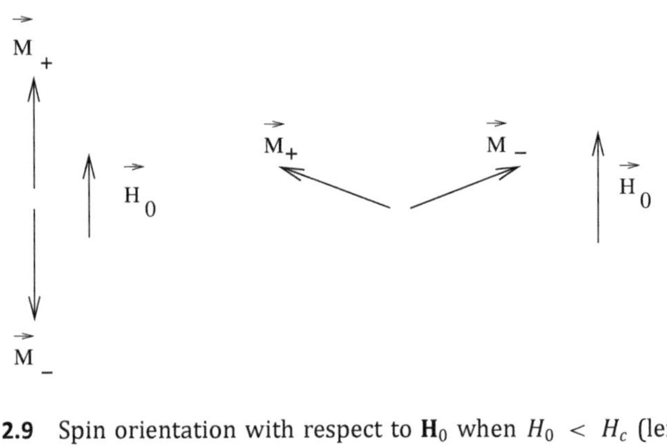

Figure 2.9 Spin orientation with respect to $\mathbf{H}_0$ when $H_0 < H_c$ (left) and $H_0 > H_c$ (right).

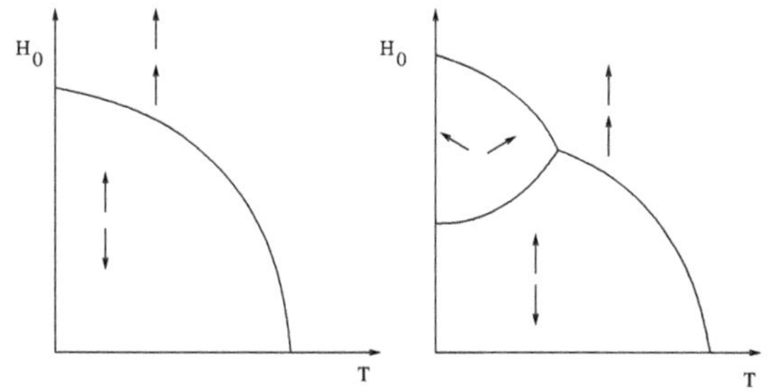

Figure 2.10 Left: Phase diagram of an antiferromagnet with Ising spins under an applied field of amplitude H_0. The line separates the antiferromagnetic and paramagnetic phases under field. Right: Phase diagram in the case of Heisenberg spins, there is a spin-flop phase.

2.2.3 Phase Transition in an Applied Magnetic Field

The results shown above were obtained at $T = 0$. We discuss now the effect of T in an antiferromagnet under a strong applied field.

To simplify the description, let us consider the Ising spin model. The field $\mathbf{H}_0$ is supposed to be parallel to **Oz**. If $H_0 < H_c$ where H_c is the critical field which is to be determined for the Ising model (see Problem 10 below), the spins remain antiparallel between them. If $H_0 > H_c$, all spins are parallel to $\mathbf{H}_0$: There is no spin-flop phase for Ising spins.

In the case of ferromagnets in a field, the magnetization is never zero, so a phase transition is impossible at any temperature. In the case of antiferromagnets, when $H_0 < H_c$ there is a possibility that the antiferromagnetic order is broken with increasing T: At high temperatures, spins excited by the temperature finish by turning themselves parallel to the field at a temperature T_c. Of course, T_c depends on H_0. We display schematically a phase diagram in Fig. 2.10. More details on the phase transition are given in Chapter 5.

2.3 Ferrimagnetism in Mean-Field Theory

Ferrimagnetic materials have complicated crystalline structures. There are often many sublattices of non equivalent spins interacting with each other via antiferromagnetic couplings. A well-known example is Fe_2O_3 (ferrites) which has three sublattices: Fe^{3+} with $\uparrow$ spin, Fe^{3+} with $\downarrow$ spin and Fe^{2+} with $\uparrow$ spin. The spin amplitude of Fe^{3+} is 5/2, that of Fe^{2+} is 2. As a consequence, the resulting magnetization comes from the spins of Fe^{2+}. For ferrimagnets, we should introduce several interaction parameters intra- and inter-sublattices. Ferrimagnets have very rich and complicated properties which are used in numerous applications such as recording devices, thanks to their very high critical temperatures of the order of 500–800°C.

We introduce here a very simple model to illustrate some remarkable properties of ferrimagnets. We consider a system of Heisenberg spins which is composed of two sublattices, sublattice A containing $\uparrow$ spins of amplitude S_A and sublattice B containing $\downarrow$ spins of amplitude S_B. The Hamiltonian is written as

$$\mathcal{H} = J_1 \sum_{(l,m)} \mathbf{S}_l \cdot \mathbf{S}_m + J_2^A \sum_{(l,l')} \mathbf{S}_l \cdot \mathbf{S}_{l'} + J_2^B \sum_{(m,m')} \mathbf{S}_m \cdot \mathbf{S}_{m'} \quad (2.72)$$

where (l, l') and (m, m') indicate the sites of A and B, respectively. The interactions are J_2^A between the neighbors belonging to the sublattice A, J_2^B between neighbors belonging to the sublattice B, and J_1 between inter-sublattice nearest neighbors. We suppose $J_1 > 0$ (antiferromagnetic). The signs of J_2^A and J_2^B can be arbitrary. For simplicity, we assume $J_2^A = J_2^B = 0$. We can start with equations (2.47) and (2.49) for two sublattices in zero applied field:

$$< S_A^z > = S_A B_{S_A}(-\beta C J_1 S_A < S_B^z >) \quad (2.73)$$

$$< S_B^z > = S_B B_{S_B}(-\beta C J_1 S_B < S_A^z >) \quad (2.74)$$

where C is the coordination number. Since the sublattices are not equivalent because $S_A \neq S_B$, we have to solve these two coupled equations by iteration.

Let us use the notations $M_A =< S_A^z >$ and $M_B =< S_B^z >$. At $T = 0$, the expansion of the functions $B_{S_A}(\cdots)$ and $B_{S_B}(\cdots)$ gives $M_A = S_A$ and $M_B = S_B$ (see Section 2.1). At low temperatures, we can obtain the solution for M_A and M_B by solving graphically Eqs. (2.73)-(2.74). However, it is more complicated to calculate the critical temperature. The high-temperature expansion similar to (2.20) gives two equations containing M_A and M_B of the form $M_A = a(S_A, T)M_B + b(S_A, T)M_B^3 + \cdots$ and $M_B = c(S_B, T)M_A + d(S_B, T)M_A^3 + \cdots$ where $a(S_A, T)$, $b(S_A, T)$, $c(S_B, T)$ and $d(S_B, T)$ are coefficients depending on S_A, S_B and T. An explicit expression of the critical temperature T_N can be obtained (see Problem 6 below). We have

$$k_B T_N = \frac{CJ_1}{3}\sqrt{S_A(S_A+1)S_B(S_B+1)} \tag{2.75}$$

This result is equivalent to (2.52) for antiferromagnets if $S_A = S_B$. Let us give a qualitative argument. We suppose that $S_A > S_B$. When M_B becomes very small M_A is still large. It induces a local field on its B neighbors, keeping them from going to zero. As long as M_A is not zero, M_B is maintained at a non zero value. However, fluctuations of M_B affect in turn M_A. Therefore, the critical temperature is somewhere between the two critical temperatures of the sublattices when they are independent, namely

$$\frac{k_B T_A}{J_1} = \frac{CS_A(S_A+1)}{3} > \frac{k_B T_N}{J_1} > \frac{k_B T_B}{J_1} = \frac{CS_B(S_B+1)}{3}$$

For $S_A = 2$, $S_B = 1$ and $C = 8$ (body-centered cubic lattice), we have $k_B T_A/J_1 = 16$, $k_B T_B/J_1 = 16/3 \simeq 5.33$, and $k_B T_N/J_1 = 8\sqrt{12}/3 \simeq 9.2376$. The numerical solution of (2.73) and (2.74) for the above values of S_A, S_B and C is shown as a function of T in Fig. 2.11.

Note that the sublattices have different characteristics: When we include J_2^A and J_2^B for example, it can happen that M_A (>0) and M_B (<0) cancel out at some T below T_N, yielding a zero net magnetization. This temperature is called "magnetization compensation point" observed in some ferrimagnetic materials such as garnets and rare-earth transition-metal alloys. Ferrimagnets may also have an angular-momentum compensation point where the net angular momentum vanishes. This compensation point is a crucial point for achieving high speed magnetization reversal in magnetic memory device [326].

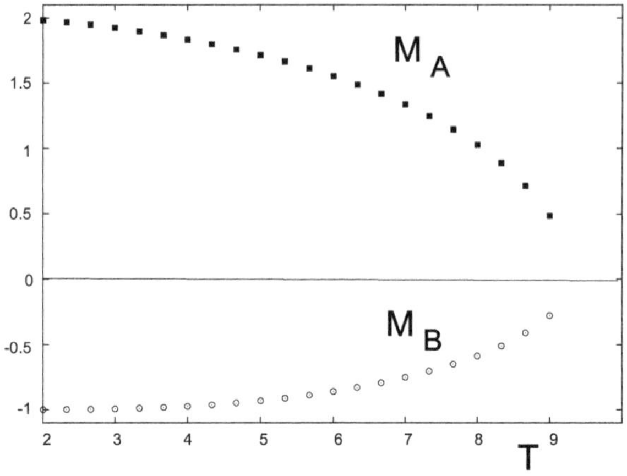

Figure 2.11 Numerical solution of (2.73) and (2.74) is shown as a function of T for $S_A = 2, S_B = 1, C = 8$ and $k_B/J_1 = 1$. See text for comments.

2.4 Conclusion

In this chapter, we have presented the mean-field theory applied to ferromagnets, antiferromagnets and ferrimagnets. This theory is considered as the first approximation used to study behaviors of systems of interacting spins. Low-temperature properties and the phase transition are described with physical quantities such as the order parameter, the critical temperature, the susceptibility and the specific heat. The theory gives qualitative results for dimension 3, but becomes incorrect in lower dimensions. It can be improved by higher-order approximations such as the Landau–Ginzburg theory (see Section 5.3) and the cluster-variation method (CVM). The CVM consists in an exact treatment of a cluster of small size embedded in a crystal: The interaction of the cluster with the crystal is treated with the mean-field theory. The larger the cluster size, the better the results. However, the calculation is very complicated for large sizes. The method with smallest clusters is sometimes called "Bethe–Peierls–Weiss" approximation (see Problem 7). To obtain more precise properties of interacting spin systems, we can use other methods such as the spin-wave theory (Chapter 3), Green's function theory (Chapter 4), the renormalization group or the Monte Carlo simulation (Chapter 5).

2.5 Problems

Problem 1. Define the order parameter of an antiferromagnetic lattice of Ising spins.

Problem 2. Consider the q-state Potts model defined by the Hamiltonian (1.66) on a square lattice.

- **(a)** Define the order parameter of the q-state Potts model.
- **(b)** Describe the ground state and its degeneracy when $J > 0$.
- **(c)** If $J < 0$, what is the ground state for $q = 2$ and $q = 3$? For $q = 3$, find ways to construct some ground states and give comments.
- **(d)** Show that the Potts model is equivalent to the Ising model when $q = 2$.

Problem 3. Domain walls:

In magnetic materials, due to several reasons, we may have magnetic domains schematically illustrated in Fig. 2.5. The spins at the interface between two neighboring domains should arrange themselves in a smooth configuration in order to make a gradual change from one domain to the other. An example of such a "domain wall" is shown in that figure. Calculate the energy of a wall of thickness of N spins.

Problem 4. Bragg–Williams approximation:

The mean-field theory presented in this chapter can be demonstrated by the Bragg–Williams approximation described in this problem.

Consider a crystal of N sites each occupied by an Ising spin at a given temperature T. The coordination number (number of nearest neighbors) is z. One supposes the periodic boundary conditions. Let N_+ and N_- be the number of up and down spins, respectively. The energy of a pair of parallel spins is $-J$ ($J > 0$) and that of a pair of antiparallel spins is $+J$. Let X be defined by $N_+ = N(1 + X)/2$. One has then $N_- = N(1 - X)/2$.

(a) Calculate the entropy S.

(b) Calculate the probability to have an up spin at a lattice site. Deduce the numbers of up-up, down-down and up-down spin pairs, as functions of X.

(c) Calculate the energy of the crystal as a functions of z, J and X.

(d) Calculate the free energy F. Deduce the expression of X at thermal equilibrium, namely at the minimum of F. Show that this expression is equivalent to the mean-field equation (2.12) with $S = \pm 1$.

(e) Give the mean-field solution for the critical temperature T_c. Calculate the entropy for $T > T_c$.

Problem 5. Binary alloys by spin language:

Consider a lattice where there are two kinds of site such as the one shown in Fig. 2.12: sites of type I (white circles) and sites of type II (black circles). There are two kinds of atoms A and B occupying the lattice sites. The number of each atom type is $N/2$. The interaction energy between two neighbors of the same kind is ϵ, that between two neighbors of different kinds is ϕ. One supposes $\epsilon > \phi$.

In the disordered phase, half of A atoms are on the white sites and the other half on the black sites. The same situation is for B atoms. We can study the ordering structure of this binary alloy by mapping the problem into a spin language: an A atom is represented by an up Ising spin and a B atom by a down Ising spin. The $A - B$ attractive interaction is replaced by an antiferromagnetic interaction.

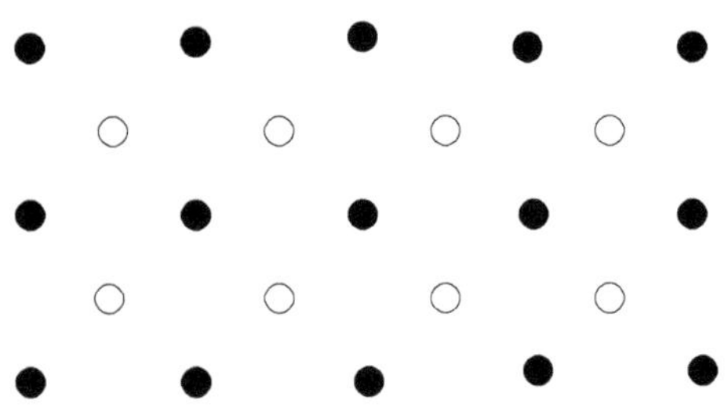

Figure 2.12 Binary alloy (see Problem 5): white and black circles represent sites of type I and II, respectively.

(a) Describe the ground state.

(b) The system energy is E. Let $N_{\uparrow,I}$ be the number of $\uparrow$-spins occupying sites of the type I. We define x by

$$N_{\uparrow,I} = N(1+x)/4 \quad (2.76)$$

- What is the value domain of x ? Which state does $x = 0$ correspond to? Calculate as a function of x the number of $\uparrow$-spins occupying sites of type II ($N_{\uparrow,II}$). The same question is for $\downarrow$-spins. One considers $x > 0$ hereafter.
- Calculate the probabilities as functions of x for a $\uparrow$-spin at a site of type I and at a site of type II, supposing that all probabilities are independent. The same question is for a $\downarrow$-spin.
- Let $N_{\uparrow,\uparrow}$, $N_{\downarrow,\downarrow}$, and $N_{\uparrow,\downarrow}$ be the numbers of $\uparrow\uparrow$, $\downarrow\downarrow$ and $\uparrow\downarrow$ spin pairs, respectively. Calculate these quantities as functions of x. Show that $N_{\uparrow,\uparrow} = N(1-x^2)/2$, $N_{\downarrow,\downarrow} = N(1-x^2)/2$, $N_{\uparrow,\downarrow} = N(1+x^2)$.
- Calculate E as functions of x, ϵ and ϕ. Show that E can be written as

$$E = N(\epsilon+\phi) - N(\epsilon-\phi)x^2 \quad (2.77)$$

- Calculate $\Omega(E)$ the number of microscopic states of energy E. Deduce the entropy S.
- Calculate temperature T. Show that

$$x = \tanh[2(\epsilon-\phi)x/(k_B T)] \quad (2.78)$$

Show that x tends to 1 at low T, and that $x = 0$ when $T > T_c = 2(\epsilon-\phi)/k_B$.

Problem 6. Critical temperature of ferrimagnet:
Using the mean-field theory, calculate the critical temperature T_N of the simple model for a ferrimagnet described in Section 2.3.

Problem 7. Improvement of mean-field theory:

(a) Two-spin problem:
Consider the following Hamiltonian

$$\mathcal{H} = -2J\,\mathbf{S}_1\cdot\mathbf{S}_2 - D[(S_1^z)^2 + (S_2^z)^2] - B(S_1^z + S_2^z) \quad (2.79)$$

where J (exchange interaction) and D (magnetic anisotropy) are positive constants and B magnitude of an applied magnetic field in the z direction. Find the eigenvalues and eigenvectors of $\mathcal{H}$ for spin one-half.

(b) Improved mean-field theory:
Consider the Heisenberg spin model:

$$\mathcal{H} = -2J \sum_{(i,j)} \mathbf{S}_i \cdot \mathbf{S}_j$$

In the first step, we treat exactly the interaction of two neighboring spins. In the second step, we use the mean-field theory to treat the interaction of the two-spin cluster embedded in the crystal. Explicitly, consider two spins $\mathbf{S}_i$ and $\mathbf{S}_j$ embedded in a crystal. The Hamiltonian is given by

$$\mathcal{H}_{ij} = -2J\,\mathbf{S}_i \cdot \mathbf{S}_j - 2(Z-1)J < S^z > (S_i^z + S_j^z) \quad (2.80)$$

where Z is the coordination number. Show that the critical temperature T_c for $S = 1/2$ is given by

$$e^{-2J/k_BT_c} + 3 - 2(Z-1)J/k_BT = 0 \quad (2.81)$$

Problem 8. Interaction between next-nearest neighbors in mean-field treatment:
Consider a centered cubic lattice where each site is occupied by an Ising spin with values ± 1. The Hamiltonian is given by

$$\mathcal{H} = -J_1 \sum_{(i,j)} \sigma_i \sigma_j - J_2 \sum_{(i,k)} \sigma_i \sigma_k \quad (2.82)$$

where σ_i is the spin at site i, J_1 (>0) exchange interaction between nearest neighbors and J_2 (>0) interaction between next-nearest neighbors. The first and second sums are made over pairs of corresponding neighbors.

(a) Describe the magnetic ordering at temperature $T = 0$.
(b) Give briefly the hypothesis of the mean-field theory.
(c) By a qualitative argument, show that the interaction between next-nearest neighbors, J_2, increases the critical temperature.

(d) Using the mean-field theory, calculate the partition function of a spin at a given temperature T. Deduce an equation which allows us to calculate $< \sigma >$, mean value of a spin, at T.

(e) Determine the critical temperature T_c as functions of J_1 and J_2.

(f) In the case where J_2 is negative (antiferromagnetic interaction), the above result is no more valid beyond a critical value of $|J_2|$. Determine that critical value J_2^c. What is the magnetic ordering when $|J_2| \gg |J_2^c|$ at $T = 0$?

Problem 9. Repeat Problem 7 in the case of an antiferromagnet.

Problem 10. Calculate the critical field H_c in the following cases

(a) a simple cubic lattice of Ising spins with antiferromagnetic interaction between nearest neighbors

(b) a square lattice of Ising spins with antiferromagnetic interaction J_1 between nearest neighbors and ferromagnetic interaction J_2 between next-nearest neighbors.

Chapter 3

Theory of Spin Waves

In this chapter, we study the properties of the spin waves excited in a system of interacting spins. We suppose that the spins are localized on the lattice sites. The exchange interaction between spins makes them rotate about the quantization axis in a collective manner: These collective excitations are called "spin waves" or "magnons" when quantized. Such collective excitations or elementary excitations are common phenomena observed in systems of interacting particles or quasi-particles in condensed matter. One can mention phonons (waves of atomic motions around the atoms' equilibrium positions in a system of interacting atoms) and plasmons (waves of the charge density). Note that there is no spin wave in systems of Ising spins because Ising spins cannot be deviated from their axis to create a wave.

Spin waves propagate in systems with a translation invariance. In systems with a broken translation, spin waves can be localized or damped such as in the presence of an impurity or a surface. Spin waves cannot be excited in disordered systems due to their wave nature since a propagating wave should have a constant amplitude and a constant phase shift in space. The same is observed with a system of interacting atoms: Phonons can be excited only at low temperatures when the system has a crystalline structure.

Physics of Magnetic Thin Films: Theory and Simulation
Hung T. Diep

ISBN 978-981-4877-42-8 (Hardcover), 978-1-003-12110-7 (eBook)
www.jennystanford.com

At low temperatures, spin waves yield the main thermal behavior of a system of ordered spins such as in a ferromagnet, as seen below. This chapter presents the spin wave theory as applied to ferromagnets, antiferromagnets and, to a lesser extent, ferrimagnets. It gives a necessary background for understanding the applications in part II of the book.

3.1 Spin Waves in Ferromagnets

3.1.1 Classical Treatment

We consider a ferromagnet where the spins are parallel, aligned along the Oz axis in the ground state. Each spin is considered as a vector of modulus S with three components. This is the classical Heisenberg spin model. As the temperature increases, the spins rotate around their z axis in a collective manner as shown in Fig. 3.1. The energy provided by the temperature is carried by the excited spin wave.

In the following, we calculate the spin wave energy for the classical Heisenberg spin model on a lattice. The spins are supposed to interact with each other via a nearest-neighbor exchange interaction J. The interaction of the spin $\mathbf{S}_l$ with its nearest neighbors is written as

$$\mathcal{H}_l = -2J\,\mathbf{S}_l \cdot \sum_{\mathbf{R}} \mathbf{S}_{l+\mathbf{R}} \tag{3.1}$$

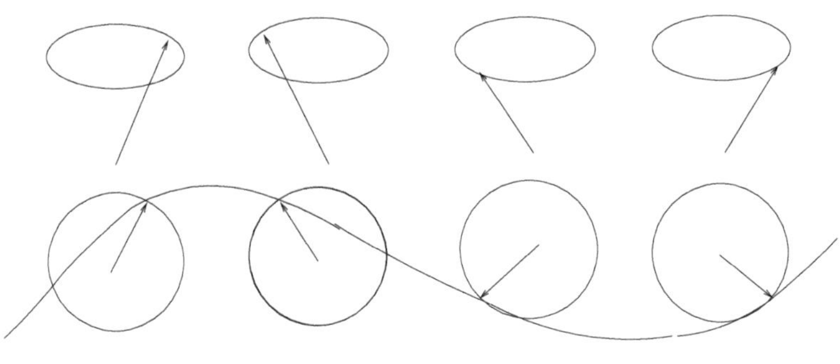

Figure 3.1 Side view and top view of a spin wave.

where $\mathbf{R}$ is the vector connecting the spin $\mathbf{S}_l$ with a nearest neighbor. The sum is performed over all nearest neighbors. Equation (3.1) can be rewritten as

$$\mathcal{H}_l = -\mathbf{M}_l \cdot \mathbf{H}_{ex} \tag{3.2}$$

where

$$\mathbf{M}_l = g\mu_B \mathbf{S}_l \tag{3.3}$$

and

$$\mathbf{H}_{ex} = \frac{2J}{g\mu_B} \sum_{\mathbf{R}} \mathbf{S}_{l+\mathbf{R}} \tag{3.4}$$

g and μ_B are the Landé factor and Bohr magneton. $\mathbf{M}_l$ and $\mathbf{H}_{ex}$ are, respectively, the magnetic moment of spin $\mathbf{S}_l$ and the field of the nearest neighbors acting on $\mathbf{S}_l$.

We write the equation of motion of the kinetic moment $\hbar\mathbf{S}_l$ as follows:

$$\hbar\frac{d\mathbf{S}_l}{dt} = [\mathbf{M}_l \wedge \mathbf{H}_{ex}] = 2J\left[\mathbf{S}_l \wedge \sum_{\mathbf{R}} \mathbf{S}_{l+\mathbf{R}}\right] \tag{3.5}$$

If $\mathbf{S}_l$ is parallel to its neighbors then the right-hand side of Eq. (3.5) is zero, one has $\hbar\frac{d\mathbf{S}_l}{dt} = 0$, i.e., $\mathbf{S}_l$ is equal to a constant vector, thus there is no spin wave.

In an excited state due to the spin rotation around the z axis, one can decompose $\mathbf{S}_l$ as shown in Fig. 3.2:

$$\mathbf{S}_l = \mathbf{S}_0 + \delta\mathbf{S}_l \tag{3.6}$$

where $\delta\mathbf{S}_l$ represents the deviation and $\mathbf{S}_0$ the z spin component. For a homogenous system and for a given spin wave mode, it is natural to suppose that $\mathbf{S}_0$ is space- and time-independent. Equation (3.5) becomes

$$\begin{aligned}\hbar\frac{d(\delta\mathbf{S}_l)}{dt} &= 2J\sum_{\mathbf{R}} [(\mathbf{S}_0 + \delta\mathbf{S}_l) \wedge (\mathbf{S}_0 + \delta\mathbf{S}_{l+\mathbf{R}})] \\ &\simeq 2J\sum_{\mathbf{R}} [(\delta\mathbf{S}_l - \delta\mathbf{S}_{l+\mathbf{R}}) \wedge \mathbf{S}_0]\end{aligned} \tag{3.7}$$

where $\delta\mathbf{S}_l$ and $\delta\mathbf{S}_{l+\mathbf{R}}$ are supposed to be small so that we neglect their second-order terms. This hypothesis is justified at low temperatures.

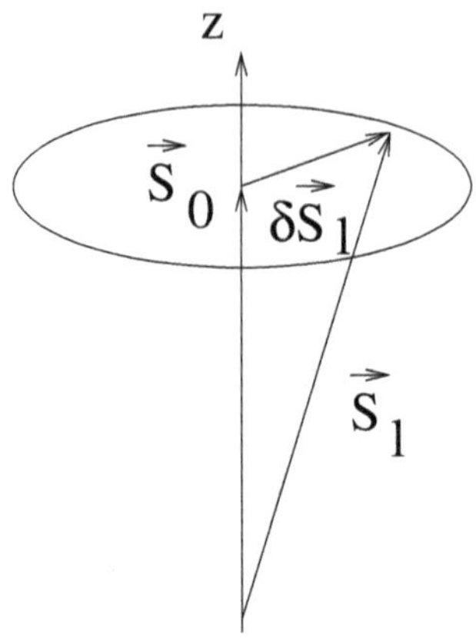

Figure 3.2 Decomposition of a spin.

We define now the spin components by: $\mathbf{S}_0 = S_l^z\hat{k} = S_0\hat{k}$, $(\delta\mathbf{S}_l)^x = S_l^x$ and $(\delta\mathbf{S}_l)^y = S_l^y$, $\hat{k}$ being the unit vector on the z axis (see Fig. 3.3).

We rewrite Eq. (3.7) as

$$\hbar\frac{dS_l^x}{dt} = 2J\,S\sum_{\mathbf{R}}\left(S_l^y - S_{l+\mathbf{R}}^y\right) \tag{3.8}$$

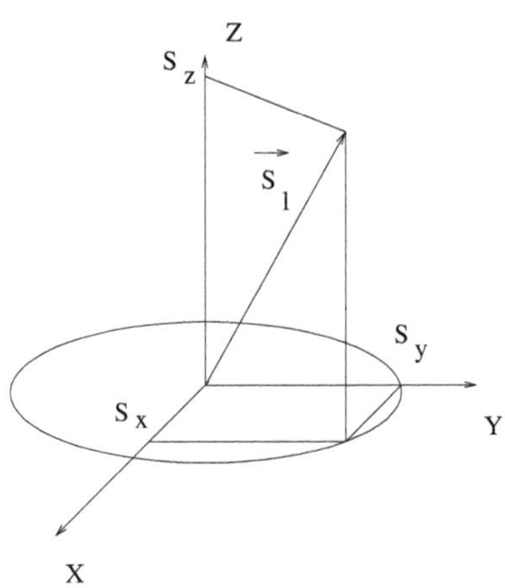

Figure 3.3 Spin components.

$$\hbar\frac{dS_l^y}{dt} = -2J\,S\sum_{\mathbf{R}}\left(S_l^x - S_{l+\mathbf{R}}^x\right) \tag{3.9}$$

$$\hbar\frac{dS_l^z}{dt} = 0 \tag{3.10}$$

The last equation indicates that S_l^z is a constant of motion. One now looks for the solutions of S_l^x and S_l^y of the form

$$S_l^x = U\,\mathrm{e}^{i(\mathbf{k}\cdot\mathbf{l}-\omega_{\mathbf{k}}t)} \tag{3.11}$$

$$S_l^y = V e^{i(\mathbf{k}\cdot\mathbf{l}-\omega_{\mathbf{k}}t)} \tag{3.12}$$

where **k** and **l** are the wave vector and the position of $\mathbf{S}_l$, respectively. Replacing (3.11)–(3.12) in (3.8)–(3.9), one obtains

$$-i\hbar\omega_{\mathbf{k}}U = 2J\,SZ\left[1-\frac{1}{Z}\sum_{\mathbf{R}}\mathrm{e}^{i\mathbf{k}\cdot\mathbf{R}}\right]V \tag{3.13}$$

$$i\hbar\omega_{\mathbf{k}}V = 2J\,SZ\left[1-\frac{1}{Z}\sum_{\mathbf{R}}\mathrm{e}^{i\mathbf{k}\cdot\mathbf{R}}\right]U \tag{3.14}$$

where Z is the coordination number (number of nearest neighbors). The non-trivial solutions of U and V verify

$$\begin{vmatrix} i\hbar\omega_{\mathbf{k}} & 2J\,SZ(1-\gamma_{\mathbf{k}}) \\ -2J\,SZ(1-\gamma_{\mathbf{k}}) & i\hbar\omega_{\mathbf{k}} \end{vmatrix} = 0$$

from which one has

$$\hbar\omega_{\mathbf{k}} = 2J\,SZ\,(1-\gamma_{\mathbf{k}}) \tag{3.15}$$

where

$$\gamma_{\mathbf{k}} = \frac{1}{Z}\sum_{\mathbf{R}}\mathrm{e}^{i\mathbf{k}\cdot\mathbf{R}} \tag{3.16}$$

Replacing (3.15)–(3.16) in the above coupled equations of U and V, one finds $V = -iU$. This relation indicates that $|U| = |V|$; therefore the spin rotation has a circular precession around the z axis.

The relation (3.15) is called "dispersion relation" of spin waves in ferromagnets.

Example: In the case of a chain of spins, one has

$$\gamma_k = \frac{1}{2}\left(\mathrm{e}^{ika}+\mathrm{e}^{-ika}\right) = \cos(ka) \tag{3.17}$$

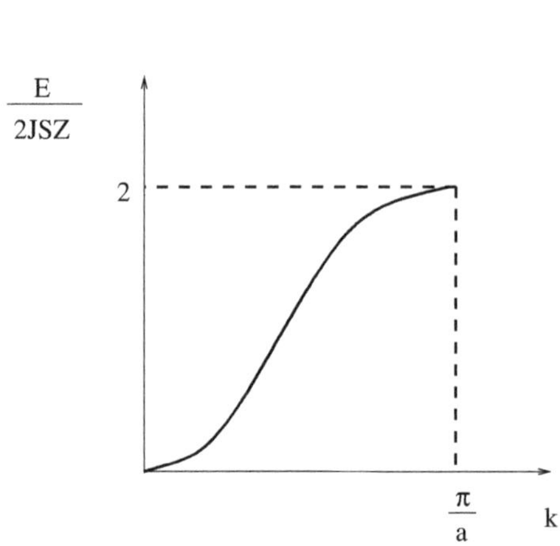

Figure 3.4 Spin wave dispersion relation of a ferromagnet in one dimension.

where a is the lattice constant. Equation (3.15) becomes

$$\hbar\omega_k = 2J\,SZ\,[1 - \cos(ka)] \tag{3.18}$$

Figure 3.4 shows $\hbar\omega_k$ versus k in the first Brillouin zone. When $ka \ll 1$, with $Z = 2$ for one dimension, one has

$$\hbar\omega_k \simeq 2J\,S(ka)^2 \tag{3.19}$$

ω_k is thus proportional to k^2 for small k (long wavelength). This behavior is also true for other dimensions. However, for antiferromagnets we will see below that $\omega_k \propto k$. Since macroscopic physical properties are calculated by averaging over the spin wave excitations, we will see the difference between ferromagnets and antiferromagnets. This difference is not observed in the mean-field theory.

Remark: Effect of the crystal symmetry is contained in the factor $\gamma_{\mathbf{k}}$. Here are a few examples:

(1) Square lattice:

$$\begin{aligned}\gamma_{\mathbf{k}} &= \frac{1}{4}\left(e^{ik_xa} + e^{-ik_xa} + e^{ik_ya} + e^{-ik_ya}\right)\\ &= \frac{1}{2}\left[\cos(k_xa) + \cos(k_ya)\right]\end{aligned} \tag{3.20}$$

(2) Simple cubic lattice:

$$\gamma_{\mathbf{k}} = \frac{1}{6}\left(e^{ik_x a} + e^{-ik_x a} + e^{ik_y a} + e^{-ik_y a} + e^{ik_z a} + e^{-ik_z a}\right)$$
$$= \frac{1}{3}\left[\cos(k_x a) + \cos(k_y a) + \cos(k_z a)\right] \quad (3.21)$$

(3) Centered cubic lattice:

$$\gamma_{\mathbf{k}} = \cos\left(\frac{k_x a}{2}\right)\cos\left(\frac{k_y a}{2}\right)\cos\left(\frac{k_z a}{2}\right) \quad (3.22)$$

(4) Face-centered cubic lattice:

$$\gamma_{\mathbf{k}} = \frac{1}{3}\left[\cos\left(\frac{k_x a}{2}\right)\cos\left(\frac{k_y a}{2}\right) + \cos\left(\frac{k_y a}{2}\right)\cos\left(\frac{k_z a}{2}\right) + \cos\left(\frac{k_z a}{2}\right)\cos\left(\frac{k_x a}{2}\right)\right] \quad (3.23)$$

3.1.2 Quantum Spin Wave Theory: Holstein–Primakoff Approximation

We consider now the case of quantum spins. The spin $\mathbf{S}_l$ at the lattice site l can be decomposed into the following spin operators: S_l^z and $S_l^{\pm} = S_l^x \pm i S_l^y$. These spin operators obey the following commutation relations:

$$\left[S_l^+, S_{l'}^-\right] = 2S_l^z\delta_{ll'} \quad (3.24)$$
$$\left[S_l^z, S_{l'}^{\pm}\right] = \pm S_l^{\pm}\delta_{ll'} \quad (3.25)$$

The Holdstein-Primakoff method consists in introducing the operators a and a^+ as follows:

$$S^z = S - a^+a \quad (3.26)$$
$$S^+ = \sqrt{2S}\, f(S)a \quad (3.27)$$
$$S^- = \sqrt{2S}a^+ f(S) \quad (3.28)$$

where

$$f(S) = \sqrt{1 - \frac{a^+a}{2S}} \quad (3.29)$$

The transformations (3.26)–(3.28) are called "Holstein–Primakoff transformations." We note that a^+a in Eq. (3.26) corresponds to the diminution of S due to the excited spin wave. a^+a is called therefore "spin wave number operator."

Let us calculate the spin wave dispersion relation using the Holstein–Primakoff transformations (3.26)–(3.28). We consider the following Heisenberg Hamiltonian

$$\mathcal{H} = -2J \sum_{(l,m)} \mathbf{S}_l \cdot \mathbf{S}_m - g\mu_B H \sum_l S_l^z \tag{3.30}$$

where for simplicity we suppose that interactions are limited to pairs of nearest neighbors (l, m) with a ferromagnetic exchange integral $J > 0$. H is the amplitude of a magnetic field applied along the z direction, g and μ_B are, respectively, the Landé factor and Bohr magneton.

Using $S^{\pm} = S^x \pm iS^y$ for $\mathbf{S}_l$ and $\mathbf{S}_m$, one rewrites $\mathcal{H}$ as

$$\mathcal{H} = -2J \sum_{(l,m)} \left[S_l^z S_m^z + \frac{1}{2}(S_l^+ S_m^- + S_l^- S_m^+) \right] - g\mu_B H \sum_l S_l^z \tag{3.31}$$

Replacing $S^{\pm}$ and S^z by (3.26)–(3.28) while keeping position indices l and m of operators a and a^+ one obtains

$$\mathcal{H} = -2J \sum_{(l,m)} [S^2 + Sa_l^+ f_l(S) f_m(S) a_m + S f_l(S) a_l a_m^+ f_m(S) - Sa_l^+ a_l$$
$$-Sa_m^+ a_m + a_l^+ a_l a_m^+ a_m] - g\mu_B H \sum_l (S - a_l^+ a_l) \tag{3.32}$$

It is impossible to find a solution of this equation because of non-linear terms such as $a_l^+ a_l a_m^+ a_m$, $f_l(S)$ and $f_m(S)$. In a first approximation, one can assume that the number of excited spin waves n is small with respect to $2S$ (namely $a^+ a \ll 2S$) so that one can expand $f_l(S)$ and $f_m(S)$ as follows:

$$f_l(S) \simeq 1 - \frac{a_l^+ a_l}{4S} + \cdots \tag{3.33}$$

$$f_m(S) \simeq 1 - \frac{a_m^+ a_m}{4S} + \cdots \tag{3.34}$$

Equation (3.32) becomes, to the quadratic order in a and a^+

$$\mathcal{H} \simeq -ZJNS^2 - g\mu_B HNS + 4JS \sum_{(l,m)} (a_l^+ a_l - a_l^+ a_m) + g\mu_B H \sum_l a_l^+ a_l \tag{3.35}$$

where one has used the following relation

$$\sum_{(l,m)} 1 = \frac{1}{2} \sum_l \sum_{\mathbf{R}} 1 = \frac{Z}{2} \sum_l 1 = \frac{Z}{2} N \tag{3.36}$$

R being the vector connecting the spin at l to one of its nearest neighbors, Z the coordination number and N the total number of spins.

The first term of (3.35) is the energy of the ground state where all spins are parallel and the second is a constant which will be omitted in the following. One introduces next the following Fourier transformations

$$a_l^+ = \frac{1}{\sqrt{N}} \sum_{\mathbf{k}} \mathrm{e}^{-i\mathbf{k}\cdot\mathbf{l}} a_{\mathbf{k}}^+ \tag{3.37}$$

$$a_l = \frac{1}{\sqrt{N}} \sum_{\mathbf{k}} \mathrm{e}^{i\mathbf{k}\cdot\mathbf{l}} a_{\mathbf{k}} \tag{3.38}$$

One can show that $a_{\mathbf{k}}$ and $a_{\mathbf{k}}^+$ obey the boson commutation relations just as real-space operators a_l and a_l^+ (see Problem 9 in Section 3.6). Putting (3.38) in (3.35) one finds

$$\mathcal{H} = \sum_{\mathbf{k}} [2ZJS(1-\gamma_{\mathbf{k}}) + g\mu_B H]\, a_{\mathbf{k}}^+ a_{\mathbf{k}} = \sum_{\mathbf{k}} \epsilon_{\mathbf{k}} a_{\mathbf{k}}^+ a_{\mathbf{k}} \tag{3.39}$$

where

$$\epsilon_{\mathbf{k}} = 2ZJS(1-\gamma_{\mathbf{k}}) + g\mu_B H \tag{3.40}$$

One sees that in the case where $H = 0$ one recovers the magnon dispersion relation (3.15) $\epsilon_{\mathbf{k}} = \hbar\omega_{\mathbf{k}}$, obtained by the classical treatment. The Holstein–Primakoff method allows, however, to go further by taking into account terms of order higher than quadratic in a^+ and a. By using expansions (3.33)–(3.34), one obtains terms of four operators, six operators, ... which represent interactions between spin waves. These terms play an important role when the temperature increases.

3.1.3 Properties at Low Temperatures

One studies here some low-temperature properties of spin waves using the dispersion relation (3.40).

3.1.3.1 Magnetization

One has seen above that a and a^+ are boson operators. The number of spin waves (or magnons) of **k** mode at temperature T is therefore given by the Bose-Einstein distribution

$$< n_{\mathbf{k}} > = < a_{\mathbf{k}}^{+} a_{\mathbf{k}} > = \frac{1}{\exp[\beta \epsilon_{\mathbf{k}}] - 1} \quad (3.41)$$

where $\beta = (k_B T)^{-1}$. The magnetization defined as the magnetic moment per unit volume is given for the volume $\Omega = 1$:

$$M = g\mu_B \sum_j < S_j^z > = g\mu_B \sum_{j=1}^{N} (S - < a_j^{+} a_j >) \quad (3.42)$$

where the sum is performed over all spins.

With (3.38), Eq. (3.42) becomes

$$M = g\mu_B \sum_j < S_j^z > = g\mu_B N \left(S - \frac{1}{N} \sum_{\mathbf{k}} < n_{\mathbf{k}} > \right) \quad (3.43)$$

where $< n_{\mathbf{k}} >$ is given by (3.41).

One shows here how to calculate M in the case of a simple cubic lattice and $H = 0$. The sum in (3.43) reads

$$\frac{1}{N} \sum_{\mathbf{k}} < n_{\mathbf{k}} > = \frac{\Omega}{(2\pi)^3} \frac{1}{N} \iiint \frac{dk_x dk_y dk_z}{\exp[\beta 2ZJ\, S(1 - \gamma_{\mathbf{k}})] - 1} \quad (3.44)$$

Using $\gamma_{\mathbf{k}}$ of (3.21), one has

$$1 - \gamma_{\mathbf{k}} = 1 - \frac{1}{3}[\cos(k_x a) + \cos(k_y a) + \cos(k_z a)] \simeq \frac{1}{6}[(k_x a)^2 + (k_y a)^2 + (k_z a)^2 - O(k^4)] \quad (3.45)$$

where one used an expansion for small $\mathbf{k}$ because at low temperatures (large β) the main contribution to the integral (3.44) comes from the region of small $\mathbf{k}$. With $\Omega = 1$, $Z = 6$ (simple cubic lattice) and (3.45), Eq. (3.44) becomes

$$\frac{1}{N}\sum_{\mathbf{k}} < n_{\mathbf{k}} > \simeq \frac{1}{(2\pi)^3}\int_0^\infty \frac{4\pi k^2 dk}{e^{\beta 2J\,S(ka)^2}-1}$$

$$= \frac{1}{(2\pi)^3}\int_0^\infty 4\pi k^2 dk \frac{e^{-\beta 2J\,S(ka)^2}}{1-e^{-\beta 2J\,S(ka)^2}}$$

$$= \frac{1}{(2\pi)^3}\int_0^\infty 4\pi k^2 dk e^{-\beta 2J\,S(ka)^2}$$

$$\sum_{l=0}^{\infty}\exp[-l\beta 2J\,S(ka)^2]$$

$$= \frac{1}{2\pi^2}\int_0^\infty \sum_{l=0}^{\infty}\exp[-(l+1)\beta 2J\,S(ka)^2]k^2 dk$$

$$= \frac{1}{2\pi^2}\int_0^\infty \sum_{m=1}^{\infty}\exp[-m\beta 2J\,S(ka)^2]k^2 dk \quad (3.46)$$

where $m = l + 1$. Note that the upper limit of the integral which is the border of the first Brillouin zone has been replaced by ∞. This is justified by the fact that important contributions are due to small k. Changing the variable $x = m\beta 2J\,S(ka)^2$, one obtains

$$\frac{1}{N}\sum_{\mathbf{k}} < n_{\mathbf{k}} > \simeq \frac{1}{(2\pi)^2}\left(\frac{k_B T}{2J\,S}\right)^{3/2}\sum_{m=1}^{\infty}\frac{1}{m^{3/2}}\int_0^\infty e^{-x}x^{1/2}dx \quad (3.47)$$

One notes that the integral of the right-hand side is equal to $\frac{\sqrt{\pi}}{2}$ and that the sum on m is the Riemann's series $\zeta(3/2)$. Finally, one arrives at

$$\frac{M}{g\mu_B N} = S - \zeta(3/2)\left(\frac{k_B}{8\pi J\,S}\right)^{3/2} T^{3/2} \quad (3.48)$$

The magnetization decreases with increasing T by a term proportional to $T^{3/2}$. This is called the Bloch's law.

As T increases further one has to take into account higher-order terms in (3.45). In doing so, one obtains

$$\frac{M}{g\mu_B N} = S - \zeta(3/2)t^{3/2} - \frac{3\pi}{4}\zeta(5/2)t^{5/2} - \frac{33\pi^2}{32}\zeta(7/2)t^{7/2} - \cdots \quad (3.49)$$

where $t = \frac{k_B T}{8\pi J S}$. This result, exact at low temperatures, has been confirmed by experiments at least up to $T^{5/2}$.

Note: $\zeta(3/2) = 2.612$, $\zeta(5/2) = 1.341$, $\zeta(7/2) = 1.127$

3.1.3.2 Energy and heat capacity

The energy of a ferromagnet is calculated in the same manner. From Eq. (3.39), one writes

$$E = \langle \mathcal{H} \rangle = -ZJNS^2 + \sum_{\mathbf{k}} \epsilon_{\mathbf{k}} < n_{\mathbf{k}} > \tag{3.50}$$

$$\simeq -ZJNS^2 + 12N\pi J S\zeta(5/2)t^{5/2} + \cdots \tag{3.51}$$

where the first term is the ground-state energy and the second term the energy of excited magnons at the quadratic order (free magnons).

The magnetic heat capacity C_V^m is thus

$$C_V^m = \frac{dE}{dT} = \frac{dE}{dt}\frac{dt}{dT} \simeq \frac{15}{4} N k_B \zeta(5/2) t^{3/2} + \cdots \tag{3.52}$$

We see that at low temperatures C_V^m is proportional to $T^{3/2}$ while the heat capacity of an electron gas C_V^e is proportional to T. It is also different from that of phonons where $C_V^p \propto T^3$. The dependence of C_V^m on T has been experimentally confirmed.

3.2 Spin Waves in Antiferromagnets

3.2.1 Dispersion Relation

Consider the following Heisenberg Hamiltonian:

$$\begin{aligned}\mathcal{H} &= J \sum_{(l,m)} \mathbf{S}_l \cdot \mathbf{S}_m - g\mu_B H \left(\sum_l S_l^z + \sum_m S_m^z \right) \\ &= J \sum_{(l,m)} \left[S_l^z S_m^z + \frac{1}{2} \left(S_l^+ S_m^- + S_l^- S_m^+ \right) \right] \\ &\quad - g\mu_B H \left(\sum_l S_l^z + \sum_m S_m^z \right)\end{aligned} \tag{3.53}$$

where interactions are limited to pairs of nearest neighbors (l, m) with an antiferromagnetic exchange integral $J > 0$. H is the amplitude of a magnetic field applied in the z direction. l and m indicate the sites belonging, respectively, to $\uparrow$ and $\downarrow$ sublattices. Note that we do not use the factor 2 in front of J in the Hamiltonian (3.53).

We use the Holstein–Primakoff method in the same manner as in the case of ferromagnets shown in (3.26)–(3.28) but with a distinction of up and down sublattices. For the up sublattice, one has

$$S_l^z = S - a_l^+ a_l \tag{3.54}$$
$$S_l^+ = \sqrt{2S} f_l(S) a_l \tag{3.55}$$
$$S_l^- = \sqrt{2S} a_l^+ f_l(S) \tag{3.56}$$

where

$$f_l(S) = \sqrt{1 - \frac{a_l^+ a_l}{2S}} \tag{3.57}$$

For the down sublattice, one defines

$$S_m^z = -S + b_m^+ b_m \tag{3.58}$$
$$S_m^+ = \sqrt{2S} b_m^+ f_m(S) \tag{3.59}$$
$$S_m^- = \sqrt{2S} f_m(S) b_m \tag{3.60}$$

where

$$f_l(S) = \sqrt{1 - \frac{b_m^+ b_m}{2S}} \tag{3.61}$$

The operators a, a^+, b and b^+ obey the commutation relations (Problem 9 in Section 3.6). Replacing operators $S^\pm$ and S^z in (3.53) by these operators, one gets

$$\mathcal{H} = J \sum_{(l,m)} [-S^2 + S f_l(S) a_l f_m(S) b_m + S a_l^+ f_l(S) b_m^+ f_m(S) + S a_l^+ a_l$$
$$+ S b_m^+ b_m - a_l^+ a_l b_m^+ b_m] - g\mu_B H [\sum_l (S - a_l^+ a_l)$$
$$- \sum_m (-S + b_m^+ b_m)] \tag{3.62}$$

In a first approximation, one supposes that the number of excited spin waves n is small with respect to $2S$, namely $a_l^+ a_l \ll 2S$ and

$b_m^+ b_m \ll 2S$, so that $f_l(S)$ and $f_m(S)$ can be expanded as follows:

$$f_l(S) \simeq 1 - \frac{a_l^+ a_l}{4S} + \cdots \tag{3.63}$$

$$f_m(S) \simeq 1 - \frac{b_m^+ b_m}{4S} + \cdots \tag{3.64}$$

Equation (3.62) becomes at the quadratic order

$$\mathcal{H} \simeq -\frac{ZJNS^2}{2} + JS \sum_{(l,m)} \left(a_l^+ a_l + b_m^+ b_m + a_l^+ b_m^+ + a_l b_m \right)$$
$$+ g\mu_B H \left(\sum_l a_l^+ a_l - \sum_m b_m^+ b_m \right) \tag{3.65}$$

where Z is the coordination number and the following relation has been used

$$\sum_{(l,m)} 1 = \sum_l \sum_{\mathbf{R}} 1 = Z \sum_l 1 = Z \frac{N}{2} \tag{3.66}$$

where $\mathbf{R}$ is the vector connecting the spin at l to a nearest neighbor belonging to the other sublattice, and $N/2$ the total number of spins in a sublattice.

The first term of (3.65) is the classical ground-state energy where neighboring spins are perfectly antiparallel (Néel ground state). One introduces now the following Fourier transformations

$$a_l^+ = \sqrt{\frac{2}{N}} \sum_{\mathbf{k}} e^{i\mathbf{k}\cdot\mathbf{l}} a_{\mathbf{k}}^+ \tag{3.67}$$

$$a_l = \sqrt{\frac{2}{N}} \sum_{\mathbf{k}} e^{-i\mathbf{k}\cdot\mathbf{l}} a_{\mathbf{k}} \tag{3.68}$$

$$b_m^+ = \sqrt{\frac{2}{N}} \sum_{\mathbf{k}} e^{-i\mathbf{k}\cdot\mathbf{m}} b_{\mathbf{k}}^+ \tag{3.69}$$

$$b_m = \sqrt{\frac{2}{N}} \sum_{\mathbf{k}} e^{i\mathbf{k}\cdot\mathbf{m}} b_{\mathbf{k}} \tag{3.70}$$

As in the case of ferromagnets, the Fourier components $a_{\mathbf{k}}$, $a_{\mathbf{k}}^+$, $b_{\mathbf{k}}$ and $b_{\mathbf{k}}^+$ obey the boson commutation relations. Using the above Fourier transforms, Eq. (3.65) becomes

$$\mathcal{H} = -\frac{ZJNS^2}{2} + ZJS \sum_{\mathbf{k}} \left[(1+h) a_{\mathbf{k}}^+ a_{\mathbf{k}} + (1-h) b_{\mathbf{k}}^+ b_{\mathbf{k}} \right.$$
$$\left. + \gamma_{\mathbf{k}} (a_{\mathbf{k}} b_{\mathbf{k}} + a_{\mathbf{k}}^+ b_{\mathbf{k}}^+) \right] \tag{3.71}$$

where

$$\gamma_{\mathbf{k}} = \frac{1}{Z}\sum_{\mathbf{R}} e^{i\mathbf{k}\cdot\mathbf{R}} \tag{3.72}$$

and

$$h = \frac{g\mu_B H}{ZJS} \tag{3.73}$$

One sees that $\mathcal{H}$ of Eq. (3.71) does not have a diagonal form of the "harmonic oscillator," namely $a^+_{\mathbf{k}}a_{\mathbf{k}}$ and $b^+_{\mathbf{k}}b_{\mathbf{k}}$, because of the existence of the term $a_{\mathbf{k}}b_{\mathbf{k}} + a^+_{\mathbf{k}}b^+_{\mathbf{k}}$. One can diagonalize $\mathcal{H}$ using the following transformation

$$\alpha_{\mathbf{k}} = a_{\mathbf{k}}\cosh\theta_k + b^+_{\mathbf{k}}\sinh\theta_k \tag{3.74}$$

$$\alpha^+_{\mathbf{k}} = a^+_{\mathbf{k}}\cosh\theta_k + b_{\mathbf{k}}\sinh\theta_k \tag{3.75}$$

$$\beta_{\mathbf{k}} = a^+_{\mathbf{k}}\sinh\theta_k + b_{\mathbf{k}}\cosh\theta_k \tag{3.76}$$

$$\beta^+_{\mathbf{k}} = a_{\mathbf{k}}\sinh\theta_k + b^+_{\mathbf{k}}\cosh\theta_k \tag{3.77}$$

where θ_k is a variable to be determined. The inverse transformation gives

$$a^+_{\mathbf{k}} = \alpha^+_{\mathbf{k}}\cosh\theta_k - \beta_{\mathbf{k}}\sinh\theta_k \tag{3.78}$$

$$a_{\mathbf{k}} = \alpha_{\mathbf{k}}\cosh\theta_k - \beta^+_{\mathbf{k}}\sinh\theta_k \tag{3.79}$$

$$b^+_{\mathbf{k}} = -\alpha_{\mathbf{k}}\sinh\theta_k + \beta^+_{\mathbf{k}}\cosh\theta_k \tag{3.80}$$

$$b_{\mathbf{k}} = -\alpha^+_{\mathbf{k}}\sinh\theta_k + \beta_{\mathbf{k}}\cosh\theta_k \tag{3.81}$$

One can verify that the new operators also obey the commutation relations (see Problem 10 of Section 3.6). Replacing (3.78)–(3.81) in (3.71), one has

$$\begin{aligned}\mathcal{H} = -\frac{ZJNS^2}{2} + ZJS\sum_{\mathbf{k}}\{&\cosh(2\theta_k) - 1 - \gamma_{\mathbf{k}}\sinh(2\theta_k)\\ &+ [\cosh(2\theta_k) - \gamma_{\mathbf{k}}\sinh(2\theta_k) + h]\alpha^+_{\mathbf{k}}\alpha_{\mathbf{k}}\\ &+ [\cosh(2\theta_k) - \gamma_{\mathbf{k}}\sinh(2\theta_k) - h]\beta^+_{\mathbf{k}}\beta_{\mathbf{k}}\\ &- [\sinh(2\theta_k) - \gamma_{\mathbf{k}}\cosh(2\theta_k)](\alpha_{\mathbf{k}}\beta_{\mathbf{k}} + \alpha^+_{\mathbf{k}}\beta^+_{\mathbf{k}})\}\end{aligned} \tag{3.82}$$

$\mathcal{H}$ is diagonal if the coefficient before the term $\alpha_{\mathbf{k}}\beta_{\mathbf{k}} + \alpha^+_{\mathbf{k}}\beta^+_{\mathbf{k}}$ is zero. This requirement allows us to determine the variable θ_k. One has

$$\tanh(2\theta_k) = \gamma_{\mathbf{k}} \tag{3.83}$$

Expressing $\sinh(2\theta_k)$ and $\cosh(2\theta_k)$ as functions of $\tanh(2\theta_k)$ then using Eq. (3.83), one obtains

$$\mathcal{H} = -\frac{ZJNS^2}{2} + ZJS\sum_{\mathbf{k}}\left[\sqrt{1-\gamma_{\mathbf{k}}^2} - 1\right] + ZJS\sum_{\mathbf{k}}\left\{\left[\sqrt{1-\gamma_{\mathbf{k}}^2} + h\right]\alpha_{\mathbf{k}}^{+}\alpha_{\mathbf{k}} + \left[\sqrt{1-\gamma_{\mathbf{k}}^2} - h\right]\beta_{\mathbf{k}}^{+}\beta_{\mathbf{k}}\right\} \tag{3.84}$$

One recognizes that for a given wave vector **k**, there are two spin wave modes corresponding to

$$\epsilon_{\mathbf{k}}^{\pm} = ZJS\left[\sqrt{1-\gamma_{\mathbf{k}}^2} \pm h\right] \tag{3.85}$$

This is the magnon dispersion relation of antiferromagnets. Without an applied field, these modes are degenerate. Note that for small k, using $\gamma_{\mathbf{k}}^2 \simeq (1 + ak^2 + \cdots)$ one obtains

$$\epsilon_{\mathbf{k}}^{\pm} \propto k \tag{3.86}$$

As said before, this result for antiferromagnets is different from $\epsilon_{\mathbf{k}}^{\pm} \propto k^2$ obtained for ferromagnets. One expects therefore that thermodynamic properties are different for the two cases in particular at low temperatures where small k modes dominate. This will be indeed seen below.

3.2.2 Properties at Low Temperatures

If one knows the dispersion relation $\epsilon_{\mathbf{k}}$ one can in principle use formulas of statistical mechanics to study properties of a system as a function of the temperature (see Appendix A). One writes the partition function $\mathbb{Z}$ as follows [see Eq. (A.9) with a change of the notation to avoid Z, the coordination number used above]:

$$\begin{aligned}\mathbb{Z} &= \mathrm{Tr}e^{-\beta\mathcal{H}} \\ &= \sum_{n_k=0}^{\infty}\sum_{n'_k=0}^{\infty}\exp\left\{-\beta\left[E_0 + \sum_{\mathbf{k}}(n_k\epsilon_{\mathbf{k}}^{+} + n'_k\epsilon_{\mathbf{k}}^{-})\right]\right\} \\ &= \mathrm{e}^{-\beta E_0}\prod_{\mathbf{k}}\left[\frac{1}{1-\mathrm{e}^{-\beta\epsilon_{\mathbf{k}}^{+}}}\frac{1}{1-\mathrm{e}^{-\beta\epsilon_{\mathbf{k}}^{-}}}\right]\end{aligned} \tag{3.87}$$

where one has used

$$n_k = \alpha_{\mathbf{k}}^+ \alpha_{\mathbf{k}} \tag{3.88}$$

$$n'_k = \beta_{\mathbf{k}}^+ \beta_{\mathbf{k}} \tag{3.89}$$

$$E_0 = -\frac{ZJ\,NS^2}{2} + ZJ\,S \sum_{\mathbf{k}} \left[\sqrt{1-\gamma_{\mathbf{k}}^2} - 1\right] \tag{3.90}$$

The free energy is written as

$$\begin{aligned} F &= -k_B T \ln \mathbb{Z} = E_0 + k_B T \sum_{\mathbf{k}} \left[\ln(1-\epsilon_{\mathbf{k}}^+) + \ln(1-\epsilon_{\mathbf{k}}^-)\right] \\ &= E_0 + 2k_B T \sum_{\mathbf{k}} \ln(1-\epsilon_{\mathbf{k}}) \quad \text{if } h = 0 \end{aligned} \tag{3.91}$$

One uses the above expression of F to calculate various thermodynamic properties as seen below.

3.2.2.1 Energy

For $h = 0$, one has

$$\begin{aligned} \mathcal{H} &= -\frac{ZJ\,NS^2}{2} + ZJ\,S\frac{N}{2} + \sum_{\mathbf{k}} \epsilon_{\mathbf{k}} (\alpha_{\mathbf{k}}^+ \alpha_{\mathbf{k}} + \beta_{\mathbf{k}}^+ \beta_{\mathbf{k}} + 1) \\ &= -\frac{ZJ\,NS(S+1)}{2} + 2\sum_{\mathbf{k}} \epsilon_{\mathbf{k}} \left(n_k + \frac{1}{2}\right) \end{aligned} \tag{3.92}$$

where one has used $\sum_{\mathbf{k}} 1 = \frac{N}{2}$ = number of microscopic states in the first Brillouin zone which is equal the number of spins in each sublattice. At $T = 0$, $n_k = n'_k = 0$ one obtains

$$E(T=0) = -\frac{ZJ\,NS(S+1)}{2} + \sum_{\mathbf{k}} \epsilon_{\mathbf{k}} \tag{3.93}$$

The second term is a correction to the classical ground-state energy $-\frac{ZJNS^2}{2}$ (Néel state). This correction is due to quantum fluctuations in analogy with the zero-point phonon energy.

At low temperatures, one calculates the magnon energy by the use of a low-temperature expansion. One gets

$$< E >= -T^2 \frac{\partial}{\partial T}(\frac{F}{T}) \simeq E(T=0) + aT^4 \tag{3.94}$$

where a is a coefficient proportional to $ZJ\,S$. One notes that the power T^4 is different from that of the ferromagnetic case [Eq. (3.51)]. This difference stems from $\epsilon_{\mathbf{k}} \propto k$ for small k in antiferromagnets.

3.2.2.2 Magnetization at low temperatures

To calculate the sublattice magnetization, one writes for the ↑ sublattice

$$\begin{aligned} M &= \sum_l (S- < a_l^+ a_l >) \\ &= \frac{N}{2}S - \frac{2}{N}\sum_{\mathbf{k}}\sum_{\mathbf{k}'} < a_{\mathbf{k}}^+ a_{\mathbf{k}'} > \sum_{\mathbf{l}} e^{i(\mathbf{k}-\mathbf{k}')\cdot \mathbf{l}} \\ &= \frac{N}{2}S - \frac{2}{N}\sum_{\mathbf{k}}\sum_{\mathbf{k}'} < a_{\mathbf{k}}^+ a_{\mathbf{k}'} > \frac{N}{2}\delta_{\mathbf{k},\mathbf{k}'} \\ &= \frac{N}{2}S - \sum_{\mathbf{k}} < a_{\mathbf{k}}^+ a_{\mathbf{k}} > \\ &= \frac{N}{2}S - \sum_{\mathbf{k}} < \cosh^2\theta_k \alpha_{\mathbf{k}}^+ \alpha_{\mathbf{k}} + \sinh^2\theta_k \beta_{\mathbf{k}}^+ \beta_{\mathbf{k}} + \sinh^2\theta_k > \end{aligned} \tag{3.95}$$

where one has used successively the Fourier transformation and relations (3.78)–(3.81). One expresses now $\cosh^2\theta_k$ and $\sinh^2\theta_k$ in terms of $\gamma_{\mathbf{k}}$ using (3.83), then one uses

$$< \alpha_{\mathbf{k}}^+ \alpha_{\mathbf{k}} >=< \beta_{\mathbf{k}}^+ \beta_{\mathbf{k}} >= \frac{1}{e^{\beta\epsilon_{\mathbf{k}}} - 1} \tag{3.96}$$

to obtain M. At low temperatures, using $\frac{1}{e^{\beta\epsilon_{\mathbf{k}}}-1}$ with $\epsilon_{\mathbf{k}} \propto k$ for small k, one gets

$$M \simeq \frac{N}{2}(S - \Delta S - AT^2) \tag{3.97}$$

where $\Delta S = \frac{2}{N}\sum_{\mathbf{k}} \sinh^2\theta_k$ is independent of T, and A a coefficient.

One sees that at $T = 0$, the magnetization is $S - \Delta S$ which is smaller than the spin magnitude S. ΔS is called the zero-point spin contraction. ΔS depends on the lattice: $\Delta S \simeq 0.197$ for an antiferromagnetic square lattice, $\Delta S \simeq 0.078$ for a cubic antiferromagnet of NaCl type.

Note that the sublattice magnetization of an antiferromagnet decreases as T^2 while the ferromagnetic magnetization decreases as $T^{3/2}$ [Eq. (3.48)].

3.3 Spin Waves in Ferrimagnets

In this section, one calculates the magnon dispersion relation in the case of a ferrimagnet. In principle, one can use the Holstein–Primakoff method as described above for antiferromagnets. However, the purpose here is to obtain the dispersion relation in a simplest manner. So, one will use the method of equation of motion as described hereafter.

One considers here a simple model of ferrimagnet which is composed of two sublattices of A and B Heisenberg spins occupying, respectively, the corner sites l and the center sites m of a body-centered cubic lattice. The A sublattice contains $\uparrow$ spins of amplitude S_A and the B sublattice contains $\downarrow$ spins of amplitude S_B. The Hamiltonian (2.72) is rewritten as

$$\begin{aligned}\mathcal{H} = J_1 \sum_{(l,m)} \left[S_l^z S_m^z + \frac{1}{2}(S_l^+ S_m^- + S_l^- S_m^+) \right] \\ + J_2^A \sum_{(l,l')} \left[S_l^z S_{l'}^z + \frac{1}{2}\left(S_l^+ S_{l'}^- + S_l^- S_{l'}^+\right) \right] \\ + J_2^B \sum_{(m,m')} \left[S_m^z S_{m'}^z + \frac{1}{2}(S_m^+ S_{m'}^- + S_m^- S_{m'}^+) \right] \qquad (3.98)\end{aligned}$$

where J_1 denotes the interaction between A and B spins (nearest neighbors), J_2^A and J_2^B denote the intra-sublattice interactions (next-nearest neighbors). To simplify the presentation, one supposes in the following $J_2^A = J_2^B = J_2$.

The Heisenberg equation of motion for the operator $S_l^-(t)$ reads

$$\begin{aligned}i\hbar\frac{dS_l^-}{dt} = \left[S_l^-, \mathcal{H}\right] \simeq J_1 \left[\sum_m < S_m^z > S_l^- - < S_l^z > S_m^+ \right] \\ + J_2 \left[\sum_{l'} < S_{l'}^z > S_l^- - < S_l^z > S_{l'}^- \right] \qquad (3.99)\end{aligned}$$

The equation of motion for S_m^+ can be obtained from this equation by exchanging $l \leftrightarrow m, l' \leftrightarrow m'$ and $S^+ \leftrightarrow S^-$. It is noted that in this equation, one has replaced, in the mean-field spirit, the operators S_m^z and S_l^z by their averaged values $< S_m^z >$ and $< S_l^z >$. Using the following Fourier transformations for S_l^- and S_m^+

$$S_l^- = \frac{1}{(2\pi)^3}\int_{-\infty}^{-\infty} d\omega e^{-i\omega t}\int_{BZ} d\mathbf{k}e^{i\mathbf{k}\cdot\mathbf{l}}U_A(\mathbf{k},\omega) \quad (3.100)$$

$$S_m^+ = \frac{1}{(2\pi)^3}\int_{-\infty}^{-\infty} d\omega e^{-i\omega t}\int_{BZ} d\mathbf{k}e^{i\mathbf{k}\cdot\mathbf{l}}U_B(\mathbf{k},\omega) \quad (3.101)$$

where BZ indicates the first Brillouin zone, one obtains for the body-centered cubic lattice the following coupled equations

$$(E + E_A)U_A = -8(1-\alpha)\gamma_1(\mathbf{k})U_B \quad (3.102)$$

$$(E - E_B)U_B = 8(1+\alpha)\gamma_1(\mathbf{k})U_A \quad (3.103)$$

where one has used

$$< S_l^z > = S_A = (1-\alpha)S \quad (3.104)$$

$$< S_m^z > = S_B = -(1+\alpha)S\ (|\alpha| < 1) \quad (3.105)$$

$$E = \frac{\hbar\omega}{J_1 S} \quad (3.106)$$

$$\varepsilon = \frac{J_2}{J_1} \quad (3.107)$$

$$E_A = 8(1+\alpha) - 6\varepsilon(1-\alpha)[1-\gamma_2(\mathbf{k})] \quad (3.108)$$

$$E_B = 8(1-\alpha) - 6\varepsilon(1+\alpha)[1-\gamma_2(\mathbf{k})] \quad (3.109)$$

$$\gamma_1(\mathbf{k}) = \frac{1}{Z_1}\sum_{\rho_1\in NN} e^{i\mathbf{k}\cdot\rho_1}$$

$$= \cos\left(\frac{k_x a}{2}\right)\cos\left(\frac{k_y a}{2}\right)\cos\left(\frac{k_z a}{2}\right) \quad (3.110)$$

$$\gamma_2(\mathbf{k}) = \frac{1}{Z_2}\sum_{\rho_2\in NNN} e^{i\mathbf{k}\cdot\rho_2}$$

$$= \frac{1}{3}\left[\cos(k_x a) + \cos(k_y a) + \cos(k_z a)\right] \quad (3.111)$$

a being the lattice constant, $Z_1 = 8$ and $Z_2 = 6$ the numbers of nearest and next-nearest neighbors. One recognizes here that ρ_1 and ρ_2 connect a site to its nearest neighbors (NN) and next-nearest neighbors (NNN), respectively.

For a non-trivial solution of (3.102)–(3.103), one imposes the secular equation

$$(E + E_A)(E - E_B) = -64(1-\alpha^2)\gamma_1^2(\mathbf{k}) \quad (3.112)$$

from which, one obtains

$$E_{\pm} = \frac{1}{2}\left[E_B - E_A \pm \sqrt{(E_A - E_B)^2 + 4[E_A E_B - 64(1-\alpha^2)\gamma_1^2(\mathbf{k})]}\right] \tag{3.113}$$

We examine some particular cases:

(1) For $k = 0$, one has $E_- = 0$ and $E_+ = -16\alpha$, independent of ϵ, namely J_2.
(2) For $k_x a = k_y a = k_z a = \pi$, one has $E_+ = 8(1-\alpha) - 12\varepsilon(1+\alpha)$ and $E_- = -8(1+\alpha) + 12\varepsilon(1-\alpha)$.
(3) If $\varepsilon = 0$, then for $k_x a = k_y a = k_z a = \pi$, one has $E_+ = 8(1-\alpha)$ and $E_- = -8(1+\alpha)$.

These results show a gap in the magnon spectrum at $\mathbf{k} = 0$ with a width proportional to α. Figure 3.5 shows the magnon spectrum versus $k_x = k_y$ for $k_z = 0$.

The energy, the heat capacity and the magnetization at low temperatures can be calculated using the method of the preceding section for antiferromagnet.

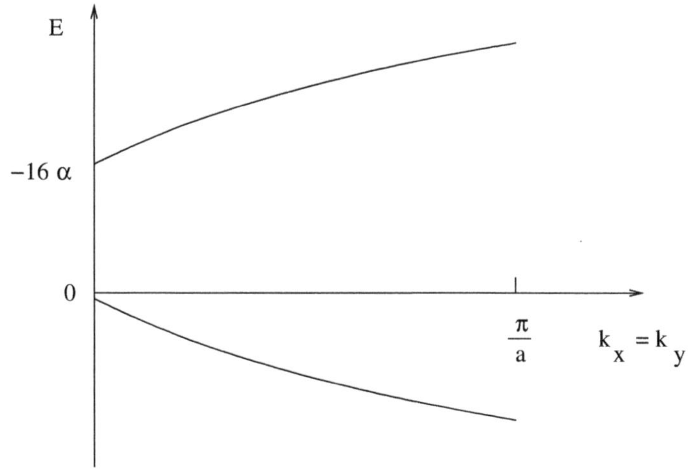

Figure 3.5 Magnon spectrum of a ferrimagnet of body-centered cubic lattice versus $k_x = k_y$ with $k_z = 0, \alpha = -1/3, \varepsilon = 0$.

3.4 Spin Waves in Helimagnets

Helimagnets are a family of materials in which the spins are not collinear in the low-T ordered phase as in other systems considered above. Due to a competition between various kinds of interaction, the neighboring spins make an angle which is neither zero nor π. Helimagnets are thus frustrated systems which present many unexpected properties [84, 85].

We consider here the simplest example of helimagnet which is a chain with a ferromagnetic interaction $J_1(> 0)$ between nearest neighbors and an antiferromagnetic interaction $J_2(< 0)$ between next-nearest neighbors. When $\varepsilon = |J_2|/J_1$ is larger than a critical value ε_c, the spin configuration of the ground state becomes non collinear. One shows that the helical configuration displayed in Fig. 3.6 is obtained by minimizing the following interaction energy

$$\begin{aligned} E &= -J_1 \sum_i \mathbf{S}_i \cdot \mathbf{S}_{i+1} + |J_2| \sum_i \mathbf{S}_i \cdot \mathbf{S}_{i+2} \\ &= S^2 \left[-J_1 \cos\theta + |J_2| \cos(2\theta)\right] \sum_i 1 \end{aligned} \tag{3.114}$$

$$\begin{aligned} \frac{\partial E}{\partial \theta} &= S^2 \left[J_1 \sin\theta - 2|J_2| \sin(2\theta)\right] \sum_i 1 = 0 \\ &= S^2 \left[J_1 \sin\theta - 4|J_2| \sin\theta \cos\theta\right] \sum_i 1 = 0 \end{aligned} \tag{3.115}$$

where one has supposed that the angle between nearest neighbors is θ.
The solutions are

- Ferromagnetic and antiferromagnetic configurations:

$$\sin\theta = 0 \longrightarrow \theta = 0, \pi$$

  Replacing the above solutions into Eq. (3.114), we see that the antiferromagnetic solution ($\theta = \pi$) corresponds to the maximum of E. It is to be discarded.
- Helical configuration:

$$\cos\theta = \frac{J_1}{4|J_2|} \longrightarrow \theta = \pm \arccos\left(\frac{J_1}{4|J_2|}\right) \tag{3.116}$$

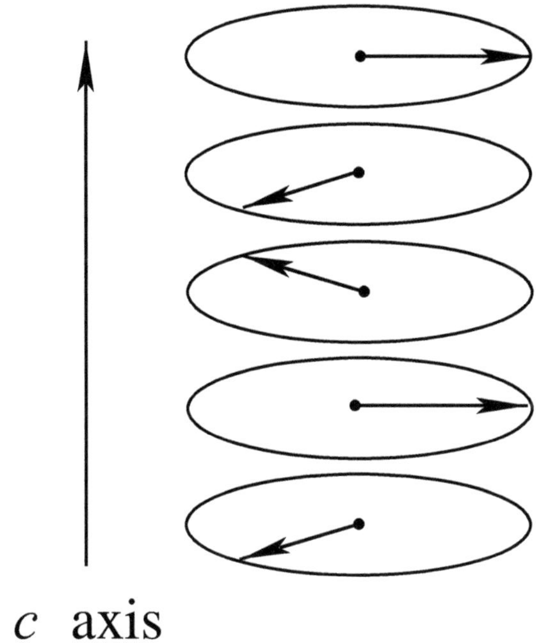

Figure 3.6 Helical configuration when $\varepsilon = |J_2|/J_1 > \varepsilon_c = 1/4$ ($J_1 > 0$, $J_2 < 0$).

The last solution is possible if $-1 \leq \cos\theta \leq 1$, i.e. $J_1/(|J_2|) \leq 4$ or $|J_2|/J_1 \geq 1/4 \equiv \varepsilon_c$. There are two degenerate configurations corresponding to clockwise and counter-clockwise turning angles. The ferromagnetic solution has an energy lower than that of the helical solution for $|J_2|/J_1 < 1/4$. It is therefore more stable in this range of parameters. For $|J_2|/J_1 > 1/4$, the helical configuration is more stable.

For the magnon spectrum, let us consider a three-dimensional body-centered cubic lattice with Heisenberg spins interacting with each other via (i) a ferromagnetic interaction $J_1 > 0$ between nearest neighbors, (ii) an antiferromagnetic interaction $J_2 < 0$ between next-nearest neighbors only along the y axis. The Hamiltonian is given by

$$\mathcal{H} = -J_1 \sum_{<i,j>} \mathbf{S}_i \cdot \mathbf{S}_j - J_2 \sum_{<i,l>} \mathbf{S}_i \cdot \mathbf{S}_l + D \sum_i (S_i^y)^2 \qquad (3.117)$$

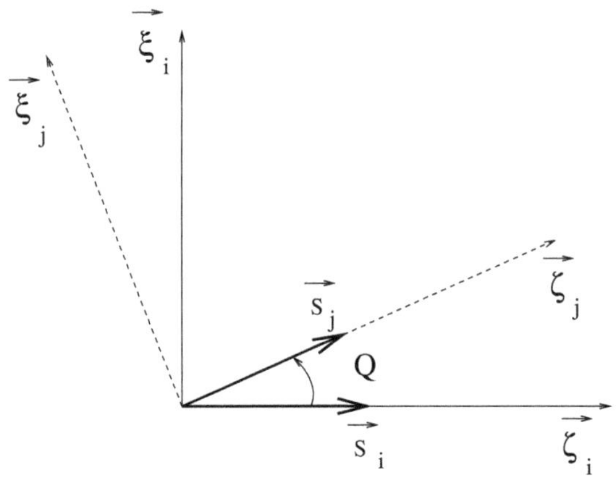

Figure 3.7 Local coordinates defined for two spins $\mathbf{S}_i$ and $\mathbf{S}_j$. The axis η is common for the two spins.

where $D > 0$ is a very small anisotropy of the type "easy-plane anisotropy" which stabilizes the spins in the xz plane. The ground state can be calculated in the same manner as in the case of a chain given above. We find $\cos\theta = \frac{J_1}{|J_2|}$ so that the helical configuration in the y direction is stable when $\varepsilon = |J_2|/J_1 > \varepsilon_c = 1$. Note that the spins in the same xz plane are parallel with each other.

Let (ξ_i, η_i, ζ_i) be the unit vectors making a direct trihedron at the site i, namely η_i is parallel to the y axis as shown in Fig. 3.7. One supposes in addition that the quantization axis of the spin $\mathbf{S}_i$ coincides with the local axis ζ_i.

One uses now the following transformation in the local coordinates associated with $\mathbf{S}_i$ and $\mathbf{S}_j$

$$\eta_j = \eta_i \tag{3.118}$$

$$\zeta_j = \cos Q\zeta_i + \sin Q\xi_i \tag{3.119}$$

$$\xi_j = -\sin Q\zeta_i + \cos Q\xi_i \tag{3.120}$$

One writes

$$\mathbf{S}_i = S_i^x\xi_i + S_i^y\eta_i + S_i^z\zeta_i \tag{3.121}$$

$$\mathbf{S}_j = S_j^x\xi_j + S_j^y\eta_j + S_j^z\zeta_j \tag{3.122}$$

Their scalar product becomes

$$\begin{aligned}
\mathbf{S}_i \cdot \mathbf{S}_j &= [S_i^x \xi_i + S_i^y \eta_i + S_i^z \zeta_i] \cdot [S_j^x(-\sin Q\zeta_i + \cos Q\xi_i) \\
&\quad + S_j^y \eta_i + S_j^z(\cos Q\zeta_i + \sin Q\xi_i)] \\
&= S_i^z(-\sin QS_j^x + \cos QS_j^z) + S_i^y S_j^y \\
&\quad + S_i^x(\cos QS_j^x + \sin QS_j^z) \\
&= \cos QS_i^z S_j^z - \sin QS_i^z \left[\frac{S_j^+ + S_j^-}{2}\right] - \frac{(S_i^+ - S_i^-)(S_j^+ - S_j^-)}{4} \\
&\quad + \cos Q\frac{(S_i^+ + S_i^-)(S_j^+ + S_j^-)}{4} + \sin Q\left[\frac{S_i^+ + S_i^-}{2}\right] S_j^z
\end{aligned} \tag{3.123}$$

To be general, the angle Q should depend on positions of $\mathbf{S}_i$ and $\mathbf{S}_j$. One defines $\cos Q = \cos(\mathbf{Q} \cdot \mathbf{R}_{ij})$ where $\mathbf{Q}$ is the vector of modulus Q, perpendicular to the plane of the angle Q, namely plane (ζ, ξ), and $\mathbf{R}_{ij}$ the vector connecting the positions of $\mathbf{S}_i$ and $\mathbf{S}_j$. One shall keep in the following $J(\mathbf{R}_{ij})$ as interaction between $\mathbf{S}_i$ and $\mathbf{S}_j$ which will be replaced by J_1 and J_2 depending on $\mathbf{R}_{ij}$ at the end of the calculation.

Equation (3.117) is rewritten as

$$\begin{aligned}
\mathcal{H} &= -\frac{1}{4}\sum_{(i,j)} J(\mathbf{R}_{ij})\{(S_i^+ S_j^- + S_i^- S_j^+)[1 + \cos(\mathbf{Q} \cdot \mathbf{R}_{ij})] \\
&\quad -(S_i^+ S_j^+ + S_i^- S_j^-)[1 - \cos(\mathbf{Q} \cdot \mathbf{R}_{ij})] + 4S_i^z S_j^z \cos(\mathbf{Q} \cdot \mathbf{R}_{ij}) \\
&\quad +2[(S_i^+ + S_i^-)S_j^z - S_i^z(S_j^+ + S_j^-)] \sin(\mathbf{Q} \cdot \mathbf{R}_{ij})\} \\
&\quad -\frac{D}{4}\sum_i (S_i^+ - S_i^-)^2
\end{aligned} \tag{3.124}$$

With this Hamiltonian, one can choose an appropriate method to calculate the magnon dispersion relation. One can use the Holstein-Primakoff method by replacing the operators $S^{\pm}$ and S^z by (3.26)–(3.29), or Green's function method (Chapter 4) or simply by the method of equation of motion (Section 3.3). Using the Holstein-Primakoff method, one obtains the magnon dispersion relation

$$\begin{aligned}
\mathcal{H} &= -NSJ(\mathbf{Q}) + \frac{S}{2}\sum_{\mathbf{k}}[A(\mathbf{k}, \mathbf{Q})(a_{\mathbf{k}} a_{\mathbf{k}}^+ + a_{\mathbf{k}}^+ a_{\mathbf{k}}) \\
&\quad +B(\mathbf{k}, \mathbf{Q})(a_{\mathbf{k}} a_{-\mathbf{k}} + a_{\mathbf{k}}^+ a_{-\mathbf{k}}^+)]
\end{aligned} \tag{3.125}$$

where

$$J(\mathbf{k}) = \sum_{\mathbf{R}_{ij}} J(\mathbf{R}_{ij}) \exp(\mathbf{k} \cdot \mathbf{R}_{ij}) \quad (\text{sum on } \mathbf{R}_{ij}) \tag{3.126}$$

$$A(\mathbf{k}, Q) = \left[2J(\mathbf{Q}) - J(\mathbf{k}) - \frac{1}{2}[J(\mathbf{k}+\mathbf{Q}) + J(\mathbf{k}-\mathbf{Q})] + D\right] \tag{3.127}$$

$$B(\mathbf{k}, \mathbf{Q}) = \left[J(\mathbf{k}) - \frac{1}{2}[J(\mathbf{k}+\mathbf{Q}) + J(\mathbf{k}-\mathbf{Q})] - D\right] \tag{3.128}$$

The Hamiltonian (3.125) can be diagonalized by introducing the new operators $\alpha_{\mathbf{k}}$ and $\alpha^+_{\mathbf{k}}$ just as in the antiferromagnetic case studied above

$$a_{\mathbf{k}} = \alpha_{\mathbf{k}} \cosh\theta_k - \alpha^+_{-\mathbf{k}} \sinh\theta_k \tag{3.129}$$

$$a^+_{\mathbf{k}} = \alpha^+_{\mathbf{k}} \cosh\theta_k - \alpha_{-\mathbf{k}} \sinh\theta_k \tag{3.130}$$

where $\alpha_{\mathbf{k}}$ and $\alpha^+_{\mathbf{k}}$ obey the boson commutation relations [see similar transformation in Eqs. (3.74)–(3.77)]. Hamiltonian (3.125) is diagonal if one takes

$$\tanh(2\theta_k) = \frac{B(\mathbf{k}, \mathbf{Q})}{A(\mathbf{k}, \mathbf{Q})} \tag{3.131}$$

One then has

$$\mathcal{H} = \frac{S}{2} \sum_{\mathbf{k}} \hbar\omega_{\mathbf{k}} [\alpha^+_{\mathbf{k}} \alpha_{\mathbf{k}} + \alpha_{\mathbf{k}} \alpha^+_{\mathbf{k}}] \tag{3.132}$$

where the energy of the magnon of mode **k** is

$$\hbar\omega_{\mathbf{k}} = \sqrt{A(\mathbf{k}, \mathbf{Q})^2 - B(\mathbf{k}, \mathbf{Q})^2} \tag{3.133}$$

In the case of the body-centered cubic lattice, one has

$$\begin{aligned} J(\mathbf{k}) &= 8J_1 \cos(k_x a/2) \cos(k_y a/2) \cos(k_z c/2) + 2J_2 \cos(k_z c) \\ &= 2J_2 \left[-4 \cos Q \cos(k_x a/2) \cos(k_y a/2) \cos(k_z c/2) + \cos(k_z c)\right] \end{aligned}$$

where $\epsilon_c = 1$ has been used, a and c being the lattice constants (one uses c for the helical axis). Figure 3.8 shows the magnon spectrum for J_2/J_1 corresponding to $Q = \pi/3$. One observes that the magnon frequency is zero not only at $k = 0$ but also at $k_z = Q$.

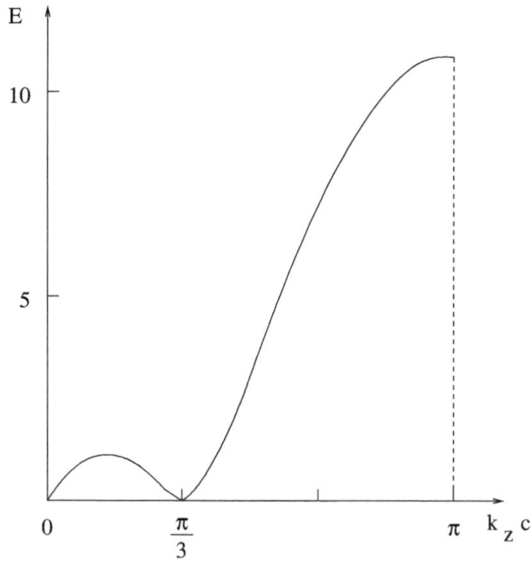

Figure 3.8 Magnon spectrum $E = \hbar\omega_{\mathbf{k}}$, Eq. (3.133), versus $k_z c$ in a helimagnet defined by the Hamiltonian (3.117), with $Q = \pi/3$, $k_x = k_y = 0$.

3.5 Conclusion

The theory of magnons presented in this chapter allows us to calculate the spin wave dispersion relation which is used to study thermodynamic properties of magnetic systems at low temperatures in a precise manner. When the temperature increases, it is necessary to take into account higher-order terms in the Hamiltonian which represent interactions between magnons. The calculation then becomes more complicated and needs other approximations to decouple chains of operators to renormalize the harmonic, or free-magnon, spectrum. One can also use Green's function method which allows us to include implicitly magnon–magnon interactions up to the transition temperature (see Chapter 4). The method involves, however, some decoupling schemes which make the results near the transition less precise.

Let us summarize some main points of this chapter. First, on the dispersion relation, results of antiferromagnets are quite different from those of ferromagnets. For instance, at small **k**, one has for antiferromagnets $\epsilon_{\mathbf{k}} \propto k$, while for ferromagnets $\epsilon_{\mathbf{k}} \propto k^2$. This difference yields different temperature-dependence of physical properties. One notes also that the completely antiparallel spin configuration (Néel state) is not the real ground state of quantum antiferromagnets. Zero-point quantum fluctuations reduce the spin amplitude at $T = 0$ in antiferromagnets. For ferrimagnets, only a simple example has been used to illustrate the magnon spectrum. A gap due to the difference of spin magnitudes is frequently found. However, it should be emphasized that real ferrimagnets often have much more complicated lattice structures.

We have also presented some aspects of helimagnets where the magnon spectrum has been shown. It is important to note that systems with non-collinear spin configurations have been and still are subject of intensive investigations since 30 years [84, 85].

Finally, to close this chapter, let us outline some aspects which are important to know:

(1) *Magnetic anisotropy*: The Heisenberg model is isotropic.It does not tell us in which direction the spins should align themselves. It is a habit to suppose that the spins are on the z axis for calculation. But all other directions are equivalent. In real materials, there often exists a preferential direction which is called "easy-magnetization axis." This magnetic "anisotropy" stems from complicated microscopic origins such as spin-orbit interaction, dipole-dipole interaction, etc. [149].

(2) *Long-range interactions*: In this chapter, one has supposed for simplicity that interactions are limited to nearest neighbors. The calculations can be of course extended to interactions up to second, third, ... nearest neighbors. However, for infinite-range interaction, specific methods should be used.

(3) *Low-dimensional systems*: In the case of one dimension, a system of spins with short-range interactions are disordered for non-zero T whatever the spin model is. In the case of two dimensions, systems of discrete spins such as Ising and Potts models have a phase transition at a finite temperature [25]. Systems of vector spins such as Heisenberg and XY models are not ordered at $T \neq 0$. This has been rigorously shown by a theorem of Mermin–Wagner [231]. One can see this in an

approximative manner: In two dimensions one replaces $4\pi k^2 dk$ in (3.46) by $2\pi k dk$, the integral then becomes divergent at the lower bound ($k \to 0$), causing undefined M except when $T = 0$.

(4) There exist other methods to study temperature-dependent properties of magnetic systems such as low- and high-temperature series expansions [95], renormalization-group analysis and numerical simulations. These are shown in Chapter 5.

3.6 Problems

Problem 1. Prove (3.51)–(3.52).

Problem 2. Chain of Heisenberg spins:

(a) Calculate the magnon spectrum $\epsilon(k)$ for a chain of Heisenberg spins of lattice constant a with ferromagnetic interactions J_1 between nearest neighbors and J_2 between next-nearest neighbors. Plot $\epsilon(k)$ versus k within the first Brillouin zone (BZ).

(b) The spectrum $\epsilon(k)$ obtained in the previous question is supposed to be valid when J_2 becomes antiferromagnetic as long as $\epsilon(k) \geq 0$. Show that the ferromagnetic order becomes unstable when $|J_2|$ is larger than a critical value. Determine this value and compare to ε_c given below Eq. (3.116).

Problem 3. Heisenberg spin systems in two dimensions:

Consider the Heisenberg spin model on a two-dimensional lattice with a ferromagnetic interaction J between nearest neighbors.

(a) Calculate the magnon spectrum $\epsilon(\mathbf{k})$ as a function of $\mathbf{k}$. Check that $\epsilon(\mathbf{k}) \propto k^2$ as $k \to 0$.

(b) Write down the formal expression connecting the magnetization M to the temperature T. Show that M is undefined as soon as T becomes non-zero. Comments.

Problem 4. Prove Eqs. (3.131)–(3.133).

Problem 5. Consider the Ising spin model on a "Union-Jack" lattice, namely the square lattice in which one square out of every two has a centered site. Define sublattice 1 containing the centered sites, and sublattice 2 containing the remaining sites (namely the cornered sites). Let J_1 be the interaction between a centered spin and its nearest neighbors, J_2 and J_3 the interactions between two nearest spins on the y and x axes of the sublattice 2, respectively. Determine the phase diagram of the ground state in the space (J_1, J_2, J_3). Indicate the phases where the centered spins are undefined (partial disorder).

Problem 6. Determine the ground-state spin configuration of a triangular antiferromagnet with XY spins interacting with each other via an exchange J_1 between nearest neighbors. Show that the spin configuration is the 120° structure shown in Fig. 18.6 (left).

Problem 7. Determine the ground state spin configuration of the 2D Villain's model with XY spins defined in Fig. 18.6 (right). Write the energy of the elementary plaquette. By minimizing this energy, determine the ground state as a function of the antiferromagnetic interaction $J_{AF} = -\eta J_F$ where η is a positive coefficient. Determine the angle between two neighboring spins as a function of η. Show that the critical value of η beyond which the spin configuration is not collinear is $\eta_c = 1/3$.

Problem 8. Uniaxial anisotropy

(a) Show that if one includes in the Heisenberg Hamiltonian for ferromagnets the following anisotropy term $-D\sum_i (S_i^z)^2$ where D is a positive constant and the sum is performed over all spins, one obtains the following magnon spectrum

$$\epsilon_{\mathbf{k}} = 2ZJS(1 - \gamma_{\mathbf{k}} + d)$$

where $d \equiv \frac{D}{2ZJS}$ [see the definitions of other notations in Eq. (3.40)].

(b) Is it possible to have a long-range magnetic ordering at finite temperature in two dimensions? (cf. Problem 3).

Problem 9. Show that the operators a^+ and a defined in the Holstein–Primakoff approximation, Eqs. (3.26)–(3.28), respect rigorously the commutation relations between the spin operators.

Problem 10. Show that the operators defined in Eqs. (3.74)–(3.77) obey the commutation relations.

Problem 11. Show that the magnon spectrum (3.113) becomes unstable when the interaction between next-nearest neighbors defined in ϵ, Eq. (3.107), is larger than a critical constant.

Chapter 4

Green's Function Theory in Magnetism

The Green's function method is a very general method in quantum field theory. It can be used for various problems in many areas such as condensed matter, nuclear physics and elementary-particle physics [115]. The general formulation is rather complicated, abstract, not suitable for a quick application. For our purpose, we present in this chapter a simplified version which, from the beginning, aims at an application of the method to systems of interacting spins. This formulation does not need a high level of knowledge in quantum field theory. A basic level in quantum mechanics is enough to understand and to apply the method.

As seen below, the Green's function method is an alternative technique to study spin waves, in addition to the theory of magnons presented in Chapter 3. We have seen that the theory of free magnons is exact at very low temperatures. However, at higher temperatures we have to take into account magnon–magnon interactions. For example, we have to treat terms of higher orders in the Holstein–Primakoff expansion [see Eqs. (3.32)–(3.34)] to modify the harmonic magnon spectrum given by Eqs. (3.39) and (3.40). This is a laborious task [91]. An alternative and, by far, simpler way to take into account some correlations is the Green's function method which can treat the whole temperature range going from

Physics of Magnetic Thin Films: Theory and Simulation
Hung T. Diep

ISBN 978-981-4877-42-8 (Hardcover), 978-1-003-12110-7 (eBook)
www.jennystanford.com

low-temperature phase up to the transition temperature T_c. This is possible because Green's function includes even at the lowest level some spin–spin correlations. We can therefore follow the evolution of the magnon spectrum with increasing temperature. However, since all correlations are not hierarchically included, its validity is still an open question. Note that there is at present no other better method to calculate the spin wave spectrum at finite temperatures.

Numerous applications of the Green's function method in surface magnetism are shown in Part II of this book.

4.1 Green's Function Method

4.1.1 Definition

Let $A(t)$ and $B(t')$ be two operators in the Heisenberg representation at times t and t', respectively. We define the retarded Green's function by

$$G^r_{AB}(t-t') = \ll A(t); B(t') \gg^r = -i\theta(t-t') < \left[A(t), B(t')\right] > \quad (4.1)$$

and the advanced Green's function by

$$G^a_{AB}(t-t') = \ll A(t); B(t') \gg^a = i\theta(t'-t) < \left[A(t), B(t')\right] > \quad (4.2)$$

where

$$\theta(t-t') = 1 \qquad \text{if } t > t' \quad (4.3)$$

$$\theta(t-t') = 0 \qquad \text{if } t < t' \quad (4.4)$$

$[A(t), B(t')]$ being the commutation relation and $< \cdots >$ denoting the thermal average. In spite of the complicated notation in its definition given above, Green's function is just a thermal average of a commutation relation between two operators. Green's function is connected, as will be seen below, to the physical properties of the system. So, depending on what we want to study, we choose the operators $A(t)$ and $B(t')$. For instance, to study an electron gas, we can choose $A(t) = a_i^+(t)$ and $B(t') = a_i(t')$ where $a_i^+(t)$ and $a_i(t')$ are creation and annihilation operators of electron state i, to study phonon excitations we can choose $A(t)$ and $B(t')$ as phonon creation and annihilation operators, and for a spin system we can choose

$A(t) = S_i^+(t)$ and $B(t') = S_i^-(t')$ where $S_i^+(t)$ and $S_i^-(t')$ are spin operators (cf. Chapter 3).

Green's function contains information on the system, in particular on the elementary-excitation spectrum. In what follows, we present the formulation of the method as it is applied to a spin system so that the reader can associate mathematical tools presented here to physical meanings of a phenomenon that has been seen in Chapter 3.

4.1.2 Formulation

We consider the following functions:

$$F_{AB}(t-t') = < A(t)B(t') > \tag{4.5}$$

$$F_{BA}(t'-t) = < B(t')A(t) > \tag{4.6}$$

The time Fourier transformation of $F_{BA}(t'-t)$ is written as

$$F_{BA}(t'-t) = \int_{-\infty}^{\infty} I(\omega)\mathrm{e}^{i\omega(t'-t)}d\omega \tag{4.7}$$

We show below that

$$< A(t)B(t') >=< B(t')A(t+i\beta) > \tag{4.8}$$

where $\beta = (k_B T)^{-1}$.

Demonstration: We use essentially the circular permutation properties of operators in $\mathrm{Tr}[\cdots]$ and $A(t) = \mathrm{e}^{i\mathcal{H}t}A\mathrm{e}^{-i\mathcal{H}t}$:

$$\begin{aligned}
< A(t)B(t') > &= \frac{\mathrm{Tr}A(t)B(t')\mathrm{e}^{-\beta\mathcal{H}}}{\mathrm{Tr}\mathrm{e}^{-\beta\mathcal{H}}} = \frac{\mathrm{Tr}\mathrm{e}^{-\beta\mathcal{H}}A(t)B(t')}{\mathrm{Tr}\mathrm{e}^{-\beta\mathcal{H}}} \\
&= \frac{\mathrm{Tr}B(t')\mathrm{e}^{-\beta\mathcal{H}}A(t)}{\mathrm{Tr}\mathrm{e}^{-\beta\mathcal{H}}} \\
&= \frac{\mathrm{Tr}\mathrm{e}^{i\mathcal{H}t'}B\mathrm{e}^{-i\mathcal{H}t'}\mathrm{e}^{-\beta\mathcal{H}}\mathrm{e}^{i\mathcal{H}t}A\mathrm{e}^{-i\mathcal{H}t}}{\mathrm{Tr}\mathrm{e}^{-\beta\mathcal{H}}} \\
&= \frac{\mathrm{Tr}\mathrm{e}^{-\beta\mathcal{H}}\mathrm{e}^{i\mathcal{H}t'}B\mathrm{e}^{-i\mathcal{H}t'}\mathrm{e}^{-\beta\mathcal{H}}\mathrm{e}^{i\mathcal{H}t}A\mathrm{e}^{-i\mathcal{H}t}\mathrm{e}^{\beta\mathcal{H}}}{\mathrm{Tr}\mathrm{e}^{-\beta\mathcal{H}}} \\
&= \frac{\mathrm{Tr}\mathrm{e}^{-\beta\mathcal{H}}\mathrm{e}^{i\mathcal{H}t'}B\mathrm{e}^{-i\mathcal{H}t'}\mathrm{e}^{i\mathcal{H}(t+i\beta)}A\mathrm{e}^{-i\mathcal{H}(t+i\beta)}}{\mathrm{Tr}\mathrm{e}^{-\beta\mathcal{H}}} \\
&= < B(t')A(t+i\beta) >
\end{aligned}$$

We have thus

$$F_{AB}(t-t') = F_{BA}(t'-t-i\beta) \tag{4.9}$$

The Fourier transformation of $F_{AB}(t-t')$ is

$$F_{AB}(t-t') = F_{BA}(t'-t-i\beta)$$
$$= \int_{-\infty}^{\infty} I(\omega)\mathrm{e}^{-i\omega(t-t')}\mathrm{e}^{\beta\omega}d\omega \qquad (4.10)$$

Using this formula for Green's function

$$G^r_{AB}(t-t') = -i\theta(t-t') < A(t)B(t') - B(t')A(t) > \qquad (4.11)$$

we obtain

$$G^r_{AB}(\omega)$$
$$= \frac{1}{(2\pi)^2}\iiint_{-\infty}^{\infty} \frac{I(\omega')(\mathrm{e}^{\beta\omega'}-1)\mathrm{e}^{i(\omega-\omega'-x)(t-t')}}{x+i\epsilon} d\omega' d(t-t')dx \qquad (4.12)$$

where we have used the following expression:

$$\theta(t-t') = \lim_{\epsilon\to 0^+} \frac{i}{2\pi}\int_{-\infty}^{\infty} \frac{\mathrm{e}^{-ix(t-t')}}{x+i\epsilon}dx \qquad (4.13)$$

Integrating on $(t-t')$ and using the formula

$$\delta(x) = \frac{1}{2\pi i}\left(\frac{1}{x-i\epsilon} - \frac{1}{x+i\epsilon}\right) \qquad (4.14)$$

we arrive at

$$G^r_{AB}(\omega) = \frac{1}{2\pi}\int_{-\infty}^{\infty} \frac{I(\omega')(\mathrm{e}^{\beta\omega'}-1)}{\omega-\omega'+i\epsilon}d\omega' \qquad (4.15)$$

In the same manner, we obtain

$$G^a_{AB}(\omega) = \frac{1}{2\pi}\int_{-\infty}^{\infty} \frac{I(\omega')(\mathrm{e}^{\beta\omega'}-1)}{\omega-\omega'-i\epsilon}d\omega' \qquad (4.16)$$

The difference between the retarded and advanced Green's functions is the sign in front of $i\epsilon$ in the denominator. We write from now on these functions without superscripts r and a, but we distinguish them by the sign in their arguments. Combining these functions we write

$$G^r_{AB}(\omega) - G^a_{AB}(\omega) = G_{AB}(\omega+i\epsilon) - G_{AB}(\omega-i\epsilon)$$
$$= \frac{1}{2\pi}\int_{-\infty}^{\infty} d\omega' I(\omega')(\mathrm{e}^{\beta\omega'}-1)$$
$$\times\left[\frac{1}{\omega-\omega'+i\epsilon} - \frac{1}{\omega-\omega'-i\epsilon}\right]$$
$$= \frac{1}{2\pi}\int_{-\infty}^{\infty} d\omega' I(\omega')(\mathrm{e}^{\beta\omega'}-1)\left[-2\pi i\delta(\omega-\omega')\right]$$
$$= -iI(\omega)(\mathrm{e}^{\beta\omega}-1) \qquad (4.17)$$

We shall use this relation to calculate the spin–spin correlation in the following sections.

4.2 Ferromagnetism by the Green's Function Method

We consider the Heisenberg spins with a ferromagnetic interaction between nearest neighbors. The Hamiltonian is given by

$$\begin{aligned}\mathcal{H} &= -J\sum_{(l,m)} \mathbf{S}_l \cdot \mathbf{S}_m \\ &= -J\sum_{(l,m)} \left[S_l^z S_m^z + \frac{1}{2}(S_l^+ S_m^- + S_l^- S_m^+)\right] \end{aligned} \tag{4.18}$$

where the spin operators obey the commutation relations

$$\left[S_l^+, S_m^-\right] = 2S_l^z \delta_{lm} \tag{4.19}$$

$$\left[S_m^z, S_l^\pm\right] = \pm S_l^\pm \delta_{lm} \tag{4.20}$$

For a ferromagnet, we define the following Green's function:

$$G_{lm}(t-t') = \ll S_l^+(t); S_m^-(t') \gg = -i\theta(t-t') < \left[S_l^+(t), S_m^-(t')\right] > \tag{4.21}$$

We set $t' = 0$ hereafter to simplify the writing.

4.2.1 Equation of Motion

The equation of motion for $G_{lm}(t)$ is written as

$$\begin{aligned} i\hbar\frac{dG_{lm}(t)}{dt} &= 2 < S_l^z > \delta_{lm}\delta(t) - \ll \left[\mathcal{H}, S_l^+\right](t); S_m^- \gg \\ &= 2 < S_l^z > \delta_{lm}\delta(t) \\ &\quad - J\sum_{\vec{\rho}} \ll S_l^z(t) S_{l+\vec{\rho}}^+(t) - S_l^+(t) S_{l+\vec{\rho}}^z(t); S_m^- \gg \end{aligned} \tag{4.22}$$

where the sum on $\vec{\rho}$ is performed on the vectors connecting spin $\mathbf{S}_l$ to its nearest neighbors.

Remark: We have used the identity $[AB, C] = [A, C]B + A[B, C]$ then (4.19) and (4.20) to calculate $\left[\mathcal{H}, S_l^+\right]$ (see Problem 2 in Section 4.6).

We see that the right-hand side of Eq. (4.22) contains Green's functions of higher order with three operators. Writing the equation of motion for these functions will generate functions of five operators. In a first approximation, we reduce higher-order Green's functions by using the so-called Tyablikov decoupling scheme [346]

$$\ll S_l^z(t)S_{l+\vec{\rho}}^+(t); S_m^- \gg \simeq < S_l^z(t) > \ll S_{l+\vec{\rho}}^+(t); S_m^- \gg \quad (4.23)$$

$$\ll S_l^+(t)S_{l+\vec{\rho}}^z(t); S_m^- \gg \simeq < S_{l+\vec{\rho}}^z(t) > \ll S_l^+(t); S_m^- \gg \quad (4.24)$$

We obtain on the right-hand side of Eq. (4.22) Green's functions of the same order as the one initially defined in (4.21). This decoupling bears the same spirit as the mean-field theory (Chapter 2): replacing an operator in a product by its average value, i.e., by a "c-number." This approximation is called sometimes "random-phase approximation" (RPA). The RPA is hierarchically higher than the mean-field theory in the sense that one operator in a three-operator product is replaced by its average value in the RPA while in the mean-field theory one operator in a two-operator product is replaced.

We notice that S_l^z is a constant of motion. Therefore, in a homogeneous system, we can suppose that

$$< S_l^z(t) > = < S_{l+\vec{\rho}}^z(t) > = < S^z > \quad \text{which is site-independent} \quad (4.25)$$

We obtain then

$$\begin{aligned} i\hbar\frac{dG_{lm}(t)}{dt} &= 2 < S^z > \delta_{lm}\delta(t) \\ &\quad -J\sum_{\vec{\rho}} < S^z > [\ll S_{l+\vec{\rho}}^+(t); S_m^- \gg \\ &\quad - \ll S_l^+(t); S_m^- \gg] \\ &= 2 < S^z > \delta_{lm}\delta(t) \\ &\quad -J < S^z > \sum_{\vec{\rho}}[G_{l+\vec{\rho},m}(t) - G_{lm}(t)] \end{aligned} \quad (4.26)$$

4.2.2 Dispersion Relation

The time Fourier transformation of Eq. (4.26) gives

$$\hbar\omega G_{lm}(\omega) = \frac{1}{\pi} < S^z > \delta_{lm} - J < S^z > \sum_{\vec{\rho}} \left[G_{l+\vec{\rho},m}(\omega) - G_{lm}(\omega)\right] \tag{4.27}$$

We use next the following spatial Fourier transformation:

$$G_{lm}(\omega) = \frac{1}{N}\sum_{\mathbf{k}} G_{\mathbf{k}}(\omega)e^{i\mathbf{k}\cdot(\mathbf{l}-\mathbf{m})} \tag{4.28}$$

where the sum on the wave vector $\mathbf{k}$ is performed in the first Brillouin zone of N states. We obtain

$$\begin{aligned}\hbar\omega G_{\mathbf{k}}(\omega) &= \frac{< S^z >}{\pi} - J < S^z > \left[G_{\mathbf{k}}(\omega)\sum_{\vec{\rho}} e^{i\mathbf{k}\cdot\vec{\rho}} - ZG_{\mathbf{k}}(\omega)\right] \\ &= \frac{< S^z >}{\pi} + J < S^z > ZG_{\mathbf{k}}(\omega)\left[1 - \frac{1}{Z}\sum_{\vec{\rho}} e^{i\mathbf{k}\cdot\vec{\rho}}\right]\end{aligned} \tag{4.29}$$

where Z is the coordination number. We get

$$\begin{aligned}G_{\mathbf{k}}(\omega) &= \frac{< S^z >}{\pi}\frac{1}{\hbar\omega - ZJ < S^z > (1-\gamma_{\mathbf{k}})} \\ &= \frac{< S^z >}{\pi}\frac{1}{\hbar\omega - \epsilon_{\mathbf{k}}}\end{aligned} \tag{4.30}$$

where

$$\gamma_{\mathbf{k}} = \frac{1}{Z}\sum_{\vec{\rho}} e^{i\mathbf{k}\cdot\vec{\rho}} \tag{4.31}$$

$$\epsilon_{\mathbf{k}} = ZJ < S^z > (1-\gamma_{\mathbf{k}}) \tag{4.32}$$

As seen in Eq. (4.30), the singularity of the Green's function is at $\hbar\omega = \epsilon_{\mathbf{k}}$. This determines the eigen-energy $\hbar\omega_{\mathbf{k}}$ of the magnon of wave vector $\mathbf{k}$:

$$\hbar\omega_{\mathbf{k}} = ZJ < S^z > (1-\gamma_{\mathbf{k}}) \tag{4.33}$$

This is the ferromagnetic magnon dispersion relation which is to be compared to Eq. (3.40) obtained by the theory of magnons:

$$\epsilon_{\mathbf{k}} = 2ZJS(1-\gamma_{\mathbf{k}})$$

We see that apart from the factor 2 due to the model defined by (3.30) the only difference is that $< S^z >$ appears in (4.33) instead of S. As $< S^z >$ depends on the temperature, the magnon spectrum varies with T. We calculate $< S^z >$ in the following.

4.2.3 Magnetization and Critical Temperature

We write

$$\mathbf{S}_l \cdot \mathbf{S}_l = S(S+1) = (S_l^z)^2 + \frac{1}{2}(S_l^+ S_l^- + S_l^- S_l^+)$$
$$= (S_l^z)^2 + (S_l^- S_l^+ + S_l^z) \tag{4.34}$$

where we have used (4.19). For $S = 1/2$, we have $(S_l^z)^2 = 1/4$ (see Pauli's matrices in Chapter 1). We get then

$$\frac{3}{4} = \frac{1}{4} + S_l^- S_l^+ + S_l^z \tag{4.35}$$

from which we have

$$< S_l^z > = \frac{1}{2} - < S_l^- S_l^+ > \tag{4.36}$$

We now calculate $< S_l^- S_l^+ >$. We have

$$< S_m^- S_l^+ > = \frac{1}{N} \sum_{\mathbf{k}} \int_{-\infty}^{\infty} I_{\mathbf{k}} e^{i\mathbf{k}\cdot(\mathbf{l}-\mathbf{m})} e^{-i\omega t} d\omega \tag{4.37}$$

from which, for $t = 0$,

$$< S_l^- S_l^+ > = \frac{1}{N} \sum_{\mathbf{k}} \int_{-\infty}^{\infty} I_{\mathbf{k}} d\omega \tag{4.38}$$

Using (4.17) we write

$$< S_l^- S_l^+ > = \frac{1}{N} \sum_{\mathbf{k}} \int_{-\infty}^{\infty} d\omega \left[\frac{G_{\mathbf{k}}(\omega + i\epsilon) - G_{\mathbf{k}}(\omega - i\epsilon)}{-i(e^{\beta\omega} - 1)} \right] \tag{4.39}$$

Replacing $G_{\mathbf{k}}(\omega + i\epsilon)$ and $G_{\mathbf{k}}(\omega - i\epsilon)$ by (4.30) and using (4.14), we obtain

$$< S_l^- S_l^+ > = -\frac{1}{N} \frac{< S^z >}{i\pi} \sum_{\mathbf{k}} \int_{-\infty}^{\infty} \frac{d\omega}{e^{\beta\omega} - 1}$$
$$\times \left[\frac{1}{\hbar\omega - \epsilon_{\mathbf{k}} + i\epsilon} - \frac{1}{\hbar\omega - \epsilon_{\mathbf{k}} - i\epsilon} \right]$$
$$= -\frac{1}{N} \frac{< S^z >}{i\pi} \sum_{\mathbf{k}} \int_{-\infty}^{\infty} \frac{d\omega}{e^{\beta\omega} - 1} [-2\pi i \delta(\hbar\omega - \epsilon_{\mathbf{k}})]$$
$$= \frac{2}{N} < S^z > \sum_{\mathbf{k}} \frac{1}{e^{\beta\hbar\omega_{\mathbf{k}}} - 1} \tag{4.40}$$

Equation (4.36) becomes

$$< S_l^z >= \frac{1}{2} - \frac{2}{N} < S^z > \sum_{\mathbf{k}} \frac{1}{e^{\beta\hbar\omega_{\mathbf{k}}} - 1} \quad (4.41)$$

This is an implicit equation for $< S_l^z >$: The right-hand side contains also $< S_l^z >$ in $\omega_{\mathbf{k}}$. Therefore, we have to solve this equation by iteration to get $< S_l^z >$.

At low temperatures, we follow the same method as that used for (3.46). We obtain the following result:

$$< S_l^z > \simeq \frac{1}{2} - a_1 T^{3/2} - a_2 T^{5/2} - a_3 T^3 - a_4 T^{7/2} - \cdots \quad (4.42)$$

where $a_i (i = 1, 2, 3, \cdots)$ are constants. We notice that the T^3 term does not exist in the low-temperature expansion (3.49) of the theory of magnons. It may be due to the Tyablikov decoupling (4.23) and (4.24), which is the only approximation used here. To improve the decoupling, the reader is referred to the references [222, 340].

At high temperatures, $< S_l^z > \rightarrow 0$. An expansion of $e^{\beta\hbar\omega_{\mathbf{k}}}$ yields

$$\begin{aligned} < S_l^z > &\simeq \frac{1}{2} - \frac{2}{N} < S^z > \sum_{\mathbf{k}} \frac{1}{1 + \beta\hbar\omega_{\mathbf{k}} + \cdots - 1} \\ &\simeq \frac{1}{2} - \frac{2}{N} < S^z > \sum_{\mathbf{k}} \frac{1}{\beta Z J < S^z > (1 - \gamma_{\mathbf{k}})} \\ &\simeq \frac{1}{2} - \frac{2}{N} \sum_{\mathbf{k}} \frac{1}{\beta Z J (1 - \gamma_{\mathbf{k}})} \end{aligned} \quad (4.43)$$

At $T = T_c$, $< S_l^z >= 0$. Equation (4.43) gives then

$$\left(\frac{k_B T_c}{J}\right)^{-1} = \frac{4}{ZN} \sum_{\mathbf{k}} \frac{1}{1 - \gamma_{\mathbf{k}}} \quad (4.44)$$

In the same manner, for spins of amplitude $S \neq 1/2$, using $(S_l^z)^2 = (\mathbf{S}_l \cdot \mathbf{S}_l)/3 = S(S+1)/3$ in Eq. (4.34) one obtains

$$< S_l^z >= \frac{2S(S+1)}{3} - < S_l^- S_l^+ > \quad (4.45)$$

The transition temperature is given by

$$\left(\frac{k_B T_c}{J}\right)^{-1} = \frac{3}{ZS(S+1)N} \sum_{\mathbf{k}} \frac{1}{1 - \gamma_{\mathbf{k}}} \quad (4.46)$$

Remark: When transforming the sum on $\mathbf{k}$ in (4.44) into an integral, we clearly see that the integral diverges at small k for dimensions $d = 1$ and 2 (see the discussion in the conclusion of Chapter 3). As a consequence, $T_c = 0$ for these low dimensions, in agreement with exact results [231] (see Chapter 5). For $d = 3$, the values of T_c numerically calculated by (4.44) are

- Simple cubic lattice: $k_B T_c/J \simeq 0.994$
- Body-centered cubic lattice: $k_B T_c/J \simeq 1.436$
- Face-centered cubic lattice: $k_B T_c/J \simeq 2.231$

These values are much lower than those given by the mean-field theory: 3, 4 and 6, respectively. The present results are better than those of the mean-field theory because in the Green's function method, fluctuations due to spin–spin correlations are partially taken into account.

4.3 Antiferromagnetism by the Green's Function Method

In a spin system with an antiferromagnetic interaction, we have to define two Green's functions, the first one concerns the correlation between spins of the same sublattice and the second one expresses the correlation between spins belonging to the two sublattices.

We consider a system of Heisenberg spins with an antiferromagnetic interaction J between nearest neighbors

$$\mathcal{H} = J \sum_{(l,m)} \left[S_l^z S_m^z + \frac{1}{2}(S_l^+ S_m^- + S_l^- S_m^+) \right] \tag{4.47}$$

where the spin operators satisfy the commutation relations (4.19) and (4.20). We define the following Green's functions:

$$\begin{aligned} G_{ll'}(t - t') &= \ll S_l^+(t); S_{l'}^-(t') \gg \\ &= -i\theta(t - t') < \left[S_l^+(t), S_{l'}^-(t')\right] > \end{aligned} \tag{4.48}$$

$$\begin{aligned} F_{ml'}(t - t') &= \ll S_m^+(t); S_{l'}^-(t') \gg \\ &= -i\theta(t - t') < \left[S_m^+(t), S_{l'}^-(t')\right] > \end{aligned} \tag{4.49}$$

where l, l' belong to sublattice of $\uparrow$ spins and m belongs to sublattice of $\downarrow$ spins. For writing simplicity, we take $t' = 0$.

The equations of motion for $G_{ll'}(t)$ and $F_{ml'}(t)$ are written as, after the Tyablikov decoupling,

$$i\hbar\frac{dG_{ll'}(t)}{dt} = 2 < S^z > \delta_{ll'}\delta(t) + J < S^z > \sum_{\vec{\rho}}[F_{l+\vec{\rho},l'}(t) + G_{ll'}(t)] \quad (4.50)$$

$$i\hbar\frac{dF_{ml'}(t)}{dt} = -J < S^z > \sum_{\vec{\rho}}[G_{m+\vec{\rho},l'}(t) + F_{ml'}(t)] \quad (4.51)$$

where we have used the fact that nearest neighbors of a l site are on m sites and vice versa, namely $l + \vec{\rho}$ is a site m of the $\downarrow$ sublattice and that $m + \vec{\rho}$ is a site l of the $\uparrow$ sublattice. As in the ferromagnetic case we have supposed $< S_l^z >=< S_{m+\vec{\rho}}^z >= - < S_m^z >= - < S_{l+\vec{\rho}}^z >=< S^z >$, independent of the lattice site. We use the Fourier transformations

$$G_{ll'}(t) = \frac{2}{N}\int_{-\infty}^{\infty}\sum_{\mathbf{k}} G_{\mathbf{k}}(\omega)\mathrm{e}^{i[\mathbf{k}\cdot(\mathbf{l}-\mathbf{l}')-\omega t]}d\omega \quad (4.52)$$

$$F_{ml'}(t) = \frac{2}{N}\int_{-\infty}^{\infty}\sum_{\mathbf{k}} F_{\mathbf{k}}(\omega)\mathrm{e}^{i[\mathbf{k}\cdot(\mathbf{m}-\mathbf{l}')-\omega t]}d\omega \quad (4.53)$$

where the factor $2/N$ comes from the fact that each sublattice has $N/2$ sites. The equations (4.50) and (4.51) become

$$(\hbar\omega - A_{\mathbf{k}})G_{\mathbf{k}}(\omega) - B_{\mathbf{k}}F_{\mathbf{k}}(\omega) = \frac{< S^z >}{\pi} \quad (4.54)$$

$$B_{\mathbf{k}}G_{\mathbf{k}}(\omega) + (\hbar\omega + A_{\mathbf{k}})F_{\mathbf{k}}(\omega) = 0 \quad (4.55)$$

where $A_{\mathbf{k}} = ZJ < S^z >$ and $B_{\mathbf{k}} = ZJ < S^z > \gamma_{\mathbf{k}}$. The solution of (4.54) and (4.55) is

$$G_{\mathbf{k}}(\omega) = \frac{< S^z >}{2\pi}\left[\frac{1 + A_{\mathbf{k}}/\epsilon_{\mathbf{k}}}{\hbar\omega - \epsilon_{\mathbf{k}}} + \frac{1 - A_{\mathbf{k}}/\epsilon_{\mathbf{k}}}{\hbar\omega + \epsilon_{\mathbf{k}}}\right] \quad (4.56)$$

$$F_{\mathbf{k}}(\omega) = -\frac{< S^z > B_{\mathbf{k}}}{2\pi\epsilon_{\mathbf{k}}}\left[\frac{1}{\hbar\omega - \epsilon_{\mathbf{k}}} - \frac{1}{\hbar\omega + \epsilon_{\mathbf{k}}}\right] \quad (4.57)$$

where

$$\epsilon_{\mathbf{k}} = \sqrt{A_{\mathbf{k}}^2 - B_{\mathbf{k}}^2} = ZJ < S^z > \sqrt{1 - \gamma_{\mathbf{k}}^2} \quad (4.58)$$

The singularities of these functions are $\pm\epsilon_{\mathbf{k}}$. The magnon mode $\mathbf{k}$ has thus two opposite precessions. This degeneracy comes from the reversal symmetry of the two sublattices. The antiferromagnetic dispersion relation is

$$\hbar\omega_{\mathbf{k}} = \pm\epsilon_{\mathbf{k}} = \pm\sqrt{A_{\mathbf{k}}^2 - B_{\mathbf{k}}^2} = \pm ZJ < S^z > \sqrt{1-\gamma_{\mathbf{k}}^2} \quad (4.59)$$

It is noted that

- for small k, one has $\epsilon_{\mathbf{k}} \propto k$ instead of k^2 of the ferromagnetic case (cf. previous paragraph and Chapter 3)
- as in the ferromagnetic case, the magnon spectrum depends on the temperature via $< S^z >$ in $\hbar\omega_{\mathbf{k}}$.

To calculate the magnetization we follow the same method used in the ferromagnetic case. We have

$$< S_{l'}^- S_l^+ > = \frac{2}{N}\int_{-\infty}^{\infty}\sum_{\mathbf{k}} I_{\mathbf{k}}(\omega)\mathrm{e}^{i[\mathbf{k}\cdot(\mathbf{l}-\mathbf{l}')-\omega t]}d\omega \quad (4.60)$$

$$< S_{l'}^- S_m^+ > = \frac{2}{N}\int_{-\infty}^{\infty}\sum_{\mathbf{k}} K_{\mathbf{k}}(\omega)\mathrm{e}^{i[\mathbf{k}\cdot(\mathbf{m}-\mathbf{l}')-\omega t]}d\omega \quad (4.61)$$

Replacing $I_{\mathbf{k}}(\omega)$ and $K_{\mathbf{k}}(\omega)$ each with its corresponding Green's function by using (4.17), we obtain

$$I_{\mathbf{k}}(\omega) = < S^z > \frac{(1+A_{\mathbf{k}}/\epsilon_{\mathbf{k}})\delta(\hbar\omega-\epsilon_{\mathbf{k}}) + (1-A_{\mathbf{k}}/\epsilon_{\mathbf{k}})\delta(\hbar\omega+\epsilon_{\mathbf{k}})}{\mathrm{e}^{\beta\hbar\omega}-1} \quad (4.62)$$

$$K_{\mathbf{k}}(\omega) = -< S^z > \frac{B_{\mathbf{k}}}{\epsilon_{\mathbf{k}}}\frac{\delta(\hbar\omega-\epsilon_{\mathbf{k}}) - \delta(\hbar\omega+\epsilon_{\mathbf{k}})}{\mathrm{e}^{\beta\hbar\omega}-1} \quad (4.63)$$

Using these relations in (4.60) and (4.61), we have

$$< S_l^- S_l^+ > = \frac{2}{N} < S^z > \sum_{\mathbf{k}}\left[-1 + \frac{A_{\mathbf{k}}}{\epsilon_{\mathbf{k}}}\coth(\beta\epsilon_{\mathbf{k}}/2)\right] \quad (4.64)$$

$$< S_l^- S_m^+ > = \frac{2}{N} < S^z > \sum_{\mathbf{k}}\frac{B_{\mathbf{k}}}{\epsilon_{\mathbf{k}}}\coth(\beta\epsilon_{\mathbf{k}}/2)\mathrm{e}^{i\mathbf{k}\cdot(\mathbf{m}-\mathbf{l})} \quad (4.65)$$

For $S = 1/2$, we replace $< S_l^- S_l^+ >$ by $1/2- < S^z >$ in (4.64). We finally arrive at

$$\frac{1}{2} = \frac{2}{N} < S^z > \sum_{\mathbf{k}}\frac{1}{\sqrt{1-\gamma_{\mathbf{k}}^2}}\coth\left[\frac{\beta ZJ < S^z > \sqrt{1-\gamma_{\mathbf{k}}^2}}{2}\right] \quad (4.66)$$

This implicit equation for $< S^z >$ should be solved by numerical iteration.

At low temperatures, expanding the right-hand side of (4.66) when $T \to 0$ ($\beta \to \infty$), one has

$$< S^z > \simeq \frac{1}{2} - \Delta S - a_2 T^2 - \cdots \qquad (4.67)$$

where ΔS is the zero-point spin contraction (at $T = 0$) (see Chapter 3). As seen, the temperature dependence of $< S^z >$ is not the same as in the ferromagnetic case. This is a consequence of the linear dependence on k of $\epsilon_{\mathbf{k}}$ at small k.

The Néel temperature T_N is calculated by letting $< S^z > \to 0$ in (4.66). One has

$$\left(\frac{k_B T_N}{J}\right)^{-1} = \frac{4}{ZN} \sum_{\mathbf{k}} \left[\frac{1}{1-\gamma_{\mathbf{k}}} + \frac{1}{1+\gamma_{\mathbf{k}}}\right] \qquad (4.68)$$

4.4 Green's Function Method for Non-Collinear Magnets

In frustrated spin systems, the competition between different kinds of interaction can give rise to ground-state spin configurations which are not collinear. The incompatibility of some lattice geometries with antiferromagnetic interaction yields also non-collinear spin configurations as those in the antiferromagnetic triangular lattice, in the antiferromagnetic face-centered cubic lattice or hexagonal-close-packed lattice. Such frustrated systems show many striking properties. The reader is referred to Ref. [85] for recent reviews on many aspects of the frustration. For non-collinear magnets, the Hamiltonian can be expressed in the local coordinates. This has been done in Eq. (3.124). We define the following Green's functions:

$$\begin{aligned} G_{k\ell}(t-t') &= \ll S_k^+(t); S_\ell^-(t') \gg \\ &= -i\theta(t-t') < \left[S_k^+(t), S_\ell^-(t')\right] > \qquad (4.69) \\ F_{k\ell}(t-t') &= \ll S_k^-(t); S_\ell^-(t') \gg \\ &= -i\theta(t-t') < \left[S_k^-(t), S_\ell^-(t')\right] > \qquad (4.70) \end{aligned}$$

With the use of the Tyablikov decoupling [346], the equations of motion of these functions are given by

$$
\begin{aligned}
i\hbar\frac{dG_{k\ell}(t-t')}{dt} &= 2 < S_k^z > \delta_{k\ell}\delta(t) \\
&\quad -\frac{1}{2}\sum_{k'} J_{k,k'}[< S_k^z >\ll S_{k'}^-; S_\ell^- \gg (\cos\theta_{k,k'} - 1) \\
&\quad + < S_k^z >\ll S_{k'}^+; S_\ell^- \gg (\cos\theta_{k,k'} + 1) \\
&\quad - 2 < S_{k'}^z >\ll S_k^+; S_\ell^- \gg \cos\theta_{k,k'}] \\
&= 2 < S_k^z > \delta_{k\ell}\delta(t) \\
&\quad -\frac{1}{2}\sum_{k'} J_{k,k'}[< S_k^z > (\cos\theta_{k,k'} - 1) \\
&\quad \times F_{k'\ell}(t-t') + < S_k^z > (\cos\theta_{k,k'} + 1) G_{k'\ell}(t-t') \\
&\quad - 2 < S_{k'}^z > \cos\theta_{k,k'} G_{k\ell}(t-t')]
\end{aligned}
\tag{4.71}
$$

$$
\begin{aligned}
i\hbar\frac{dF_{k\ell}(t-t')}{dt} &= \frac{1}{2}\sum_{k'} J_{k,k'}[< S_k^z >\ll S_{k'}^+; S_\ell^- \gg (\cos\theta_{k,k'} - 1) \\
&\quad + < S_k^z >\ll S_{k'}^-; S_\ell^- \gg (\cos\theta_{k,k'} + 1) \\
&\quad -2 < S_{k'}^z >\ll S_k^-; S_\ell^- \gg \cos\theta_{k,k'}] \\
&= \frac{1}{2}\sum_{k'} J_{k,k'}[< S_k^z > (\cos\theta_{k,k'} - 1) G_{k'\ell}(t-t') \\
&\quad + < S_k^z > (\cos\theta_{k,k'} + 1) F_{k'\ell}(t-t') \\
&\quad -2 < S_{k'}^z > \cos\theta_{k,k'} F_{k\ell}(t-t')]
\end{aligned}
\tag{4.72}
$$

Note that the sinus terms in Eq. (3.124) are canceled out upon summing over symmetric neighbors at each lattice site (inversion symmetry) in the above equations. The next steps are similar to those used in the antiferromagnetic case in the previous section: Using the Fourier transformations, the solution of the resulting coupled equations gives the dispersion relation which allows us to calculate the magnetization and the transition temperature. Some applications are given as problems in Section 4.6.

4.5 Conclusion

We have presented in this chapter the Green's function method, which is used to investigate spin waves in ferromagnetic, antiferromagnetic and non-collinear magnets. The method gives the temperature-dependence of the magnon spectrum and a compact expression which allows us to calculate the magnetization up to the transition temperature. To keep the method simple and tractable, we have used the Tyablikov decoupling scheme to reduce higher-order Green's functions. There is room for improving it but we will lose the simplicity of the method. The dispersion relation can be explicitly obtained in simple cases. However, for complicated systems where we have to define several Green's functions, we obtain a system of coupled linear equations which can be numerically solved to obtain the dispersion relation which is used to compute the temperature dependence of various physical quantities. Such applications are treated in part II of this book.

We note that the method, though efficient, is not accurate enough to allow us to calculate critical exponents of the phase transition. However, it can detect first-order transitions as well as multiple phase transitions of the system at different temperatures as seen in Ref. [285] and in part II.

4.6 Problems

Problem 1. Give proofs of the formula (4.13).

Problem 2. Give the demonstration of Eq. (4.22).

Problem 3. Helimagnet by Green's function method:

Consider a crystal of simple cubic lattice with Heisenberg spins of amplitude 1/2. The interaction J_1 between nearest neighbors is ferromagnetic. Suppose that along the y axis there exists an antiferromagnetic interaction J_2 between next nearest neighbors, in addition to J_1.

(a) Follow the method in Section 3.4, show that the ground state is helimagnetic in the y direction if $|J_2|/J_1$ is larger than a critical value α_c. Determine α_c.

(b) Let θ be the helical angle between two nearest neighboring spins in the y direction. Express the Hamiltonian in terms of θ.

(c) Define two Green's functions which allow us to calculate the spin wave spectrum using the RPA decoupling scheme. Calculate that spectrum.

(d) Show that this spectrum is reduced to the ferromagnetic and antiferromagnetic spectra when $J_2 = 0$ and $J_1 \gtrless 0$.

Problem 4. Apply the Green's function method to a system of Ising spins $S = \pm 1$ in one dimension, supposing a ferromagnetic interaction between nearest neighbors under an applied magnetic field.

Problem 5. Apply the Green's function method to a system of Heisenberg spins on a simple cubic lattice, supposing ferromagnetic interactions between nearest neighbors and between next-nearest neighbors.

Problem 6. Magnon spectrum in Heisenberg triangular antiferromagnet: Green's function method

Calculate the magnon spectrum of a triangular lattice with spins 1/2 interacting with each other via an antiferromagnetic interaction between nearest neighbors. Estimate numerically the zero-point spin contraction. Guide: Use the spin configuration obtained in Problem 6 of Section 3.6.

Chapter 5

Theory of Phase Transitions and Critical Phenomena

When a system changes its symmetry under the effect of an external parameter such as temperature, pressure, applied magnetic or electric field, we say it undergoes a phase transition. Phase transitions in systems of interacting particles are subjects of intensive investigations in modern physics.

In the present chapter, we show the theory of phase transition using some systems of interacting spins. Note that various systems of different nature can be mapped into spin systems and solved using the spin language. There are many approximations and methods with various degrees of precision for the study of phase transitions. Among these approximations, the mean-field approximation is by far the simplest one as we have seen in Chapter 2. However, the mean-field theory gives some artifacts at low dimensions and cannot determine with precision the critical exponents which characterize the nature of a phase transition (see Section 5.1.5). These critical exponents are intimately related to the microscopic interaction between the particles and the symmetry of the system.

In the following, we first present Beth's approximation and the Landau–Ginzburg theory, which improve the mean-field theory.

Physics of Magnetic Thin Films: Theory and Simulation
Hung T. Diep

ISBN 978-981-4877-42-8 (Hardcover), 978-1-003-12110-7 (eBook)
www.jennystanford.com

The concept of the renormalization group is presented next. The renormalization group was formulated by K. G. Wilson [361, 362] in the early 1970s. It provides an accurate insight into the mechanism of a phase transition and has been applied with success in many fields of physics ranging from condensed matter to quantum field theory. Among the most remarkable results, we can mention the notion of universality. Other well-known successes in condensed matter include the Kondo problem [363] and the Kosterlitz–Thouless transition [191]. The chapter ends with the presentation of some other methods dealing with the phase transition.

5.1 Introduction

We present here some fundamental notions necessary to understand a phase transition. There exist a huge number of reviews and books specialized in this field [10, 54, 95, 380].

All systems do not have obligatorily a phase transition. In general, the existence of a phase transition depends on a few general parameters such as the space dimension, the nature of the interaction between particles and the system symmetry. For spin systems, one can give a brief summary here. In one dimension, in general there is no phase transition at a non-zero temperature for systems of short-range interactions regardless of the spin model. The long-range ordering at $T = 0$ is destroyed as soon as $T \neq 0$. However, in two dimensions discrete spin models such as the Ising and Potts models have a phase transition at a finite temperature T_c, while continuous spin models such as the Heisenberg model do not have a transition at a finite temperature [231]. The XY spin model in two dimensions is a very particular case: In spite of the absence of a long-range order at finite temperatures, there is a phase transition of a special kind called the "Kosterlitz–Thouless" transition [191]. In three dimensions, all known spin models have in general a phase transition at $T_c \neq 0$. Note that this summary is for non-frustrated systems. Frustrated spin systems have to be separately considered, they often do not follow these observations (see reviews in Ref. [85]).

5.1.1 Symmetry Breaking: Order Parameter

A transition from one phase to another may take place when an external parameter varies. Such a parameter can be the temperature or an external applied magnetic field. At the transition point, the system changes from one symmetry to another. The most studied type of transition is no doubt the order-disorder transition. A well-known transition of this kind is the loss of magnetic attraction of a permanent magnet with increasing temperature: This is a transition from a magnetically ordered phase to a magnetically disordered (or paramagnetic) phase.

In order to measure the degree of ordering, one defines an order parameter which depends on the system symmetry. A good order parameter should be non-zero in one phase and zero in the other phase, signaling thus the symmetry breaking when the system changes its phase. Let us present in the following the order parameters defined for some systems.

For a ferromagnetic system of N Ising spins, the order parameter is defined as

$$P = \frac{1}{N}\left|\sum_i \sigma_i\right| \tag{5.1}$$

where $\sigma_i = \pm 1$ is the spin at the site i. It is seen that in the ground state where all spins are parallel, one has $P = 1$, and in the disordered state (or paramagnetic state) where there is a random mixing of up and down spins with the same number, one has $P = 0$.

For an antiferromagnetic system, the order parameter is the so-called staggered magnetization. For example, in one dimension with the Ising model, the staggered magnetization is given by

$$P = \frac{1}{N}\left|\sum_i (-1)^i \sigma_i\right| \tag{5.2}$$

where $(-1)^i$ is the parity of the site i. One can verify that in the antiferromagnetic ground state one has $P = 1$, and in the disordered phase $P = 0$.

For the ferromagnetic Potts model with q states, namely $\sigma_j = 1, \ldots, q$, the order parameter is defined as

$$P = \frac{qM_{\text{max}} - 1}{q - 1} \tag{5.3}$$

where

$$M_{\max} = \frac{\max(M_1, M_2, \ldots, M_q)}{N} \tag{5.4}$$

with

$$M_i = \sum_j \delta_{\sigma_j, i} \quad (i = 1, 2, \ldots, q) \tag{5.5}$$

where the sum on j is performed over all sites of the system. It is observed that in the ground state where there is only one kind of σ_j one has $P = 1$. In the disordered phase where all values of spin are equally present, namely $M_i = N/q$ for any $i = 1, \ldots, q$, one has $P = 0$.

For the Heisenberg spins of amplitude 1 in a ferromagnet, the order parameter is the magnetization defined by

$$M = \frac{1}{N}\left|\sum_i \mathbf{S}_i\right| \tag{5.6}$$

One can verify that in the ground state, where all spins are parallel, M is 1 whatever the orientation of spins with respect to the crystal axes. In the disordered state, each spin has a random spatial orientation so that $\sum_i \mathbf{S}_i = 0$, namely $M = 0$.

5.1.2 Order of a Phase Transition

A phase transition takes place when physical quantities of the system undergo an anomaly. In order to define properly a phase transition, one should examine various physical quantities at the transition temperature T_c. If physical quantities which are second derivatives of the free energy F, such as the specific heat C_V and the susceptibility χ, diverge (see Appendix A), then the corresponding phase transition is a "phase transition of second order." In this case, the first derivatives of F such as the average energy $\overline{E}$ and the average magnetization $\overline{M}$ are continuous functions at T_c. On the other hand, in a first-order phase transition, these first-derivative quantities undergo a discontinuity at the transition point. The reader is referred to Section 6.3 for schematic illustrations of these two kinds of phase transition.

5.1.3 Correlation Function: Correlation Length

An important function in the study of phase transitions is the correlation function defined by

$$G(\mathbf{r}) =< \mathbf{S}(0) \cdot \mathbf{S}(\mathbf{r}) > \tag{5.7}$$

where $\mathbf{S}(0)$ is the spin at a site chosen as the origin of the coordinates, $\mathbf{S}(\mathbf{r})$ the spin at the position $\mathbf{r}$, and $< \cdots >$ denotes the thermal average. In an isotropic system, $G(\mathbf{r})$ depends on r.

In a phase where $\mathbf{S}(0)$ and $\mathbf{S}(\mathbf{r})$ are independent, namely their fluctuations are not correlated, $G(\mathbf{r})$ is zero due to the thermal average. This is the case of a point in the paramagnetic phase well above the transition temperature T_c and at a large distance r.

When the transition temperature T_c is approached from the high-temperature side, a correlation resulting from the interaction between spins sets in, $G(r)$ becomes non-zero for spins at short distances. One can define the "correlation length" ξ as the distance beyond which $G(r)$ is no more significant. The fluctuations of two spins at a distance $r < \xi$ are said "correlated." The correlation length ξ is written as

$$G(r) =< \mathbf{S}(0) \cdot \mathbf{S}(r) >= A\frac{\exp(-r/\xi)}{r^{(d-1)/2}} \tag{5.8}$$

where d is the space dimension and A a constant. In a second-order transition, the correlation length diverges at the transition, namely all spins are correlated at the transition regardless of their distance. On the contrary, at a first-order transition, the correlation length is finite and there is a coexistence of the two phases at the transition point.

5.1.4 Critical Exponents

When the transition is of second order, one can define in the vicinity of T_c the following critical exponents:

$$C_V = A\left|\frac{T - T_c}{T_c}\right|^{-\alpha} \tag{5.9}$$

$$\overline{M} = B\left[\frac{T_c - T}{T_c}\right]^{\beta} \tag{5.10}$$

$$\chi = C\left|\frac{T - T_c}{T_c}\right|^{-\gamma} \tag{5.11}$$

$$\xi \propto \left[\frac{T - T_c}{T_c}\right]^{-\nu} \tag{5.12}$$

$$\overline{M} = H^{1/\delta} \tag{5.13}$$

Note that the same α is defined for $T > T_c$ and $T < T_c$ but the coefficient A is different for each side of T_c. This is also the case for γ. However, β is defined only for $T < T_c$ because $\overline{M} = 0$ for $T \geq T_c$. The definition of δ is valid only at $T = T_c$ when the system is under an applied magnetic field of amplitude H. Finally, at $T = T_c$, one defines exponent η of the correlation function by

$$G(r) \propto \frac{1}{r^{d-2+\eta}} \tag{5.14}$$

We note that there is another exponent called "dynamic exponent" z defined via the relaxation time τ of the spin system for $T \geq T_c$:

$$\tau \propto \xi^z \propto \left[\frac{1}{T - T_c}\right]^{z\nu} \tag{5.15}$$

Only the six exponents α, β, γ, δ, ν and η are critical exponents. It is observed below that there are four relations between them (see Section 5.4). Therefore, only two of them are to be determined.

It is obvious that the above exponents are not defined in a first-order transition because there is no divergence of physical quantities at T_c. For this reason, one says that first-order transitions are not critical. To be precise, we shall call "critical temperature" for second-order transitions and "transition temperature" for first-order transitions.

5.1.5 Universality Class

Phase transitions of second order are distinguished by their "universality class." Phase transitions having the same values of critical exponents belong to the same university class. Renormalization group analysis shows that the universality class depends only on a few very general parameters such as the space dimension, the symmetry of the order parameter and the nature of the interaction. So, for example, Ising spin systems with short-range ferromagnetic interaction in two dimensions belong to the same universality class whatever the lattice structure is. Of course, the critical temperature T_c is not the same for square, hexagonal, rectangular, honeycomb, ..., lattices, but T_c is not a universal quantity. It depends on the interaction value, the coordination number, ... but these quantities do not affect the values of the critical exponents.

Table 5.1 shows the critical exponents of some known universality classes.

Table 5.1 Critical exponents of some known universality classes

Class	Symmetry	α	β	γ	ν	η
2d Ising	Z_2	0	1/8	7/4	1	1/4
2d Potts (q = 3)	Z_3	1/3	1/9	13/9	5/6	4/15
2d Potts (q = 4)	Z_4	2/3	1/12	7/6	2/3	1/4
3d Ising	Z_2	0.11	0.325	1.241	0.63	0.031
3d XY	$O(2)$	−0.007	0.345	1.316	0.669	0.033
3d Heisenberg	$O(3)$	0.115	0.3645	1.386	0.705	0.033
Mean-field		0	1/2	1	1/2	0

Note that when a system is invariant by the following local transformation ($J \rightarrow -J$, $\mathbf{S}_i \rightarrow -\mathbf{S}_i$), where J is the interaction of $\mathbf{S}_i$ with its neighbors, the universality class of the new system does not change. This is understood immediately if one looks at the partition function: Such a local transformation does not change the argument of the exponential of the partition function. By consequence, physical properties do not change. As an example, let us consider a ferromagnetic square lattice. If one operates the local transformation on one spin out of every two in the square lattice (see Fig. 5.1), the system changes from a ferromagnetic crystal into an antiferromagnetic one. However, as said above, this local transformation does not change the physical properties of the system: One concludes that a ferromagnetic crystal and its

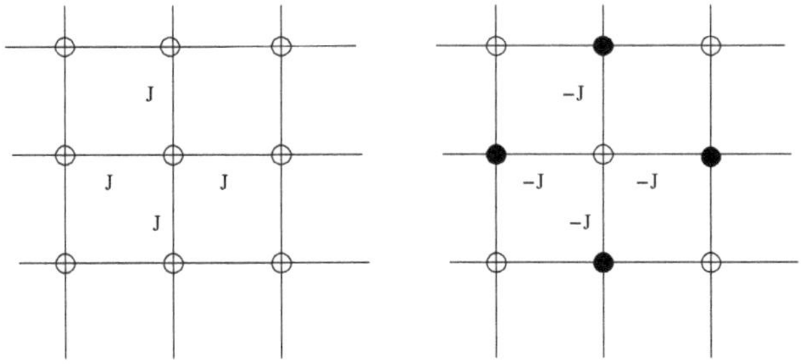

Figure 5.1 Local transformation $J \rightarrow -J$, $\mathbf{S}_i \rightarrow -\mathbf{S}_i$ operated on one spin out of every two changes the square ferromagnetic lattice (left) into an antiferromagnetic lattice (right). White and black circles indicate $\uparrow$ spins and $\downarrow$ spins, respectively.

antiferromagnetic counterpart have the same critical temperature and the same critical exponents.

We emphasize that there exist many systems in which it is impossible to operate local transformation without changing the argument of the partition function. One of these systems is the triangular lattice: It is impossible to find a spin configuration to satisfy all interactions if one changes J into $-J$ everywhere. The energy of the system is not conserved, so the partition function changes, giving rise to a new system with different properties. Such systems are called "frustrated systems" which are discussed in Section 5.7.3 and studied in Ref. [85].

In Chapter 2, we have presented the mean-field theory by using the Heisenberg model for ferromagnets, antiferromagnets and ferrimagnets. The mean-field theory neglects instantaneous fluctuations of spins. Due to this approximation, it overestimates the critical temperature T_C. Fluctuations favor disorder, so when taken into account, fluctuations cause a transition at a temperature lower than T_c given by the mean-field theory [see (2.24)], or even destroy the magnetic long-range order at any finite temperature in low dimensions $d = 1$ and $d = 2$. The mean-field theory can be improved by several methods. In the following, we present the method proposed by Bethe. The treatment of fluctuations with the Landau–Ginzburg theory is presented in Section 5.3 where the mean-field theory is shown to be exact for dimension $d > 4$.

5.2 Improved Mean-Field Theory: Bethe's Approximation

We consider the case of Ising spins. The mean-field theory can be improved by using the partition function of an approximate Hamiltonian constructed as follows. We separate the whole Hamiltonian into two parts: a cluster containing a spin, say σ_0 and its z neighbors, and the remaining crystal. The method consists in treating exactly the interactions of σ_0 with its surrounding z neighbors, but treating the interaction of the cluster with the remaining crystal by the use of the mean-field theory. The Hamiltonian is written as

$$\mathcal{H} = -J\sum_{i=1}^{z}\sigma_0\sigma_i - \mu_B B\sigma_0 - \mu_B(B+H)\sum_{i=1}^{z}\sigma_i \tag{5.16}$$

where B is an applied magnetic field. The spin σ_0 and its z surrounding nearest neighbors σ_i form a cluster which is embedded in the crystal. H is the molecular field acting on σ_i by its $(z-1)$ neighbors outside of the cluster. H is thus given by $H = (z-1)J < \sigma > /\mu_B$. The mean-field equation will be obtained at the end by setting $< \sigma_0 >=< \sigma_i >\equiv< \sigma >$. One has

$$\begin{aligned} Z &= \sum_{\sigma_0=\pm1}\sum_{\sigma_1=\pm1}\cdots\sum_{\sigma_z=\pm1} e^{\beta[J\sum_{i=1}^{z}\sigma_0\sigma_i+\mu_B B\sigma_0+\mu_B(B+H)\sum_{i=1}^{z}\sigma_i]} \\ &= \sum_{\sigma_0=\pm1}\sum_{\sigma_1=\pm1}\cdots\sum_{\sigma_z=\pm1} e^{a\sigma_0\sum_{i=1}^{z}\sigma_i+b\sigma_0+(b+c)\sum_{i=1}^{z}\sigma_i} \end{aligned} \tag{5.17}$$

where $a = J/k_BT$, $b = \mu_B B/k_BT$ and $c = \mu_B H/k_BT = (z-1)J < \sigma > /k_BT$. Summing on $\sigma_0 = \pm1$ and factorizing the other sums, one gets

$$\begin{aligned} Z &= Z_+ + Z_- \quad \text{where} \\ Z_\pm &= \sum_{\sigma_1=\pm1}\cdots\sum_{\sigma_z=\pm1} e^{\pm a\sum_{i=1}^{z}\sigma_i\pm b+(b+c)\sum_{i=1}^{z}\sigma_i} \\ &= e^{\pm b}[2\cosh(\pm a+b+c)]^z \end{aligned} \tag{5.18}$$

The averaged $< \sigma_0 >$ and $< \sigma_i >$ $(i = 1,\ldots,z)$ are given by

$$< \sigma_0 > = \frac{Z_+ - Z_-}{Z} \tag{5.19}$$

$$< \sigma_i > = \frac{1}{z}\sum_{i=1}^{z} < \sigma_i > = \frac{1}{z}\frac{\partial Z/\partial c}{Z}$$
$$= [Z_+ \tanh(a+b+c) + Z_- \tanh(-a+b+c)]/Z \tag{5.20}$$

Setting $< \sigma_0 > = < \sigma_i > \equiv < \sigma >$, one has

$$Z_+[1 - \tanh(a+b+c)] = Z_-[1 + \tanh(-a+b+c)] \tag{5.21}$$

Replacing $Z_\pm$ by (5.18), one obtains

$$\left[\frac{\cosh(a+b+c)}{\cosh(-a+b+c)}\right]^{z-1} = e^{2c} \tag{5.22}$$

This equation is used to determine self-consistently c, namely $< \sigma >$. In zero field ($B = 0$), one has $b = 0$ so that

$$\frac{\cosh(a+c)}{\cosh(-a+c)} = e^{2c/(z-1)}$$
$$\frac{c}{z-1} = \frac{1}{2}\ln\frac{\cosh(a+c)}{\cosh(-a+c)} \tag{5.23}$$

It is observed that $c = 0$ is a solution of the last equation. However, there is a non-zero solution by making an expansion at small c (small $< \sigma >$) to the third order:

$$\frac{c}{z-1} \simeq \frac{1}{2}\ln\frac{\cosh a + c\sinh a + (1/2)c^2\cosh a + \cdots}{\cosh a - c\sinh a + (1/2)c^2\cosh a + \cdots}$$
$$= \left[c - \frac{1}{3}\frac{c^3}{\cosh^2 a} + \cdots\right]\tanh a$$
$$\frac{1}{z-1} = \left[1 - \frac{1}{3}\frac{c^2}{\cosh^2 a} + \cdots\right]\tanh a \tag{5.24}$$

from which one has

$$c^2 = 3\frac{\cosh^3 a}{\sinh a}\left[\tanh a - \frac{1}{z-1} + \cdots\right] \tag{5.25}$$

Since the left-hand side is positive, this equation admits a non-zero solution if $[\tanh a - \frac{1}{z-1}] > 0$. This means that the non-zero solution exists if $T < T_c$ where T_c is given by

$$\tanh(J/k_B T_c) = \frac{1}{z-1} \tag{5.26}$$

It is interesting to note that for $z = 1$, namely a cluster of two spins, there is no solution for T_c, and for $z = 2$ (one-dimensional chain) one has $T_c = 0$. This corresponds to the results from the exact solutions (see paragraphs 5.4.2 and 5.6, and Problems 1). We recall that the mean-field theory incorrectly yields $T_c \neq 0$ for one dimension [see Eq. (2.23)].

5.3 Landau–Ginzburg Theory

It has been shown above that the mean-field theory suffers a serious flaw due to the fact that instantaneous local spin fluctuations have been neglected in this theory. For example, the mean-field theory can give rise to a phase transition where there is none in low dimensions. In addition, in the critical region neglecting fluctuations modifies the behavior of the phase transition as we will see when comparing the mean-field critical exponents with the exact ones. There exist several more efficient theories such as the high- and low-temperature series expansions [95], the Landau–Ginzburg theory and the renormalization group. In what follows, we present the Landau–Ginzburg theory and the concepts of the renormalization group.

The Landau–Ginzburg theory is an extension of the mean-field theory which includes a great part of fluctuations so far neglected near the transition. The main idea is to start from an expansion of the free energy per spin f in the vicinity of T_c when the magnetization m is sufficiently small:

$$f = -\frac{k_B T}{N} \ln Z \tag{5.27}$$

$$\begin{aligned} f_{MF} &= -\frac{k_B T}{N} \ln Z_{MF} \\ &\simeq C + A\left(T - T_c^{MF}\right) m^2 + Bm^4 + Dhm + \cdots \end{aligned} \tag{5.28}$$

where C, A, B and D are constants, h is an applied magnetic field and the suffix MF denotes quantities coming from the mean-field theory. The form of this expansion and the sign of B reflect the system symmetry. In the case where $h = 0$ and $B > 0$, f_{MF} presents a minimum at $m = 0$ for $T > T_c^{MF}$ and two symmetric minima at $\pm m_0$ for $T < T_c^{MF}$ (see Fig. 5.2), indicating two degenerate ordered states. This degeneracy is removed when $h \neq 0$ as shown in Fig. 5.3.

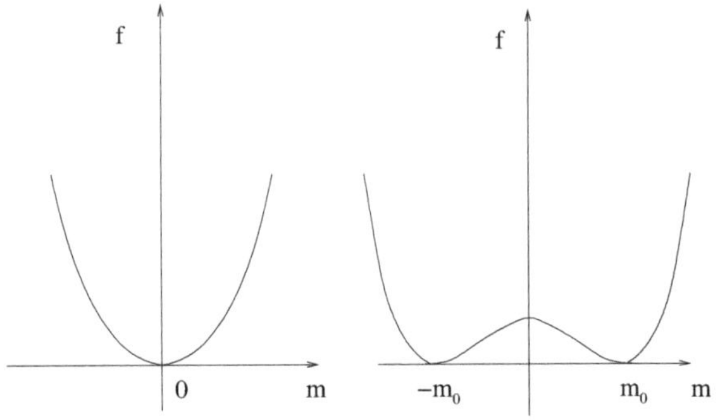

Figure 5.2 Mean-field free energy at $T > T_c$ (left) and at $T < T_c$ (right).

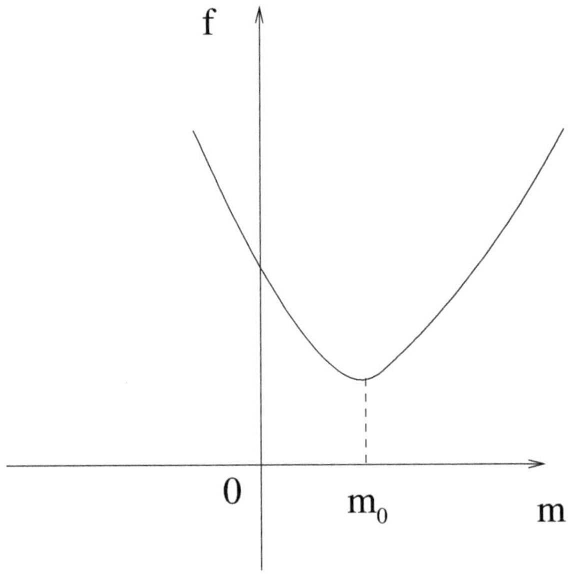

Figure 5.3 Mean-field free energy at $T < T_c$ in an applied magnetic field.

When $B < 0$, a first-order transition is possible. At $T = T_c$, there are three equivalent minima of f_{MF} at 0, $\pm m_0$ (see Fig. 5.4) contrary to the case $B > 0$. This means that at the transition the three phases $m = 0$ (paramagnetic state) and $\pm m_0$ (ordered states) coexist. The energy distribution at $T = T_c$ is thus bimodal, the peak at low

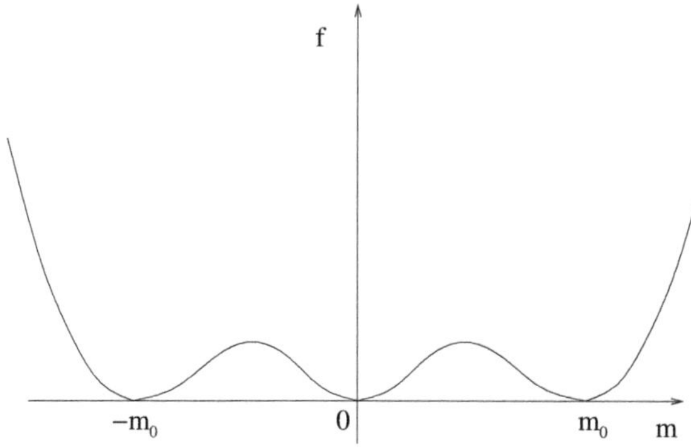

Figure 5.4 Mean-field free energy at a first-order transition.

energy corresponds to the energy of the ordered phase while that at high energy corresponds to the energy of the disordered state. The distance between the two peaks is the latent heat which is observed at a first-order transition.

When an m^3 term is present in the expansion (5.28), the transition is always of first order.

5.3.1 Mean-Field Critical Exponents

The critical exponents calculated by the mean-field theory are (see Chapter 2) $\beta = 1/2$ [see (2.24)], $\gamma = 1$ [see (2.38)]. One can find them again here by using (5.28). Putting $t = (T - T_c^{MF})/T_c^{MF}$, one has

- When $t < 0$ and $h = 0$, f_{MF} is minimum at $m_0 \propto (-t)^{\frac{1}{2}}$, so that $\beta = 1/2$.
- When $h \neq 0$ and $t \geq 0$, one has $m \propto h/t$ so that $\chi \propto t^{-1}$, hence $\gamma = 1$.
- At $t = 0$, f_{MF} is minimum at $m \propto (\frac{Dh}{B})^{1/3}$, hence $\delta = 3$.
- For $t > 0$, the minimum of f_{MF} is equal to C and for $t < 0$, it is equal to $C + \mathcal{O}(\frac{A^2}{B}t^2)$. One has $C_V \propto \frac{\partial^2 f}{\partial T^2}$ which is 0 for $t > 0$ and equal to $\frac{A^2}{B}$ for $t < 0$. C_V is thus independent of t. The discontinuity of C_V at $t = 0$ is an artifact of the mean-field theory. When fluctuations are included, one finds $\alpha = 0$ as will be seen below.

5.3.2 Correlation Function

This subsection considers a system of Ising spins for simplicity. A simplified notation has been used in this paragraph: r instead of $\mathbf{r}$. The correct notation will be recovered when necessary. The Hamiltonian is given by

$$\mathcal{H} = -\sum_{(r',r'')} J(r'-r'')S(r')S(r'') \tag{5.29}$$

The correlation function is calculated as follows:

$$\begin{aligned} G(r) &= < S(0)S(r) > \\ &= \frac{\mathrm{Tr}S(0)S(r)\exp(\frac{1}{2}\beta\sum_{r',r''} J(r'-r'')S(r')S(r''))}{\mathrm{Tr}\exp(\frac{1}{2}\beta\sum_{r',r''} J(r'-r'')S(r')S(r''))} \\ &= \frac{\mathrm{Tr}S(r)\exp(\frac{1}{2}\beta\sum_{r',r''} J(r'-r'')S(r')S(r''))}{\mathrm{Tr}\exp(\frac{1}{2}\beta\sum_{r',r''} J(r'-r'')S(r')S(r''))} \\ &= < S(r) >= \tanh\left[\beta\sum_{r'} J(r-r')S(r')\right] \end{aligned} \tag{5.30}$$

where one has taken the spin at the origin $S(0) = 1$. The trace was taken over all configurations. In the mean-field spirit, one replaces $S(r')$ on the right-hand side of (5.30) by its average value $< S(r') >$, then one makes an expansion around T_c when $< S(r') >$ is small, one obtains

$$m(r) \simeq \beta\sum_{r'} J(r-r')m(r') \tag{5.31}$$

using the notation $m(r) =< S(r) >$. The Fourier transform of this equation gives

$$m(k) = \beta J(k)m(k) + C \tag{5.32}$$

where C is a constant. For small k, one has

$$\begin{aligned} J(k) &= \sum_{r=-\infty}^{\infty} J(r)\exp(ikr) \simeq \sum_{r=-\infty}^{\infty} J(r)[1+ikr+(ikr)^2] \\ &\simeq \sum_{r=-\infty}^{\infty} J(r)[1-k^2r^2] \simeq \sum_{r=-\infty}^{\infty} J(r) - k^2\sum_{r=-\infty}^{\infty} r^2 J(r) \\ &\simeq \sum_{r=-\infty}^{\infty} J(r) - k^2\sum_{r'=-\infty}^{\infty} J(r')\sum_{r=-\infty}^{\infty} r^2 J(r)/\sum_{r'=-\infty}^{\infty} J(r') \\ &\simeq \sum_{r}\tilde{J}[1-k^2R^2] \end{aligned} \tag{5.33}$$

where the sum on the term $J(r)r$ of the second equality is zero because $J(r)r$ is an odd function of r [$J(r)$ is an even function: $J(r) = J(-r)$]. $\tilde{J}$ is defined by

$$\tilde{J} = \sum_r J(r) \tag{5.34}$$

and R^2 is defined by

$$R^2 = \frac{\sum_r r^2 J(r)}{\sum_r J(r)} \tag{5.35}$$

R^2 is thus the order of the interaction range. One obtains from (5.32) and (5.33)

$$m(k) = \frac{C}{1 - \beta\tilde{J}(1 - R^2k^2)} \tag{5.36}$$

One recalls here that in the mean-field theory $k_BT_c = \tilde{J}$. One writes thus

$$\begin{aligned} m(k) &= \frac{C}{1 - \beta k_BT_c(1 - R^2k^2)} = \frac{C}{1 - \frac{T_c}{T}(1 - R^2k^2)} \\ &= \frac{C}{t + \frac{T_c}{T}R^2k^2} \simeq \frac{CR^{-2}}{tR^{-2} + k^2} \end{aligned} \tag{5.37}$$

where one has taken $T \simeq T_c$. Putting

$$\xi^{-2} = tR^{-2} \tag{5.38}$$

and using $m(r) = G(r)$ of (5.30), one finally arrives at

$$G(k) = \frac{CR^{-2}}{\xi^{-2} + k^2} \tag{5.39}$$

The inverse Fourier transform of (5.39) gives

$$G(r) \propto \frac{\exp(-r/\xi)}{r^{(d-1)/2}} \tag{5.40}$$

This form of $G(r)$ justifies the fact that ξ is called the "correlation length" in the mean-field theory. The expression (5.39) is called "Ornstein–Zernike correlation function."

From (5.38), one sees that $\xi \propto t^{-1/2}$; therefore, $\nu = 1/2$ in this mean-field theory. In addition, at $t = 0$, ξ tends to ∞ (see the following paragraph), $G(k)$ of (5.39) is then proportional to $1/k^2$. The inverse Fourier transform of this function is $\frac{1}{r^{d-2}}$ for large r, indicating therefore that exponent η of the mean-field theory is equal to 0 [see definition (5.14)].

5.3.3 Corrections to Mean-Field Theory

One decomposes the following term [see (12.16)]:

$$\begin{aligned} J(r-r')\mathbf{S}_r \cdot \mathbf{S}_{r'} &= J(r-r')[\mathbf{S}_r^z + \delta\vec{S}_r] \cdot [\mathbf{S}_{r'}^z + \delta\mathbf{S}_{r'}] \\ &\simeq J(r-r') < \mathbf{S}_r^z > \cdot < \mathbf{S}_{r'}^z > \\ &\simeq J(r-r')m(r)m(r') \end{aligned} \tag{5.41}$$

where the average values of the linear terms in $\delta\mathbf{S}_r$ and $\delta\mathbf{S}_{r'}$ are zero by symmetry (rotating vectors in the xy plane). In the last equality, one has neglected, in the mean-field spirit, the following term:

$$J(r-r')\delta\mathbf{S}_r \cdot \delta\mathbf{S}_{r'} \tag{5.42}$$

This term, however, when taken into account, will improve the mean-field theory as seen below. Using the correlation function

$$G(r-r') =< \mathbf{S}_r \cdot \mathbf{S}_{r'} >= \text{constant} + < \delta\vec{S}_r \cdot \delta\mathbf{S}_{r'} >$$

one writes (5.42) as

$$\sum_{r'} J(r-r') < \delta\mathbf{S}_r \cdot \delta\mathbf{S}_{r'} >= \sum_{r'} J(r-r')G(r-r') \tag{5.43}$$

One is interested now in a long-distance behavior, namely at small k. Using the Fourier transform of (5.39) for small k, one writes

$$\begin{aligned} \sum_{r'} J(r-r') < \delta\mathbf{S}_r \cdot \delta\mathbf{S}_{r'} > &\simeq \sum_{r'} J(r-r')\frac{C}{R^2}\int_{ZB}\frac{d^d k}{k^2+\xi^{-2}} \\ &= \tilde{J}\frac{C}{R^2}\int_{ZB}\frac{d^d k}{k^2+\xi^{-2}} \end{aligned} \tag{5.44}$$

where the integral is performed in the first Brillouin zone. One decomposes this integral as follows:

$$\int_{ZB}\frac{d^d k}{k^2+\xi^{-2}} = \int_{ZB}\frac{d^d k}{k^2} - \xi^{-2}\int_{ZB}\frac{d^d k}{k^2(k^2+\xi^{-2})} \tag{5.45}$$

The first integral does not depend on ξ, namely independent of T. It contributes to a shift of the value of T_c calculated by the mean-field theory. The second integral depends on ξ thus on T: It converges if $k = \pm\pi/a \to \infty$, namely $a \to 0$ (continuum limit), and if d (space dimension) < 4. By a simple dimension analysis, one sees that this integral is proportional to ξ^{2-d} at the limit $\xi \to \infty$. The second term is thus proportional to $\frac{\tilde{J}}{R^2}\xi^{2-d}$. This term has been neglected before in the mean-field theory because it was wrongly considered as always small with respect to the term coming from the mean field. Let us examine the condition when it can be neglected:

$$\frac{\tilde{J}}{R^2}\xi^{2-d} \ll \tilde{J}\,t \tag{5.46}$$

where $t = \frac{T-T_c}{T}$. One knows that $\xi = Rt^{-1/2}$ [see (5.38)], expression (5.46) can be thus rewritten as

$$\xi^{4-d} \ll R^4 \tag{5.47}$$

This condition for neglecting the second term of (5.45) is called "Ginzburg's criterion." One sees that when $d < 4$ this criterion is not satisfied near T_c where $\xi \to \infty$. By consequence, the mean-field theory which neglects fluctuations characterized by ξ^{4-d} is not valid in the critical region for $d < 4$. The dimension $d = 4$ is called "upper critical dimension" for the Ising model with short-range interaction.

5.4 Renormalization Group

5.4.1 Transformation of the Renormalization Group: Fixed Point

The central idea of the renormalization group is to replace the set of parameters which define the system by another set of parameters while conserving the essential physical ingredients, in particular the system symmetry. This set of parameter is simpler to deal with.

In the study of the phase transition, this transformation consists in dividing the system into blocks of spins and replacing each block by a single spin. This "new" spin interacts with the others by the renormalized interactions calculated while decimating the block. The distances between the new spins are measured with a new lattice constant. This procedure is repeated for the system of new spins to obtain the next generation of spins which is used to generate the following generation, and so on. At each iteration, also called decimation or transformation, one has a relation between the new interaction $K' = \beta J'$ and the previous one $K = \beta J$:

$$K' = f(K) \tag{5.48}$$

The new correlation length ξ which is a function of K' is equal to the previous correlation length divided by the factor called "dilatation" b defined as the ratio between the new and old lattice constants. One has

$$\xi(K') = \frac{\xi(K)}{b} \tag{5.49}$$

At the phase transition, ξ becomes infinite, the measuring distance unit is no more important. The interaction constants K' and K become identical $K' = K = K^*$. This point is called the "fixed point" in the renormalization group language. The fixed point is thus determined by

$$K^* = f(K^*) \tag{5.50}$$

The one-dimensional case is simple to proceed as seen in paragraph 5.4.2 below. However, in the case of a general dimension $d > 1$, relation (5.48) is rather complicated. It is often impossible to find a solution of (5.50). One then has to take into account physical considerations in order to find appropriate approximations.

One considers a point K in the proximity of the fixed point K^*. If the iteration process takes K away from the fixed point, K^* is an unstable fixed point. This is a "run away" case. On the other hand, in the case where K tends to K^* by iteration, K^* is a stable fixed point (see Problem 5.4.2 below). The map of these trajectories near a fixed point with indicated moving directions is called a "flow diagram." An example is shown in Fig. 5.5. This figure corresponds to the case of a square lattice of Ising spins with interactions $K_1 = \frac{J_1}{k_B T}$ in the x direction and $K_2 = \frac{J_2}{k_B T}$ in the y direction. P is the fixed point. For a given ratio K_2/K_1, namely one follows the discontinued line: Intersection P_1 with the line separating regions of different flows is the critical point corresponding to that ratio K_2/K_1. The line of flow separation shown by the heavy solid line is the critical line. One sees that P_1 runs toward P on the critical line; therefore, P_1 and P belong to the same universality class: The universality class does not thus depend on the ratio K_2/K_1.

One can calculate the critical exponents using the renormalization group if one knows how K' depends on K in the vicinity of a fixed point K^*. One makes then an expansion of their relation (5.48)

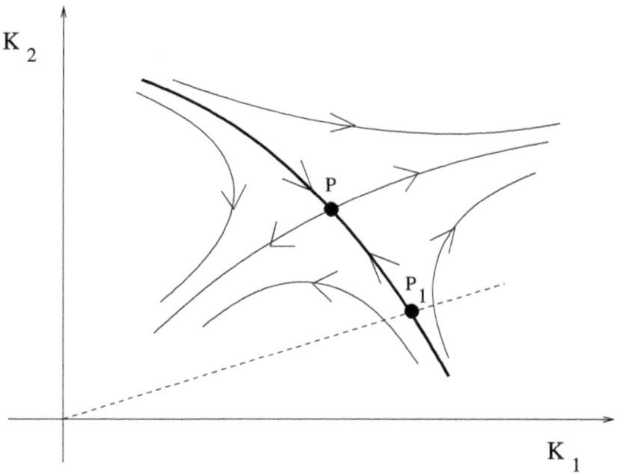

Figure 5.5 Flow diagram for a square lattice of Ising spins with nearest-neighbor interactions K_1 along the x axis and K_2 along the y axis. See text for comments.

around K^*:

$$\begin{aligned} K' &\simeq f(K^*) + (K - K^*) f'(K^*) \\ &\simeq K^* + b^y (K - K^*) \end{aligned} \tag{5.51}$$

where one has replaced $f(K^*)$ by K^* using (5.50), and one has used the following notation:

$$y = \frac{\ln f'(K^*)}{\ln b}.$$

Now, near K^* one knows that $\xi \propto (K - K^*)^{-\nu}$ (definition of ν). Therefore, by using (5.49) and (5.51) one obtains

$$(K - K^*)^{-\nu} = b(K' - K^*)^{-\nu} = b[b^y (K - K^*)]^{-\nu} \tag{5.52}$$

By identifying the two sides of the above equation, one gets

$$\nu = 1/y \tag{5.53}$$

This example shows that the critical exponent ν is a derivative of the renormalization group equation (5.48). The other critical exponents, defined in (5.9), (5.10), (5.11), (5.13) and (5.14), are calculated by the use of the free energy. Details are given elsewhere, for example, in Refs. [10, 54, 380]. They are connected by the following scaling relations:

$$\alpha + 2\beta + \gamma = 2 \tag{5.54}$$

$$\alpha + \beta(1 + \delta) = 2 \tag{5.55}$$

In addition, the two exponents concerning the spin–spin correlation, ν and η, are connected via the following "hyperscaling relations":

$$\alpha = 2 - d\nu \tag{5.56}$$

$$\gamma = \nu(2 - \eta) \tag{5.57}$$

In summary, there are six critical exponents and four relations between them. It suffices to determine two among six exponents. The other four can be then calculated using the above four relations.

To close this section, one emphasizes that, in addition to the results shown above, another important result of the renormalization group is the relations connecting the system size to the critical exponents: Physical quantities calculated at a finite system size are shown to depend on powers of the linear system size. These powers are simple functions of critical exponents. They are very useful for the determination of critical exponents by Monte Carlo simulation (see, for example, Refs. [33, 87, 199]).

5.4.2 Renormalization Group Applied to an Ising-Spin Chain

One applies in the following the renormalization group to the case of a chain of Ising spins with a ferromagnetic interaction between nearest neighbors. One will show that there is no phase transition at finite temperature.

One considers the following Hamiltonian:

$$\mathcal{H} = -K \sum_n \sigma_n \sigma_{n+1} \tag{5.58}$$

where $K = J/k_B T$, J being a ferromagnetic interaction between nearest neighbors. The partition function is given by

$$Z = Tr \exp(-\mathcal{H}) \tag{5.59}$$

To study this spin chain, one uses the decimation method. One divides the system into three-spin blocks as shown in Fig. 5.6.

One writes for the two blocks in the figure the corresponding factors in Z

$$\exp(K\sigma_2\sigma_3)\exp(K\sigma_3\sigma_4)\exp(K\sigma_4\sigma_5) \tag{5.60}$$

Using the following equality for the case $\sigma_n = \pm 1$:

$$\exp(K\sigma_2\sigma_3) = \cosh K(1 + x\sigma_2\sigma_3) \tag{5.61}$$

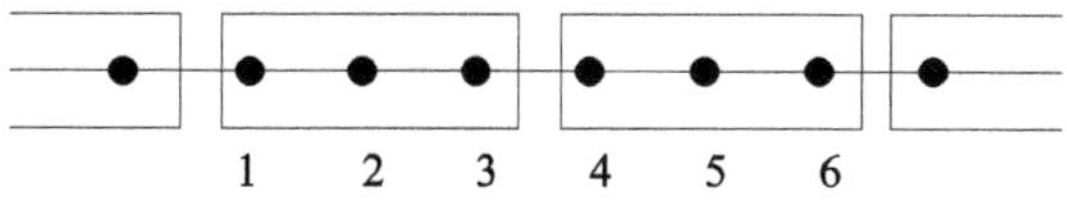

Figure 5.6 Blocks of three spins used for the decimation.

where $x = \tanh K$, one rewrites (5.60) as

$$\cosh^3 K(1 + x\sigma_2\sigma_3)(1 + x\sigma_3\sigma_4)(1 + x\sigma_4\sigma_5) \tag{5.62}$$

Writing explicitly each term of the product and making the sum on $\sigma_3 = \pm 1$ and $\sigma_4 = \pm 1$ (decimation of two spins at the block border), one sees that the odd terms of these variables give zero contributions. There remains

$$2^2 \cosh^3 K(1 + x^3\sigma_2\sigma_5)$$

One can rewrite it as

$$2^2 \cosh^3 K(1 + x^3\sigma_2\sigma_5) = \exp[K'\sigma_2\sigma_5 + C] \tag{5.63}$$

where C is a constant. If one forgets C, the right-hand side is similar to the initial Hamiltonian with a new interaction K' between the remaining spins σ_2 and σ_5. To calculate K' one writes

$$\begin{aligned}\exp[K'\sigma_2\sigma_5 + C] &= \exp(C)\exp(K'\sigma_2\sigma_5)\\ &= \exp(C)\cosh K'(1 + x'\sigma_2\sigma_5)\end{aligned} \tag{5.64}$$

By identifying this with the left-hand side of (5.63), one obtains

$$x' = \tanh K' = x^3 \quad \text{where} \quad K' = \tanh^{-1}[\tanh^3 K]$$

and

$$\begin{aligned}\exp(C)\cosh K' &= 2^2 \cosh^3 K\\ \exp(C) &= \frac{2^2 \cosh^3 K}{\cosh K'}\\ C &= -\ln\left[\frac{\cosh^3 K}{\cosh K'}\right] - 2\ln 2\end{aligned} \tag{5.65}$$

One renumbers the spins after the first decimation as follows: $\sigma'_1 = \sigma_2, \sigma'_2 = \sigma_5, \ldots$ (every three old spins). The new Hamiltonian is thus

$$\mathcal{H}' = -K'\sum_n \sigma'_n\sigma'_{n+1} - C\frac{N}{3} \tag{5.66}$$

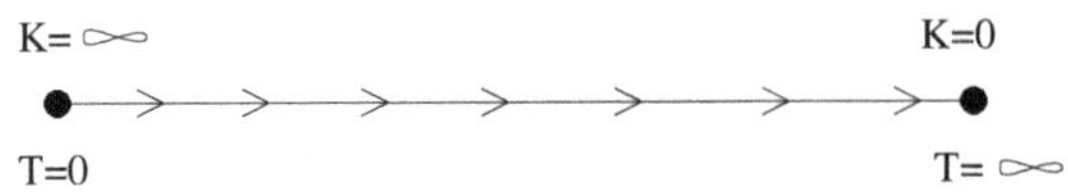

Figure 5.7 Flow diagram of a chain of Ising spins.

where $N/3$ is the number of three-spin blocks (N: initial number of spins).

The equation of the renormalization group is thus

$$K' = \tanh^{-1}[\tanh^3 K] \tag{5.67}$$

From the above equation, one sees that $K' = K$ if $K = 0$ and $K = \infty$:

(1) At high T, $K \to 0^+$, $\tanh^3 K < 1$, hence $K' \to 0$ after successive decimations. The fixed point at $T = \infty$ ($K = 0$) is thus stable.
(2) At low T, $K \to \infty$, $\tanh^3 K \to 1^-$, hence $K' \to 0$ after successive decimations, namely a "run away" flow. The fixed point at $T = 0$ ($K = \infty$) is thus unstable.

The flow diagram is shown in Fig. 5.7. Any point between $K = 0$ and $K = \infty$ moves to $K = 0$ after successive decimations. The nature of any point between these limits is therefore the same as that of $K = 0$ ($T = \infty$), namely it belongs to the paramagnetic phase. The one-dimensional chain is thus disordered at any finite temperature.

5.5 Migdal–Kadanoff Decimation Method: Migdal–Kadanoff Bond-Moving Approximation

In the same spirit as the renormalization concept presented above, the Migdal–Kadanoff decimation method consists of decimating one spin out of every two on a chain to reduce the number of spins. The scaling parameter is thus $b = 2$. Let us describe how it works.

One writes the partition function for an Ising spin chain in an applied magnetic field h:

$$Z = \sum_{\{\sigma_i = \pm 1\}} \exp\left\{\sum_i [K\sigma_i\sigma_{i+1} + h\sigma_i]\right\}$$

where the first sum runs over all spin configurations. Now, instead of dividing the chain into blocks as in the renormalization group described in the previous section, one decimates one spin out of every two as follows: One considers three consecutive spins i, k and j and one writes the corresponding part of the Hamiltonian with the sum on the middle spin σ_k:

$$\begin{aligned}
-\beta H(i,j) &= \frac{h}{2}\sum_{\sigma_k=\pm 1}(\sigma_i+\sigma_k+\sigma_j) \\
&\quad + \ln\sum_{\{\sigma_k=\pm 1\}}\exp\left[K(\sigma_i\sigma_k+\sigma_k\sigma_j)+h\sigma_k\right] \\
&= \frac{h}{2}(\sigma_i+\sigma_j)+\ln 2+\ln\cosh\left[K(\sigma_i+\sigma_j)+h\right] \\
&\simeq \frac{h}{2}(\sigma_i+\sigma_j)+\ln 2+\ln\cosh\left[K(\sigma_i+\sigma_j)\right] \\
&\quad + h\tanh\left[K(\sigma_i+\sigma_j)\right]
\end{aligned} \tag{5.68}$$

where in the last line one has assumed a small h. Since $\sigma_i^2 = 1$, one has the following expressions:

$$\begin{aligned}
(\sigma_i+\sigma_j)^{2n} &= 2^{2n-1}(1+\sigma_i\sigma_j) \\
(\sigma_i+\sigma_j)^{2n+1} &= 2^{2n}(\sigma_i+\sigma_j)
\end{aligned}$$

With these, one can write for any even and odd functions of $K(\sigma_i + \sigma_j)$:

$$\begin{aligned}
f_e[K(\sigma_i+\sigma_j)] &= \sum_{n=0}\frac{a_n K^{2n}(\sigma_i+\sigma_j)^{2n}}{n!} \\
&= \sum_{n=0}\frac{a_n K^{2n}2^{2n-1}(1+\sigma_i\sigma_j)}{n!} \\
&= \frac{1}{2}f_e(2K)(1+\sigma_i\sigma_j)
\end{aligned} \tag{5.69}$$

$$\begin{aligned}
f_o[K(\sigma_i+\sigma_j)] &= \sum_{n=0}\frac{b_n K^{2n+1}(\sigma_i+\sigma_j)^{2n+1}}{n!} \\
&= \sum_{n=0}\frac{b_n K^{2n+1}2^{2n}(\sigma_i+\sigma_j)}{n!} \\
&= \frac{1}{2}f_o(2K)(\sigma_i+\sigma_j)
\end{aligned} \tag{5.70}$$

Using these expressions, one gets from (5.68)

$$\begin{aligned}-\beta H(i,j) &= \frac{1}{2}\ln\cosh(2K)\sigma_i\sigma_j + \frac{1}{2}h\,[1+\tanh(2K)]\,(\sigma_i+\sigma_j)\\ &\quad +\ln 2 + \frac{1}{2}\ln\cosh(2K)\\ &= K'\sigma_i\sigma_j + h'(\sigma_i+\sigma_j) \qquad (5.71)\end{aligned}$$

where

$$K' = \frac{1}{2}\ln\cosh(2K) \qquad (5.72)$$

$$h' = h\,[1+\tanh(2K)] + O(h^2) \qquad (5.73)$$

Note that all quantities independent of spins have been omitted since they do not affect expectation values of physical quantities. Equation (5.71) has the same form as if the spins σ_i and σ_j are neighbors but with the parameters K' and h' instead of the original K and h in the initial Hamiltonian. If one looks for a fixed point one has to solve (5.72) when $K' = K = K^*$ with $h = 0$. But the equation $f(x) = \frac{1}{2}\ln\cosh(2x) = x$ has no solution for $0 < x < \infty$. There is a solution at $x = 0$ corresponding to the stable fixed point at $T = \infty$ as found in the previous section. There is an unstable fixed point at $K = \infty$ as seen by the following expansion:

$$\begin{aligned}K' &= \frac{1}{2}\ln[\exp(2K)+\exp(-2K)] - \frac{\ln 2}{2}\\ &= K + \frac{1}{2}\ln[1+\exp(-4K)] - \frac{\ln 2}{2}\\ &\simeq K - \frac{\ln 2}{2} + O[\exp(-4K)] \qquad (5.74)\end{aligned}$$

One sees that K varies very slowly starting with the first iteration where $K = K_0 = J/T \gg 1$ ($T \sim 0$): Between two iterations, K diminishes by $dK = -\frac{\ln 2}{2}$. On the other hand, the scaling parameter after n iterations is $b = 2^n$ so that $\ln b = n\ln 2$. For one step, one has $d(\ln b) = \ln 2$. Therefore,

$$dK = -\frac{d\ln b}{2} \Rightarrow \frac{dK}{d\ln b} = -\frac{1}{2}.$$

This yields an exponential law as seen below, instead of the power law in the case where there is a fixed point at a finite K. Integrating the above equation, one has

$$K(b) = K_0 - \frac{1}{2}\ln b \qquad (5.75)$$

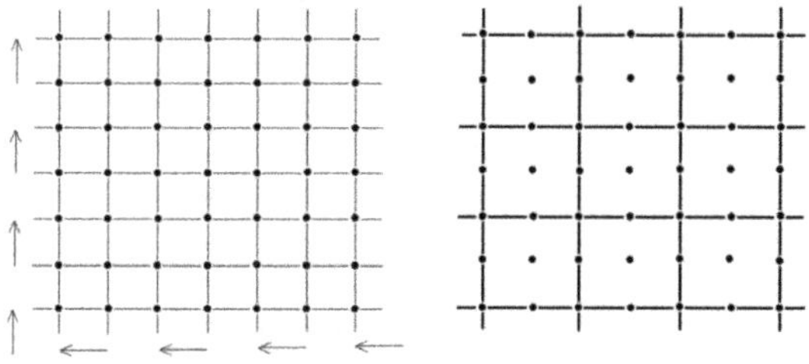

Figure 5.8 Moving one horizontal line and one vertical line out of every two lines as indicated by arrows: The left lattice becomes the right lattice. Black circles denote the spins.

The correlation length is measured in unit of b: One has the following scaling relation:

$$\xi(K) = b\xi[K(b)]$$

where, under renormalization, $K(b)$ tends to zero. Since $\xi(0) \sim 1$ (paramagnetic state), one has $\xi(K) \simeq b$. Equation (5.75) becomes at the limit $K(b) \sim 0$

$$\ln b = 2K_0 = \frac{2J}{T}$$

$$\Rightarrow b \simeq \exp(2K_0) = \exp(2J/T) \Rightarrow \xi(K) \simeq \exp(2J/T)$$

The Migdal–Kadanoff decimation shown above is exact for one dimension.

In two dimensions, one uses the so-called bond-moving approximation which consists of moving one horizontal bond line and one vertical bond line every two lines as shown in Fig. 5.8. In doing so, the number of lattice cells is reduced to a half and the spins left behind are free spins, they do not participate in the collective properties of the system. Each remaining bond has a new strength $2J$. Note that the spins on the new bonds (not at the crossings) have each two neighbors: One can decimate these spins by the Migdal–Kadanoff decimation shown above with the result

$$K' = \frac{1}{2} \ln \cosh(4K) \tag{5.76}$$

$$h' = h[1 + \tanh(4K)] + O(h^2) \tag{5.77}$$

The fact that K becomes $2K$ changes a lot of things: The equation for the fixed points ($h = 0$) is

$$K^* = \frac{1}{2} \ln \cosh(4K^*) \tag{5.78}$$

which admits a solution at a finite value of K. One sees this by examining the two limits:

$$\frac{1}{2}\ln\cosh(4K) \simeq \begin{cases} 4K^2 \ll K & \text{if} \quad K \ll 1 \\ 2K > K & \text{if} \quad K \gg 1 \end{cases} \tag{5.79}$$

From this one sees that the function $K - \frac{1}{2}\ln\cosh(4K)$ should change its sign somewhere between 0 and ∞. The phase transition occurs thus at a finite temperature. In the case of the square lattice considered here, one has $K^* \simeq 0.30469$, or $T_c/J \simeq 3.282$ which is below the mean-field value 4, but above the exact value 2.2692. Our conclusion is that the bond moving improves the mean-field theory but the moving procedure is not justified. In spite of this, the bond-moving approximation gives the values of critical exponents not very bad.

To compute the critical exponents, one expands the recursion relation (5.78) around its fixed point. Putting $K = K^* + t$ where t is the reduced temperature, one has $t' = 1.6786t = b^{\lambda_t}t$ where one has used $b = 2$ and

$$\lambda_t = \frac{\ln 1.678}{\ln 2} \simeq 0.74674$$

This value is better than the mean-field value $\nu = 0.5$ but still smaller than the exact value $\nu = 1$. Note that the bond moving approximation becomes exact for some hierarchical lattices which correspond to fractal spatial dimensions [131].

5.6 Transfer-Matrix Method

The transfer-matrix method is very useful when the system can be divided into subsystems, each of which interacts only with its adjacent nearest neighboring subsystems. For example, the simple cubic lattice can be considered as composed of planes each of which interacts only with its neighboring planes. In the case of periodic boundary conditions, the partition function can be written as a product of partition functions of its N subsystems:

$$Z = \text{trace}\left[\prod_{i=1}^{N} W_i\right]$$

where W_i is the "transfer matrix" of dimension $n \times n$ representing the interaction connection between two adjacent blocks (subsystems). The trace of Z is the sum of the eigenvalues of Z. If the system is homogeneous, then all W_i are identical: Each eigenvalue of the product of identical matrices W_i is equal the product of the corresponding eigenvalue of W_i (properties of trace). Let $z_1, z_2, \cdots$, z_n be the eigenvalues of W_i, one writes $Z = z_1^N + z_2^N + \cdots + z_n^N$. If $N \to \infty$, then $Z = z_{\text{max}}^N$ where z_{max} is the largest eigenvalue among $z_1, z_2, \cdots, z_n$.

One applies in the following the transfer-matrix method to the case of a chain of Ising spins using the periodic boundary condition. Let N be the total number of spins. The Hamiltonian is given by

$$\mathcal{H}_0 = -J \sum_{n=1}^{N-1} \sigma_n \sigma_{n+1} - J\, \sigma_N \sigma_1 \tag{5.80}$$

where the last term expresses the periodic boundary condition. One has $\sigma_i = \pm 1$. One can define new variables $\alpha_n = \sigma_n \sigma_{n+1}$. α_n takes the values ± 1 as σ_i. One can rewrite $\mathcal{H}$ as

$$\mathcal{H}_0 = -J \sum_{n=1}^{N-1} \alpha_n - J\, \alpha_N \tag{5.81}$$

The partition function is then

$$\begin{aligned} Z &= \text{Tr} \exp\left[\beta J \sum_{n=1}^{N} \alpha_n\right] = Tr \prod_{n=1}^{N} \exp(\beta J\, \alpha_n) \\ &= \prod_{n=1}^{N} [\exp(\beta J) + \exp(-\beta J)] = [2 \cosh \beta J\,]^N \end{aligned} \tag{5.82}$$

This result is the same as that obtained by the exact method shown in Problem 1. The average energy is calculated by $\overline{E} = -\partial \ln Z / \partial \beta$ [see (A.10)]:

$$\overline{E} = -\frac{\partial \ln Z}{\partial \beta} = -NJ \tanh(\beta J) \tag{5.83}$$

One obtains the following heat capacity:

$$C_V = dE/dT = Nk_B \left[\frac{k_B T}{J} \cosh \frac{J}{k_B T}\right]^{-2} \tag{5.84}$$

In an applied magnetic field H, one proceeds as follows:

$$\mathcal{H} = \mathcal{H}_0 - H \sum_{n=1}^{N} \sigma_n \tag{5.85}$$

where $\mathcal{H}_0$ is given by (5.80). The partition function is $Z = \prod_{n=1}^{N} V_n$ where

$$V_n = \exp[\beta(J\,\sigma_n\sigma_{n+1} + H\sigma_n)] \tag{5.86}$$

$$V_N = \exp[\beta(J\,\sigma_N\sigma_1 + H\sigma_N)] \tag{5.87}$$

The matrix elements V_n, of dimension 2x2, depend on σ_n and σ_{n+1}, namely

$$V_n(1,1) = \exp[\beta(J + H)] \quad (\sigma_n = 1, \sigma_{n+1} = 1)$$

$$V_n(1,2) = \exp[\beta(-J + H)] \quad (\sigma_n = 1, \sigma_{n+1} = -1)$$

$$V_n(2,1) = \exp[\beta(-J - H)] \quad (\sigma_n = -1, \sigma_{n+1} = 1)$$

$$V_n(2,2) = \exp[\beta(J - H)] \quad (\sigma_n = -1, \sigma_{n+1} = -1) \tag{5.88}$$

The matrix V_n is called "transfer matrix." Note that all V_n ($n = 1, \cdots, N$) have the same elements, say V. One thus has $Z = \mathrm{Tr}V^N$. Let z_1 and z_2 be the eigenvalues of V obtained by diagonalizing V, using (5.88). One has

$$z_1 = \exp(\beta J)\cosh(\beta H) + \sqrt{\exp(2\beta J)\cosh^2(\beta H) - 2\sinh(2\beta J)}$$

$$z_2 = \exp(\beta J)\cosh(\beta H) - \sqrt{\exp(2\beta J)\cosh^2(\beta H) - 2\sinh(2\beta J)}$$

One obtains then

$$Z = z_1^N + z_2^N = z_1^N(1 + \exp[-N\ln(z_1/z_2)] \tag{5.89}$$

where z_1 denotes the larger eigenvalue. When $N \to \infty$, one has $Z = z_1^N$.

The susceptibility is calculated by $\chi = (dM/dH)_{H\to 0}$ where $M = -\partial F/\partial H$ with $F = -k_B T \ln Z$. One obtains

$$\chi \simeq \frac{1}{T}\exp[2J/k_B T] \tag{5.90}$$

This result shows that there is no phase transition in one dimension (absence of anomaly of χ with varying T as seen in Fig. 5.9), confirming the results shown in the two previous sections.

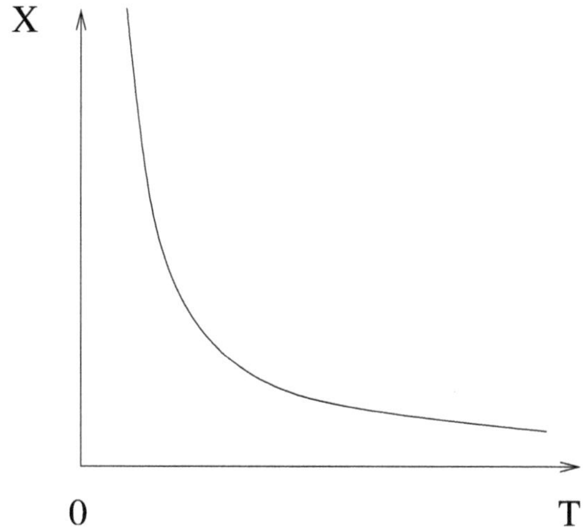

Figure 5.9 χ versus T [Eq. (5.90)].

5.7 Phase Transition in Particular Systems

One has seen so far various methods used to study phase transitions and critical phenomena. In general, the nature of a phase transition depends on the symmetry of the order parameter, the spatial dimension and the nature of the interaction (short or long range). Standard methods presented above can be used to determine it with satisfactory precision. However, in some particular systems one needs special methods. Some of these remarkable systems are presented hereafter.

5.7.1 Exactly Solved Spin Systems

There are several families of systems in one or two dimensions with short-range non-crossing interactions which can be exactly solved. The spin models in those solvable systems are often Ising and Potts models. One needs exact solutions in such simple systems to test approximations conceived for more complicated systems or systems in three dimensions. Methods for searching exact solutions are lengthy to present here. The reader is referred to the book by R. J. Baxter [25] for the general methods and examples of exactly solved models. For some exactly solved frustrated spin systems, the reader is referred to the review by Diep and Giacomini [86]. To summarize, to find an exact solution, the most frequently used method is to transform the system under study into a vertex model where the solutions for the critical surfaces are known. Among the most popular models, one can mention the 8-, 16- and 32-vertex models [25].

5.7.2 Kosterlitz–Thouless Transition

One considers the XY spins on a two-dimensional lattice with a ferromagnetic interaction between nearest neighbors. In the ground state, the spin configuration is a perfect ferromagnetic state, namely all spins are parallel. However, this system does not have a normal order-disorder transition at a finite temperature: There is no long-range ordering as soon as the temperature is not zero, following the Mermin–Wagner theorem [231] valid for two-dimensional systems with continuous spins (see discussion in Chapter 3). Kosterlitz and Thouless [191] have shown that this system has a special phase transition due to the unbinding of vortex-antivortex pairs at a finite temperature below (above) which the correlation function decays as a power law (exponential law) with increasing distance. This transition, called Kosterlitz–Thouless (KT) or Kosterlitz–Thouless–Berezinskii transition, is of infinite order. For the reader interested in this special transition, an appendix in Ref. [88] gives the main points explaining the mechanism lying behind the KT transition.

5.7.3 Frustrated Spin Systems

A system is said "frustrated" when the interaction bonds between a spin with its neighbors cannot be fully satisfied. An example is the triangular antiferromagnet: (i) in the case of Ising spin model, the three spins on a triangle cannot find orientations to satisfy the three antiferromagnetic bonds, ii) in the case of XY or Heisenberg spin models, the spins make a "compromise" to form a non-collinear configuration in order to partially satisfy each bond as shown in Fig. 18.6 in Problems 6 and 7 of Chapter 3. The ground-state spin configuration of a helimagnet has been given in Section 3.4.

Effects due to the frustration are numerous and spectacular. One can mention a few of them: (i) high ground-state degeneracy, (ii) non-collinear spin configuration, (iii) multiple phase transitions, (iv) reentrance phenomenon, (v) disorder lines, (vi) partial ordering at equilibrium, (vii) difficulty in determining the nature of phase transitions in several systems, etc. Some of these spectacular effects (iii)–(vi) have been observed in exactly solved two-dimensional systems [16, 67, 68, 76, 86]. It is believed that these effects persist in three-dimensional systems and in other more complicated unsolved models. For advanced reviews on frustrated systems, the reader is referred to Refs. [85, 87].

5.8 Conclusion

In this chapter, we introduced basic notions as well as some fundamental methods which are widely used in the field of phase transitions. The mean-field theory presented in Chapter 2 paves the way for other improving methods such as the Bethe's approximation and the Ginzburg's criterion shown in this chapter. The renormalization group has been shown with simple examples to illustrate its concepts. In particular, the notion of universality class and the relations between the critical exponents have been discussed. The Migdal–Kadanoff decimation method and bond-moving technique have been explained. An example of the transfer-matrix method has been treated, and some complementary methods such as canonical and micro-canonical methods are also introduced as problems which are given below. More advanced methods, such as quantum phase transitions of low-dimensional systems [202, 235, 301] and the Hubbard model [108] are not included here to keep the contents of the book suitable for lectures in a graduate course.

5.9 Problems

Problem 1. Chain of Ising spins by exact method:

Consider a chain of N Ising spins with a ferromagnetic interaction between nearest neighbors, maintained at temperature T, with the following Hamiltonian:

$$\mathcal{H} = -J \sum_{i=1}^{N} \sigma_i \sigma_{i+1} \tag{5.91}$$

One supposes the periodic boundary condition $\sigma_{N+1} = \sigma_1$. Calculate exactly the partition function of the system. Find the free energy, the average energy and the heat capacity, as functions of T. Show that there is no phase transition at finite temperature.

Problem 2. Chain of Ising spins by micro-canonical method:

Consider a chain of N Ising spins interacting with each other via a nearest-neighbor coupling $J > 0$. The system is isolated with the Hamiltonian

$$\mathcal{H} = -J \sum_{i=1}^{N} \sigma_i \sigma_{i+1} \tag{5.92}$$

One supposes that the periodic boundary condition $\sigma_{N+1} = \sigma_1$ applies and N is even.

(a) Calculate the energy of the ground state and its degeneracy.

(b) Find the energy of the lowest excited state where there are two unsatisfied bonds, namely one reversed spin or two antiparallel spin pairs. Find its degeneracy. Deduce the energy $E(2n)$ of a state in which there are $2n$ unsatisfied bonds. Indicate the maximum energy of the system and its degeneracy.

(c) For a given $E(2n)$ with $n \gg 1$, calculate the entropy and the micro-canonical temperature. Find the percentage x of unsatisfied bonds with respect to the total number of bonds. Find its low- and high-temperature limits.

Problem 3. Chain of Ising spins by canonical method:

Consider again the system in the preceding problem but put it now in the canonical situation at temperature T.

(a) Calculate the partition function of the system using the Newton binomial relations: $(1+u)^N = \sum_{n=0}^{N} C_N^n u^n$ and $(1-u)^N = \sum_{n=0}^{N}(-1)^n C_N^n u^n$.

(b) Calculate the system energy.

(c) Calculate the average percentage $\bar{x}$ of unsatisfied bonds with respect to the total number of bonds. Find its low- and high-temperature limits. Compare these results to those of Problem 2.

Problem 4. Low- and high-temperature expansions of the Ising model on the square lattice:

The low- and high-temperature expansions are useful not only for studying physical properties of a spin system in these temperature regions, but also for introducing a new concept called duality which allows to map a system of weak coupling into a system of strong coupling, as seen in the problem below.

Consider N Ising spins on a square lattice with the Hamiltonian

$$\mathcal{H} = -J \sum_{<i,j>} \sigma_i \sigma_j$$

where the sum is performed over nearest neighbors and $\sigma_{i(j)} = \pm 1$. The periodic boundary conditions are used.

(a) Write the partition function Z. Calculate Z for the ground state (GS).

(b) Low-temperature expansion: Consider the GS in which one reverses one spin, two nearest neighboring spins, a block of three nearest spins, ... Count for each case the number of "broken" links, namely links between the reversed spins with the remaining spins. Calculate the degeneracy of each case. Write Z with the first excited states.

(c) Draw a path P crossing the broken links around each reversed spin cluster. Verify that each cluster has an even number of broken links and P is always a closed path. Let $\ell(P)$ the number of broken links crossed by the path. Show that

$$Z = 2e^{N_b K} \sum_P e^{-2K\ell(P)} \qquad (5.93)$$

where N_b is the total number of links and $K = J/(k_B T)$.

(d) High-temperature expansion: Using Eq. (18.118) or Eq. (5.61) show that the partition function is written as

$$Z = \sum_{\sigma_1=\pm 1,\cdots} \prod_{<ij>} (\cosh K + \sigma_i \, \sigma_j \sinh K)$$
$$= (\cosh K)^{N_b} \sum_{\sigma_1=\pm 1,\cdots} \prod_{<ij>} (1 + \sigma_i \, \sigma_j \tanh K) \quad (5.94)$$

Expand the product in the last equation and show that

$$Z = 2^N (\cosh K)^{N_b} \sum_P (\tanh K)^{\ell(P)} \quad (5.95)$$

(e) Duality: The partition function Z in (5.93) and (5.95) has the same structure: the prefactors are non-singular, the summations over the paths determine the singularity of Z. Show that the two Z have the same critical behavior if

$$K^* = -\frac{1}{2} \ln \tanh K \quad (5.96)$$

where K^* corresponds to the low-T phase and K to the high-T phase. The relation (5.96) is called the "duality" condition which connects the low- and high-T phases.

(f) Deduce the critical temperature of the Ising model on the square lattice.

Problem 5. Critical temperatures of the triangular lattice and the honeycomb lattice by duality:
Consider the triangular lattice with Ising spins with a ferromagnetic interaction between nearest neighbors. Construct its dual lattice. Calculate the partition functions of the two lattices. Deduce the critical temperature of each of them by following the method outlined in the previous problem.

Chapter 6

Monte Carlo Simulation: Principle and Implementation

6.1 Principle of Monte Carlo Simulation

The method of Monte Carlo simulation for the study of thermodynamic properties of matter is based on statistical physics with the use of a computer programming language. It was introduced for the first time in 1953 by Metropolis et al. [233]. However, simulations became popular only from the early 1990s when several new methods of simulation of high precision were introduced and, at the same time, powerful computers became accessible for a large number of researchers [33]. Today, numerical simulations are considered as an important investigation method very efficient and complementary to theory and experiment in the study of properties of matter. Numerical simulations allow us to test the validity of theoretical approximations, to compare quantitatively simulated results with experimental data, and to propose interpretations. In addition, real systems have many parameters but often only a few of them govern their main properties. Theories and experiments cannot test all of them for the simple reason that experimental realizations and theoretical calculations take time and cost to carry

Physics of Magnetic Thin Films: Theory and Simulation
Hung T. Diep

ISBN 978-981-4877-42-8 (Hardcover), 978-1-003-12110-7 (eBook)
www.jennystanford.com

out. Simulations can help identify relevant mechanisms before realizing experimental setups or constructing theoretical models. In many cases, numerical simulations are the only way to study very complex systems that theory and experiment cannot investigate properly.

It is obvious that numerical simulations also have particular difficulties. The first kind of difficulty is related to the capacity of computers: limited memory and speed. The second kind of difficulty concerns the efficiency of the simulation method to get good statistical averages, to treat correctly particular physical effects, to reduce errors, ... Of course, there are remedies to improve and to get rid of most of these difficulties as seen below. A good simulation requires a great care in every step from the choice of model to a deep analysis of results. We recall that simulation is a method to study a problem: a good knowledge on theoretical background and experimental data prior to the simulation is necessary.

In the following, we present the principle of the Monte Carlo simulation using the Metropolis algorithm. Although Monte Carlo simulations can be performed using the micro-canonical and grand-canonical statistical descriptions (see Appendix A), we employ in the following the canonical description for illustration because this description is most frequently used in simulations. The system we consider is kept at a constant temperature T. The internal variables such as energy and magnetization are free to fluctuate at T. Of course, to show a quantity versus temperature using the Metropolis algorithm we have to perform independent simulations at several discrete temperatures in the temperature region of interest. However, when we use advanced Monte Carlo techniques, we can calculate physical quantities as continuous functions of temperature, as seen in Section 6.5.

To illustrate the principle of the Monte Carlo simulation using the Metropolis algorithm, we consider the Ising model with ferromagnetic interaction J between nearest neighbors. This simple example, however, does not cause a loss of generality of the method.

In a simulation, we wish to calculate average values of physical quantities such as the average energy $< E >$, the heat capacity C_V, the average magnetization $< M >$ and the susceptibility χ. The average value of a physical quantity A is defined by

$$< A >= \frac{1}{Z(T)} \sum_s A(s) \exp[-\beta E(s)] \tag{6.1}$$

where $Z(T)$ is the partition function at the temperature T, $E(s)$ and $A(s)$ are the system energy and the value of A in the microscopic state s. For a spin system, s is a spin configuration.

In principle, we have to sum over all spin configurations s. The total number of spin configurations in a system of N Ising spins is 2^N. This is a huge number when N is large for a reasonable value of N. It is impossible to take into account all microscopic states. However, we can overcome this difficulty by two following ways:

6.1.1 Simple Sampling

We generate a number C of random spin configurations by giving randomly a value $+1$ or -1 to each spin. We then calculate $< A >$ using these C states

$$< A >= \frac{\sum_{s=1}^{C} A(s) \exp[-\beta E(s)]}{\sum_{s=1}^{C} \exp[-\beta E(s)]} \tag{6.2}$$

It is obvious that the precision on the obtained average value depends on the number C of configurations: The larger C, the more precise $< A >$. This simple sampling suffers from a more serious problem: the randomly generated spin configurations correspond to disordered states at high temperatures. So, for low T, we have to generate random configurations with a larger concentration of up spins with respect to down spins (or vice versa). But then, how do we know which concentration corresponds to which temperature? The average value so calculated contains certainly uncontrolled errors. We can use simulations at constant concentration instead of constant temperature, but the description will be micro-canonical.

6.1.2 Importance Sampling

The importance sampling is based on the following principle: We generate spin configurations s which are most probable at T, namely we generate microscopic states using the canonical probability. Once these states are generated, the average value $< A >$ is calculated by a simple addition

$$< A >= \frac{1}{C} \sum_{s=1}^{C} A(s) \tag{6.3}$$

The question is "how to generate these microscopic states according to the canonical probability at T in a convenient way for simulation"?

An answer to this question is to use the following algorithm. Instead of generating these states in an independent manner, we can generate a series of states called "Markov chain" in which the $(i+1)$-th state, called s_{i+1}, is obtained from the precedent state s_i by an appropriate transition probability $w(s_i \rightarrow s_{i+1})$ between these two states. The choice of $w(s_i \rightarrow s_{i+1})$ should obey the following probability at equilibrium (see Appendix A):

$$P(s_i) = \frac{1}{Z(T)} \exp[-\beta E(s_i)] \tag{6.4}$$

This is possible if we impose on $w(s_i \rightarrow s_{i+1})$ the following "principle of detailed balance":

$$P(s_i)w(s_i \rightarrow s_k) = P(s_k)w(s_k \rightarrow s_i) \tag{6.5}$$

Using (6.4), we get the detailed balance for a system at equilibrium

$$\frac{w(s_i \rightarrow s_k)}{w(s_k \rightarrow s_i)} = \exp(-\beta \Delta E) \tag{6.6}$$

where $\Delta E = E(s_k) - E(s_i)$.

If the above relation for equilibrium is obeyed, then there is in principle no problem for the system to reach equilibrium. The choices frequently used to respect the detailed balance (6.6) are

$$w(s_i \rightarrow s_k) = \frac{\exp(-\beta \Delta E)}{1 + \exp(-\beta \Delta E)} \tag{6.7}$$

and

$$\begin{aligned} w(s_i \rightarrow s_k) &= \exp(-\beta \Delta E) && \text{if } \Delta E \geq 0 \\ &= 1 && \text{if } \Delta E < 0 \end{aligned} \tag{6.8}$$

The second choice will be used in the following to write a Monte Carlo program for the Ising spin model.

Remarks:

(1) The state s_{i+1} can be different from the state s_i by just the state of a single spin. This facilitates the simulation task: The difference of energy ΔE of the two states is equal to the difference of energy of the single spin under consideration.
(2) The power of the Monte Carlo method resides in the facility to make the system transit from the state s_i to the state s_{i+1}: We just reverse one single spin or a block of spins. The first choice is called "single-spin flip algorithm" or "Metropolis algorithm," and the second choice is called "cluster-flip algorithm" which is shown in Section 6.5.

6.2 Implementation: Construction of a Computer Program

We implement now the principle of the Monte Carlo simulation presented in the previous section. To simplify the presentation, we use the Metropolis algorithm hereafter. Let us consider the Ising model on a square lattice with a ferromagnetic interaction J between nearest neighbors. The Hamiltonian is given by

$$\mathcal{H} = -J \sum_{(k,m)} S(\mathbf{R}_k)S(\mathbf{R}_m) \tag{6.9}$$

where the sum is made over pairs of nearest neighbors. $S(\mathbf{R}_k)$ indicates the spin at the position $\mathbf{R}_k$. We suppose the periodic boundary conditions. Next, we proceed to the following steps:

(1) Initial spin configuration:
In the case of the Ising model, we create a square lattice of dimension $L \times L$ where at each site we attribute a spin. For the square lattice, each site is defined by two Cartesian indices (i, j) corresponding to the position of the site (see Fig. 6.1). The spin occupying this site is $S(i, j)$ with an attributed initial value (1 or -1).
(2) Choice of physical parameters:
Here the parameters of the model are J and T. We take $J = 1$ as the energy unit.

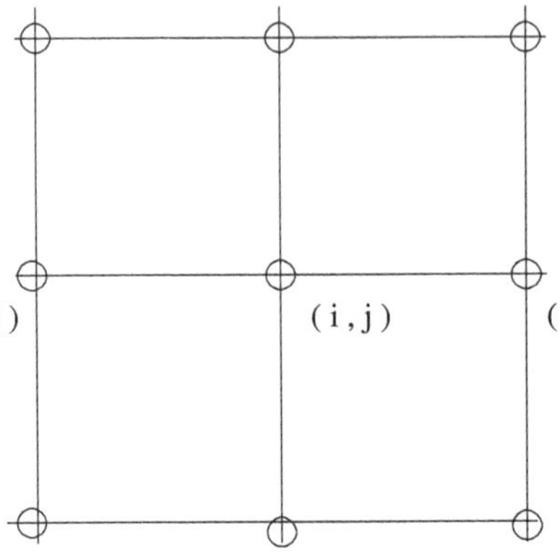

Figure 6.1 Coding of the square lattice. Each site is defined by two Cartesian indices.

(3) Calculation of the energy of spin $S(i, j)$, using the periodic boundary conditions if it is at an edge: The best and fast way to identify the neighbors of $S(i, j)$ is to call its left neighbor by (im, j), its right neighbor by (ip, j), the one on its top by (i, jp) and the one below it by (i, jm) (see Fig. 6.1). One writes

$$im = i - 1 + (1/i) * L \tag{6.10}$$

$$ip = i + 1 - (i/N) * L \tag{6.11}$$

$$jm = j - 1 + (1/j) * L \tag{6.12}$$

$$jp = j + 1 - (j/N) * L \tag{6.13}$$

where we see that $im = i - 1$ as long as $i \neq 1$ and $ip = i + 1$ as long as $i \neq L$ because the integer divisions $(1/i)$ and (i/L) give zero (with Fortran). When $i = 1$, we have $im = L$ and when $i = L$ we have $ip = 1$. This automatically respects the periodic condition in the x direction: The left neighbor of the spin $S(1, j)$ is $S(L, j)$, and the right neighbor of $S(L, j)$ is $S(1, j)$. The same explanation is for the y direction. The energy of the spin $S(i, j)$ is thus written as

$$E1 = -J * S(i, j) * (S(im, j) + S(ip, j) + S(i, jm) + S(i, jp)) \tag{6.14}$$

(4) Updating the spin $S(i, j)$ and calculating its new energy:
For an Ising spin, the new state of $S(i, j)$ is obtained by changing its sign. Its new energy is thus $E2 = -E1$. If $E2 < E1$, the new orientation of $S(i, j)$ is accepted using (6.8). If $E2 > E1$, the new orientation is accepted with a probability $e^{-(E2-E1)/T}$ (we take $k_B = 1$ for simplicity). This step is called "spin update."

(5) Taking another spin and repeating steps 3 and 4:
We continue until all spins have been updated: We say we have made one Monte Carlo step per spin.

(6) Equilibrating the system with N_1 Monte Carlo steps:
We have to make many Monte Carlo steps to equilibrate the system at T since the initial configuration does not in general correspond to an equilibrium state at the temperature we make the simulation. The repetition of spin updates is thus necessary to bring the system to equilibrium at T.

(7) Averaging physical quantities with N_2 Monte Carlo steps:
During N_2 Monte Carlo steps, we calculate average values of physical quantities of interest such as the energy and the magnetization

$$< E > = \frac{1}{N_2} \sum_{t=N_1+1}^{N_1+N_2} E(t) \tag{6.15}$$

where

$$E(t) = -\frac{1}{2}\frac{J}{L \times L} \sum_{i,j=1}^{L} S(i, j) * (S(im, j) + S(ip, j) + S(i, jm) + S(i, jp)) \tag{6.16}$$

$$M = \frac{1}{N_2} \sum_{t=N_1+1}^{N_1+N_2} M(t) \tag{6.17}$$

where

$$M(t) = \frac{1}{L \times L} \sum_{i,j=1}^{L} S(i, j) \tag{6.18}$$

where t is the Monte Carlo "time," $E(t)$, $M(t)$ and $S(i, j)$ are the energy per spin, the magnetization and the spin at t. The factor $1/2$ in (6.16) is to avoid the double counting.

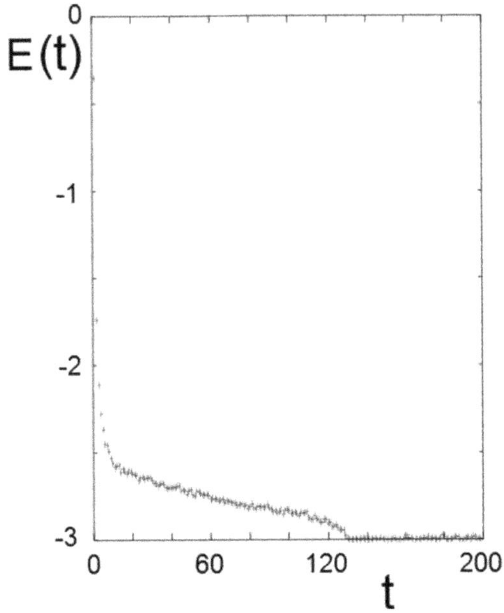

Figure 6.2 Example of time evolution (in unit of Monte Carlo step), during equilibrating, of the energy per spin $E(t)$ (in unit of J), at $k_BT/J = 1$ for the ferromagnetic Ising model on the triangular lattice with a random initial spin configuration.

Remarks:

(i) During the N_1 Monte Carlo steps for equilibrating, we can record the instantaneous energy and magnetization at each Monte Carlo step in order to observe their time evolution. We can then see the time necessary to get equilibrium. An example is shown in Fig. 6.2.

(ii) At the end of step 7, we record all average values in a file, and then restart the simulation at another temperature. We have at the end average values for several temperatures so that we can plot average values of physical quantities versus T. From those curves, we can recognize the phase transition as well as other properties of the system. We show some examples in the next section.

6.3 Phase Transition as Seen in Monte Carlo Simulations

We perform a Monte Carlo simulation for a system of linear size L at various temperatures as described in the previous section. We examine the curves of the average values of different physical quantities versus T. A phase transition is recognized by the anomalies of these quantities at some temperature.

In a second-order transition, the internal energy changes its curvature at the transition temperature T_c, the heat capacity and the susceptibility have a peak at T_c, the magnetization falls to zero but with a small tail above T_c due to a finite-size effect. These are schematically shown in Fig. 6.3.

In a first-order transition, for a sufficiently large L, the energy and the magnetization are discontinued at the transition; the heat capacity and the susceptibility therefore cannot be defined at the transition. We schematically show E and M in Fig. 6.4.

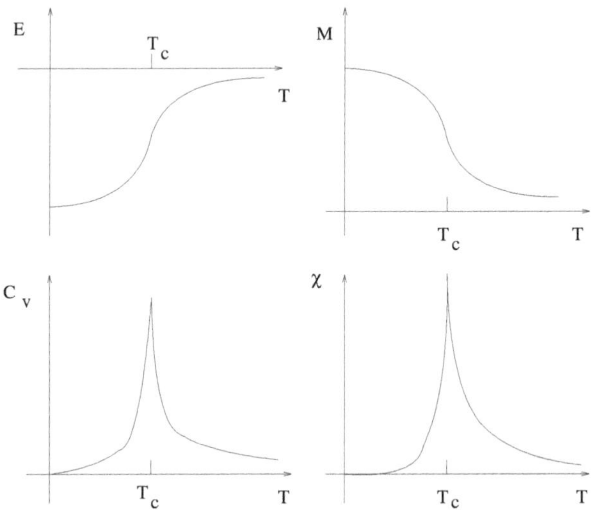

Figure 6.3 Physical quantities of a second-order phase transition as seen in a simulation with a linear size L: energy E, heat capacity C_V, magnetization M, susceptibility χ.

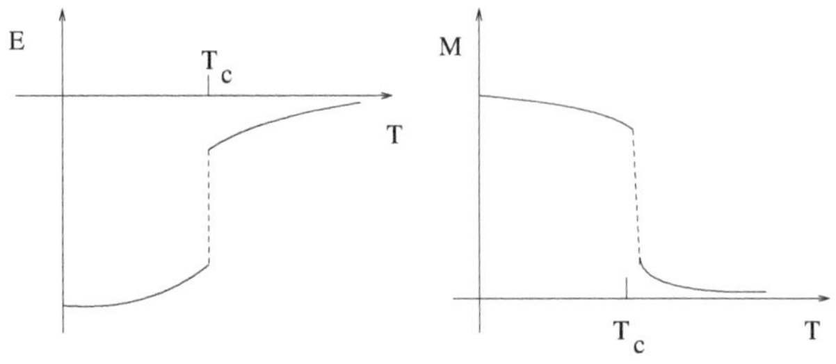

Figure 6.4 First-order transition as seen in a simulation for a large size L: Discontinuities of E and M are observed at the transition (discontinued vertical lines).

Several remarks are in order:

(1) We have to pay attention to size effects while analyzing the results. A first-order transition can look like a second-order transition if L is not larger than the correlation length at the transition (see paragraph 5.1.3). If we cannot take large-enough L because of the computer-memory limit and/or of a too long CPU time, then we have to use a finite-size scaling analysis shown below to determine the order of the transition.

(2) Some systems show a maximum of the heat capacity at a temperature but do not have a phase transition. One of these systems is a chain of Ising spins with a ferromagnetic interaction between nearest neighbors. Therefore, we have to verify the existence of a phase transition by several methods. For example, if it is a real transition, the heights of the maxima of C_V and χ as well as of other quantities (see below) should depend on the system size.

6.4 Finite-Size Scaling Laws

Finite-size effects on thermodynamic quantities and on spin–spin correlation have been shown by a renormalization group analysis [21]. The calculations are too complicated to be reproduced here. We just recall some results which are very helpful while analyzing data obtained from Monte Carlo simulations.

Let L be the linear dimension of the system under consideration. We distinguish two cases:

6.4.1 Second-Order Phase Transition

- The height of the maximum of the heat capacity depends on L as follows:

$$C_V^{\max}(L) \propto L^{\alpha/\nu} \tag{6.19}$$

- The height of the maximum of the susceptibility behaves as

$$\chi^{\max}(L) \propto L^{\gamma/\nu} \tag{6.20}$$

- The magnetization at the transition depends on L through

$$M_{T_c}(L) \propto L^{-\beta/\nu} \tag{6.21}$$

- The moment of n-th order is defined as

$$V_n =< (\ln M^n)' >= \frac{< \partial \ln M^n >}{\partial \beta} \tag{6.22}$$

where $\beta = \frac{1}{k_B T}$. We can show that

$$V_1 = \frac{< ME > - < E >}{< M >} \tag{6.23}$$

$$V_2 = \frac{< M^2 E > - < M^2 >< E >}{< M^2 >} \tag{6.24}$$

The finite-size effects on the maxima of these moments are given by [21]

$$V_1^{\max}, \; V_2^{\max} \propto L^{1/\nu} \tag{6.25}$$

- The Binder cumulant [33] is defined by

$$U_E = 1 - \frac{< E^4 >}{3 < E^2 >^2} \tag{6.26}$$

In a second-order phase transition, we have

$$U_E(L) - U_E(\infty) \propto L^{-\theta} \tag{6.27}$$

where $U_E(\infty) = 2/3$ and $\theta < d$.

- The critical temperature depends on L through the relation

$$T_c(L) - T_c(\infty) \propto L^{-1/\nu} \tag{6.28}$$

We use the above relations to determine the critical exponents by realizing simulations with many sizes L. Note that L has to be chosen so as the system sizes are different by at least two orders.

6.4.2 First-Order Phase Transition

When the system size is not sufficiently large, a first-order phase transition can have aspects of a second-order one, namely E and M are continuous at the transition with a maximum of C_V and χ. In the case of a first-order transition, we should have

- The height of the maximum of C_V and χ are proportional to the system volume L^d:

$$C_V^{\max}, \chi^{\max} \propto L^d \tag{6.29}$$

- The Binder cumulant depends on L as

$$U_E(L) - U_E^*(\infty) \propto L^{-d} \tag{6.30}$$

  where $U_E^*(\infty) \neq 2/3$.

Note that finite-size effects are seen in the transition region around T_c because only in this region that the correlation length may become as large as the system size. This is the reason why the size effects depend on the critical exponents. They act therefore in different manners on different physical quantities. Note that at a finite size, the "pseudo" transition temperature where C_V is maximum is not that where χ is maximum. Only at the infinite size that these maxima occur at the same temperature which is the "real" critical temperature. This is schematically shown in Fig. 6.5.

6.4.3 Error Estimations

One of the problems encountered in Monte Carlo simulations is the estimation of errors due to a number of causes: simulation time, system size, artificial procedures used to accelerate the convergence toward equilibrium, etc. There are two types of principal errors (i) statistical errors due to autocorrelation and finite-size effects on these errors (ii) errors due to the fitting of raw data with some laws. Errors of the second type are directly given by the computer according to the chosen fitting procedure (mean least squares, for example). Errors of the first type become less and less important because computers are more and more powerful, rapid in execution with huge available memories. Simulation time nowadays is much longer than autocorrelation time and the relaxation time in most

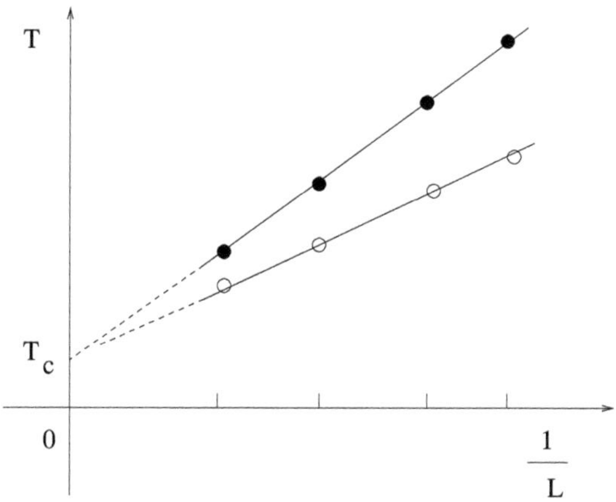

Figure 6.5 Finite-size effects on the temperatures corresponding to the maxima of C_V (black circles) and χ (white circles). These temperatures coincide only when $L \rightarrow \infty$ (extrapolated by discontinued lines).

systems. Therefore, errors are extremely small. Most of Monte Carlo works at the present time do not show any more errors since they are often smaller than the size of the presented data points. We give anyway in the following some notions on error sources and show how to calculate errors.

6.4.4 Autocorrelation

The autocorrelation function of a quantity A at the time t with itself at $t = 0$ is defined by

$$\phi(t) = \frac{< A(0)A(t) > - < A >^2}{< A^2 > - < A >^2} \tag{6.31}$$

where $< \cdots >$ is the thermal average taken between $t = 0$ and t, $A(t)$ the instantaneous value of A. By definition, we have $\phi(0) = 1$ and $\phi(\infty) = 0$. In a simulation, we can calculate $\phi(t)$ and obtain the integrated autocorrelation time $\overline{\tau}$ by

$$\overline{\tau} = \int_0^\infty \phi(t)dt \tag{6.32}$$

The autocorrelation time τ is defined by

$$\phi(t) = C \exp(-t/\tau) \tag{6.33}$$

We then have

$$\int_0^\infty \phi(t)dt = C \int_0^\infty \exp(-t/\tau)dt = C\tau \tag{6.34}$$

Comparing to (6.32), we get

$$\tau = \overline{\tau}/C \tag{6.35}$$

The error on A is given by its variance

$$\begin{aligned} < (\delta A)^2 > &= < \frac{1}{N}\sum_{n=1}^{N}(A(t_n) - < A >)^2 > \\ &= \frac{< A^2 > - < A >^2}{N} \times \\ &\quad \left[1 + \frac{2}{\Delta t}\int_0^{t_N}\left(1 - \frac{t}{t_N}\right)\phi(t)dt\right] \end{aligned} \tag{6.36}$$

where N is the total number of measures and $A(t_n)$ the instantaneous value of A measured at t_n. The second equality has been obtained by expanding the square of the first equality and then replacing the sum by an integral using (6.31). Δt is the interval between two measures. To de-correlate successive values $A(t_n)$ we should not measure A at each Monte Carlo step. We can, for example, take a measure once every ten steps, namely $\Delta t = 10$. Since $\phi(t) \simeq$ constant for $t \gg \tau$, we replace the upper limit t_N by ∞. In addition, we can neglect t/t_N with respect to 1. We then obtain

$$< (\delta A)^2 > \simeq \left(< A^2 > - < A >^2\right)\frac{2\tau + \Delta t}{N\Delta t} \tag{6.37}$$

where $N\Delta t = N_{MC}$ is the total number of Monte Carlo steps used in the simulation.

The number of independent measures among N measures is

$$n = \frac{N\Delta t}{2\tau + \Delta t} \tag{6.38}$$

The relative error on A can be estimated by

$$\rho = \frac{< (\delta A)^2 >^{1/2}}{< A >} \simeq \left[\frac{< A^2 > - < A >^2}{< A >^2}\frac{2\tau}{N_{MC}}\right]^{1/2} \tag{6.39}$$

where we neglected Δt while comparing to 2τ. Since τ is given by (6.35), $< A^2 >$ and $< A >$ are known from the simulation, we obtain ρ by the above formula.

6.4.5 Size Effects on Errors

In the vicinity of the transition, the relaxation time τ is very long. Spins are correlated even at very long distances. This phenomenon is called "critical slowing-down." The first error due to the finite system size comes from the error on the relaxation time. We have [154]

$$\tau \propto \xi^z \propto \left(\frac{T_c}{T - T_c}\right)^{z\nu} \tag{6.40}$$

where ξ is the correlation length, z the dynamic exponent (see previous sections). It has been shown that with the Metropolis algorithm (single-spin flip) we have $z \simeq 2$ for many systems. In addition, theoretically ξ diverges at the transition of second order, but in simulation the transition takes place when $\xi \simeq L$ which is the infinite limit due to the periodic boundary conditions. For this reason, in a second-order transition, τ depends on the size L through

$$\tau(L) \propto L^z \tag{6.41}$$

In a first-order transition, we have [21, 33]

$$\tau(L) \simeq L^a \exp(2\sigma L^{d-1}) \tag{6.42}$$

where a depends on the algorithm and $(2\sigma L^{d-1})$ is the height of the barrier of the free energy at the transition. We have $a \simeq 2.1$ obtained by the Metropolis algorithm for the 10-state Potts model, but $a \simeq 1.5$ obtained by the Swendsen–Wang algorithm (see description in Section 6.5.1) for the same model.

Replacing $\tau(L)$ of (6.41) and (6.42) in (6.39), we obtain for a second-order transition

$$\rho(L) \simeq \frac{1}{N_{MC}^{1/2}} L^{(z-x)/2} \tag{6.43}$$

and for a first-order transition, we have

$$\rho(L) \simeq \frac{1}{N_{MC}^{1/2}} L^{(a-x)/2} \exp(\sigma L^{d-1}) \tag{6.44}$$

where x is a fitting parameter.

The systematic error is defined by

$$\sigma^2(n) \equiv (< A^2 > - < A >^2)\left(1 - \frac{1}{n}\right) \tag{6.45}$$

The relative error on the variance $< (\delta A)^2 >$ is

$$\epsilon = \frac{< A^2 > - < A >^2 - \sigma^2(n)}{< A^2 > - < A >^2} = \frac{1}{n} = \frac{2\tau + \Delta t}{N\Delta t} \tag{6.46}$$

For an error of $\epsilon = 1\%$, we see that the simulation time should be 200 times τ ($\Delta t \ll \tau$). As said at the beginning of this section, with the power of today's computers, such a simulation time is not a problem. Also, the necessity for large-enough system sizes is no longer a real problem in Monte Carlo simulations with ever-increasing computer capacity.

6.5 Advanced Techniques in Monte Carlo Simulations

In simulations, we have to solve two practical problems: (i) to shorten the waiting time, (ii) to find better algorithms to study physical phenomena which cannot be treated with precision by the standard Metropolis algorithm.

To solve the first kind of problems, mainly we have to find ways to accelerate convergence to equilibrium so that equilibrating time is shortened and the quality of physical quantities during a given averaging time is better. One of the best methods proposed so far is the cluster-flip method due to Wolff [364] and Swendsen–Wang [337] described below. For the second kind of problems, we can mention difficulties encountered by the Metropolis algorithm in the calculation of critical exponents and in the detection of extremely weak first-order transition. To deal with such difficulties, histogram and multiple-histogram techniques as well as flat-histogram methods have been proposed in the literature [110, 351]. We will describe them below.

6.5.1 Cluster-Flip Algorithm

The central idea comes from the following observation. In the vicinity of the transition, the correlation length is very long (see previous sections). Large clusters of parallel spins are formed just above T_c (we are in the case of a ferromagnetic crystal). The

Metropolis algorithm which updates spin after spin will take a long time to consider unnecessarily all spins in a cluster because, except for a small fraction at spins at the boundary, the spins in a cluster do not need to change orientation (they are ready for the ferromagnetic state at T_c). Therefore, to perform single-spin flips for all spins in a cluster near T_c is a loss of time. In addition, due to the existence of large clusters near T_c, the relaxation time is very long [see Eq. (6.40)]. This phenomenon is called "critical slowing-down." Wolff [364] and Swendsen and Wang [337] have proposed to update simultaneously all spins of a cluster by flipping the entire cluster (we have in mind the case of the Ising model). The method is simple to implement. To simplify the description, we consider a system of Ising spins with a ferromagnetic interaction J between nearest neighbors. The principle of the algorithm is the following:

(1) For a given spin configuration, we consider a spin S_i and we "construct" a "cluster" around S_i as follows. We examine the neighboring spins: If a neighbor in one direction is parallel to S_i, then it belongs to the cluster with a probability $p = 1 - \exp(-2\beta J)$ where $\beta = (k_B T)^{-1}$. We consider the spin next to that spin in that direction, and we continue the cluster construction. The limit of the cluster in the considered direction is where the cluster encounters an antiparallel spin or if a random number taken between 0 and 1 is larger than p. We have to go to all directions to determine the boundary of the cluster. Note that there is a very efficient algorithm for the cluster construction proposed by Hoshen and Kopelman [156].

(2) We flip the cluster and we calculate ΔE the difference in energy with the previous state: $\Delta E = 2J\,[C(++) - C(-+)]$ where $C(++)$ the number of broken parallel links along the boundary of the cluster, and $C(-+)$ the number of antiparallel links along the boundary. An example is shown in Fig. 6.6. The new orientation of the cluster is accepted or not, following the Metropolis criterion as for a single spin.

(3) We take another spin outside the above cluster and we begin again another cluster construction, etc.

The difference between the method by Wolff and that by Swendsen–Wang is the following. In the former, we flip only large

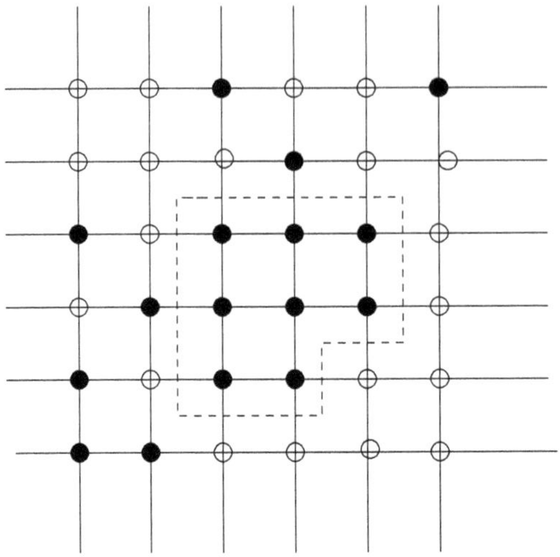

Figure 6.6 Example of a cluster, limited by the discontinued contour, constructed by cluster-construction algorithm. Black circles = $\uparrow$ spins, white circles = $\downarrow$ spins. The number of broken parallel links $C(++)$ is 1, that of antiparallel spin links $C(-+)$ is 11.

clusters, while in the latter one we flip all clusters. It is true that near T_c, the two methods are equivalent because most of clusters are large. However, a little bit further from T_c where there are many small clusters, we spend much time to flip small clusters in the Swendsen–Wang method, this does not significantly improve the result.

In practice, we can combine the Metropolis algorithm and the cluster-flip algorithm in one simulation: We use the latter from time to time because the cluster construction takes time. We need it only very near T_c to get rid of the critical slowing-down. The result is striking: The dynamic exponent $z \simeq 2$ as obtained by the Metropolis algorithm alone becomes $z \simeq 0.5$ using the cluster-flip method near T_c.

6.5.2 Histogram Method

In standard Metropolis Monte Carlo simulations, we calculate average physical quantities at discrete temperatures. We extrapolate results between discrete temperatures. However, near the transition temperature, extrapolation is impossible because physical quantities diverge at a precise single temperature value. Practically, we cannot find the exact location of the transition temperature using discrete temperatures. If the chosen temperature is not exactly T_c, we will not have the precision on the critical exponents calculated with the heights of C_V and χ which are not at T_c.

To avoid this difficulty, Ferrenberg and Swendsen [110] have proposed the histogram method which consists of making a simulation at a temperature T_0 as close as possible to T_c and recording the instantaneous system energy E during the simulation as long as possible to establish a histogram $H(E)$. Using this histogram we can calculate the canonical probabilities $P(T, E)$ at neighboring temperatures T around T_0 at as many points as we wish. Using these probabilities $P(T, E)$, we can calculate average values of physical quantities as continuous functions of T around T_0, by the formulas of the canonical description (see Appendix A), not by simulations. It suffices to choose T_0 close to T_c (not necessarily at T_c since T_c is not known before the simulation), we can find the exact location of the maximum of C_V and χ, for example. How to choose T_0? The answer is we have to make a preliminary run with many discrete temperatures using the Metropolis algorithm. From these preliminary results, we take T_0 at the maximum of C_v or χ for the histogram run.

The method is described in the following. We consider the partition function at T_0 (see Appendix A):

$$Z(T_0) = \sum_s \exp[-\beta_0 E(s)] \tag{6.47}$$

$$= \sum_E W(E) \exp(-\beta_0 E) \tag{6.48}$$

where the sum is taken over all energies of microscopic states, $W(E)$ denotes the degeneracy of the energy level E independent of T_0. The probability of the level E at T_0 is then

$$P(T_0, E) = \frac{W(E) \exp[-\beta_0 E(s)]}{Z(T_0)} \tag{6.49}$$

Now, by simulation we obtain the energy histogram $H(E)$ at T_0. The probability $P(T_0, E)$ is nothing but

$$P(T_0, E) = \frac{H(E)}{N_{MC}} \tag{6.50}$$

where N_{MC} is the number of Monte Carlo steps used to establish $H(E)$. By comparison of (6.50) to (6.49) we have

$$H(E) = N_{MC} P(T_0, E) = N_{MC} \frac{W(E) \exp[-\beta_0 E]}{Z(T_0)} \tag{6.51}$$

We consider now a temperature T near T_0. The probability $P(T, E)$ is written by

$$\begin{aligned} P(T, E) &= \frac{W(E) \exp[-\beta E]}{Z(T)} \\ &= \frac{Z(T_0) H(E) \exp[(\beta_0 - \beta)E]}{N_{MC} Z(T)} \\ &= \frac{H(E) \exp[(\beta_0 - \beta)E]}{\sum_E H(E) \exp[(\beta_0 - \beta)E]} \end{aligned} \tag{6.52}$$

where we have used (6.51) to replace $W(E)$ and the following relation to replace $Z(T)$:

$$\begin{aligned} Z(T) &= \sum_E W(E) \exp[-\beta E] \\ &= \frac{Z(T_0)}{N_{MC}} \sum_E H(E) \exp[(\beta_0 - \beta)E] \end{aligned} \tag{6.53}$$

The histogram $H(E)$ established for T_0 is thus used to calculate the probability at another temperature T by (6.52). Using this probability we can calculate without simulation the average value of a quantity A by the formula

$$< A >= \sum_E A P(T, E) \tag{6.54}$$

Remarks:

(1) To get a precise $H(E)$, we have to use a very large number of Monte Carlo steps N_{NC} in order to include as many as possible microscopic states in the sum.
(2) Since $H(E)$ is obtained with an importance sampling at T_0, if T is rather far from T_0, the probability $P(T, E)$ calculated by (6.52) using $H(E)$ is not precise. It is therefore very important to verify the form of each $P(T, E)$ by looking at its plot before using it. In general, we have a Gaussian form as shown in

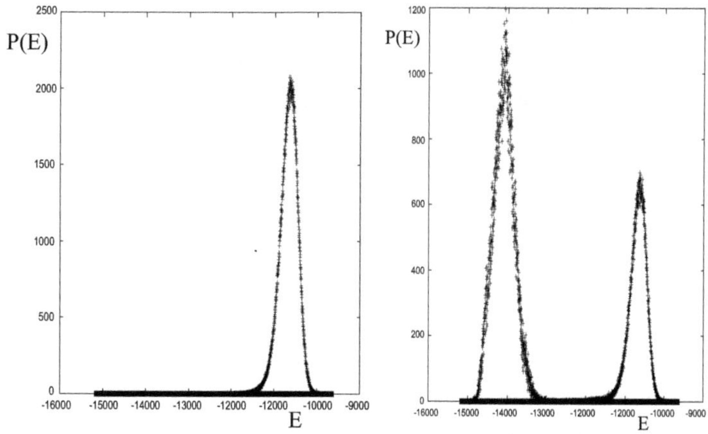

Figure 6.7 Probability $P(E)$ obtained by simulation in the case of a face-centered cubic antiferromagnet with interaction J between nearest neighbors. Left: Energy histogram taken at temperature $T_0 = 1.76J$ ($k_B = 1$), just above the transition temperature. Right: Histogram taken at the transition temperature $T_0 = T_c = 1.755J$ E is the total energy of the system.

Fig. 6.7 (left). If T is far from T_0, $P(T, E)$ calculated has an irregular, asymmetric form. We should not use it [110].

(3) To determine with precision the transition temperature $T_c(L)$, we have to choose T_0 in the critical region, as close as possible to the presumed $T_c(L)$. To have a good choice, we have to make several simulations with Metropolis algorithm to detect a good value of T_0.
(4) When the transition is of first order, $P(T_0, E)$ presents a double peak if T_0 coincides or very close to $T_c(L)$. This is shown in Fig. 6.7 (right) where the energies at the peaks E_1 and E_2 correspond to energies of the ordered and disordered phases, which coexist at T_0. The histogram between two peaks is almost zero, indicating an energy discontinuity, namely the latent heat of the system.

The determination of the critical exponents with the histogram method is very precise (see, for example, the original papers of Ferrenberg and Swendsen [110]).

6.5.3 Multiple-Histogram Technique

The multiple-histogram technique is known to reproduce with very high accuracy the critical exponents of second-order phase transitions [50, 111]. It is more complicated to be implemented because we have to realize many histograms in independent simulations, but it gives much better results in difficult cases.

The principle consists of the following steps [111]:

(1) First, we realize independent simulations at n temperatures T_i ($i = 1, \cdots, n$). For each temperature T_i, the number of Monte Carlo steps is N_i. The histogram taken during the simulation at that temperature is $H(E, T_i)$: One has $\sum_E H(E, T_i) = N_i$.

(2) Second, we calculate the density of states $\rho(E)$ by

$$\rho(E) = \frac{\sum_{i=1}^{n} H(E, T_i)}{\sum_{i=1}^{n} N_i Z(T_i)^{-1} e^{-E/k_B T_i}} \tag{6.55}$$

where the partition function $Z(T_i)$ is

$$Z(T_i) = \sum_E \rho(E) e^{-E/k_B T_i} \tag{6.56}$$

We see here that $\rho(E)$ and $Z(T_i)$ should be calculated self-consistently. The choice of neighboring temperatures $T_1, T_2, \cdots, T_n$ should be guided as the choice of T_0 discussed in the single histogram technique shown above.

(3) Once $\rho(E)$ and $Z(T)$ are obtained, we can calculate the thermal average of any physical quantity A at T by

$$\langle A(T) \rangle = \frac{\sum_E A(E) \rho(E) e^{-E/k_B T}}{Z(T)} \tag{6.57}$$

Thermal averages of physical quantities are thus calculated as continuous functions of T. The results are valid over a much wider range of temperature than for any single histogram. The calculation of the critical exponents is more precise than with a single histogram technique.

6.5.4 Wang–Landau Flat-Histogram Method

Wang and Landau [351] have proposed a Monte Carlo algorithm for classical statistical models which allowed us to study systems with difficultly accessed microscopic states. In particular, it permits to detect with efficiency weak first-order transitions [250, 251]. The algorithm uses a random walk in the energy space in order to obtain an accurate estimate for the density of states $g(E)$ which is defined as the number of spin configurations for any given E. This method is based on the fact that a flat energy histogram $H(E)$ is produced if the probability for the transition to a state of energy E is proportional to $g(E)^{-1}$.

We summarize how this algorithm is implemented here. At the beginning of the simulation, the density of states is set equal to one for all possible energies, $g(E) = 1$. We begin a random walk in energy space (E) by choosing a site randomly and flipping its spin with a probability proportional to the inverse of the temporary density of states. In general, if E and E' are the energies before and after a spin is flipped, the transition probability from E to E' is

$$p(E \to E') = \min\left[g(E)/g(E'), 1\right] \tag{6.58}$$

Each time an energy level E is visited, the density of states is modified by a modification factor $f > 0$ whether the spin is flipped or not, i.e., $g(E) \to g(E)f$. At the beginning of the random walk, the modification factor f can be as large as $e^1 \simeq 2.7182818$. A histogram $H(E)$ records the number of times a state of energy E is visited. Each time the energy histogram satisfies a certain "flatness" criterion, f is reduced according to $f \to \sqrt{f}$ and $H(E)$ is reset to zero for all energies. The reduction process of the modification factor f is repeated several times until a final value f_{final} which is close enough to one. The histogram is considered as flat if

$$H(E) \geq x\% \cdot \langle H(E) \rangle \tag{6.59}$$

for all energies, where $x\%$ is chosen between 70% and 95% and $\langle H(E) \rangle$ is the average histogram.

Thermodynamic quantities [49, 351] can be evaluated using $g(E)$. For example,

$$\langle E^n \rangle = \frac{1}{Z} \sum_E E^n g(E) \exp(-E/k_B T) \tag{6.60}$$

$$C_v = \frac{\langle E^2 \rangle - \langle E \rangle^2}{k_B T^2} \tag{6.61}$$

$$\langle M^n \rangle = \frac{1}{Z} \sum_E M^n g(E) \exp(-E/k_B T) \tag{6.62}$$

$$\chi = \frac{\langle M^2 \rangle - \langle M \rangle^2}{k_B T} \tag{6.63}$$

where Z is the partition function defined by

$$Z = \sum_E g(E) \exp(-E/k_B T) \tag{6.64}$$

The canonical distribution at any temperature can be calculated simply by

$$P(E, T) = \frac{1}{Z} g(E) \exp(-E/k_B T) \tag{6.65}$$

In practice, we have to choose an energy range of interest [223, 311] $(E_{\text{min}}, E_{\text{max}})$. We divide this energy range into R subintervals, the minimum energy of each subinterval is E^i_{min} for $i = 1, 2, \cdots, R$, and the maximum of the subinterval i is $E^i_{\text{max}} = E^{i+1}_{\text{min}} + 2\Delta E$, where ΔE can be chosen large enough for a smooth boundary between two subintervals. The Wang–Landau algorithm is used to calculate the relative density of states of each subinterval $(E^i_{\text{min}}, E^i_{\text{max}})$ with the modification factor $f_{\text{final}} = \exp(10^{-9})$ and flatness criterion $x\% = 95\%$. We reject the suggested spin flip and do not update $g(E)$ and the energy histogram $H(E)$ of the current energy level E if the spin-flip trial would result in an energy outside the energy segment. The density of states of the whole range is obtained by joining the density of states of each subinterval $(E^i_{\text{min}} + \Delta E, E^i_{\text{max}} - \Delta E)$. Numerous examples of applications of this method can be found in the literature and in part II of this book.

6.6 Conclusion

We have presented above the principle of Monte Carlo simulation and shown how to implement it to study properties of spin systems, in particular at the phase transition. Several advanced techniques have also been described to help solve a number of problems encountered in Monte Carlo simulations to investigate phase transitions. Numerical simulations are considered today as important as theory and experiment in the study of materials.

To make a good simulation, we have to know results from previous simulations, theories and experiments. This knowledge allows us to decide ingredients to include and ingredients not to include in the model. We should bear in mind that simulation is just a method of investigation. Simulation results come from the model and should be interpretable. A deep understanding of difficulties encountered during simulations helps us modify technical details or introduce new techniques to circumvent those obstacles. For example, we have shown above that to get rid of the critical slowing-down, we can use cluster-flips, to detect weak first-order transitions we have to use new methods such as the Wang–Landau technique, and to calculate critical exponents with precision we should use high-performance techniques such as the multiple-histogram method. To succeed in simulations, we should have, beyond a numerical skill, a good background in theories in the field under study. This helps us understand and interpret what comes out from the computer after a long waiting time.

Part II

Magnetism of Thin Films

Chapter 7

Exactly Solved Frustrated Models in Two Dimensions

In this chapter, we show properties of exactly solved frustrated two-dimensional (2D) spin systems. We would like to emphasize the physics near a phase boundary where interesting phenomena can occur due to competing interactions of the two phases around the boundary. Two-dimensional systems are, in fact, the limiting case of thin films with a monolayer. We give examples of frustrated 2D Ising systems that we can exactly solve by transforming them into vertex models. We show that these simple systems contain already most of the striking features of frustrated systems such as the high degeneracy of the ground state (GS), many phases in the GS phase diagram in the space of interaction parameters, the reentrance occurring near the boundaries of these phases, the disorder lines in the paramagnetic phase and the partial disorder coexisting with the order at equilibrium.

We discuss a number examples of thin films and 3D systems where we see some phenomena observed in the above exactly found solutions such as the reentrance and the partial disorder at equilibrium.

The results shown in this chapter are taken from Refs. [16, 67, 68, 76, 77, 153, 285, 305, 307].

Physics of Magnetic Thin Films: Theory and Simulation
Hung T. Diep

ISBN 978-981-4877-42-8 (Hardcover), 978-1-003-12110-7 (eBook)
www.jennystanford.com

7.1 Introduction

Extensive investigations on materials have been carried out during the past three decades. This is due to an enormous number of industrial applications which drastically change our life style. The progress in experimental techniques, the advance on theoretical understanding and the development of high-precision simulation methods together with the rapid increase of computer power have made possible the rapid development in materials science. Today, it is difficult to predict what will be discovered in this research area in 10 years.

The purpose of this chapter is to recall important results found by solving exactly several frustrated 2D Ising systems which help understand recent results in other non-solvable models such as frustrated thin films studied in the following chapters of the present book. The physics of frustrated spin systems at low dimensions, 2D systems and magnetic thin films, attracts an enormous number of investigations during the past decades due to many industrial applications. We would like to connect these results, published over a long period of time, on a line of thoughts: physics near a phase boundary. A boundary between two phases of different orderings is determined as a compromise of competing interactions each of which favors one kind of ordering. The frustration is thus minimum on the boundary (see reviews on many aspects of frustrated spin systems in Ref. [85]). When an external parameter varies, this boundary changes and we will see in this review that many interesting phenomena occur in the boundary region. We will concentrate ourselves in the search for interesting physics near the phase boundaries in various frustrated spin systems in this review.

In the 1970s, statistical physics with the Renormalization Group analysis has greatly contributed to the understanding of the phase transition from an ordered phase to a disordered phase [88, 380]. We will show methods to study the phase transition in magnetic thin films where surface effects when combined with frustration effects give rise to many new phenomena. Physical properties of solid surfaces, thin films and superlattices have been intensively studied due to their many applications [18, 22, 36, 87, 113, 134, 374].

Section 7.2 is devoted to the definition of the frustration and to the determination of the ground states of a few popular models. We show in Section 7.3 the decimation method and how to find the disorder line and the reentrance phase. Section 7.4 shows several exactly solved models with their striking properties. Section 7.5 shows some other 2D exactly solved decorated models. As seen, many interesting phenomena such as partial disorder, reentrance, disorder lines and multiple phase transitions are exactly uncovered. Only exact mathematical techniques can allow us to reveal such beautiful phenomena which occur around the boundary separating two phase of different ground-state orderings. These exact results permit to understand similar behaviors in systems that cannot be solved, some of these systems are shown and discussed in Sections 7.6 and 7.7.

7.2 Frustration

7.2.1 Definition

Since the 1980s, frustrated spin systems have been subjects of intensive studies [85]. The word "frustration" has been introduced to describe the fact that a spin cannot find an orientation to *fully* satisfy all interactions with its neighbors, namely the energy of a bond is not the lowest one [344, 348]. This will be seen below for Ising spins where at least one among the bond with the neighbors is not satisfied. For vector spins, frustration is shared by all spins so that all bonds are only partially satisfied, i.e., the energy of each bond is not minimum. Frustration results either from the competing interactions or from the lattice geometry such as the triangular lattice with antiferromagnetic nearest-neighbor (NN) interaction, the face-centered cubic (FCC) antiferromagnet and the antiferromagnetic hexagonal-close-packed (HCP) lattice (see [85]).

Note that real magnetic materials have complicated interactions and there are large families of frustrated systems such as the heavy lanthanides metals (holmium, terbium and dysprosium) [381, 382], helical MnSi [330], pyrochore antiferromagnets [124], and spin-ice materials [47]. Exact solutions on simpler systems may help understand qualitatively real materials. Besides, exact results can be used to validate approximations.

We recall in the following some basic arguments leading to the definition of the frustration.

The interaction energy of two spins $\mathbf{S}_i$ and $\mathbf{S}_j$ interacting with each other by J is written as $E = -J\,(\mathbf{S}_i \cdot \mathbf{S}_j)$. If J is ferromagnetic ($J > 0$) then the minimum of E is $-J$ corresponding to $\mathbf{S}_i$ parallel to $\mathbf{S}_j$. If J is antiferromagnetic ($J < 0$), E is minimum when $\mathbf{S}_i$ is antiparallel to $\mathbf{S}_j$. One sees that in a crystal with NN ferromagnetic interaction, the ground state (GS) of the system is the configuration where all spins are parallel: The interaction of every pair of spins is "fully" satisfied, namely the bond energy is equal to $-J$. This is true for any lattice structure. If J is antiferromagnetic, the GS depends on the lattice structure: (i) For lattices containing no elementary triangles, i.e., bipartite lattices (such as square lattice, simple cubic lattices, ...) in the GS, each spin is antiparallel to its neighbors, i.e., every bond is fully satisfied, its energy is equal to $-|J|$; (ii) for lattices containing elementary triangles such as the triangular lattice, the FCC lattice and the HCP lattice, one cannot construct a GS where all bonds are fully satisfied (see Fig. 7.1). The GS does not correspond to the minimum interaction energy of every spin pair: The system is frustrated.

Let us formally define the frustration. We consider an elementary lattice cell which is a polygon formed by faces called "plaquettes." For example, the elementary cell of the simple cubic lattice is a cube with six square plaquettes, the elementary cell of the FCC lattice is a tetrahedron formed by four triangular plaquettes. According to the definition of Toulouse [344] the plaquette is frustrated if the parameter P defined below is negative

$$P = \prod_{\langle i,j \rangle} \mathrm{sign}(J_{i,j}) \tag{7.1}$$

where $J_{i,j}$ is the interaction between two NN spins of the plaquette and the product is performed over all $J_{i,j}$ around the plaquette.

We show two examples of frustrated plaquettes in Fig. 7.1, a triangle with three antiferromagnetic bonds and a square with three ferromagnetic bonds and one antiferromagnetic bond. P is negative

Figure 7.1 Two frustrated cells are shown. The thin (heavy) lines denote the ferromagnetic (antiferromagnetic) bonds. Up and down spins are shown by green and red circles, respectively. Question marks indicate undetermined spin orientation. Choosing an orientation for the spin marked by the question mark will leave one of its bonds unsatisfied (frustrated bond with positive energy).

in both cases. If one tries to put Ising spins on those plaquettes, at least one of the bonds around the plaquette will not be satisfied. For vector spins, we show below that the frustration is equally shared by all bonds so that in the GS, each bond is only partially satisfied.

One sees that for the triangular plaquette, the degeneracy is three, and for the square plaquette it is four. Therefore, the degeneracy of an infinite lattice for these cases is infinite, unlike the non-frustrated case.

The frustrated triangular lattice with NN interacting Ising spins was studied in 1950 [356].

7.2.2 Non-Collinear Ground-State Spin Configurations

For vector spins, non-collinear configurations due to competing interactions were found in 1959 independently by Yoshimori [369], Villain [350] and Kaplan [175].

We emphasize that the frustration may be due to the competition between a Heisenberg exchange model which favors a collinear spin configuration and the Dzyaloshinskii-Moriya interaction $E = -\mathbf{D} \cdot (\mathbf{S}_i \wedge \mathbf{S}_j)$ [99, 240] which favors the perpendicular configuration. We will return to this interaction in Chapters 13, 14 and 15.

We show below how to determine the GS of some frustrated systems and discuss some of their properties.

We consider the plaquettes shown in Fig. 7.1 with XY spins. The GS configuration corresponds to the minimum of the energy E of the plaquette. In the case of the triangular plaquette, suppose that spin $\mathbf{S}_i$ ($i = 1, 2, 3$) of amplitude S makes an angle θ_i with the **Ox** axis. One has

$$\begin{aligned} E &= -J\left(\mathbf{S}_1 \cdot \mathbf{S}_2 + \mathbf{S}_2 \cdot \mathbf{S}_3 + \mathbf{S}_3 \cdot \mathbf{S}_1\right) \\ &= -J\,S^2\left[\cos(\theta_1 - \theta_2) + \cos(\theta_2 - \theta_3) + \cos(\theta_3 - \theta_1)\right] \end{aligned} \quad (7.2)$$

where $J < 0$ (antiferromagnetic). Minimizing E with respect to 3 angles θ_i, we find the solution $\theta_1 - \theta_2 = \theta_2 - \theta_3 = \theta_3 - \theta_1 = 2\pi/3$.

One can also write

$$E = -J\left(\mathbf{S}_1 \cdot \mathbf{S}_2 + \mathbf{S}_2 \cdot \mathbf{S}_3 + \mathbf{S}_3 \cdot \mathbf{S}_1\right) = +\frac{3}{2}J\,S^2 - \frac{J}{2}\left(\mathbf{S}_1 + \mathbf{S}_2 + \mathbf{S}_3\right)^2$$

J is negative, the minimum thus corresponds to $\mathbf{S}_1 + \mathbf{S}_2 + \mathbf{S}_3 = 0$ which gives the 120° structure. This is true also for the Heisenberg spins.

For the frustrated square plaquette, we suppose that the ferromagnetic bonds are J and the antiferromagnetic bond is $-J$) connecting the spins $\mathbf{S}_1$ and $\mathbf{S}_4$ (see Fig. 7.2). The energy minimization gives

$$\theta_2 - \theta_1 = \theta_3 - \theta_2 = \theta_4 - \theta_3 = \frac{\pi}{4} \text{ and } \theta_1 - \theta_4 = \frac{3\pi}{4} \quad (7.3)$$

If the antiferromagnetic interaction is $-\eta J$ ($\eta > 0$), the angles are [27]

$$\cos\theta_{21} = \cos\theta_{32} = \cos\theta_{43} \equiv \theta = \frac{1}{2}\left[\frac{\eta + 1}{\eta}\right]^{1/2} \quad (7.4)$$

and $|\theta_{14}| = 3|\theta|$, where $\cos\theta_{ij} \equiv \cos\theta_i - \cos\theta_j$. This solution exists if $|\cos\theta| \leq 1$, namely $\eta > \eta_c = 1/3$. One recovers when $\eta = 1$, $\theta = \pi/4, \theta_{14} = 3\pi/4$.

The GS spin configurations of the frustrated triangular and square lattices are displayed in Fig. 7.2 with XY spins. We see that the frustration is shared by all bonds: The energy of each bond is $-0.5J$ for the triangular lattice, and $-\sqrt{2}J/2$ for the square lattice. Thus, the bond energy in both cases is not fully satisfied, namely not equal to $-J$, as we said above when defining the frustration.

At this stage, we note that the GS found above have a two-fold degeneracy resulting from the equivalence of clockwise and

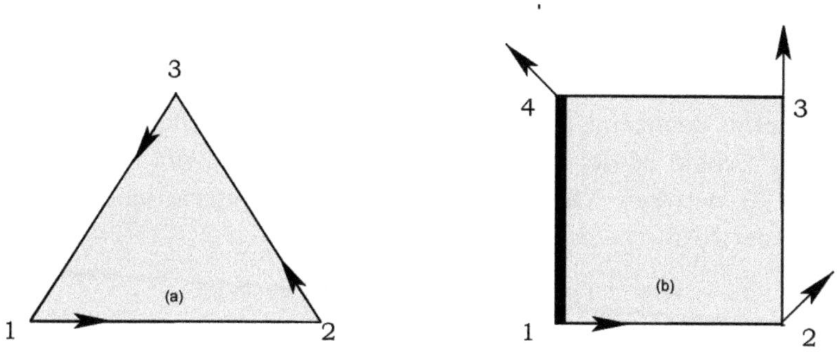

Figure 7.2 Ground state of XY spins on frustrated triangular and square cells: non-collinear spin arrangements. The thin lines denote the ferromagnetic interaction, the thick line is the antiferromagnetic one.

counter-clockwise turning angle (noted by $+$ and $-$ in Fig. 7.3) between adjacent spins on a plaquette in Fig. 7.2. Therefore, the symmetry of these spin systems is of Ising type $O(1)$, in addition to the symmetry $SO(2)$ due to the invariance by global spin rotation in the plane.

From the GS symmetry, one expects that the respective breaking of $O(1)$ and $SO(2)$ symmetries would behave, respectively, as the 2D Ising universality class and the Kosterlitz–Thouless transition [380].

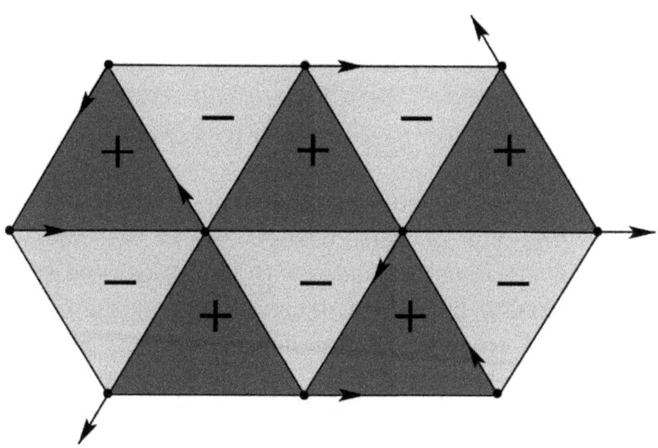

Figure 7.3 Triangular antiferromagnet with XY spins: The left (right) chirality is indicated by $+$ ($-$). See text.

However, the question of whether the two phase transitions would occur at the same temperature and the nature of their universality remains an open question [27, 45].

Let us determine the GS of a helimagnet. Consider the simplest case: a chain of Heisenberg spins with ferromagnetic interaction $J_1(>0)$ between NN and antiferromagnetic interaction $J_2(<0)$ between NNN. The interaction energy is

$$
\begin{aligned}
E &= -J_1 \sum_i \mathbf{S}_i \cdot \mathbf{S}_{i+1} + |J_2| \sum_i \mathbf{S}_i \cdot \mathbf{S}_{i+2} \\
&= S^2 \left[-J_1 \cos\theta + |J_2| \cos(2\theta)\right] \sum_i 1 \\
\frac{\partial E}{\partial \theta} &= S^2 \left[J_1 \sin\theta - 2|J_2| \sin(2\theta)\right] \sum_i 1 = 0 \\
&= S^2 \left[J_1 \sin\theta - 4|J_2| \sin\theta \cos\theta\right] \sum_i 1 = 0 \qquad (7.5)
\end{aligned}
$$

where one has supposed that the angle between NN spins is θ. The first solution is

$$\sin\theta = 0 \longrightarrow \theta = 0 \text{ which is the ferromagnetic solution}$$

and the second one is

$$\cos\theta = \frac{J_1}{4|J_2|} \longrightarrow \theta = \pm \arccos\left(\frac{J_1}{4|J_2|}\right) \qquad (7.6)$$

This solution is possible when $-1 \leq \cos\theta \leq 1$, i.e., when $J_1/(4|J_2|) \leq 1$ or $|J_2|/J_1 \geq 1/4 \equiv \varepsilon_c$. An example of configuration is shown in Fig. 7.4. Note that there are two degenerate configurations of clockwise and counter-clockwise turning angles as the other examples shown above.

Note that the two frequently studied frustrated spin systems are the FCC and HCP antiferromagnets. These two magnets are constructed by stacking tetrahedra with four frustrated triangular faces. Frustration by the lattice structure such as these cases are called "geometry frustration." Another 3D popular model which has been extensively studied since 1984 is the system of stacked antiferromagnetic triangular lattices (satl). The phase transition of this system with XY and Heisenberg spins was a controversial subject for more than 20 years. The controversy was ended with our works: The reader is referred to Refs. [187, 250, 251] for the history.

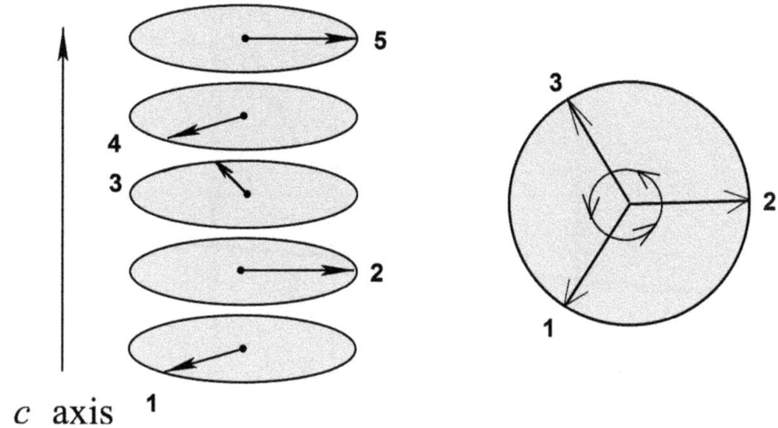

Figure 7.4 Example of a helimagnetic configuration using $\varepsilon = |J_2|/J_1 > \varepsilon_c = 1/4$ ($J_1 > 0, J_2 < 0$), namely $\theta = 2\pi/3$. Left: 3D view. Right: top view.

In short, we found that in known 3D frustrated spin systems (FCC, HCP, satl, helimagnets, ...) with Ising, XY or Heisenberg spins, the transition is of first order [83, 152].

Another subject which has been much studied since the 1980s is the phenomenon called "order by disorder": We have seen that the ground state of frustrated spin systems is highly degenerate and often infinitely degenerate (entropy not zero at temperature $T = 0$). However, it has been shown in many cases that when T is turned on the system chooses a state which has the largest entropy, namely the system chooses its order by the largest disorder. We call this phenomenon "order by disorder" or "order by entropic selection" (see references cited in Section III.B of Ref. [83]).

We will not discuss these subjects in this review, which is devoted to low-dimensional frustrated spin systems.

7.3 Exactly Solved Frustrated Models

Any 2D Ising model with non-crossing interactions can be exactly solved. To avoid the calculation of the partition function one can transform the model into a 16-vertex model or a 32-vertex model. The resulting vertex model is exactly solvable. We have applied this

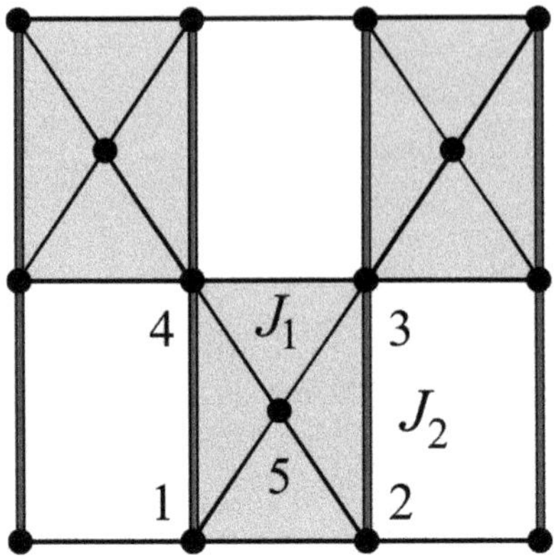

Figure 7.5 Kagomé lattice: Diagonal and horizontal bonds are NN antiferromagnetic interactions J_1, vertical double lines indicate the NNN interactions J_2.

method to search for the exact solution of several Ising frustrated 2D models with non-crossing interactions shown in Figs. 7.5–7.7.

Details have been given in Ref. [86]. We outline below a simplified formulation of a model for illustration. The aim is to discuss the results. As we will see, these models possess striking phenomena due to the frustration.

7.3.1 Example of the Decimation Method

We take the case of the centered honeycomb lattice with the following Hamiltonian:

$$H = -J_1 \sum_{(ij)} \sigma_i \sigma_j - J_2 \sum_{(ij)} \sigma_i \sigma_j - J_3 \sum_{(ij)} \sigma_i \sigma_j \tag{7.7}$$

where $\sigma_i = \pm 1$ is an Ising spin at the lattice site i. The first, second, and third sums are performed on the spins interacting via J_1, J_2 and J_3 bonds, respectively (see Fig. 7.7). The case $J_2 = J_3 = 0$ corresponds to the honeycomb lattice, and the case $J_1 = J_2 = J_3$ to the triangular lattice.

Let σ be the central spin of the lattice cell shown in Fig. 7.7. Other spins are numbered from σ_1 to σ_6. The Boltzmann weight of

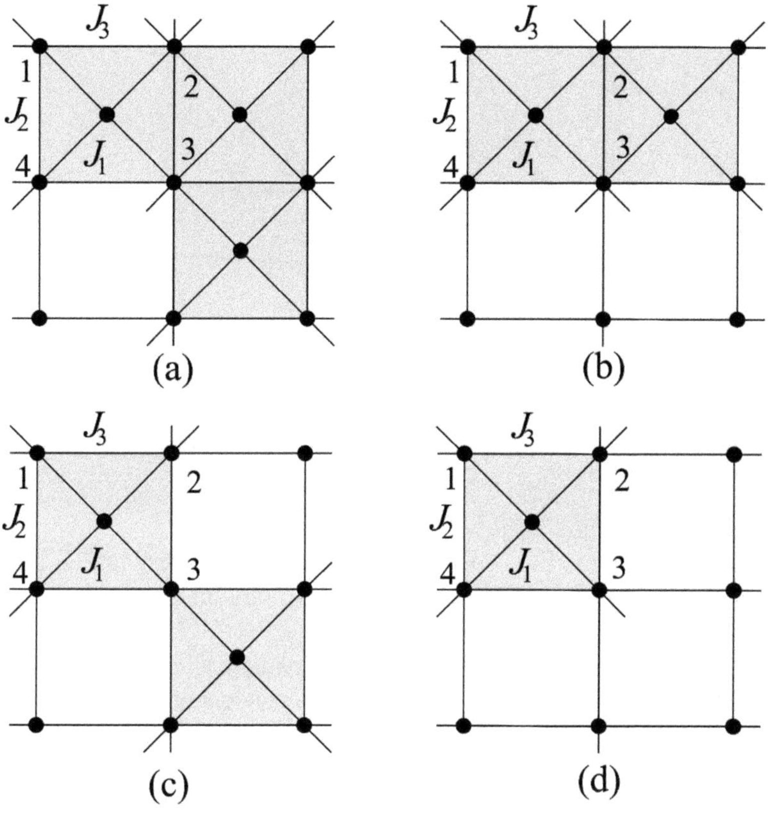

Figure 7.6 Exactly solved dilute centered square lattices: Interactions along diagonal, vertical and horizontal spin pairs are noted by J_1, J_2, and J_3, respectively.

the elementary cell is written as

$$W = \exp[K_1(\sigma_1\sigma_2 + \sigma_2\sigma_3 + \sigma_3\sigma_4 + \sigma_4\sigma_5 + \sigma_5\sigma_6 + \sigma_6\sigma_1) + K_2\sigma(\sigma_1 + \sigma_2 + \sigma_4 + \sigma_5) + K_3\sigma(\sigma_3 + \sigma_6)] \tag{7.8}$$

where $K_i \equiv \frac{J_i}{k_B T}$ $(i = 1, 2, 3)$. The partition function reads

$$Z = \sum_\sigma \prod_c W \tag{7.9}$$

where the sum is taken over all spin configurations and the product over all elementary cells of the lattice. One imposes the periodic boundary conditions. The above model is exactly solvable. To that

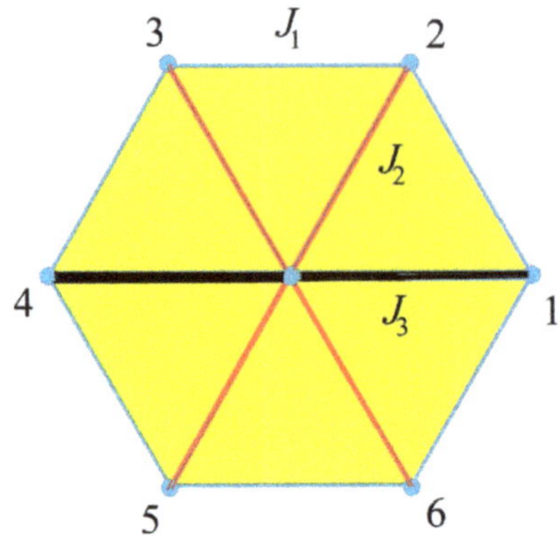

Figure 7.7 Centered honeycomb lattice. Spins are numbered for decimation demonstration (see text). Blue, red and black bonds denote interactions J_1, J_2 and J_3, respectively.

end, we decimate the central spin of every elementary lattice cell. We finally get a honeycomb Ising model (without centered spins) with multispin interactions.

After decimation of the central spin, namely after summing the values of the central spin σ, the Boltzmann weight of an elementary cell reads

$$W' = 2\exp[K_1(\sigma_1\sigma_2 + \sigma_2\sigma_3 + \sigma_3\sigma_4 + \sigma_4\sigma_5 + \sigma_5\sigma_6 + \sigma_6\sigma_1)] \\ \times \cosh[K_2(\sigma_1 + \sigma_2 + \sigma_4 + \sigma_5) + K_3(\sigma_3 + \sigma_6)] \quad (7.10)$$

We show below that this model is, in fact, a case of the 32-vertex model on the triangular lattice which has an exact solution.

We consider the dual triangular lattice of the honeycomb lattice obtained above [25]. The sites of the dual triangular lattice are at the center of each elementary honeycomb cell with bonds perpendicular to the honeycomb ones, as illustrated in Fig. 7.8.

Let us define the conventional arrow configuration for each site of the dual triangular lattice: If all six spins of the honeycomb cell are parallel, then the arrows, called "standard configuration," are shown in Fig. 7.9. From this "conventional" configuration, antiparallel spin pairs on the two sides of a triangular lattice bond will have its corresponding arrow change the direction.

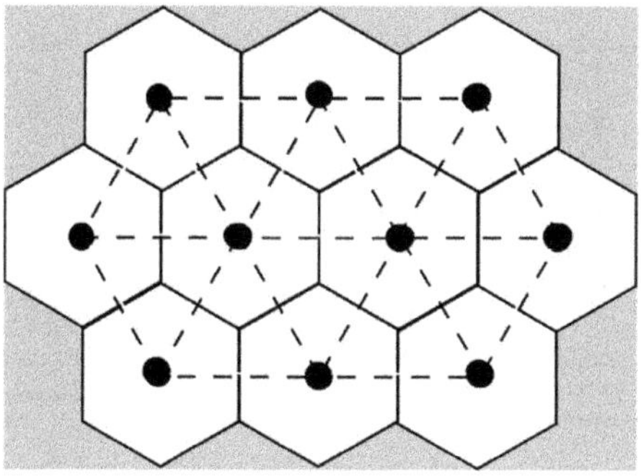

Figure 7.8 The dual triangular lattice, shown by discontinued lines, of the honeycomb lattice.

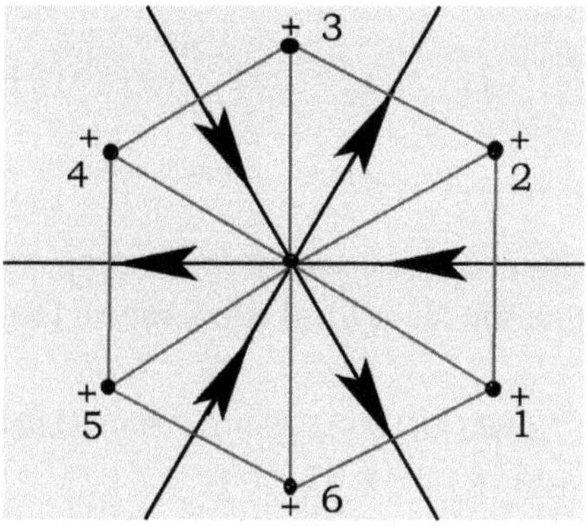

Figure 7.9 The conventional definition of the "standard" arrows for NN around a site of the triangular lattice: Spins are numbered so the arrows can be recognized in examples shown in Fig. 7.10. Note that the configuration of all down spins has the same arrow configuration. See text.

As examples, two spin configurations on the honeycomb lattice and their corresponding arrow configurations on the triangular lattice are displayed in Fig. 7.10.

Counting all arrow configurations, we obtain 32. To each of these 32 vertices one associates the Boltzmann weight $W'(\sigma_1, \sigma_2, \sigma_3, \sigma_4, \sigma_5, \sigma_6)$ given by Eq. (7.10). Let us give explicitly a few of them:

$$\omega_1 = W'(+, -, -, -, +, +) = 2e^{2K_1} \tag{7.11}$$

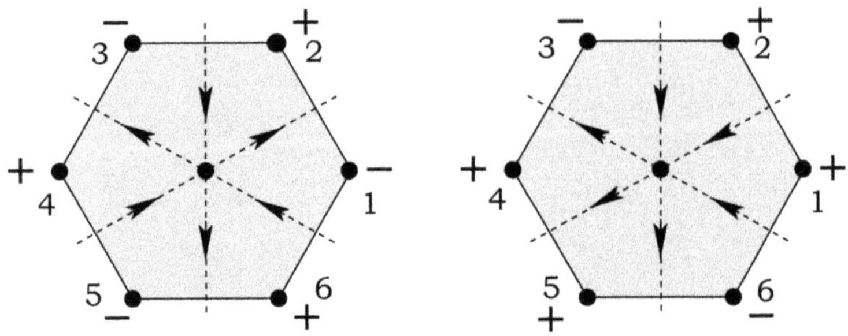

Figure 7.10 Two examples of spin configurations and their corresponding arrow configurations. To understand, compare with the standard arrows defined in Fig. 7.9. See text.

$$\omega_2 = W'(+,+,-,+,+,-) = 2e^{-2K_1}\cosh(4K_2 - 2K_3) \quad (7.12)$$

$$\omega_3 = W'(+,-,+,-,+,+) = 2e^{-2K_1}\cosh(2K_3) \quad (7.13)$$

$$\omega_4 = W'(+,+,+,+,+,-) = 2e^{2K_1}\cosh(4K_2) \quad (7.14)$$

$$\ldots$$

Using the above expressions of the 32-vertex model, one finds the following equation for the critical temperature (see details in Ref. [76]):

$$\begin{aligned} & e^{2K_1} + e^{-2K_1}\cosh(4K_2 - 2K_3) + 2e^{-2K_1}\cosh(2K_3) + 2e^{2K_1} \\ & \quad + e^{6K_1}\cosh(4K_2 + 2K_3) + e^{-6K_1} \\ & = 2\max\{e^{2K_1} + e^{-2K_1}\cosh(4K_2 - 2K_3); \\ & \quad e^{2K_1} + e^{-2K_1}\cosh(2K_3); e^{6K_1}\cosh(4K_1 + 2K_3) + e^{-6K_1}\} \end{aligned} \quad (7.15)$$

The solutions of this equation are given in 7.4.2 below for some special cases.

Following the case studied above, we can study the 2D models shown in Figs. 7.5–7.6: After decimation of the central spin in each square, these models can be transformed into a special case of the 16-vertex model which yields the exact solution for the critical surface (see details in Ref. [86]).

Before showing some results in the space of interaction parameters, let us introduce the definitions of disorder line and reentrant phase.

7.3.2 Disorder Line, Reentrance

It is not the purpose of this review to enter technical details. We would rather like to describe the physical meaning of the disorder line and the reentrance. A full technical review has been given in Ref. [86].

Disorder solutions exist in the paramagnetic region which separate zones of fluctuations of different nature. They are where the short-range pre-ordering correlations change their nature to allow for transitions in the phase diagrams of anisotropic models. They imply constraints on the analytical behavior of the partition function of these models.

To obtain the disorder solution, one makes a certain local decoupling of the degrees of freedom. This yields a dimension reduction: A 2D system then behaves on the disorder line as a 1D system. This local decoupling is made by a simple local condition imposed on the Boltzmann weights of the elementary cell [329].

This is very important while interpreting the system behavior: On one side of the disorder line, pre-ordering fluctuations have correlation different from those of the other side. Crossing the line, the system pre-ordering correlation changes. The dimension reduction is often necessary to realize this.

Note that disorder solutions may be used in the study of cellular automata as it has been shown in Ref. [297].

Let us give now a definition for the reentrance. A reentrant phase lies between two ordered phases. For example, at low temperature (T) the system is in an ordered phase I. Increasing T, it undergoes a transition to a paramagnetic phase R, but if one increases further T, the system enters another ordered phase II before becoming disordered at a higher T. Phase R is thus between two ordered phases I and II. It is called "reentrant paramagnetic phase" or "reentrant phase."

How physically is it possible? At a first sight, it cannot be possible because the entropy of an ordered phase is smaller than that of an disordered phase so that the disordered phase R cannot exist at lower T than the ordered phase II. In reality, as we will see below, phase II has always a partial disorder which compensates the loss of entropy while going from R to II. The principle that entropy increases with T is thus not violated.

7.4 Phase Diagram

7.4.1 Kagomé Lattice

7.4.1.1 Model with NN and NNN interactions

The Kagomé lattice shown in Fig. 7.5 has attracted much attention not only by its great interest in statistical physics but also in real materials [124]. The Kagomé Ising model with only NN interaction J_1 has been solved a long time ago [173]. No phase transition at finite T when J_1 is antiferromagnetic. Taking into account the NNN interaction J_2, we have solved [16] this model by transforming it into a 16-vertex model which satisfies the free-fermion condition. The equation of the critical surface is

$$\begin{aligned}
&\frac{1}{2}\left[\exp(2K_1 + 2K_2)\cosh(4K_1) + \exp(-2K_1 - 2K_2)\right] \\
&\quad + \cosh(2K_1 - 2K_2) + 2\cosh(2K_1) \\
&= 2\max\left\{\frac{1}{2}\left[\exp(2K_1 + 2K_2)\cosh(4K_1)\right.\right. \\
&\quad \left.\left. + \exp(-2K_1 - 2K_2)\right];\ \cosh(2K_2 - 2K_1); \cosh(2K_1)\right\}
\end{aligned} \tag{7.16}$$

We are interested in the region near the phase boundary between two phases IV (partially disordered) and I (ferromagnetic) in Fig. 7.11 (top). We show in Fig. 7.11 (bottom) the small region near the boundary $\alpha = J_2/J_1 = -1$ which has the reentrant paramagnetic phase and a disorder line.

We note that only near the phase boundary such a reentrant phase and a disorder line can exist.

7.4.1.2 Generalized Kagomé lattice

If we suppose that all interactions J_1, J_2 and J_3 in the model shown in Fig. 7.5 are different, the phase diagram becomes very rich [67]. For instance, the reentrance can occur in an *infinite region*par

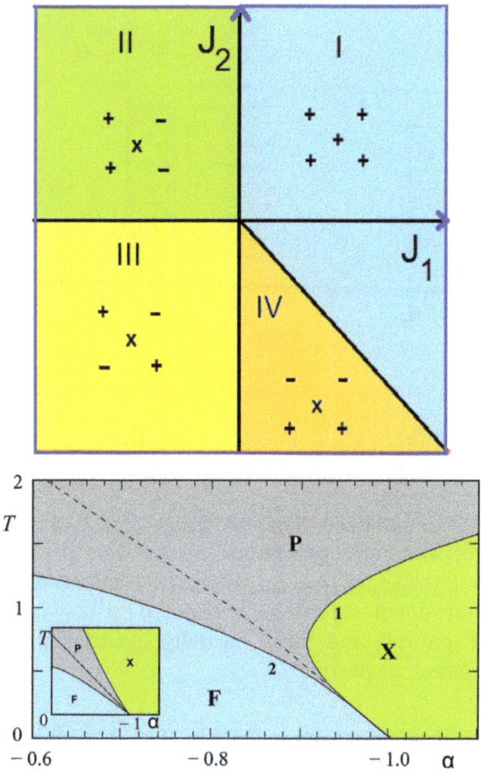

Figure 7.11 Top: Each color represents the ground-state configuration in the space (J_1, J_2) where $+$, $-$ and x denote up, down and free spins, respectively. Bottom: Phase diagram in the space $(\alpha = J_2/J_1, T)$ with $J_1 > 0$. T is in the unit of J_1/k_B. Solid lines are critical lines, dashed line is the disorder line. P, F and X stand for paramagnetic, ferromagnetic and partially disordered phases, respectively. The inset shows schematically the enlarged region near the critical value $J_2/J_1 = -1$.

of interaction parameters and several reentrant phases can occur for a given set of interactions when T varies.

The Hamiltonian reads

$$H = -J_1 \sum_{(ij)} \sigma_i \sigma_j - J_2 \sum_{(ij)} \sigma_i \sigma_j - J_3 \sum_{(ij)} \sigma_i \sigma_j \tag{7.17}$$

where σ_i is the Ising spin occupying the lattice site i, and the sums are performed over the spin pairs connected by J_1, J_2 and J_3, respectively.

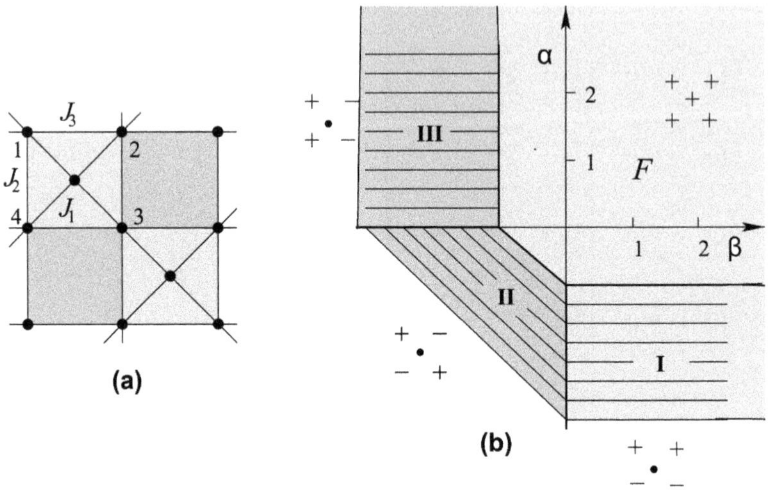

Figure 7.12 (a) Generalized Kagomé lattice: J_1, J_2 and J_3 denote the diagonal, vertical and horizontal bonds, respectively. (b) The ground-state phase diagram in the space $(\alpha = J_2/J_1, \beta = J_3/J_1)$. Each phase is displayed by a color with up, down and free spins denoted by +, − and o, respectively. I, II, III and F indicate the three partially disordered phases and the ferromagnetic phase, respectively.

The phase diagram at temperature $T = 0$ is shown in Fig. 7.12 in the space $(\alpha = J_2/J_1, \beta = J_3/J_1)$, supposing $J_1 > 0$. The spin configuration of each phase is indicated. The three partially disordered phases (I, II, and III) have free central spins. With $J_1 < 0$, it suffices to reverse the central spin in the F phase of Fig. 7.12. In addition, the permutation of J_2 and J_3 will not change the system, because it is equivalent to a $\pi/2$ rotation of the lattice.

We examine now the temperature effect. We have seen above that a partially disordered phase lies next to the ferromagnetic phase in the ground state gives rise to the reentrance phenomenon. We expect, therefore, similar phenomena near the phase boundary in the present model. As it turns out, we find below a new and richer behavior of the phase diagram.

We use the decimation of central spins described in Ref. [86], we get then a checkerboard Ising model with multispin interactions. This corresponds to a symmetric 16-vertex model which satisfies the free-fermion condition [123, 336, 366]. The critical temperature

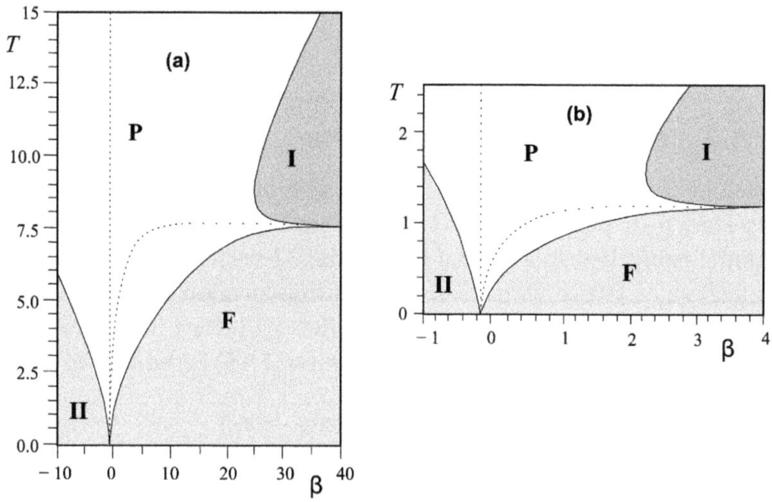

Figure 7.13 Phase diagram in the ($\beta = J_3/J_1, T$) space for negative $\alpha = J_2/J_1$. (a) $\alpha = -0.25$; (b) $\alpha = -0.8$. Partially disordered phases of type I and II and F are defined in Fig. 7.12. The disorder lines are shown by dotted lines.

is the solution of the following equation:

$$\cosh(4K_1)\exp(2K_2 + 2K_3) + \exp(-2K_2 - 2K_3)$$
$$= 2\cosh(2K_3 - 2K_2) \pm 4\cosh(2K_1) \quad (7.18)$$

Note the invariance of Eq. (7.18) with respect to changing $K_1 \to -K_1$ and interchanging K_2 and K_3. Let us show just the solution near the phase boundary in the plane ($\beta = J_3/J_1, T$) for two values of $\alpha = J_2/J_1$.

It is interesting to note that in the interval $0 > \alpha > -1$, there exist three critical lines. Two of them have a common horizontal asymptote as β tends to infinity. They limit a reentrant paramagnetic phase between the F phase and the partially disordered phase I for β between β_2 and infinite β (see Fig. 7.13). Such an infinite reentrance has never been found before in other models. With decreasing α, β_2 tends to zero and the F phase is reduced (comparing Figs. 7.13a and 7.13b). For $\alpha < -1$, the F phase and the reentrance no longer exist.

We note that for $-1 < \alpha < 0$, the model possesses two disorder lines (see equations in Ref. [67]) starting from a point near the phase boundary $\beta = -1$ for α close to zero; this point position moves to $\beta = 0$ as α tends to -1 (see Fig. 7.13).

7.4.2 Centered Honeycomb Lattice

We use the decimation of the central spin of each elementary cell as shown in paragraph 7.3.1. After the decimation, we obtain a model equivalent to a special case of the 32-vertex model [299] on a triangular lattice which satisfies the free-fermion condition. The general treatment has been given in Ref. [76]. Here we show the result of the case where $K_2 = K_3$. Equation (7.15) is reduced to

$$\exp(3K_1)\cosh(6K_2) + \exp(-3K_1) = 3[\exp(K_1) + \exp(-K_1)\cosh(2K_2)] \tag{7.19}$$

When $K_2 = 0$, Eq. (7.15) gives the critical line

$$\exp(3K_1)\cosh(2K_3) + \exp(-3K_1) = 3[\exp(K_1) + \exp(-K_1)\cosh(2K_3)] \tag{7.20}$$

When $K_3 = 0$, we observe a reentrant phase. The critical lines are given by

$$\cosh(4K_2) = \frac{\exp(4K_1) + 2\exp(2K_1) + 1}{[1 - \exp(4K_1)]\exp(2K_1)} \tag{7.21}$$

$$\cosh(4K_2) = \frac{3\exp(4K_1) + 2\exp(2K_1) - 1}{[\exp(4K_1) - 1]\exp(2K_1)} \tag{7.22}$$

The phase diagram obtained from Eqs. (7.21) and (7.22) near the phase boundary $\alpha = -0.5$ is displayed in Fig. 7.14. One observes here that the reentrant zone goes down to $T = 0$ at the boundary $\alpha = -0.5$ separating the GS phases II and III (see Fig. 7.14b).

Note that phase II has the antiferromagnetic ordering on the hexagon and the central spin free to flip, while phase III is the ordered phase where the central spin is parallel to 4 diagonal spins (see Fig. 2 of Ref. [76]). Therefore, if $-0.6 < \alpha < -0.5$ (reentrant region, see Fig. 7.14b), when one increases T from $T = 0$, ones goes across successively the ordered phase III, the narrow paramagnetic

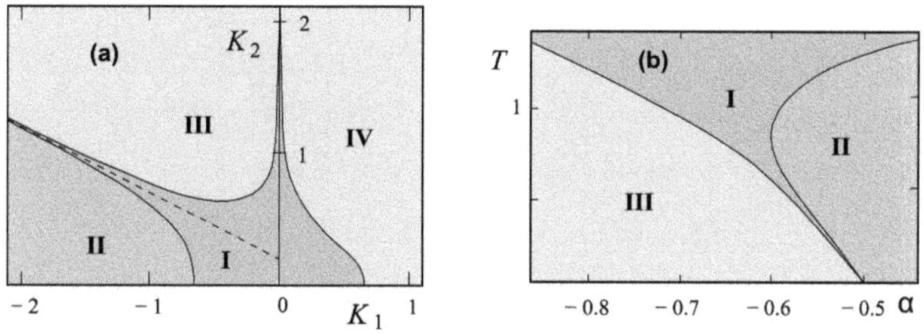

Figure 7.14 Centered honeycomb lattice: (a) Phase diagram in the space (K_1, K_2), discontinued line is the asymptote; (b) Reentrance in the space $(T, \alpha = K_2/K_1)$. I, II, III phases denote paramagnetic, partially disordered and ordered phases, respectively.

reentrant phase and the partially disordered phase II. Two remarks are in order: (i) The reentrant phase occurs here between an ordered phase and a partially disordered phase. However, as will be seen below, we discover in the three-center square lattice, reentrance can occur between two partially disordered phase; (ii) In any case, we find reentrance between phases when and only when there are free spins in the ground state. The entropy of the high-T partially disordered phase is higher than that of the low-T one. The second thermodynamic principle is not violated.

It is noted that the present honeycomb model does not possess a disorder solution with a reduction of dimension as the Kagomé lattice shown earlier.

7.4.3 Centered Square Lattices

In this paragraph, we study several centered square Ising models by mapping them onto 8-vertex models that satisfy the free-fermion condition. The exact solution is then obtained for each case. Let us anticipate that in some cases, for a given set of parameters, up to five transitions have been observed with varying temperature. In addition, there are two reentrant paramagnetic phases going to infinity in the space of interaction parameters, and there are two additional reentrant phases found, each in a small zone of the phase space [68, 77].

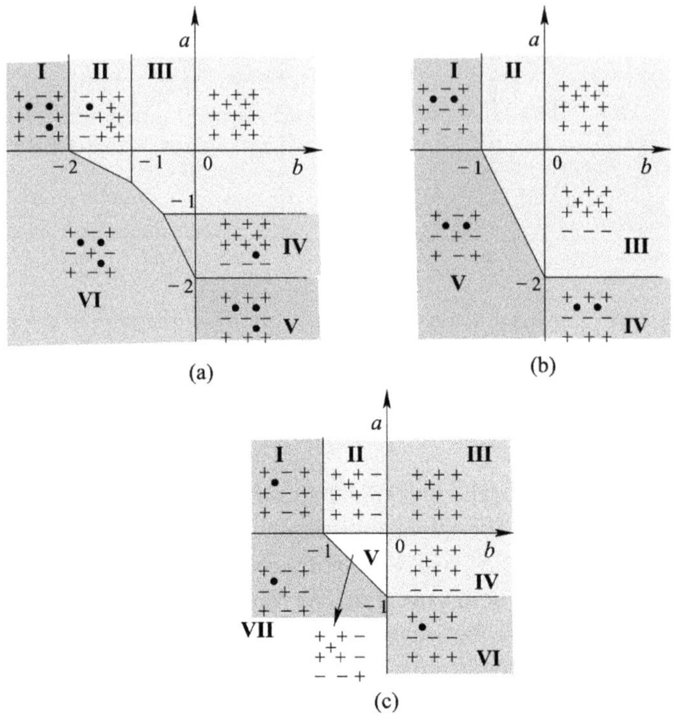

Figure 7.15 Ground-state phase diagram in the space ($a = J_2/J_1$, $b = J_3/J_1$) for (a) three-center square lattice; (b) two-adjacent center case; (c) and one-center case. Phase boundaries are indicated by heavy lines. Each phase is numbered and the spin configuration is displayed (+, −, and o are up, down, and free spins, respectively).

We consider the dilute centered square lattices shown in Fig. 7.6. The Hamiltonian of these models reads

$$H = -J_1 \sum_{(ij)} \sigma_i \sigma_j - J_2 \sum_{(ij)} \sigma_i \sigma_j - J_3 \sum_{(ij)} \sigma_i \sigma_j \tag{7.23}$$

where σ_i is an Ising spin at the lattice site i. The sums are performed over the spin pairs interacting by J_1, J_2 and J_3 bonds (diagonal, vertical and horizontal bonds, respectively).

Figure 7.15 shows the ground-state phase diagrams of the models displayed in Figs. 7.6a, 7.6b and 7.6d, where $a = J_2/J_1$ and $b = J_3/J_1$. The spin configurations in different phases are also displayed. The model in Fig. 7.15a has six phases (numbered from

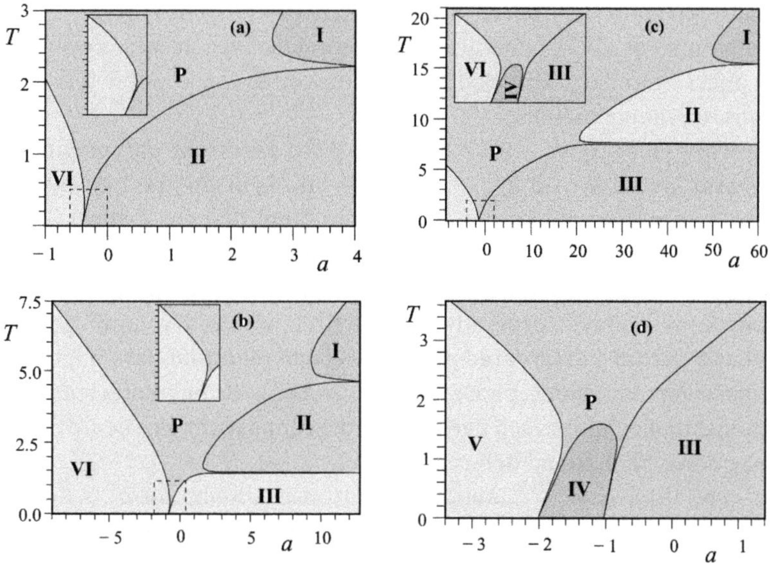

Figure 7.16 Three-center model: phase diagram in the space $(T, a = J_2/J_1)$ for several values of $b = J_3/J_1$: (a) $b = -1.25$, (b) $b = -0.75$, (c) $b = -0.25$, (d) $b = 0.75$. Reentrant regions indicated by discontinued lines are enlarged in the insets. A number indicates the corresponding spin configuration shown in Fig. 7.15a. P is paramagnetic phase.

I to VI), five of which (I, II, IV, V and VI) are partially disordered (at least one centered spin being free), the model in Fig. 7.15b has five phases, three of which (I, IV, and V) are partially disordered, and the model in Fig. 7.15c has seven phases with three partially disordered ones (I, VI and VII).

It is interesting to note that each model shown in Fig. 7.6 possesses the reentrance along most of the phase boundary lines when the temperature is turned on. This striking feature of the centered square Ising lattices has not been observed in other known models.

Let us show in Fig. 7.16 the results of the three-center model of Fig. 7.6a, in the space ($a = J_2/J_1$, T) for typical values of $b = J_3/J_1$.

For $b < -1$, there are two reentrances as seen in Fig. 7.16a for $b = -1.25$. The phase diagram is shown using the same numbers of corresponding ground state configurations of Fig. 7.15. Note that the centered spins disordered at $T = 0$ in phases I, II and VI (Fig. 7.15a) remain so at all T. Note also that the reentrance occurs always at a phase boundary. This point is emphasized in this chapter through various shown models.

For $-1 < b < -0.5$, there are three reentrant paramagnetic phases as shown in Fig. 7.16b, two of them on the positive a are so narrow while a goes to infinity. Note that the critical lines in these regions have horizontal asymptotes. For a large value of a, one has five transitions with decreasing T: paramagnetic phase–partially disordered phase I–first reentrant paramagnetic phase–partially disordered phase II–second reentrant paramagnetic phase–ferromagnetic phase (see Fig. 7.16b). To our knowledge, a model that exhibits such five phase transitions with two reentrances has never been found before.

For $-0.5 < b \leq 0$, another reentrance is found for $a < -1$ as seen in the inset of Fig. 7.16c. With increasing b, the ferromagnetic phase III in the phase diagram becomes large, reducing phases I and II. At $b = 0$, only the ferromagnetic phase remains.

For positive b, we have two reentrances for $a < 0$, ending at $a = -2$ and $a = -1$ when $T = 0$ as seen in Fig. 7.16d.

In conclusion, we summarize that in the three-center square lattice model shown in Fig. 7.6a, we found two reentrant phases occurring on the temperature scale at a given set of interaction parameters. A new feature found here is that a reentrant phase can occur between two partially disordered phases, unlike in other models such as the Kagomé Ising lattice where a reentrant phase occurs between an ordered phase and a partially disordered phase.

7.5 Other Exactly Solved Models

There is a number of papers dealing with exactly solved frustrated models published by J. Strečka and coworkers since 2006. These models are essentially decorated Ising models in one or two dimensions.

In Ref. [331] ground-state and finite-temperature properties of the mixed spin-1/2 and spin-S Ising–Heisenberg diamond chains are examined within an exact analytical approach based on the generalized decoration-iteration map. A particular emphasis is laid on the investigation of the effect of geometric frustration, which is generated by the competition between Heisenberg- and Ising-type exchange interactions. It is found that an interplay between the geometric frustration and quantum effects gives rise to several quantum ground states with entangled spin states in addition to some semiclassically ordered ones. Among the most interesting results to emerge from our study, one could mention a rigorous evidence for quantized plateaux in magnetization curves, an appearance of the round minimum in the thermal dependence of susceptibility times temperature data, double-peak zero-field specific heat curves, or an enhanced magneto-caloric effect when the frustration comes into play. The triple-peak specific heat curve is also detected when applying small external field to the system driven by the frustration into the disordered state.

In Ref. [332], the geometric frustration of the spin-1/2 Ising–Heisenberg model on the triangulated kagomé "triangles-in-triangles" lattice is investigated within the framework of an exact analytical method based on the generalized star-triangle mapping transformation. Ground-state and finite-temperature phase diagrams are obtained along with other exact results for the partition function, Helmholtz free energy, internal energy, entropy, and specific heat, by establishing a precise mapping relationship to the corresponding spin-1/2 Ising model on the kagomé lattice. It is shown that the residual entropy of the disordered spin liquid phase for the quantum Ising–Heisenberg model is significantly lower than for its semiclassical Ising limit ($S_0/NTk_B = 0.2806$ and 0.4752, respectively), which implies that quantum fluctuations partially lift a macroscopic degeneracy of the ground-state manifold in the frustrated regime.

In Ref. [333], spin-1/2 Ising model with a spin–phonon coupling on decorated planar lattices partially amenable to lattice vibrations is examined using the decoration-iteration transformation and harmonic approximation. It is shown that the magneto-elastic coupling gives rise to an effective antiferromagnetic next-nearest-neighbor interaction, which competes with the nearest-neighbor interaction and is responsible for a frustration of decorating spins. A strong enough spin–phonon coupling consequently leads to an appearance of striking partially ordered and partially disordered phase, where a perfect antiferromagnetic alignment of nodal spins is accompanied with a complete disorder of decorating spins.

In the above works, the decorations are local couplings of an Ising spin to decorated Heisenberg spins or to phonons which can be summed up to renormalize the interactions between Ising spins. This leaves the systems solvable by star-triangle transformations, vertex models or other methods. Their results are interesting. We find again in these models striking features shown above in the present chapter for 2D solvable frustrated Ising models. In particular, partially disordered systems, multiple phase transitions and the reentrant phase due to the frustration are shown to exist in the phase diagrams. For details, the reader is referred to Refs. [331–333].

7.6 Evidence of Partial Disorder and Reentrance in Non-Solvable Frustrated Systems

The previous sections show interesting phenomena due to the frustration. What to be retained is the fact that those phenomena occur around the boundary of two phases of different ground states, namely different symmetries. These phenomena include the following:

(1) The partial disorder at equilibrium: disorder is not equally shared on all particles as usually the case in non-frustrated systems,
(2) The reentrance: this occurs around the phase boundary when T increases $\rightarrow$ the phase with larger entropy will win at finite T. In other words, this is a kind of selection by entropy.
(3) The disorder line: this line occurs in the paramagnetic phase. It separates the pre-ordering zones between two nearby ordered phases.

The partial disorder and the reentrance which occur in exactly solved Ising systems shown above are expected to occur also in models other than the Ising one as well as in some three-dimensional systems. Unfortunately, these systems cannot be exactly solved. One has to use approximations or numerical simulations to study them. This renders difficult the interpretation of the results. Nevertheless, in the light of what has been found in exactly solved systems, we can introduce the necessary ingredients into the model under study if we expect the same phenomenon to occur.

As seen above, the most important ingredient for a partial disorder and a reentrance to occur at low T in the Ising model is the existence of a number of free spins in the ground state.

In three dimensions, apart from a particular exactly solved case [155] showing a reentrance, a few Ising systems such as the fully frustrated simple cubic lattice [30, 81], a stacked triangular Ising antiferromagnet [31, 244] and a body-centered cubic (bcc) crystal [17] exhibit a partially disordered phase in the ground state. We believe that reentrance should also exist in the phase space of such systems though evidence is found numerically only for the bcc case [17].

In two dimensions, a few non-Ising models show also evidence of a reentrance. For the q-state Potts model, evidence of a reentrance is found in a recent study of the two-dimensional frustrated Villain lattice (the so-called piled-up domino model) by a numerical transfer matrix calculation [119, 120]. It is noted that the reentrance occurs near the fully frustrated situation, i.e. $\alpha_c = J_{AF}/J_F = -1$ (equal antiferromagnetic and ferromagnetic bond strengths), for q between $\simeq$ 1.0 and $\simeq$ 4. Note that there is no reentrance in the case $q = 2$. Below (above) this q value, the reentrance occurs above (below) the fully frustrated point α_c as shown in Figs. 7.17 [120] and 7.18 [120]. For q larger than $\simeq$ 4, the reentrance disappears [120].

A frustrated checkerboard lattice with XY spins shows also evidence of a paramagnetic reentrance [45].

In vector spin models such as the Heisenberg and XY models, the frustration is shared by all bonds so that no free spins exist in the ground state. However, one can argue that if there are several kinds of local field in the ground state due to several kinds of interaction, then there is a possibility that a subsystem with weak local field is disordered at low T while those of stronger local field stay ordered up to higher temperatures. This conjecture has been verified in a number of recent works on classical [307] and quantum spins

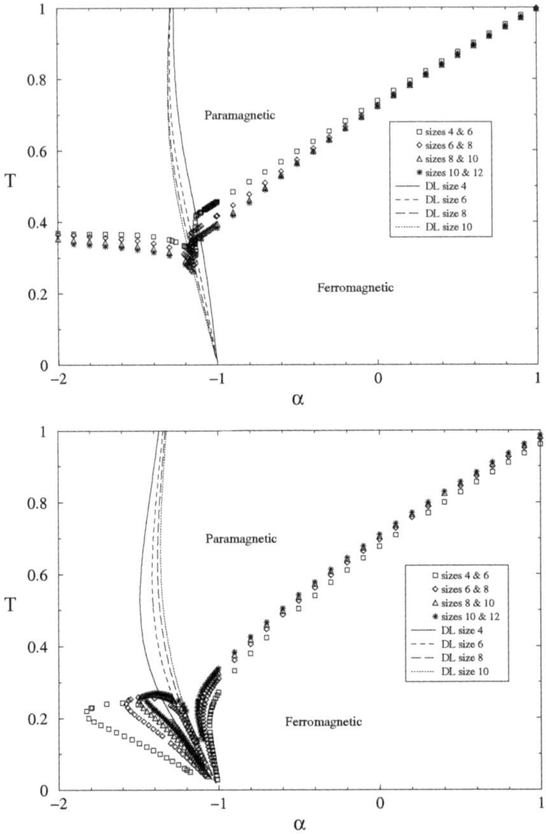

Figure 7.17 Phase diagram for the Potts piled-up-domino model with $q = 3$: periodic boundary conditions (top) and free boundary conditions (bottom). The disorder lines are shown as lines, and the phase boundaries as symbols. The numerical uncertainty is smaller than the size of the symbols.

[285, 305]. Consider, for example, Heisenberg spins $\mathbf{S_i}$ on a bcc lattice with a unit cell shown in Fig. 7.19 [307].

For convenience, let us call sublattice 1 the sublattice containing the sites at the cube centers and sublattice 2 the other sublattice. The Hamiltonian reads

$$H = -\frac{1}{2}\sum_{<i,j>_1} \mathbf{S_i}.\mathbf{S_j} - \frac{1}{2}\sum_{<i,j>_2} \mathbf{S_i}.\mathbf{S_j}$$

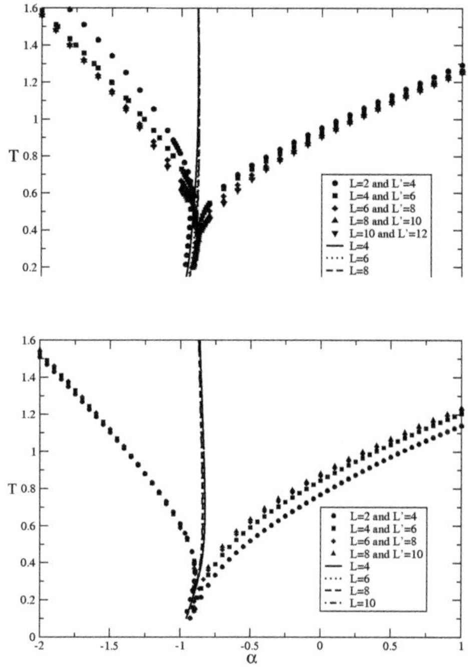

Figure 7.18 The phase diagram for $q = 1.5$ found using transfer matrices and the phenomenological renormalization group with periodic (top) and free boundary conditions (bottom). The points correspond to finite-size estimates for T_c, whilst the lines correspond to the estimates for the disorder line, ($\alpha = J_2/J_1$).

where $\sum_{<i,j>_1}$ indicates the sum over the NN spin pairs with exchange coupling J_1, while $\sum_{<i,j>_2}$ is limited to the NNN spin pairs belonging to sublattice 2 with exchange coupling J_2. It is easy to see that when J_2 is antiferromagnetic the spin configuration is non-collinear for $J_2/|J_1| < -2/3$. In the non-collinear case, one can verify that the local field acting a center spin (sublattice 1) is weaker in magnitude than that acting on a corner spin (sublattice 2). The partial disorder is observed in Fig. 7.20: The sublattice of black spins (sublattice 1) is disordered at a low T.

The same argument is applied for quantum spins [285]. Consider the bcc crystal as shown in Fig. 7.19, but the sublattices are supposed now to have different spin magnitudes, for example, $S_A = 1/2$

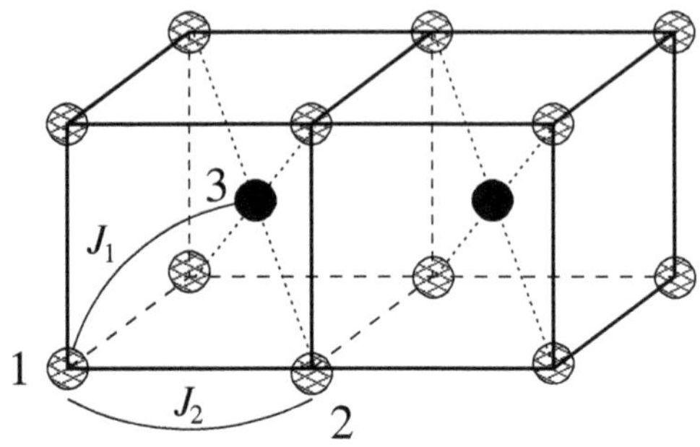

Figure 7.19 bcc lattice. Spins are shown by gray and black circles. Interaction between NN (spins numbered 1 and 3) is denoted by J_1 and that between NNN (spins 1 and 2) by J_2. Note that there is no interaction between black spins.

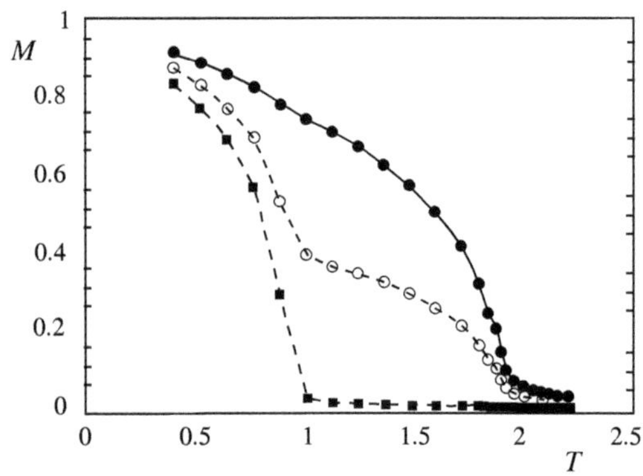

Figure 7.20 Monte Carlo results for sublattice magnetizations vs T in the case $J_1 = -1, J_2 = -1.4$: Black squares and black circles are for sublattices 1 and 2, respectively. Void circles indicate the total magnetization.

(sublattice 1) and $S_B = 1$ (sublattice 2). In addition, one can include NNN interactions in both sublattices, namely J_{2A} and J_{2B}. The Green's function technique is then applied for this quantum system [285]. We show in Fig. 7.21 the partial disorder observed in two cases: Sublattice 1 is disordered (Fig. 7.21a) or sublattice 2 is disordered (Fig. 7.21b). In each case, one can verify, using the corresponding parameters, that in the ground state the spin of the

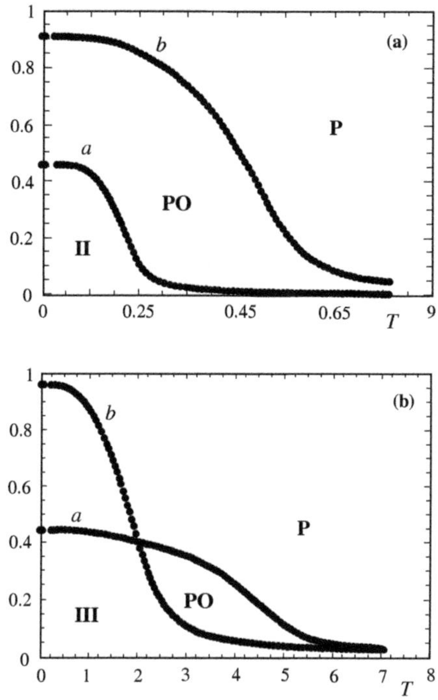

Figure 7.21 Sublattice magnetizations vs T in the case $S_A = 1/2, S_B = 1$: (a) Curve a (b) is the sublattice-1(2) magnetization in the case $J_{2A}/|J_1| = 0.2$ and $J_{2B}/|J_1| = 0.9$ (b) curve a (b) is the sublattice-1(2) magnetization in the case $J_{2A}/|J_1| = 2.2$ and $J_{2B}/|J_1| = 0.1$. P is the paramagnetic phase, PO the partial order phase (only one sublattice is ordered), II and III are non collinear spin configuration phases. See text for comments.

disordered sublattice has an energy lower than a spin in the other sublattice.

We show in Fig. 7.22 the specific heat versus T for the parameters used in Fig. 7.21. One observes the two peaks corresponding to the two phase transitions associated with the loss of sublattice magnetizations.

The necessary condition for the occurrence of a partial disorder at finite T is thus the existence of several kinds of site with different energies in the ground state. This has been so far verified in a number of systems as shown above.

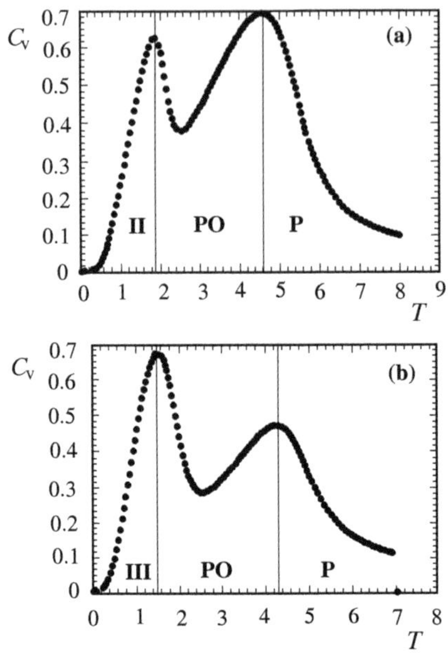

Figure 7.22 Specific heat versus T of the parameters used in Fig. 7.21, $S_A = 1/2$, $S_B = 1$: (a) $J_{2A}/|J_1| = 0.2$ and $J_{2B}/|J_1| = 0.9$ (b) $J_{2A}/|J_1| = 2.2$ and $J_{2B}/|J_1| = 0.1$. See the caption of Fig. 7.21 for the meaning of P, PO, II and III. See text for comments.

7.7 Re-Orientation Transition in Molecular Thin Films: Potts Model with Dipolar Interaction

We show in below the case of a 2D system and a thin film where there is the competition between a dipolar interaction of strength D which makes the spins lie in the plane and a surface perpendicular anisotropy A which favors the perpendicular spin configuration [153]. The dipolar Hamiltonian is written as

$$\mathcal{H}_d = D \sum_{(i,j)} \left\{ \frac{\mathbf{S}(\sigma_i) \cdot \mathbf{S}(\sigma_j)}{r_{i,j}^3} - 3 \frac{[\mathbf{S}(\sigma_i) \cdot \mathbf{r}_{i,j}][\mathbf{S}(\sigma_j) \cdot \mathbf{r}_{i,j}]}{r_{i,j}^5} \right\} \tag{7.24}$$

where $\mathbf{r}_{i,j}$ is the vector of modulus $r_{i,j}$ connecting the site i to the site j. One has $\mathbf{r}_{i,j} \equiv \mathbf{r}_j - \mathbf{r}_i$. In Eq. (7.24), D is a positive constant depending on the material, the sum $\sum_{(i,j)}$ is limited at pairs of spins within a cut-off distance r_c, and $\mathbf{S}(\sigma_i)$ is given by the following three-component pseudo vector representing the spin state:

$$\mathbf{S}(\sigma_i) = (s_x(i), 0, 0) \text{ if } \sigma_i = 1 \tag{7.25}$$

$$\mathbf{S}(\sigma_i) = (0, s_y(i), 0) \text{ if } \sigma_i = 2 \tag{7.26}$$

$$\mathbf{S}(\sigma_i) = (0, 0, s_z(i)) \text{ if } \sigma_i = 3 \tag{7.27}$$

where s_α ($\alpha = x, y, z$) is the α component with values ± 1. The perpendicular anisotropy is introduced by the following term:

$$\mathcal{H}_a = -A \sum_i s_z(i)^2 \tag{7.28}$$

where A is a constant. Note that the dipolar interaction as applied in our Potts model is not similar to that used in the vector spin model where $\mathbf{S}(\sigma_i)$ is a true vector. In our model, each spin can only choose to lie on one of three axes, pointing in positive or negative direction. We use $J = 1$ as the unit of energy. The temperature T is expressed in the unit of J/k_B where k_B is the Boltzmann constant.

7.7.1 Two-Dimensional Case

In the case of 2D, for a given A, the steepest-descent method gives the "critical value" D_c of D above (below) which the GS is the in-plane (perpendicular) configuration. D_c depends on r_c. Let us take $A = 0.5$ and make vary D and r_c. The GS numerically obtained is shown in Ref. [153] for several sets of (D, r_c). For instance, when $r_c = \sqrt{6} \simeq 2.449$, we have $D_c = 0.100$.

The phase diagrams obtained from the histogram Monte Carlo technique (see Chapter 6) in the 2D case are shown in Fig. 7.23.

It is interesting to compare the present system using the 3-state Potts model with the same system using the Heisenberg spins [306]. In that work, the re-orientation transition line is also of first order but it tilts on the left of D_c, namely the re-orientation transition of type I occurring in a small region *below* D_c, unlike the re-orientation of type II found *above* D_c for the present Potts model. To explain the "left tilting" of the Heisenberg case, we have used the following entropy argument: The Heisenberg in-plane configuration has a spin wave entropy larger than that of the perpendicular configuration

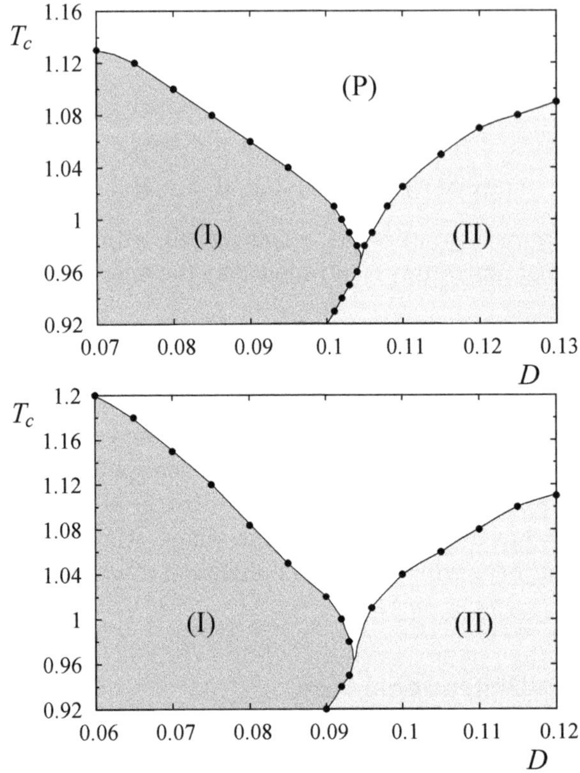

Figure 7.23 (Color online) Phase diagram in 2D: Transition temperature T_C versus D, with $A = 0.5$, $J = 1$, $r_c = \sqrt{6}$ (top) and $r_c = 4$ (bottom). Phase (I) is the perpendicular spin configuration, phase (II) the in-plane spin configuration and phase (P) the paramagnetic phase. See text for comments.

at finite T, so the re-orientation occurs in "favor" of the in-plane configuration, it goes from perpendicular to in-plane ordering with increasing T. Obviously, this argument for the Heisenberg case does not apply to the Potts model because we have here the inverse re-orientation transition. We think that, due to the discrete nature of the Potts spins, spin waves cannot be excited, so there is no spin wave entropy as in the Heisenberg case. The perpendicular anisotropy A is thus dominant at finite T for D slightly larger than D_c.

Experimentally, the "right tilting" of the re-orientation line from D_c in Fig. 7.23 has been observed in Fe/Gd [14]. So we emphasize that re-orientation depends on the system, it can be of type I or type II. Our simple model shows this possibility. Of course, to compare quantitatively our results with experimental data of Fe/Gd we need to take into account details of the real system such as Fe and Gd lattice structures and magnetic interactions.

There is another important point which is worth to mention: the scenario of the re-orientation transition. Arnold et al. [14] have suggested a two-step transition with an intermediate phase. Theoretically this scenario is possible: We have found in several exactly solved models [67, 86] that the transition between two ordered phases can be assisted by a very small disordered phase between them which is called "reentrant phase." In this extremely small region, one can have a so-called "disorder line" with dimension reduction to help the system go from one symmetry to another one. Numerically and experimentally the reentrance region is hardly detected due to its smallness. If the size of the reentrance region is large enough, we can have two second-order lines departing from D_c (see discussion on several exactly solved models in Ref. [86]). If the size is so small, these two lines look like a single line within numerical and/or experimental resolution, then the re-orientation on the "single" line has a first-order nature which is necessary to allow a transition between two different symmetries [380]. We think that this latter hypothesis corresponds to our finding of the first-order re-orientation shown in Fig. 7.24.

7.7.2 Thin Films

The case of thin films with a thickness L_z where L_z goes from a few to a dozen atomic layers has a very similar re-orientation transition as that shown above for the 2D case. Changing the film thickness results in changing the dipolar energy at each lattice site. Therefore, the critical value D_c will change accordingly. We note the periodic layered structures at large D and r_c for both cases. In the case $L_z = 4$, for $r_c = \sqrt{6}$ the critical value D_c above which the GS changes from the perpendicular to the in-plane configuration is $D_c = 0.305$.

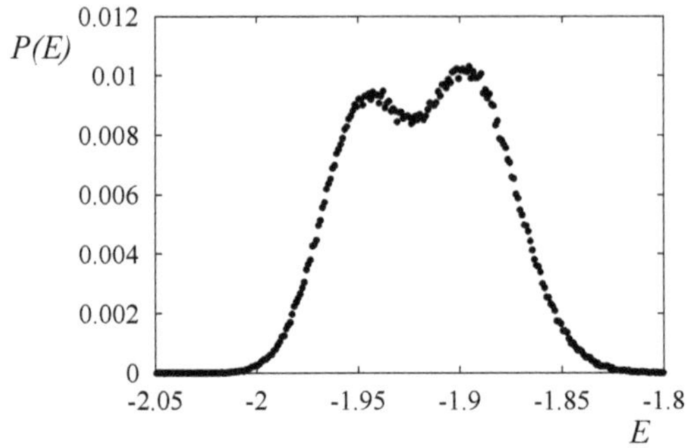

Figure 7.24 Energy histogram P versus energy E at the re-orientation transition temperature $T = 0.930$, for $D = 0.101$, $A = 0.5$, $J = 1$, $r_c = \sqrt{6}$.

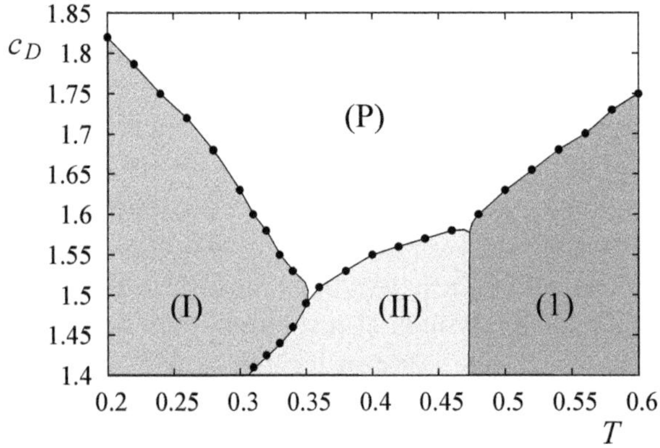

Figure 7.25 (Color online) Phase diagram in thin film of 4-layer thickness: Transition temperature T_C versus D, with $A = 0.5$, $J = 1$ and $L = 24$. The number (I) stands for the perpendicular configuration, the number (II) for the in-plane configuration (spins pointing along x or y axis), the number (1) for alternately one layer in x and one layer in y direction (periodic single-layered structure), P is paramagnetic phase. See text for comments.

The whole phase diagram is shown in Fig. 7.25. Note that the line separating the uniform in-plane phase (II) and the periodic single-layered phase (1) is vertical.

Again here, the line separating the perpendicular configuration (I) and the in-plane one (II) is a first-order line. However, we would like to emphasize that the same remark as that given above for the

2D case: We cannot exclude the possibility of a very small reentrance phase between two phases (I) and (II).

7.7.3 Effect of Surface Exchange Interaction

We have calculated the effect of J_s by taking its values far from the bulk value ($J = 1$) for several values of D. In general, when J_s is smaller than J the surface spins become disordered at a temperature T below the temperature where the interior layers become disordered. This case corresponds to the soft surface (or magnetically "dead" surface layer) [75]. On the other hand, when $J_s > J$, we have the inverse situation: The interior spins become disordered at a temperature lower that of the surface disordering. We have here the case of a magnetically hard surface. We show in Fig. 7.26 an example of a hard surface in the case where $J_s = 3$ for $D = 0.6$ with $L_z = 4$. The same feature is observed for $D = 0.4$. Note that the surface and bulk transitions are seen by the respective peaks in the specific heat and the susceptibility. In the re-orientation

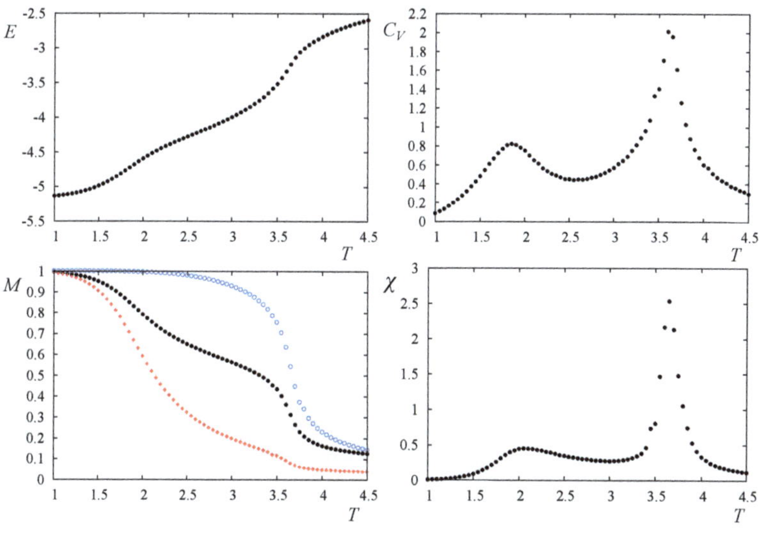

Figure 7.26 E, C_V, M and χ of a 4-layer film versus T for D =0.6 with $J_s = 3$. The surface magnetization is shown by blue void circles, the bulk magnetization by red diamonds and the total curves by black solid circles.

region, the situation is very complicated as expected because the surface transition occurs in the re-orientation zone.

Let us discuss about experimental data of Fe/Gd. It has been found that the overall transition to the paramagnetic phase in the experiment on Fe/Gd, driven by the Gd substrate [14], is a first-order transition which occurs at room temperature. To simulate the Fe/Gd system, one can proceed by changing J_s and the nature of the surface spins to represent Fe atoms, one then can perform simulations in order to compare quantitatively with the experiment.

7.8 Conclusion

In this review, we have shown a number of studied cases on the frustration effects in two dimensions. We have discussed some properties of periodically frustrated Ising systems and limited the discussion to exactly solved models which possess at least a reentrant phase. Other frustrated Ising systems which used approximations are discussed in the chapter by Nagai et al. in Ref. [85] and in the book by Liebmann [212].

The main purpose of the review is to show some frustrated magnetic systems which present a number of common interesting features. These features are discovered by solving exactly some 2D Ising frustrated models. They occur near the frontier of two competing phases of different ground-state orderings. Without frustration, such frontiers do not exist. Among the striking features, one can mention the "partial disorder," namely a number of spins stay disordered in coexistence with ordered spins at equilibrium, the "reentrance," namely a paramagnetic phase exists between two ordered phases in a small region of temperature, and "disorder lines," namely lines on which the system loses one dimension to allow for a symmetry change from one side to the other. Such beautiful phenomena can only be uncovered and understood by means of exact mathematical solutions.

Let us emphasize that simple models having no bond disorder like those presented in this chapter can possess complicated phase diagrams due to the frustration generated by competing interactions. Many interesting physical phenomena such as successive phase transitions, disorder lines, and reentrance are found. In particular, a reentrant phase can occur in an infinite region of parameters. For a given set of interaction parameters in this region, successive phase transitions take place on the temperature scale, with one or two paramagnetic reentrant phases.

The relevance of disorder solutions for the reentrance phenomena has also been pointed out. An interesting finding is the occurrence of two disorder lines which divide the paramagnetic phase into regions of different kinds of fluctuations (see Section 7.4.1). Therefore, care should be taken in analyzing experimental data such as correlation functions, susceptibility, etc. in the paramagnetic phase of frustrated systems.

Although the reentrance is found in the models shown above by exact calculations, there is no theoretical explanation why such a phase can occur. In other words, what is the necessary and sufficient condition for the occurrence of a reentrance? We have conjectured [16, 68, 77] that the necessary condition for a reentrance to take place is the existence of at least a partially disordered phase next to an ordered phase or another partially disordered phase in the ground state. The partial disorder is due to the competition between different interactions.

The existence of a partial disorder yields the occurrence of a reentrance in most of known cases [16, 62, 68, 76, 77, 239, 347], except in some particular regions of interaction parameters in the centered honeycomb lattice (Section 7.4.2): The partial disorder alone is not sufficient to make a reentrance, the finite zero-point entropy due to the partial disorder of the three ground states is the same, i.e., $S_0 = \log(2)/3$ per spin [76], but only one case yields a reentrance. Therefore, the existence of a partial disorder is a necessary, but not sufficient, condition for the occurrence of a reentrance.

The anisotropic character of the interactions can also favor the occurrence of the reentrance. For example, the reentrant region is enlarged by anisotropic interactions as in the centered square lattice [62], and becomes infinite in the generalized Kagomé model (Section 7.4.1). But again, this alone cannot cause a reentrance as seen in the centered honeycomb case [76]: Only in one case a reentrance does occur. The presence of a reentrance may also require a coordination number at a disordered site large enough but below a limit to influence the neighboring ordered sites.

Finally, let us emphasize that when a phase transition occurs between states of different symmetries which have no special group-subgroup relation, it is generally accepted that the transition is of first order. However, the reentrance phenomenon is a symmetry breaking alternative which allows one ordered phase to change into another incompatible ordered phase by going through an intermediate reentrant phase. A question which naturally arises is, under which circumstances does a system prefer an intermediate reentrant phase to a first-order transition? In order to analyze this aspect we have generalized the centered square lattice Ising model into three dimensions [17]. This is a special bcc lattice. We have found that at low T the reentrant region observed in the centered square lattice shrinks into a first-order transition line which is ended at a multi-critical point from which two second order lines emerge forming a narrow reentrant region [17].

Let us mention that although the exactly solved systems shown in this chapter are models in statistical physics, we believe that the results obtained in this work have qualitative bearing on real frustrated magnetic systems. In view of the simplicity of these models, we believe that the results found here will have several applications in various areas of physics.

We have also studied frustrated magnetic systems close to the 2D solvable systems, namely thin magnetic thin films with Ising or Heisenberg spin models that are not exactly solvable. Guided by the insights of exactly solvable systems, we have introduced ingredients in the Hamiltonian to find some striking phenomena mentioned above: We have seen in thin films partial disorder (surface disorder coexisting with bulk order), reentrance at phase boundaries in face-centered cubic antiferromagnetic films. Thin films have their own interest such as surface spin rearrangement (helimagnetic films) and surface effects on their thermodynamic properties. Those systems will be presented in Chapters 9 and 10.

To conclude, we would like to say that investigations on the subjects discussed above continue intensively today. Note that there is an enormous number of investigations of other researchers on the above subjects and on other subjects concerning frustrated magnetic thin films. We have mentioned these works in our original papers, but we did not present them here. Also, for the same reason, we have cited only a limited number of experiments and applications in this review.

Chapter 8

Spin Wave Theory for Thin Films

We have introduced in Chapters 3 and 4 basic methods to study general bulk properties of spin waves excited in systems of interacting spins. We have taken advantage of the periodic crystalline structure to use the Fourier transforms and the periodic boundary conditions which allowed us to simplify calculations via the sum rules and the crystal symmetry.

When the invariance by translation is broken because of the presence of impurities, defects or surfaces, the calculation becomes more complicated. Often we have to modify methods established for the bulk and to introduce new techniques. Physically, the loss of the spatial periodicity causes a change in bulk properties of materials. The physics of disorder has been and still is a major research domain for more than 40 years. We can mention some well-studied disordered systems such as spin glasses, amorphous compounds and doped semiconductors. There is another domain which has been intensively developed in the past decades: the physics of nanomaterials. Nanomaterials such as ultrafine particles, ultrathin films and nanoribbons do not have periodic structures due to their nanometric dimension. These tiny objects have been used in many industrial applications which rapidly change our

Physics of Magnetic Thin Films: Theory and Simulation
Hung T. Diep

ISBN 978-981-4877-42-8 (Hardcover), 978-1-003-12110-7 (eBook)
www.jennystanford.com

daily life: high-performance computers, smart phones, high-speed internet, . . .

In this chapter, we present a few standard methods to study a family of simple magnetic systems where the loss of translational invariance is caused by the existence of surfaces or interfaces such as semi-infinite crystals, thin films, or multilayers. More complicated systems such as systems of non-collinear spin configurations are presented in the following chapters.

Surface effects are very important because they modify drastically properties of a bulk material when the so-called aspect ratio, i.e., the ratio of the number of surface atoms to the total number of atoms, becomes important as it is the case in small particles or in films of a few atomic layers. Surface physics has been intensively developed during the last 30 years. Among the main reasons for that rapid and successful development we can mention the interest in understanding the physics of low-dimensional systems and an immense potential of industrial applications of thin films [36, 72, 374]. In particular, theoretically it has been shown that systems of continuous spins (XY and Heisenberg) in two dimensions (2D) with a short-range interaction cannot have a long-range order at finite temperatures [231]. In the case of thin films, it has been shown that low-lying localized spin waves can be found at the film surface [281] and effects of these localized modes on the surface magnetization at finite temperatures (T) and on the critical temperature have been investigated by the Green's function technique [75, 78]. Experimentally, objects of nanometric size such as ultrathin films and nanoparticles have also been intensively studied because of numerous and important applications in industry. An example is the so-called giant magnetoresistance used in data storage devices, magnetic sensors, etc. [18, 22, 134, 345]. Recently, much interest has been attracted towards practical problems such as spin transport, spin valves and spin-transfer torques, due to numerous applications in spintronics. A family of topological magnetic structures called "skyrmions" are currently under intensive investigations. They have already important applications in various domains (see Chapters 14 and 15).

8.1 Surface Effects

All crystals terminate in space by a surface. It can have various geometries and situations. The simplest one is a clean surface which is a perfect crystalline atomic plane such as (100) and (111) planes. It can be spherical as in the case of spherical aggregates. In general, the surface can include impurities, dislocations, vacancies, islands, steps, ... A surface can also chemisorb or physisorb alien atoms. Chemisorbed atoms form a strong chemical binding with surface atoms while physisorbed atoms are physically bound to surface atoms by weak potentials such as the long-range van der Waals interaction. They are sometimes called "adatoms." It is obvious that the more the surface is disordered, the more it is difficult to study. Spectacular effects are often observed with well controlled and well characterized surfaces. Today, sophisticated techniques allow to create a surface with desired characteristics.

At the surface, atoms do not have the same environment as those inside the crystal. Due to the lack of neighbors and various neighboring defects and geometries, electronic states of surface atoms are modified in one way or in another, giving rise to changes in their effective interactions with neighboring atoms. The density of states shows often surface states which modify the filling of electronic bands, the position of the Fermi level and the magnetic moment of surface atoms. In addition, the surface anisotropy and surface exchange interaction can be very different from those of the interior atoms. We mention here some remarkable observations:

(i) For a thin film, the dipolar interaction favors an in-plane spin configuration. However, when the film thickness becomes very small a perpendicular anisotropy comes into play to favor a spin configuration perpendicular to the film surface at low temperatures [36]. In addition, the sign and amplitude of the surface exchange may also change with decreasing thickness. These modifications can cause interesting surface behaviors such as magnetic ordering reconstructions near the surface and localized surface spin waves. Also, the competition between the dipolar interaction and the perpendicular surface anisotropy can give rise to the so-called "re-orientation transition" which turns the perpendicular configuration into the in-plane one at a finite temperature.

(ii) Perturbations in the cohesive interaction which binds surface atoms can yield a modification of the lattice constant at the surface (contraction or dilatation) and even a reconstruction of surface geometry.

(iii) Perturbations in electronic energy bands can give rise to anomalies in electronic (charge and/or spin) transport observed in multilayers and near the surface.

It is not the purpose of this chapter to give experimental data on problems raised above. There exist a great number of handbooks and reviews [35, 36, 72, 73, 374] for that purpose. In this chapter, we present some fundamental aspects which are well understood at present. Our purpose is to provide a theoretical framework to understand microscopic mechanisms which lead to macroscopic surface effects such as low surface magnetization, low transition temperature, surface phase transition and surface spin-configuration instability. We will concentrate our attention to simple methods which allow us to study properties of magnetic semi-infinite crystals and thin films.

8.2 Surface Effects in Magnetism

We consider a ferromagnetic thin film of N_T atomic layers. The surface is denoted by index $n = 1$ and the last layer by $n = N_T$. We suppose the **Oz** axis is perpendicular to the film surface.

8.2.1 Surface Magnons

We have studied the bulk spin waves in Chapter 3. The amplitude of a bulk spin wave mode does not vary in space. In general, near magnetic perturbation sources such as magnetic impurities and surfaces, spin waves can be spatially localized. Such modes are called "surface spin wave modes" or "surface-localized modes." The amplitude of a surface-localized mode decays when it propagates

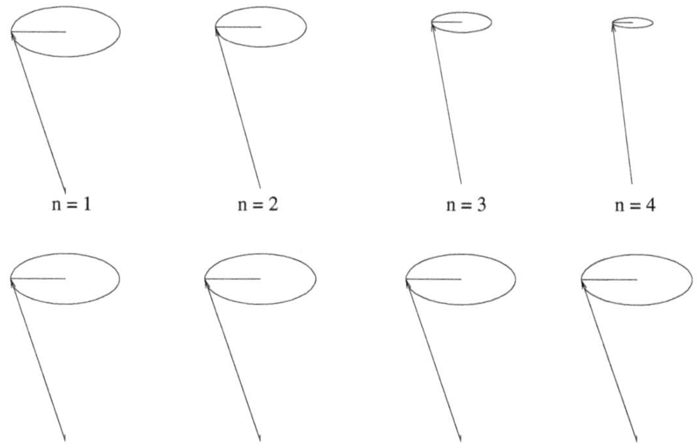

Figure 8.1 A surface spin wave mode (top) and a bulk mode (bottom).

from the surface into the bulk. We show in Fig. 8.1 a surface mode and a bulk mode, for comparison.

We express the amplitude of a surface mode by $U_n(\mathbf{k}) = Ae^{i\mathbf{k}\cdot\mathbf{r}_n}$ where n denotes the index of the layer and $\mathbf{r}_n = \mathbf{r}_{\parallel} + \mathbf{z}$ is the position of a lattice site of the n-th layer at the z position $z = na$ on the **Oz** axis, a being the distance between two successive layers in the z direction. In such a notation, a surface mode corresponds to a complex wave vector $\mathbf{k} = k_1 + ik_2$ where k_2 is non-zero. Its amplitude $U_n(\mathbf{k}) \propto e^{-k_2na}$, therefore, diminishes while propagating into the crystal interior (increasing n). We will show some examples in the following.

8.2.2 Reconstruction of Surface Magnetic Ordering

As we said above, surface exchange interaction and surface anisotropy may suffer from modifications not only in their magnitudes but also in their signs. These modifications may result in a rearrangement of the magnetic ordering at the surface to minimize the system energy. In the case of Heisenberg spin model, we can have a non-collinear spin configuration in the vicinity of the surface as will be seen below.

8.2.3 Surface Phase Transition

Basic methods leading to fundamental properties of phase transitions in the bulk have been presented in Chapter 5. Phase transition is a collective phenomenon which takes place when the system changes its symmetry. At the transition, the system spins become strongly correlated at a macroscopic scale. In a system with a finite size such as fine particles, the theoretical definition of the phase transition is not rigorously obeyed. For example, the correlation length in a second-order phase transition cannot go to infinity in a finite system. Nevertheless, anomalies in physical quantities can be observed in small systems. Finite-size scaling relations have been established [10, 95, 380] to allow us to obtain properties of the phase transition such as universality class at the infinite-size limit. In thin films, the infinite dimension of the film planes makes transitions possible. The characteristics of the phase transition in films depend on the surface conditions: If the film thickness is important then the influence of surface parameters is small since the aspect ratio is small. However, for ultrathin films, surface parameters become dominant making the surface phase transition very different from the bulk one.

There exist a large number of books on the surface phase transition. The reader is referred to, for example, reviews given in references [35, 54, 72, 87] for more details. One of the remarkable results is the existence of surface critical exponents and surface scaling laws which are different from the bulk ones [54].

8.3 Semi-Infinite Solids

One examines the case of a semi-infinite crystal. One calculates the spin wave spectrum and shows the existence of surface modes in this section. The simplest way to do is to use the method of equation of motion described in Chapter 3. To give a simple example, one considers the same system illustrated there, namely a semi-infinite ferrimagnetic crystal of body-centered cubic lattice but with the inclusion of a surface. One uses the same Hamiltonian (3.98) and the same equations of motion (3.99) with the same hypothesis

$J_2^A = J_2^B \equiv J_2$. Note that this system is a body-centered cubic antiferromagnet if $S_A = S_B$. Now, for a semi-infinite crystal, one uses the following Fourier transforms only in the xy plane which is still periodic:

$$S_l^- = \frac{1}{(2\pi)^3} \int_{-\infty}^{-\infty} d\omega e^{-i\omega t} \int_{BZ} d\mathbf{k}_\parallel e^{i\mathbf{k}_\parallel \cdot \mathbf{l}} U_n(\mathbf{k}_\parallel, \omega) \tag{8.1}$$

$$S_m^+ = \frac{1}{(2\pi)^3} \int_{-\infty}^{-\infty} d\omega e^{-i\omega t} \int_{BZ} d\mathbf{k}_\parallel e^{i\mathbf{k}_\parallel \cdot \mathbf{l}} U_{n'}(\mathbf{k}_\parallel, \omega) \tag{8.2}$$

where $\mathbf{k}_\parallel$ is a wave vector parallel to the surface, n and n' denote respectively the indices of the layers to which the spins l and m belong, BZ stands for the first Brillouin zone in the xy plane.

To simplify the presentation, we take the body-centered cubic lattice with the surface plane (001). We suppose that the sublattice of A spins (↑) occupies the planes of even indices and the sublattice of B spins (↓) takes the planes of odd indices as indicated in Fig. 8.2.

The equations of motion for S_l^- and S_m^+ give two following coupled equations:

$$\begin{aligned}(E + E_A)U_{2n} &= -(1-\alpha)[4\gamma_1(\mathbf{k}_\parallel)(U_{2n-1} + U_{2n+1}) \\ &\quad + \epsilon(U_{2n-2} + U_{2n+2}) \quad (n \geq 2)\end{aligned} \tag{8.3}$$

$$\begin{aligned}(E - E_B)U_{2n+1} &= (1+\alpha)[4\gamma_1(\mathbf{k}_\parallel)(U_{2n} + U_{2n+2}) \\ &\quad + \epsilon(U_{2n-1} + U_{2n+3}) \quad (n \geq 1)\end{aligned} \tag{8.4}$$

For the first two layers, we have

$$(E - E_1)U_1 = (1+\alpha)\left[4\gamma_1(\mathbf{k}_\parallel)U_2 + \epsilon U_3\right] \tag{8.5}$$

$$(E + E_2)U_2 = -(1-\alpha)\left[4\gamma_1(\mathbf{k}_\parallel)(U_1 + U_3) + \epsilon U_4\right] \tag{8.6}$$

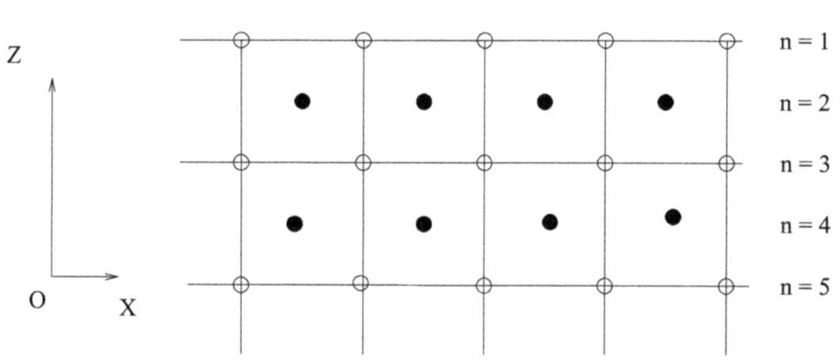

Figure 8.2 Semi-infinite crystal of body-centered cubic lattice with a (001) surface (side view). The sublattices ↑ and ↓ are denoted by black and white circles, respectively. The surface has index $n = 1$.

where we have used, at $T = 0$, $< S_A^z > \simeq S_A = (1-\alpha)S$, $< S_B^z > \simeq S_B = -(1+\alpha)S$ with $|\alpha| < 1$ and S a constant. The other notations used in (8.3)–(8.6) are

$$E = \frac{\hbar\omega}{J_1 S} \tag{8.7}$$

$$\epsilon = \frac{J_2}{J_1} \tag{8.8}$$

$$E_A = 8(1+\alpha) - 6\epsilon(1-\alpha)[1-\gamma_2(\mathbf{k}_\parallel)] \tag{8.9}$$

$$E_B = 8(1-\alpha) - 6\epsilon(1+\alpha)[1-\gamma_2(\mathbf{k}_\parallel)] \tag{8.10}$$

$$E_2 = 8(1+\alpha) - \epsilon(1-\alpha)[5-4\gamma_2(\mathbf{k}_\parallel)] \tag{8.11}$$

$$E_1 = 4(1-\alpha) - \epsilon(1+\alpha)[5-4\gamma_2(\mathbf{k}_\parallel)] \tag{8.12}$$

$$\gamma_1(\mathbf{k}_\parallel) = \cos\left(\frac{k_x a}{2}\right)\cos\left(\frac{k_y a}{2}\right) \tag{8.13}$$

$$\gamma_2(\mathbf{k}_\parallel) = \frac{1}{2}\left[\cos(k_x a) + \cos(k_y a)\right] \tag{8.14}$$

a being the lattice constant. The factor $\gamma_1(\mathbf{k}_\parallel)$ couples the nearest neighbors belonging to the adjacent planes while $\gamma_2(\mathbf{k}_\parallel)$ connects the neighbors belonging to the same sublattice (namely, next-nearest neighbors, by distance).

We take the following forms for the bulk spin wave amplitudes U_{2n} and U_{2n+1}:

$$U_{2n} = U_e \exp(ik_z na) \tag{8.15}$$

$$U_{2n+1} = U_o \exp(ik_z(1+1/2)na) \tag{8.16}$$

where k_z is the real wave vector in the z direction. Replacing these in (8.3) and (8.4), we obtain the following secular equation for a non-trivial solution:

$$(E+E_A)(E-E_B) = -64(1-\alpha^2)\gamma_1(\mathbf{k}_\parallel)^2 \tag{8.17}$$

We deduce the dispersion relation for bulk modes

$$E_\pm = \frac{1}{2}\left[E_B - E_A \pm \sqrt{(E_A-E_B)^2 + 4[E_A E_B - 64(1-\alpha^2)\gamma_1(\mathbf{k}_\parallel)^2]}\right] \tag{8.18}$$

We are now interested in finding the surface modes. We look for solutions of the form

$$U_{2n} = U_2\phi^{n-1} \tag{8.19}$$

$$U_{2n+1} = U_1\phi^n \tag{8.20}$$

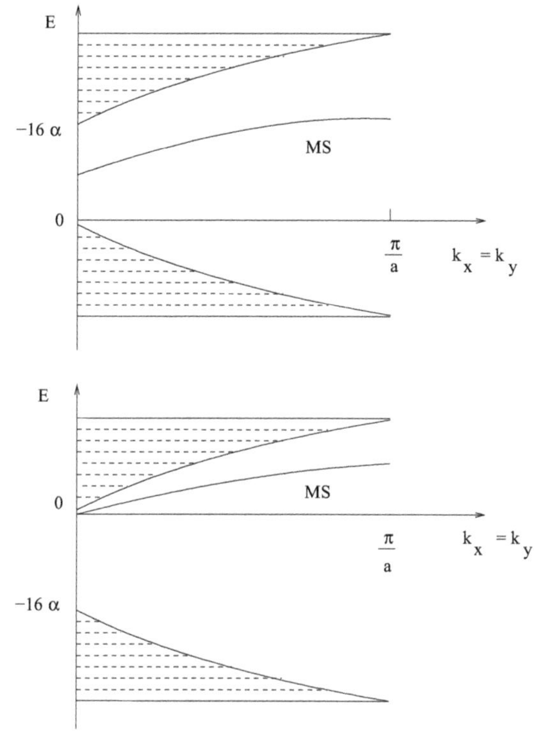

Figure 8.3 Spin wave spectrum of a semi-infinite ferrimagnet versus $k_x = k_y$ in the case where $\epsilon = 0, \alpha = -1/3$ (top) and $\alpha = 1/3$ (bottom). Surface spin wave branches are indicated by MS. The hachured bands are the bulk continuum. The upper limit of each band corresponds to $k_z = \pi/a$ and the lower limit to $k_z = 0$.

where ϕ is a real factor defined by $\phi^n = \mathrm{e}^{-k_2 na}$ where k_2 is the imaginary part of k_z. For a decaying wave, $\phi < 1$. ϕ is called "decay factor." Replacing these amplitudes in (8.3)–(8.6) we obtain a system of coupled equations. Surface modes correspond to solutions $\phi < 1$. We examine a particular case where $k_x = k_y = 0$. In this case, the following solution for a surface mode is found:

$$E_s = -8\alpha + \frac{4\alpha\epsilon(1+\alpha)}{2(1-\alpha)-\epsilon} \quad (8.21)$$

$$\phi = 1 + \frac{4\alpha}{2(1-\alpha)-\epsilon} \quad (8.22)$$

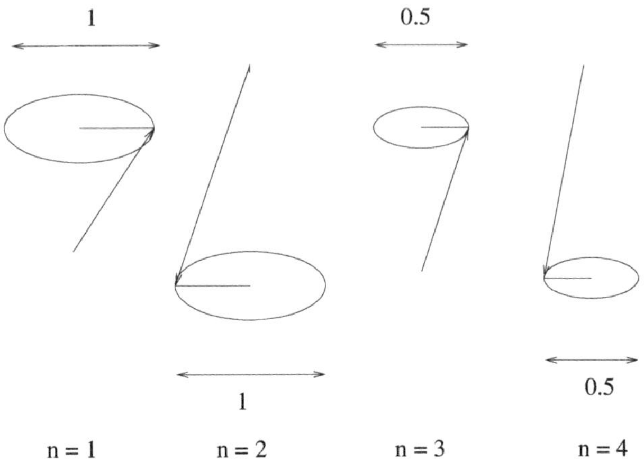

Figure 8.4 Variation of the amplitude of the surface mode near the surface. $k_x = k_y = 0, \epsilon = 0, \alpha = -1/3$. Surface spins are B spins (for convenience, B spins are drawn as up spins).

We see that in order to have $\phi < 1$, we should have $\alpha < 0$ for the case $k_x = k_y = 0$. For k_x, $k_y \neq 0$, there exists for $\alpha < 0$ and $\alpha > 0$ a surface spin wave branch in the spectrum as shown in Fig. 8.3.

Note: In the case of an antiferromagnet, we just put $\alpha = 0$ in the above equations. There is no surface mode for $k_x = k_y = 0$.

The gap at $k_x = k_y = 0$ in the ferrimagnet is proportional to α. We show in Fig. 8.3 the spin wave spectrum versus $k_x = k_y$.

We show in Fig. 8.4 the spatial variation of the amplitude of the surface mode at $k_x = k_y = 0$ with $\alpha = -1/3$. The decay factor calculated by (8.22) is equal to 0.5.

We discuss now the effect of the interaction between the next-nearest neighbors. The interaction between nearest neighbors J_1 is antiferromagnetic [$J_1 > 0$ as seen in the definition of the Hamiltonian (3.98)]. If J_2 is ferromagnetic, i.e., $\epsilon < 0$, the magnetic order is antiferromagnetic between the two sublattices. On the other hand, if J_2 is antiferromagnetic (>0), there is a competition between J_1 and J_2. When $|J_2|$ is large enough, the collinear antiferromagnetic configuration is no more stable for $\epsilon = \frac{J_2}{J_1} > \epsilon_c$. The determination of ϵ_c with a surface is more complicated than in the bulk case because there may exist a non-uniform spin configuration near the

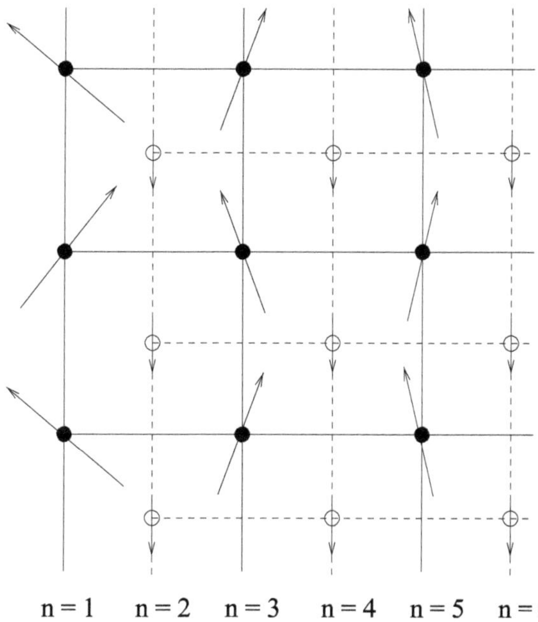

Figure 8.5 Non-collinear spin configuration near the surface (side view) for $\epsilon > \epsilon_c$, with $\alpha > 0$. Surface spins are B spins (for convenience, B spins are drawn as up spins).

surface. In the present case, the critical value ϵ_c depends also on α. We show in Fig. 8.5 the spin configuration near the surface for a value of $\epsilon > \epsilon_c$.

8.4 Spin Wave Theory in Ferromagnetic Films

We show in this section that localized surface spin wave modes affect strongly thermodynamic behaviors of ferromagnetic thin films. In particular, we will show that low-lying localized modes diminish the surface magnetization and the Curie temperature with respect to the bulk ones. These quantities depend, of course, on the surface interaction parameters and the film thickness.

The method which can cover correctly a large region of temperature is no doubt the Green's function method (cf. Chapter 4).

We shall use here that method to study properties of thin films from $T = 0$ up to the phase transition.

We consider a thin film of N_T layers with the Heisenberg quantum spin model. The Hamiltonian is written as

$$\mathcal{H} = -2\sum_{<i,j>} J_{ij}\mathbf{S}_i \cdot \mathbf{S}_j - 2\sum_{<i,j>} D_{ij} S_i^z S_j^z$$
$$= -2\sum_{\langle i,j\rangle} J_{ij}\left(S_i^z S_j^z + \frac{1}{2}(S_i^+ S_j^- + S_i^- S_j^+)\right) - 2\sum_{<i,j>} D_{ij} S_i^z S_j^z \tag{8.23}$$

where J_{ij} is positive (ferromagnetic) and $D_{ij} > 0$ denotes an exchange anisotropy. When D_{ij} is very large with respect to J_{ij}, the spins have an Ising-like behavior. The factor 2 in front of the terms is used for historical reasons [see (1.57)].

8.4.1 Method

The Green's function method has been formulated in detail in Chapter 4. It is useful to summarize here the main steps in its application to thin films.

We define one Green's function for each layer, numbering the surface as the first layer. We write next the equation of motion for each of Green's functions. We obtain a system of coupled equations. We linearize these equations to reduce higher-order Green's functions by using the Tyablikov decoupling scheme. We are then ready to make the Fourier transforms for all Green's functions in the xy planes. We obtain a system of equations in the space $(\mathbf{k}_{xy}, \omega)$ where $\mathbf{k}_{xy}$ is the wave vector parallel to the xy plane and ω the spin wave frequency (pulsation). Solving this system we obtain Green's functions and ω as functions of $\mathbf{k}_{xy}$. Using the spectral theorem, Eq. (4.39), we calculate the layer magnetization.

Let us define the following Green's function for two spins $\mathbf{S}_i$ and $\mathbf{S}_j$:

$$G_{i,j}(t,t') = \langle\langle S_i^+(t); S_j^-(t')\rangle\rangle \tag{8.24}$$

The equation of motion of $G_{i,j}(t,t')$ is written as

$$i\hbar\frac{dG_{i,j}(t,t')}{dt} = (2\pi)^{-1}\langle[S_i^+(t), S_j^-(t')]\rangle + \langle\langle[S_i^+;\mathcal{H}](t); S_j^-(t')\rangle\rangle \tag{8.25}$$

where $[\cdots]$ is the boson commutator and $\langle\cdots\rangle$ the thermal average in the canonical ensemble defined as

$$\langle F\rangle = \text{Tr}e^{-\beta\mathcal{H}}F/\text{Tr}e^{-\beta\mathcal{H}} \tag{8.26}$$

with $\beta = 1/k_B T$.

The commutator of the right-hand side of Eq. (8.25) generates functions of higher orders. In a first approximation, we can reduce these functions with the help of the Tyablikov decoupling [43, 346] as follows:

$$\langle\langle S^z_m S^+_i; S^-_j\rangle\rangle \simeq \langle S^z_m\rangle\langle\langle S^+_i; S^-_j\rangle\rangle \tag{8.27}$$

We obtain then the same kind of Green's function defined in Eq. (8.24). As the system is translation invariant in the xy plane, we use the following Fourier transforms:

$$G_{i,j}(t,t') = \frac{1}{\Delta}\int\int d\mathbf{k}_{xy}\frac{1}{2\pi}\int_{-\infty}^{+\infty} d\omega\, e^{-i\omega(t-t')}\, g_{n,n'}(\omega,\mathbf{k}_{xy}) \times e^{i\mathbf{k}_{xy}\cdot(\mathbf{R}_i-\mathbf{R}_j)} \tag{8.28}$$

where ω is the spin wave pulsation (frequency), $\mathbf{k}_{xy}$ the wave vector parallel to the surface, $\mathbf{R}_i$ the position of the spin at the site i, n and n' are respectively the indices of the planes to which i and j belong ($n = 1$ is the index of the surface plane). The integration on $\mathbf{k}_{xy}$ is performed within the first Brillouin zone in the xy plane. Let Δ be the surface of that zone. Equation (8.25) becomes

$$(\hbar\omega - A_n)g_{n,n'} + B_n(1-\delta_{n,1})g_{n-1,n'} + C_n(1-\delta_{n,N_T})g_{n+1,n'} = 2\delta_{n,n'} < S^z_n > \tag{8.29}$$

where the factors $(1-\delta_{n,1})$ and $(1-\delta_{n,N_T})$ are added to ensure that there are no C_n and B_n terms for the first and the last layer. The coefficients A_n, B_n and C_n depend on the crystalline lattice of the film. We give here some examples:

8.4.1.1 Film of stacked triangular lattices

$$\begin{aligned} A_n &= -2J_n < S^z_n > C\gamma_k + 2C(J_n + D_n) < S^z_n > \\ &\quad +2(J_{n,n+1} + D_{n,n+1}) < S^z_{n+1} > \\ &\quad +2(J_{n,n-1} + D_{n,n-1}) < S^z_{n-1} > \end{aligned} \tag{8.30}$$

$$B_n = 2J_{n,n-1} < S^z_n > \tag{8.31}$$

$$C_n = 2J_{n,n+1} < S^z_n > \tag{8.32}$$

where the following notations have been used:

(i) J_n, D_n are the interactions in the layer n,
(ii) $J_{n,n\pm1}$ and $D_{n,n\pm1}$ are the interactions between a spin in the layer n and a spin in the layer $(n \pm 1)$. Of course, $J_{n,n-1}, D_{n,n-1} = 0$ if $n = 1$, and $J_{n,n+1}, D_{n,n+1} = 0$ if $n = N_T$,
(iii) $\gamma_k = [2\cos(k_x a) + 4\cos(k_x a/2)\cos(k_y a\sqrt{3}/2)]/C$
(iv) $C = 6$ is the coordination number in the xy plane.

8.4.1.2 Film of simple cubic lattice

$$\begin{aligned} A_n &= -2J_n < S_n^z > C\gamma_k + 2C(J_n + D_n) < S_n^z > \\ &\quad +2(J_{n,n+1} + D_{n,n+1}) < S_{n+1}^z > \\ &\quad +2(J_{n,n-1} + D_{n,n-1}) < S_{n-1}^z > \end{aligned} \tag{8.33}$$

$$B_n = 2J_{n,n-1} < S_n^z > \tag{8.34}$$

$$C_n = 2J_{n,n+1} < S_n^z > \tag{8.35}$$

where $C = 4$ and $\gamma_k = \frac{1}{2}[\cos(k_x a) + \cos(k_y a)]$.

8.4.1.3 Film of body-centered cubic lattice

$$\begin{aligned} A_n &= 8(J_{n,n+1} + D_{n,n+1}) < S_{n+1}^z > \\ &\quad +8(J_{n,n-1} + D_{n,n-1}) < S_{n-1}^z > \end{aligned} \tag{8.36}$$

$$B_n = 8J_{n,n-1} < S_n^z > \gamma_k \tag{8.37}$$

$$C_n = 8J_{n,n+1} < S_n^z > \gamma_k \tag{8.38}$$

where $\gamma_k = \cos(k_x a/2)\cos(k_y a/2)$.

We go back to Eq. (8.29). Writing it for $n = 1, 2, \cdots, N_T$, we obtain a system of N_T equations which can be rewritten in a matrix form

$$\mathbf{M}(\omega)\mathbf{g} = \mathbf{u} \tag{8.39}$$

where $\mathbf{u}$ is a column matrix whose n-th element is $2\delta_{n,n'} < S_n^z >$.

For a given $\mathbf{k}_{xy}$ the spin wave dispersion relation $\hbar\omega(\mathbf{k}_{xy})$ can be obtained by solving the secular equation $det|\mathbf{M}| = 0$. There are N_T eigenvalues $\hbar\omega_i$ ($i = 1, \cdots, N_T$) for each $\mathbf{k}_{xy}$. It is obvious that ω_i depends on all $\langle S_n^z \rangle$ contained in the coefficients A_n, B_n and C_n.

To calculate the thermal average of the magnetization of the layer n in the case where $S = \frac{1}{2}$, we use the following relation: (see Chapter 4):

$$\langle S_n^z \rangle = \frac{1}{2} - \langle S_n^- S_n^+ \rangle \tag{8.40}$$

where $\langle S_n^- S_n^+ \rangle$ is given by the following spectral theorem [see (4.39)]:

$$\langle S_i^- S_j^+ \rangle = \lim_{\epsilon \to 0} \frac{1}{\Delta} \iint d\mathbf{k}_{xy} \int_{-\infty}^{+\infty} \frac{i}{2\pi} \left[g_{n,n'}(\omega + i\epsilon) - g_{n,n'}(\omega - i\epsilon) \right] \times \frac{d\omega}{e^{\beta\omega} - 1} e^{i\mathbf{k}_{xy}\cdot(\mathbf{R}_i - \mathbf{R}_j)}. \tag{8.41}$$

ϵ being an infinitesimal positive constant. Equation (8.40) becomes

$$\langle S_n^z \rangle = \frac{1}{2} - \lim_{\epsilon \to 0} \frac{1}{\Delta} \iint d\mathbf{k}_{xy} \int_{-\infty}^{+\infty} \frac{i}{2\pi} \left[g_{n,n}(\omega + i\epsilon) - g_{n,n}(\omega - i\epsilon) \right] \times \frac{d\omega}{e^{\beta\hbar\omega} - 1} \tag{8.42}$$

where Green's function $g_{n,n}$ is obtained by the solution of Eq. (8.39):

$$g_{n,n} = \frac{|\mathbf{M}|_n}{|\mathbf{M}|} \tag{8.43}$$

$|\mathbf{M}|_n$ is the determinant obtained by replacing the n-th column of $|\mathbf{M}|$ by $\mathbf{u}$.

To simplify the notations we put $\hbar\omega_i = E_i$ and $\hbar\omega = E$ in the following. By expressing

$$|\mathbf{M}| = \prod_i (E - E_i) \tag{8.44}$$

we see that E_i $(i = 1, \cdots, N_T)$ are the poles of Green's function. We can, therefore, rewrite $g_{n,n}$ as

$$g_{n,n} = \sum_i \frac{f_n(E_i)}{E - E_i} \tag{8.45}$$

where $f_n(E_i)$ is given by

$$f_n(E_i) = \frac{|\mathbf{M}|_n(E_i)}{\prod_{j \neq i}(E_i - E_j)} \tag{8.46}$$

Replacing Eq. (8.45) in Eq. (8.42) and making use of the following identity:

$$\frac{1}{x - i\eta} - \frac{1}{x + i\eta} = 2\pi i \delta(x) \tag{8.47}$$

we obtain

$$\langle S_n^z \rangle = \frac{1}{2} - \frac{1}{\Delta} \iint dk_x dk_y \sum_{i=1}^{N_T} \frac{f_n(E_i)}{e^{\beta E_i} - 1} \tag{8.48}$$

where $n = 1, \cdots, N_T$.

As $< S_n^z >$ depends on the magnetizations of the neighboring layers via $E_i (i = 1, \cdots, N_T)$, we should solve by iteration the equations (8.48) written for all layers, namely for $n = 1, \cdots, N_T$, to obtain the layer magnetizations at a given temperature T.

The critical temperature can be calculated in a self-consistent manner by iteration, letting all $< S_n^z >$ tend to zero.

In the case where the surface parameters are not different from the bulk ones, it is not exaggerated to calculate the critical temperature by supposing that all layer magnetizations are equal to a unique average value M which is to be determined self-consistently. The value of M is defined from $< S_n^z >$ of Eq. (8.48) as follows:

$$\begin{aligned} M &= \frac{1}{N_T} \sum_{n=1}^{N_T} \langle S_n^z \rangle \\ &= \frac{1}{2} - \frac{1}{N_T} \frac{1}{\Delta} \iint dk_x dk_y \sum_{n=1}^{N_T} \sum_{i=1}^{N_T} \frac{f_n(E_i)}{e^{\beta E_i} - 1} \end{aligned} \tag{8.49}$$

Replacing all $< S_n^z >$ in the matrix elements by M, we see that $\sum_n f_n(E_i) = 2M$ by using Eq. (8.46). We deduce

$$M = \frac{1}{2} - \frac{1}{N_T} \frac{1}{\Delta} \iint dk_x dk_y \sum_{i=1}^{N_T} \frac{2M}{e^{\beta E_i} - 1} \tag{8.50}$$

When $T \to T_c$, $M \to 0$. We can then make an expansion of the exponential of the denominator of Eq. (8.50). We obtain

$$\left[\frac{J}{k_B T_c}\right]^{-1} = \frac{2}{\Delta} \iint dk_x dk_y \frac{1}{N_T} \sum_{i=1}^{N_T} \frac{2M}{E_i} \tag{8.51}$$

We notice that all matrix elements A_n, B_n and C_n are proportional to M in the above hypothesis of uniform layer magnetization. We see that E_i is proportional to M. The right-hand side of Eq. (8.51), therefore, does not depend on M.

8.4.2 Results

To calculate numerically the above equations, one must first determine the first Brillouin zone according to the lattice structure. For the iteration process, one starts in general at a very low T with input values of $< S_n^z >$ close to the spin amplitude. Next, one uses the solutions of $< S_n^z >$ as inputs for a temperature not far from the previous T in order to facilitate the convergence. Of course, if we use the hypothesis of uniform layer magnetizations, i.e., $< S_n^z >= M$ for all n, then there is only one solution M to find at a given T.

At each T, using input values for $< S_n^z >$, one calculates the eigenvalues of the spin wave energy E_i by solving $det|\mathbf{M}| = 0$ for each value of $\mathbf{k}_{xy}$. Using the values so obtained of E_i one calculates the output values of $< S_n^z >$ $(n = 1, \cdots, N_T)$. If the output values are equal to the input values within a given precision, one stops the iteration. In general, a few iterations suffice at low T for a solution with a precision of 1%. For a temperature close to T_c, one needs a few dozens to a few hundreds of iterations. We show here the results of a few cases for comparison.

8.4.2.1 Spin wave spectrum

We show in Figs. 8.6 and 8.6 the spin wave spectra of ferromagnetic films of simple cubic and body-centered cubic lattices with a thickness $N_T = 8$. For simplicity, we suppose all exchange interactions are identical and equal to J. Also, all anisotropy constants D are equal to $0.01J$. We see in Figs. 8.6 and 8.6 that there is no surface mode in the simple cubic case while there are two branches of surface modes in the body-centered lattice. Two branches result from the interaction of the surface modes coming from the two surfaces of the film. If the thickness is thick enough (longer than the penetration length of the surface mode), then two branches are degenerate due to symmetrical surface conditions. This is not the case for $N_T = 8$ shown in Fig. 8.7.

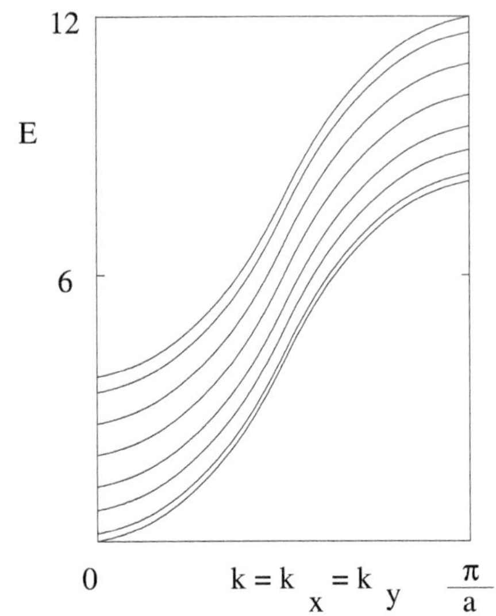

Figure 8.6 Magnon spectrum $E = \hbar\omega$ of a ferromagnetic film with a simple cubic lattice versus $k \equiv k_x = k_y$ for $N_T = 8$ and $D/J = 0.01$. No surface mode is observed for this case.

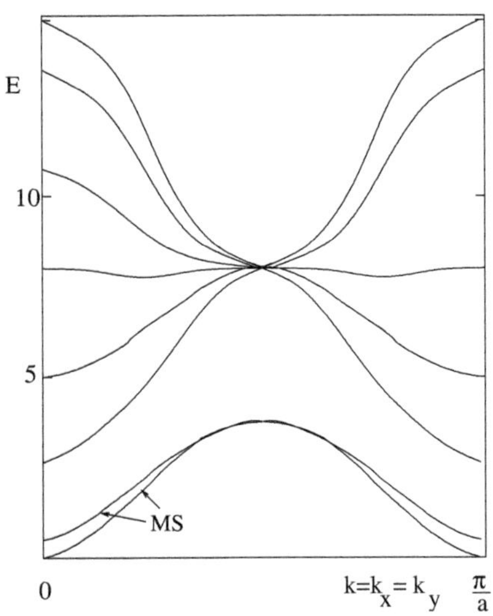

Figure 8.7 Magnon spectrum $E = \hbar\omega$ of a ferromagnetic film with a body-centered cubic lattice versus $k \equiv k_x = k_y$ for $N_T = 8$ and $D/J = 0.01$. The branches of surface modes are indicated by MS.

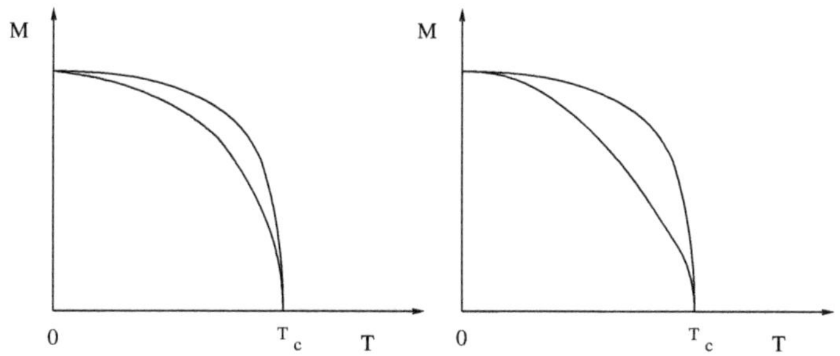

Figure 8.8 Ferromagnetic films of simple cubic lattice (left) and body-centered cubic lattice (right): magnetizations of the surface layer (lower curve) and the second layer (upper curve), with $N_T = 4$, $D = 0.01J$, $J = 1$.

8.4.2.2 Layer magnetizations

Figure 8.8 shows the results of the layer magnetizations for the first two layers in the cases considered above with $N_T = 4$. One observes that the layer magnetization at the surface is smaller than that of the second layer. This difference is larger in the case where there exists a surface mode as in the body-centered cubic lattice because surface modes are localized at the surface, making a larger deviation for surface spins.

One also observes that the critical temperature is strongly decreased in the case where surface modes of low-lying energy exist (acoustic surface modes).

8.5 Antiferromagnetic Films

We can adapt the Green's function method presented in Chapter 4 for bulk antiferromagnets to the case of antiferromagnetic thin films. We consider the following Hamiltonian:

$$\begin{aligned}\mathcal{H} &= 2\sum_{<i,j>} J_{ij}\mathbf{S}_i \cdot \mathbf{S}_j + 2\sum_{<i,j>} D_{ij} S_i^z S_j^z \\ &= 2\sum_{\langle i,j\rangle} J_{ij}\left[S_i^z S_j^z + \frac{1}{2}(S_i^+ S_j^- + S_i^- S_j^+)\right] + 2\sum_{<i,j>} D_{ij} S_i^z S_j^z \qquad (8.52)\end{aligned}$$

where $J_{ij} > 0$ and $D_{ij} > 0$. We divide the lattice into two sublattices: sublattice A contains $\uparrow$ spins and sublattice B $\downarrow$ spins. We define the following Green's functions:

$$G_{j,j'}(t-t') = \ll S_j^+(t); S_{j'}^-(t') \gg$$
$$= -i\theta(t-t') < \left[S_j^+(t), S_{j'}^-(t')\right] > \qquad (8.53)$$
$$F_{i,j'}(t-t') = \ll S_i^+(t); S_{j'}^-(t') \gg$$
$$= -i\theta(t-t') < \left[S_i^+(t), S_{j'}^-(t')\right] > \qquad (8.54)$$

where j and j' belong to sublattice A, and i to sublattice B.

8.5.1 Films of Simple Cubic Lattice

We write the equations of motion for $G_{j,j'}(t)$ and $F_{i,j'}(t)$, then we use the Tyablikov decoupling [43] and the following Fourier transforms:

$$G_{j,j'}(t,t') = \frac{1}{\Delta}\iint d\mathbf{k}_{xy}\frac{1}{2\pi}\int_{-\infty}^{+\infty} d\omega\, \mathrm{e}^{-i\omega(t-t')}\, g_{n,n'}(\omega, \mathbf{k}_{xy})$$
$$\times \mathrm{e}^{i\mathbf{k}_{xy}.(\mathbf{R}_j-\mathbf{R}_{j'})} \qquad (8.55)$$
$$F_{i,j'}(t,t') = \frac{1}{\Delta}\iint d\mathbf{k}_{xy}\frac{1}{2\pi}\int_{-\infty}^{+\infty} d\omega\, \mathrm{e}^{-i\omega(t-t')}\, f_{n,n'}(\omega, \mathbf{k}_{xy})$$
$$\times \mathrm{e}^{i\mathbf{k}_{xy}.(\mathbf{R}_i-\mathbf{R}_{j'})} \qquad (8.56)$$

where we use the same notations as in the above ferromagnetic case. We obtain

$$Eg_{n,n'}(\omega) = 2 < S_n^z > \delta_{n,n'}$$
$$+8\left[\gamma_k < S_n^z > f_{n,n'}(\omega) + (1+d_e) < S_n^z > g_{n,n'}(\omega)\right]$$
$$+2\left[< S_n^z > f_{n+1,n'}(\omega) + (1+d) < S_{n+1}^z > g_{n,n'}(\omega)\right]$$
$$\times(1-\delta_{N_T,n})$$
$$+2\left[< S_n^z > f_{n-1,n'}(\omega) + (1+d) < S_{n-1}^z > g_{n,n'}(\omega)\right]$$
$$\times(1-\delta_{1,n}) \qquad (8.57)$$

$$Ef_{n,n'}(\omega) = -8\left[\gamma_k < S_n^z > g_{n,n'}(\omega) + (1+d_e) < S_n^z > f_{n,n'}(\omega)\right]$$
$$-2\left[< S_n^z > g_{n+1,n'}(\omega) + (1+d) < S_{n+1}^z > f_{n,n'}(\omega)\right]$$
$$\times(1-\delta_{N_T,n})$$
$$-2\left[< S_n^z > g_{n-1,n'}(\omega) + (1+d) < S_{n-1}^z > f_{n,n'}(\omega)\right]$$
$$\times(1-\delta_{1,n}) \qquad (8.58)$$

where $n, n' = 1, \cdots, N_T$ and the following notations have been used:

- $J_{i,j} = J$ for all pairs (i, j),
- $d = D_{i,j}/J$ for all pairs (i, j) except when (i, j) are both on the surface ($n = 1$ and $n = N_T$) where $d = d_s$,
- $E = \hbar\omega/J$,
- $d_e = d(1 - \delta_{1,n})(1 - \delta_{N_T,n}) + d_s(\delta_{1,n} + \delta_{N_T,n})$ (this complicated notation was used in order to include all cases in the same formula),
- $\gamma_k = \cos(k_x a)\cos(k_y a)$ (for convenience, we used the distance between the two neighbors of the same sublattice in the xy plane equal to $2a$ and k_x and k_y oriented along the axes of one sublattice).

It is noted that in Eqs. (8.57)–(8.58) we have changed the sign of the average values of B spins ($< S_n^z > \to - < S_n^z >$) so that all $< S_n^z >$ are positive in Eqs. (8.57)–(8.58).

8.5.2 Films of Body-Centered Cubic Lattice

In the case of a film of body-centered cubic lattice with a (001) surface, we suppose that ↑ sublattice and ↓ sublattice occupy even- and odd-index layers, respectively. Using the following Fourier transforms:

$$G_{j,j'}(t,t') = \frac{1}{\Delta}\iint d\mathbf{k}_{xy}\frac{1}{2\pi}\int_{-\infty}^{+\infty} d\omega\, \mathrm{e}^{-i\omega(t-t')}\, g_{2n,2n'}(\omega,\mathbf{k}_{xy})$$
$$\times \mathrm{e}^{i\mathbf{k}_{xy}.(\mathbf{R}_j-\mathbf{R}_{j'})}$$
$$F_{i,j'}(t,t') = \frac{1}{\Delta}\iint d\mathbf{k}_{xy}\frac{1}{2\pi}\int_{-\infty}^{+\infty} d\omega\, \mathrm{e}^{-i\omega(t-t')}\, f_{2n+1,2n'}(\omega,\mathbf{k}_{xy})$$
$$\times \mathrm{e}^{i\mathbf{k}_{xy}.(\mathbf{R}_i-\mathbf{R}_{j'})}$$

we obtain

$$\begin{aligned} Eg_{2n,2n'}(\omega) = {} & 2 < S_{2n}^z > \delta_{n,n'} + 8[\gamma_k < S_{2n}^z > f_{2n+1,2n'}(\omega) \\ & +(1+d) < S_{2n+1}^z > g_{2n,2n'}(\omega)](1-\delta_{n,N}) \\ & +8[\gamma_k < S_{2n}^z > f_{2n-1,2n'}(\omega) \\ & +(1+d_f) < S_{2n-1}^z > g_{2n,2n'}(\omega)] \end{aligned} \quad (8.59)$$

$$\begin{aligned} Ef_{2n-1,2n'}(\omega) = & -8[\gamma_k < S^z_{2n-1} > g_{2n,2n'}(\omega) \\ & +(1+d_f) < S^z_{2n} > f_{2n-1,2n'}(\omega)] \\ & -8[\gamma_k < S^z_{2n-1} > g_{2n-2,2n'}(\omega) \\ & +(1+d) < S^z_{2n-2} > f_{2n-1,2n'}(\omega)](1-\delta_{1,n}) \end{aligned} \tag{8.60}$$

where $n, n' = 1, \cdots, N$ with $N = N_T/2$ (N=number of layers in each sublattice), $d_f = d(1-\delta_{1,n})(1-\delta_{N,n}) + d_s(\delta_{1,n}+\delta_{N,n})$, $\gamma_k = \cos(k_x a/2)\cos(k_y a/2)$. As before, we have redefined $< S^z_{2n\pm1} > \rightarrow - < S^z_{2n\pm1} >$ of B spins.

The next steps of the calculation are the same as in the ferromagnetic case: We calculate spin wave energy eigenvalues E_i by solving $det|\mathbf{M}| = 0$. The layer magnetization is calculated by Eq. (8.48). The value of the spin in the layer n at $T = 0$ is calculated by (see Chapter 3)

$$\langle S^z_n \rangle (T=0) = \frac{1}{2} + \frac{1}{\Delta} \int\int dk_x dk_y \sum_{i=1}^{N_T} f_n(E_i) \tag{8.61}$$

where the sum is performed over negative values of E_i (for positive values the Bose–Einstein factor is equal to 0 at $T = 0$).

The numerical results for $S = 1/2$, $N_T = 4$ and $d = 0.01$ are $S^z_1 = 0.44075$, $S^z_2 = 0.41885$. We conclude that the zero-point spin contraction (cf. Chapter 3) is smaller for surface spins. This is not surprising because zero-point fluctuations are due to the antiferromagnetic interaction acting on a spin: The weaker the antiferromagnetic interaction, the smaller the zero-point spin contraction. Due to the lack of neighbors, the antiferromagnetic local field acting on a surface spin is weaker than that on an interior spin [87].

The Néel temperature T_N is obtained in the same manner as for T_c. We have the following N_T coupled equations to solve by iteration:

$$\left[\frac{J}{k_B T_N}\right]^{-1} = \frac{1}{\Delta} \iint dk_x dk_y \sum_{i=1}^{N_T} \frac{f_n(E_i)}{E_i} \tag{8.62}$$

where $n = 1, \cdots, N_T$. The uniform layer-magnetization approximation presented above for ferromagnetic films [see (8.49)–(8.51)] yields for the antiferromagnetic case the following expression:

$$\left[\frac{J}{k_B T_N}\right]^{-1} = \frac{1}{\Delta} \iint dk_x dk_y \frac{1}{N_T} \sum_{i=1}^{N_T} \frac{2M}{E_i} \tag{8.63}$$

Table 8.1 Values of Néel temperature T_N and Curie temperature T_c for thicknesses $N_T = 2$ and 20, calculated with several values of anisotropy d

	$N_T = 2$	$N_T = 2$	$N_T = 2$	$N_T = 20$	$N_T = 20$	$N_T = 20$
d_s	0.01	0.1	0.2	0.01	0.1	0.2
$k_B T_N/J$	1.37	1.91	2.25	1.86	2.02	2.14
$k_B T_c/J$	1.40	1.93	2.27	2.03	2.10	2.14

We show in Table 8.1 the values of T_N and T_c calculated for films of simple cubic lattice with two thicknesses $N_T = 2$ and $N_T = 20$. We see that the antiferromagnetic film has the critical temperature lightly but systematically smaller than that of the ferromagnetic film, except when there is a strong anisotropy which suppresses more or less antiferromagnetic fluctuations (last column). The critical temperature increases with increasing thickness as expected. It reaches the bulk value at a few dozens of layers.

8.6 Conclusion

We have shown in this chapter how to calculate the spin wave spectrum and properties due to the spin waves in thin ferromagnetic and antiferromagnetic films with several lattice structures. The aim of this chapter is to provide basic theoretical methods to deal with simple models of clean surfaces. Physical results obtained here such as origin of low surface magnetization and low critical temperature are explained by the existence of localized surface spin wave modes which depends on the lattice structure, the surface orientation, the surface exchange interaction, etc. Many aspects of these results remain in real magnetic films where surface conditions are much more complicated. We did not show here all types of surface modes. In general, we can have acoustic surface modes which lie in the low-energy region below the bulk band, and optical surface modes which lie in the high energy region above the bulk band. The low-lying acoustic modes lower the surface magnetization and the critical temperature as can be seen by examining the formulas shown above. The low surface magnetization corresponds what was called in the 1980s "magnetically dead surface." On the other hand, high-energy optical surface modes make the surface magnetization larger than the bulk one. This situation occurs when the surface exchange interactions are much larger than the bulk values. The surface in this case was called "hard surface" [374].

So far, we have studied thin films with collinear spin ordering. In the following chapters, we consider the effect of the frustration in thin films. The situation in frustrated thin films is more complicated because of the non-collinear spin configuration.

8.7 Problems

Problem 1. Surface spin wave modes:

Calculate the surface spin wave modes in the case of a semi-infinite ferromagnetic crystal of body-centered cubic lattice for $k_x = k_y = 0, \pi/a$ in using the method presented in Section 8.3.

Problem 2. Critical next-nearest-neighbor interaction:

Calculate the critical value of ϵ defined in Section 8.3 for an infinite crystal.

Problem 3. Uniform magnetization approximation:

Show that with the hypothesis of uniform layer-magnetization [Eq. (8.51)], the energy eigenvalue E_i is proportional to M.

Problem 4. Multilayers: critical magnetic field

One considers a system composed of three films A, B and C, of Ising spins with respective thicknesses N_1, N_2 and N_3. The lattice sites are occupied by Ising spins pointing in the $\pm z$ direction perpendicular to the films. The interaction between two spins of the same film is ferromagnetic. Let J_1, J_2 and J_3 be the magnitudes of these interactions in the three films. One supposes that the interactions at the interfaces $A - B$ and $B - C$ are antiferromagnetic and both equal to J_s. One applies a magnetic field along the z direction. Determine the critical field above which all spins are turned into the field direction. For simplicity, consider the case $J_1 = J_2 = J_3$.

Problem 5. Mean-field theory of thin films:

Calculate the layer magnetizations of a 3-layer film by the mean-field theory (cf. Chapter 2). One supposes the Ising spin model with values $\pm 1/2$ and a ferromagnetic interaction J for all pairs of nearest neighbors.

Problem 6. Holstein–Primakoff method:

Using the Holstein–Primakoff method of Chapter 3 for a semi-infinite crystal with the Heisenberg spin model, write the expression which allows us to calculate the surface magnetization as a function of temperature. Show that a surface mode of low energy (acoustic surface mode) diminishes the surface magnetization.

Problem 7. Frustrated surface: surface spin rearrangement

Consider a semi-infinite system of Heisenberg spins composed of stacked triangular lattices. Suppose that the interaction between nearest neighbors J is everywhere ferromagnetic except for the spins on the surface: They interact with each other via an antiferromagnetic interaction J_s. Determine the ground state of the system as a function of J_s/J.

Problem 8. Ferrimagnetic film:

Write the equations of motion for a four-layer ferrimagnetic film of body-centered cubic lattice, using the model and the method presented in Section 8.3. Consider the cases $k_x = k_y = 0, \pi/a$. Solve numerically these equations to find surface and bulk magnons.

Chapter 9

Frustrated Thin Films of Antiferromagnetic FCC Lattice

In this chapter, we show the effects of frustration in an antiferromagnetic film of face-centered cubic (FCC) lattice with Heisenberg spin model including an Ising-like anisotropy. Monte Carlo (MC) simulations have been used to study thermodynamic properties of the film. We show that the presence of the surface reduces the ground state (GS) degeneracy found in the bulk. The GS is shown to depend on the surface in-plane interaction J_s with a critical value at which ordering of type I coexists with ordering of type II. Near this value, a reentrant phase is found. Various physical quantities such as layer magnetizations and layer susceptibilities are shown and discussed. We study here how physical properties vary as the surface bond strength changes at a fixed film thickness. The nature of the phase transition is also studied by the histogram technique.

We have also used the Green's function (GF) method for the quantum counterpart model. The results at low-T show interesting effects of quantum fluctuations. Results obtained by the GF method at high T are compared to those of MC simulations. A good agreement is observed.

Physics of Magnetic Thin Films: Theory and Simulation
Hung T. Diep

ISBN 978-981-4877-42-8 (Hardcover), 978-1-003-12110-7 (eBook)
www.jennystanford.com

Some results shown in this chapter has been published in Ref. [249].

9.1 Introduction

Effects of the frustration in spin systems have been extensively investigated during the past 30 years. Frustrated spin systems are shown to have unusual properties such as large ground state (GS) degeneracy, additional GS symmetries, successive phase transitions with complicated nature. Frustrated systems still challenge theoretical and experimental methods. For recent reviews, the reader is referred to Ref. [85].

On the other hand, during the same period physics of surfaces and objects of nanometric size have also attracted an immense interest. This is due to important applications in industry [35, 36, 72, 374]. In this field, results from laboratory research are often immediately used for industrial applications, without waiting for a full theoretical understanding. An example is the so-called giant magneto-resistance (GMR) used in data storage devices, magnetic sensors, ... [18, 22, 134, 345]. In parallel to these experimental developments, much theoretical effort has also been devoted to the search of physical mechanisms lying behind new properties found in nanometric objects such as ultrathin films, ultrafine particles, quantum dots, spintronic devices, etc. This effort aimed not only at providing explanations for experimental observations but also at predicting new effects for future experiments [248, 252].

The aim of this chapter is to show the effect of the presence of a film surface in a system which is known to be very frustrated, namely the FCC antiferromagnet. The bulk properties of this material have been largely studied as we will show below. We would like to see here in particular how the frustration effects on the nature of the phase transition in 3D are modified in thin films and how the surface conditions affect the magnetic phase diagram. We emphasize that we do not try to address how physical properties vary as the film thickness changes. Rather we work with only one film thickness and address the question how a change in the surface bond strength changes the physical properties. The effects obtained here certainly remain generic when the film thickness varies. We shall use in the following Monte Carlo (MC) simulations and the Green's function method for qualitative comparison.

In the following section, we describe the model and recall the properties of the 3D counterpart model in order to better appreciate properties of thin films. A determination of its GS properties is also given. In Section 9.3, we show our results obtained by MC simulations as functions of temperature T. The surface exchange interaction J_s is made to vary. A phase diagram in the space (T, J_s) is shown and discussed. In general, the surface transition is found to be distinct from the transition of interior layers. An interesting reentrant region is observed in the phase diagram. We also show in this section the results on the critical exponents obtained by MC multi-histogram technique. A detailed discussion on the nature of the phase transition is given. Section 9.4 is devoted to a study of the quantum version of the same model by the use of the GF method. We find interesting effects of quantum fluctuations at low T. The phase diagram (T, J_s) is established and compared to that obtained by MC simulations for the classical model.

9.2 Model and Classical Ground-State Analysis

It is known that the antiferromagnetic (AF) interaction between nearest-neighbor (NN) spins on the FCC lattice causes a very strong frustration. This is due to the fact that the FCC lattice is composed of corner-sharing tetrahedra each of which has four equilateral triangles. It is well-known [86] that it is impossible to fully satisfy simultaneously the three AF bond interactions on each triangle.

The analytical determination of the GS of systems of classical spins with competing interactions is a fascinating subject. For a recent review, the reader is referred to Ref. [175]. For the bulk FCC antiferromagnet, the Heisenberg spins on a tetrahedron form a configuration characterized by two arbitrary angles [262]. The ground state (GS) degeneracy is, therefore, infinite. This is also found in fully frustrated simple cubic lattice with classical Heisenberg spins [197]: The GS is also characterized by two random continuous parameters. To give an idea about the GS of the bulk FCC antiferromagnet [262], let us imagine two planes, xz and ψ, where ψ intersects the xz plane along the z axis and makes an angle ϕ with the x axis. Two of the four spins make an angle θ in the xz plane symmetric with respect to the z axis. The other two spins make also the same angle, symmetric with respect to the z axis, but in the plane ψ. It has been shown [262] that the two angles θ and ϕ are arbitrary between 0 and π. Note that when $\theta = 0$ the spin configuration is collinear with two spins along the $+z$ axis and the other two along the $-z$ one. The phase transition of the bulk frustrated FCC Heisenberg antiferromagnet has been studied [82, 137]. In particular, the transition is found to be of the first order as in the Ising case [272, 278, 334]. Other similar frustrated antiferromagnets such as the HCP antiferromagnet show the same behavior [83].

Let us consider a film of FCC lattice structure with (001) surfaces. To avoid the absence of long-range order of isotropic non-Ising spin model at finite temperature (T) when the film thickness is very small, i.e., quasi two-dimensional system [231], we add in the Hamiltonian an Ising-like uniaxial anisotropy term. The Hamiltonian is given by

$$\mathcal{H} = -\sum_{\langle i,j\rangle} J_{i,j}\mathbf{S}_i \cdot \mathbf{S}_j - \sum_i D_i (S_i^z)^2 \tag{9.1}$$

where $\mathbf{S}_i$ is the Heisenberg spin at the lattice site i, $\sum_{\langle i,j\rangle}$ indicates the sum over the NN spin pairs $\mathbf{S}_i$ and $\mathbf{S}_j$.

In the following, the interaction between two NN surface spins is denoted by J_s, while all other interactions are supposed to be antiferromagnetic and all equal to $J = -1$ for simplicity. Note that the case of pure Ising model on the simple cubic lattice has been studied by MC simulation with various surface conditions [34, 200].

We first determine the GS configuration by using the steepest descent method : Starting from a random spin configuration, we calculate the magnetic local field at each site and align the spin of the site in its local field. In doing so for all spins and repeat until the convergence is reached, we obtain easily the GS configuration without metastable states. The result is shown in Fig. 9.1.

We observe that there is a critical value $J_s^c = -0.5$. For $J_s < J_s^c$, the spins in each yz plane are parallel while spins in adjacent

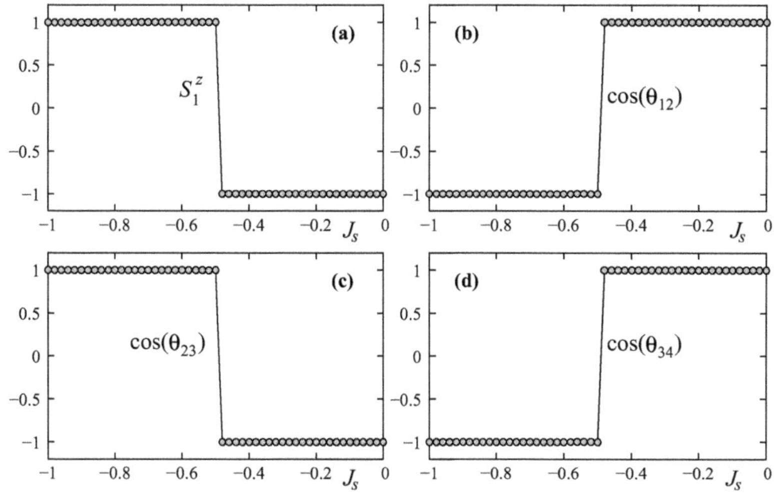

Figure 9.1 A ground state configuration of single plaquette (a) S_1^z is S^z of sublattice 1, (b) $\cos\theta_{12}$, (c) $\cos\theta_{23}$, (d) $\cos\theta_{34}$. $\cos\theta_{ij}$ is the cosine of the angle between the two spins of sublattices i and j.

yz planes are antiparallel (Fig. 9.2a). This ordering will be called hereafter "ordering of type I": In the x direction, the ferromagnetic planes are antiferromagnetically coupled as shown in this figure. Of course, there is a degenerate configuration where the ferromagnetic planes are antiferromagnetically ordered in the y direction. Note that the surface layer has an AF ordering for both configurations. The degeneracy of type I is, therefore, 4, including the reversal of all spins.

For $J_s > J_s^c$, the spins in each xy plane is ferromagnetic. The adjacent xy planes have an AF ordering in the z direction perpendicular to the film surface. This will be called hereafter "ordering of type II." Note that the surface layer is then ferromagnetic (Fig. 9.2b). The degeneracy of type II is 2 due to the reversal of all spins.

Without using a general method [175, 262], let us calculate analytically the GS configuration in a simple manner for the present model.

Consider a tetrahedron with the spins numbered as in Fig. 9.2: S_1, S_2, S_3 and S_4 are the spins in the surface FCC cell (first cell). The interaction between S_1 and S_2 is set to be equal to J_s $(-1 \leq J_s \leq 0)$

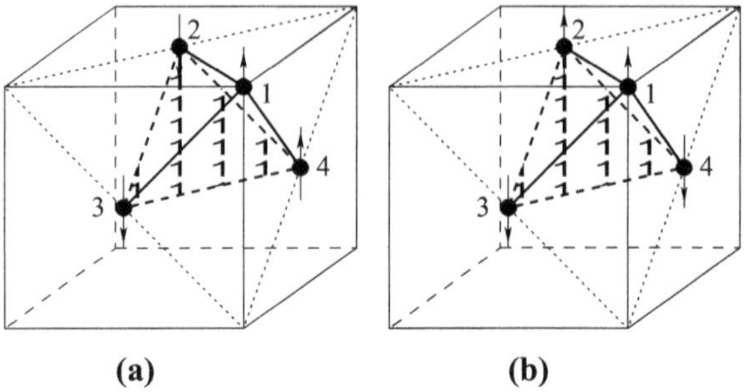

Figure 9.2 The ground state spin configuration of the FCC cell at the film surface: (a) ordering of type I for $J_s < -0.5$, (b) ordering of type II for $J_s > -0.5$.

and all others are taken to be equal to J (<0), and all $D_i = D$ for simplicity. The Hamiltonian for the cell is written as

$$H_p = -\frac{1}{2}\Big\{8J_s(\mathbf{S}_1 \cdot \mathbf{S}_2) + 8J\,(\mathbf{S}_3 \cdot \mathbf{S}_4) + 6J\,[\mathbf{S}_1 \cdot \mathbf{S}_3 + \mathbf{S}_1 \cdot \mathbf{S}_4 + \mathbf{S}_2 \cdot \mathbf{S}_3 + \mathbf{S}_2 \cdot \mathbf{S}_4] + 2D\left[(S_1^z)^2 + (S_2^z)^2 + (S_3^z)^2 + (S_4^z)^2\right]\Big\} \quad (9.2)$$

Let us decompose each spin into two components: an xy component, which is a vector, and a z component $\mathbf{S}_i = (\mathbf{S}_i^{\parallel}, S_i^z)$. The numerical results shown above indicate that the spins have only z component. Taking advantage of this, we suppose that the xy vector components of the spins are all equal to zero. The angles θ_i of $\mathbf{S}_i$ with the z axis are then

$$\begin{cases} \theta_1 = 0, \quad \theta_3 = \pi \\ \theta_2 = \beta_1, \; \theta_4 = \beta_2 \end{cases}$$

The total energy of the cell (9.2), with $S_i = \frac{1}{2}$, can be rewritten as

$$H_p = -\frac{D}{2} + \frac{3J}{4} + \left(\frac{3J}{4} - J_s - \frac{D}{4}\cos\beta_1\right)\cos\beta_1 + \frac{1}{4}(J - D\cos\beta_2)\cos\beta_2 - \frac{3J}{4}\cos\beta_1\cos\beta_2 \quad (9.3)$$

By a variational method, the minimum of the cell energy corresponds to

$$\frac{\partial H_p}{\partial \beta_1} = J_s \sin\beta_1 + \frac{D}{2}\cos\beta_1 \sin\beta_1 - \frac{3J}{4}\sin\beta_1 + \frac{3J}{4}\cos\beta_2 \sin\beta_1 = 0 \quad (9.4)$$

$$\frac{\partial H_p}{\partial \beta_2} = \left[\frac{3J}{4}\cos\beta_1 - \frac{J}{4} + \frac{D}{2}\cos\beta_2\right]\sin\beta_2 = 0 \quad (9.5)$$

The solutions of Eq. (9.4) and Eq. (9.5) corresponding to the minimal energy are

$$\begin{cases} \cos\beta_1 = -\cos\beta_2 = -1 \text{ for } |J_s| > 0.5 \\ \cos\beta_1 = -\cos\beta_2 = 1 \quad \text{for } |J_s| < 0.5 \end{cases} \quad (9.6)$$

Note that these solutions do not depend on D. The GS energy per spin is

$$\begin{cases} H_p = -D + J + J_s \quad \text{for } |J_s| > 0.5, \\ H_p = -D + 2J - J_s \text{ for } |J_s| < 0.5. \end{cases} \quad (9.7)$$

We see that the solution (9.6) agrees with the numerical result.

9.3 Monte Carlo Results

In this paragraph, we show the results obtained by MC simulations with the Hamiltonian (9.1). The spins are the classical Heisenberg model of magnitude $S = 1$.

The film size is $L \times L \times N_z$ where N_z is the number of FCC cells along the z direction (film thickness). Note that each cell has two atomic planes. We use here L = 12, 18, 24, 30, 36 and N_z = 4. Periodic boundary conditions are used in the xy planes. The equilibrating time is about 10^6 MC steps per spin and the averaging time is 2×10^6 MC steps per spin. $|J| = 1$ is taken as unit of energy in the following.

Before showing the results, let us adopt the following notations. The sublattice 1 of the first cell belongs to the surface layer, while the sublattice 3 of the first cell belongs to the second layer. The sublattices 1 and 3 of the second cell belong, respectively, to the third and fourth layers. In our simulations, we used four cells, $N_z = 4$, i.e., 8 atomic layers. The symmetry of the two film surfaces imposes the equivalence of the first and fourth cells and that of the second and third cells. It suffices then to show the results of the first two cells, i.e., four first layers. In addition, in each atomic layer the two sublattices are equivalent by symmetry. Therefore, we choose to show in the following the results of the sublattices 1 and 3 of the first two cells, i.e., results of the first four layers.

Let us show in Fig. 9.3 the magnetizations and the susceptibilities of sublattices 1 and 3 of the first two cells, in the case where $J_s = -1$.

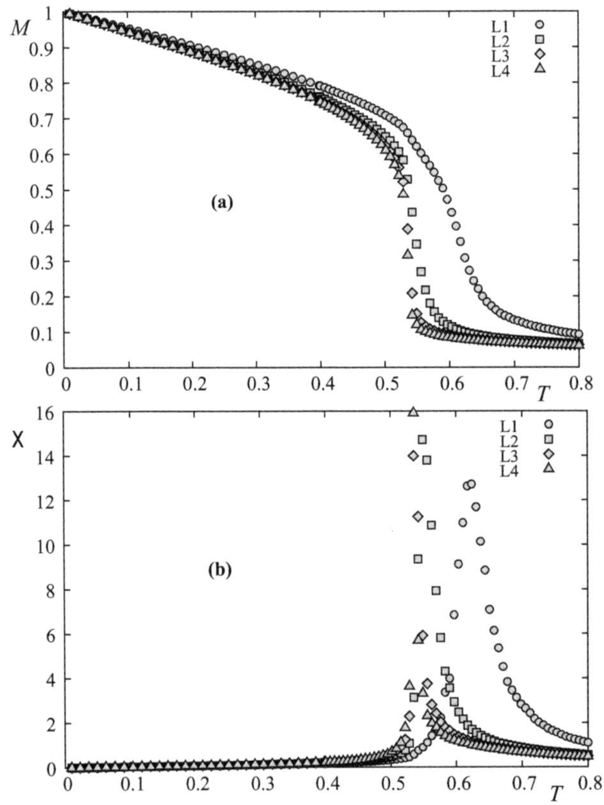

Figure 9.3 Magnetizations and susceptibilities of sublattices 1 and 3 first two cells vs temperature for $J_s = -1.0$ with $L = 24$ and $D = 0.1$. L_j denotes the sublattice magnetization of layer j.

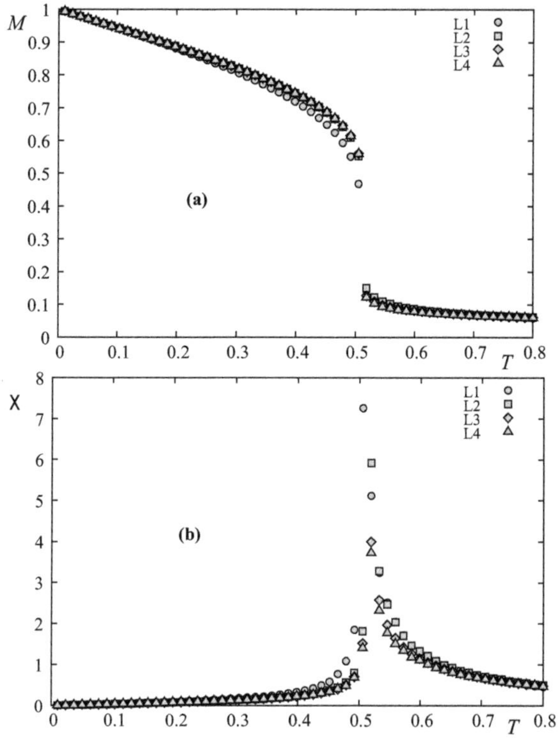

Figure 9.4 Magnetizations and susceptibilities of sublattices 1 and 3 of first two cells vs temperature for $J_s = -0.8$ with $L = 24$ and $D = 0.1$. L_j denotes the sublattice magnetization of layer j.

It is interesting to note that the surface layer has largest magnetization followed by that of the second layer, while the third and fourth layers have smaller magnetizations. This is not the case for non-frustrated films where the surface magnetization is always smaller than the interior ones because of the effects of low-lying energy surface-localized magnon modes [75, 78]. One explanation can be advanced: Due to the lack of neighbors, surface spins suffer fluctuations due to the frustration less than the interior spins so they maintain their ordering up to a higher temperature. Let us decrease the J_s strength. The surface spins then have smaller local field, so thermal fluctuations will reduce their ordering to a lower temperature. Figures 9.4 and 9.5 show, respectively, the cases where

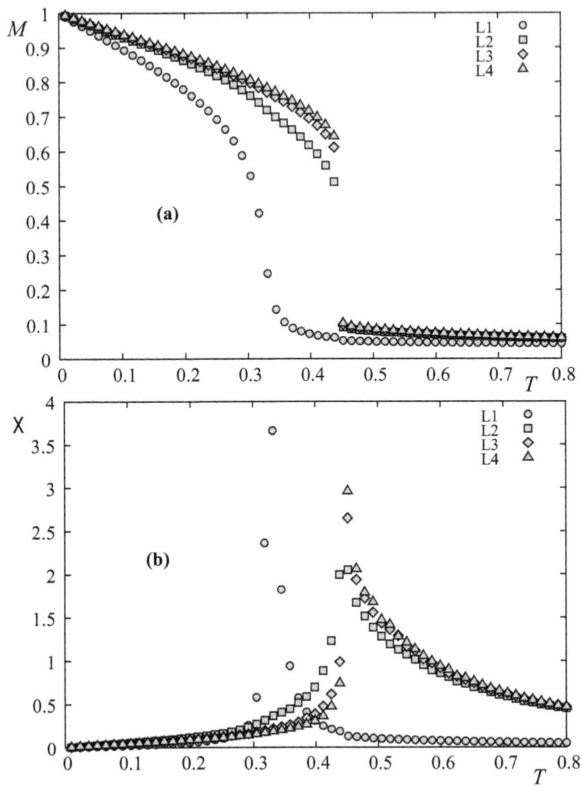

Figure 9.5 Magnetizations and susceptibilities of first two cells vs temperature for $J_s = -0.5$ with $L = 24$ and $D = 0.1$. L_j denotes the sublattice magnetization of layer j. The susceptibility of sublattice 1 of the first cell is divided by a factor 5 for presentation convenience.

$J_s = -0.8$ and -0.5. Near $J_s = -0.8$ the crossover takes place: The surface magnetization becomes smaller than the interior ones for $J_s > -0.8$. Note that the magnetizations of second, third and fourth layers undergo a discontinuity at the transition temperature for $J_s = -0.8$ and -0.5. This suggests that the phase transitions for interior layers are of first order as it has been found for bulk FCC antiferromagnet [82].

For weak $|J_s|$, there is only one transition for all layers. An example is shown in Fig. 9.6 for $J_s = -0.1$. Note that the first-

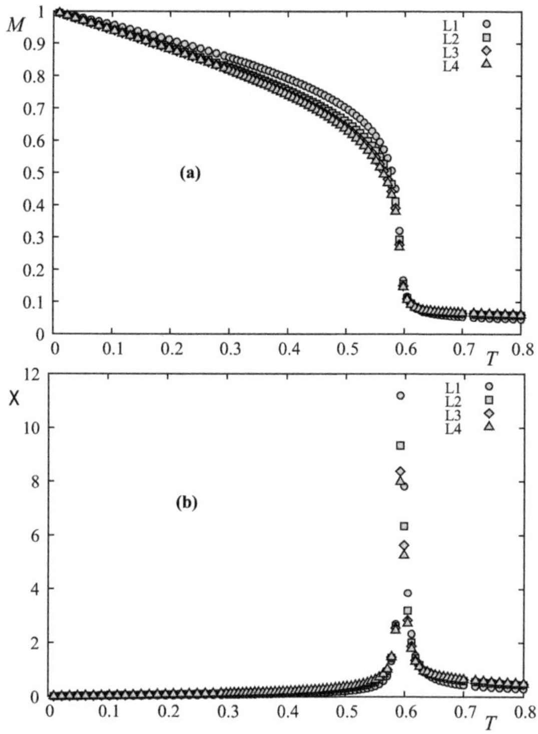

Figure 9.6 Magnetization and susceptibility of first two cells vs temperature for $J_s = -0.1$ with $L = 24$ and $D = 0.1$. L_j denotes the sublattice magnetization of layer j.

order character disappears: There is no discontinuity of layer magnetizations at the transition temperature.

In the region $-0.5 < J_s < -0.45$, there is an interesting reentrant phenomenon. To facilitate the description of this phenomenon, let us show the phase diagram in the space (J_s, T_c) in Fig. 9.7. In the region $-0.5 < J_s < -0.45$, the GS is of type II as seen above. According to the phase diagram, we see that when the temperature increases from zero, the system goes through the phase of type II, undergoes a transition to enter the phase of type I before making a second transition to the paramagnetic phase at high temperature. This kind of behavior is termed as reentrant phenomenon which has been found by exact solutions in a number of very frustrated systems

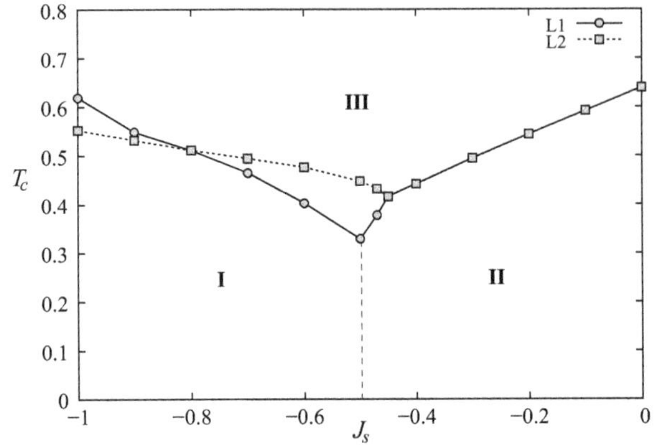

Figure 9.7 Critical temperature vs J_s with $L = 24$ and $D = 0.1$. L_j denotes data points for the maximum of the sublattice magnetization of layer j. I and II denote ordering of type I and II defined in Fig. 9.2. III is paramagnetic phase. The discontinued vertical line is a first-order line. Errors are smaller than symbol sizes. See text for comments.

[67, 76]. For a complete review on these exactly solved systems, the reader is referred to the chapter by Diep and Giacomini [86] in Ref. [85]. We note here that the reentrance is often found near the frontier where two phases coexist in the GS [86]. This is the case at $J_s = J_s^c = -0.5$.

The discontinued vertical line at $J_s = -0.5$ is a first-order line separating phases I and II. The coexistence of these two phases which do not have the same symmetry explains the first-order character of this line. To show it explicitly, we have calculated at $T = 0.15$ the magnetization M and the staggered magnetization M_{st} of the first layer with varying J_s across -0.5. From the GS configurations shown in Fig. 9.2, M should be zero in phase I and finite in phase II, and vice versa for M_{st}. This is observed at $T = 0.15$ as shown in Fig. 9.8. The large discontinuity of M and M_{st} at $J_s = -0.5$ shows a very strong first-order character across the vertical line in Fig. 9.7.

Let us discuss on finite-size effects in the transitions observed in Fig. 9.3 to Fig. 9.6. This is an important question because it is known that some apparent transitions are artifacts of small system sizes. We have checked with $L = 36$. The results do not change.

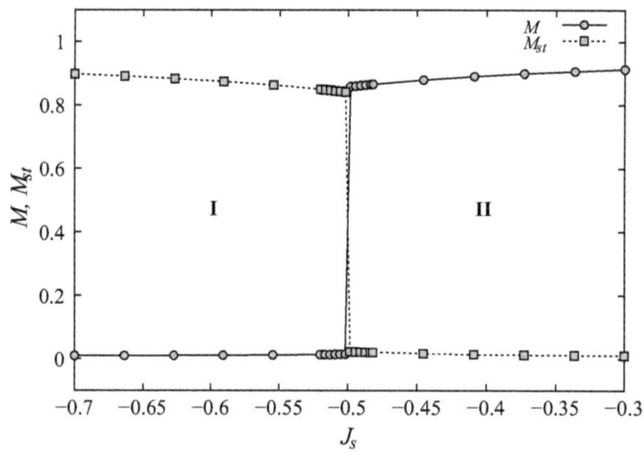

Figure 9.8 The magnetization M and the staggered magnetization M_{st} of first layer versus J_s are shown, at $T = 0.15$, with $L = 24$ and $D = 0.1$. I and II denote ordering of type I and II defined in Fig. 9.2. III is paramagnetic phase. See text for comments.

Of course, we may think that these sizes are still small to change the shape of the phase diagram, especially near $J_s = -0.5$, where finite-size effects may be strong because of the frontier between two phases. But the results from the Green's function method which are for infinite L show, as will be seen later, the similar shape near $J_s = -0.5$. So, we believe that the results in Fig. 9.7 remain for the infinite size.

To confirm further the observed transitions, we have made a study of finite-size effects on the layer susceptibilities at some chosen values of J_s by using the accurate MC multi-histogram technique [110–112] (see Chapter 6). At this point, let us recall that transitions in bulk Ising frustrated systems, unlike unfrustrated counterparts, have different natures: The antiferromagnetic FCC and HCP Ising lattices have strong first-order transition [272, 278, 334], while the stacked antiferromagnetic triangular lattice has a controversial nature (see references in Ref. [277]). The model studied here is the frustrated FCC film where surface effects can modify the strong first-order observed in its bulk counterpart.

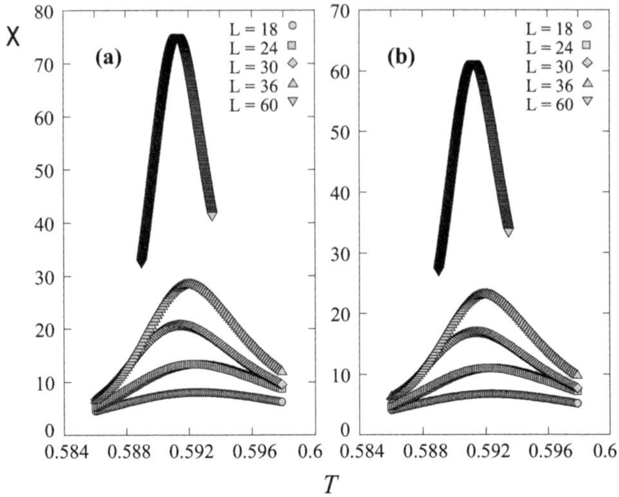

Figure 9.9 Susceptibilities of layer 1 (left) and 2 (right) are shown for various sizes L as a function of temperature for $J_s = -0.1$ and $D = 0.1$.

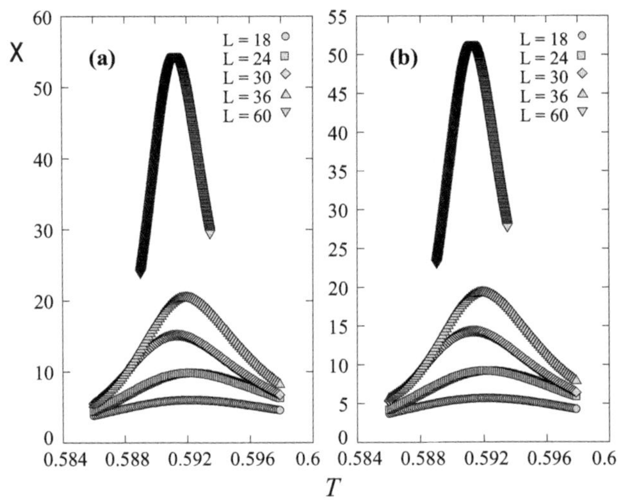

Figure 9.10 Susceptibilities of layer 3 (left) and 4 (right) are shown for various sizes L as a function of temperature for $J_s = -0.1$ and $D = 0.1$.

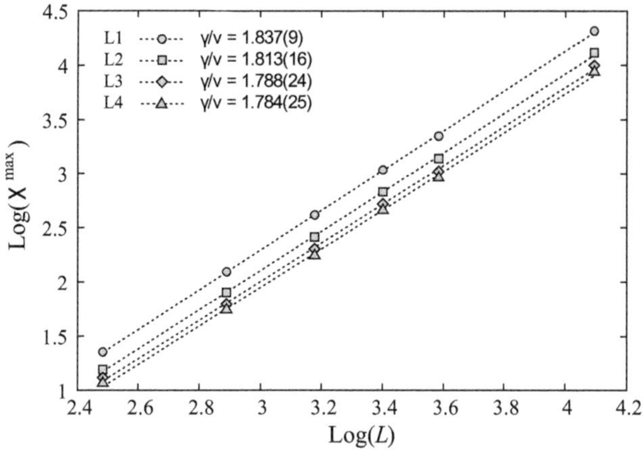

Figure 9.11 Maximum sublattice susceptibility $\chi^{\max}$ versus L in the $\ln-\ln$ scale, for $J_s = -0.1$ and $D = 0.1$. L_j denotes the sublattice magnetization of layer j. The slopes of these lines give the ratios of exponents γ/ν.

Our results show that transitions at $J_s = -1$ and $J_s = -0.1$ are real second-order transitions obeying some scaling law. Figure 9.9 shows the size effects on the maximum of the susceptibilities of the first and second layers for $J_s = -0.1$, while Fig. 9.10 shows that of the third and fourth layers. As seen, the maximum of the susceptibilities $\chi^{\max}$ increases with increasing L.

Using the scaling law $\chi^{\max} \propto L^{\gamma/\nu}$ (see Chapter 6), we plot $\ln\chi^{\max}$ versus $\ln L$ in Fig. 9.11. The ratio of the critical exponents γ/ν is obtained by the slope of the straight line connecting the data points of each layer.

Within errors the third and fourth layers have the same value of γ/ν which is neither 2D nor 3D Ising universality classes, 1.75 and 2, respectively. The same holds for the values of the first and second layers. The exponent ν can be obtained as follows. We calculate as a function of T the magnetization derivative with respect to $\beta = (k_B T)^{-1}$: $V_1 = \langle (\ln M)' \rangle = \langle E \rangle - \langle ME \rangle / \langle M \rangle$ where E is the system energy and M the sublattice order parameter. We identify the maximum of V_1 for each size L. From the finite-size scaling we know that $V_1^{\max}$ is proportional to $L^{1/\nu}$ [112]. We plot in Fig. 9.12 $\ln V_1^{\max}$ as a function of $\ln L$ for $J_s = -0.1$. The slope of each line gives $1/\nu$. For the case $J_s = -0.1$, we obtain $\nu = 0.822 \pm 0.020$, $0.795 \pm$

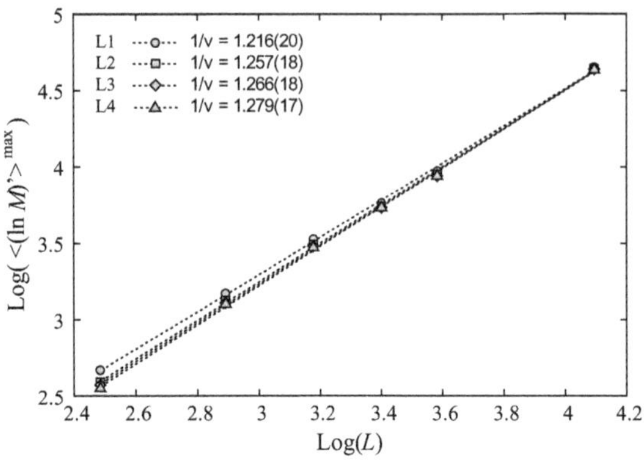

Figure 9.12 The maximum value of $\langle(\ln M)'\rangle = \langle E\rangle - \langle ME\rangle / \langle M\rangle$ versus L in the ln-ln scale for $J_s = -0.1$, where M is the sublattice order parameter. The slope of each line gives $1/\nu$. L_j denotes the sublattice magnetization of layer j.

0.020, 0.790 ± 0.020, 0.782 ± 0.020 for the first, second, third and fourth layers. These values are far from the 2D value ($\nu = 1$). We deduce $\gamma = 1.510 \pm 0.010, 1.442 \pm 0.015, 1.412 \pm 0.025, 1.395 \pm 0.025$. The values of ν and γ are decreased when one goes from the surface to the interior of the film.

We show in Fig. 9.13 and Fig. 9.14 the maximum of sublattice magnetizations and their derivatives for the first two layers in the case of $J_s = -1$. We find $\nu_1 = 0.794 \pm 0.022$, $\nu_2 = 0.834 \pm 0.027$, $\gamma_1 = 1.524 \pm 0.0.040$, and $\gamma_2 = 1.509 \pm 0.022$.

Let us discuss on the values of the critical exponents obtained above. These values do not correspond neither to 2D nor 3D Ising models ($\gamma_{2D} = 1.75$, $\nu_{2D} = 1$, $\gamma_{3D} = 1.241$, $\nu_{3D} = 0.63$). There are multiple reasons for those deviations. Apart from numerical precisions and the modest sizes we used, there may be deep physical origins.

A first question which naturally arises is the effect of the frustration. The 3D version of this model, as said above, has a first-order transition, with a very strong character for the Ising case [272, 278, 334] and somewhat less strong for the continuous spin models [82, 83]. It has been shown that at finite temperature, the phenomenon called "order by disorder" occurs leading to a

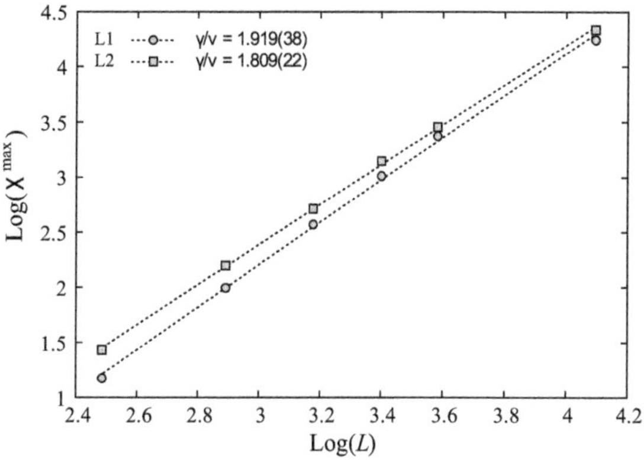

Figure 9.13 Maximum sublattice susceptibility $\chi^{\max}$ versus L in the ln − ln scale, for $J_s = -1$ and $D = 0.1$. L_j denotes the sublattice magnetization of layer j. The slopes of these lines give the ratios of exponents γ/ν.

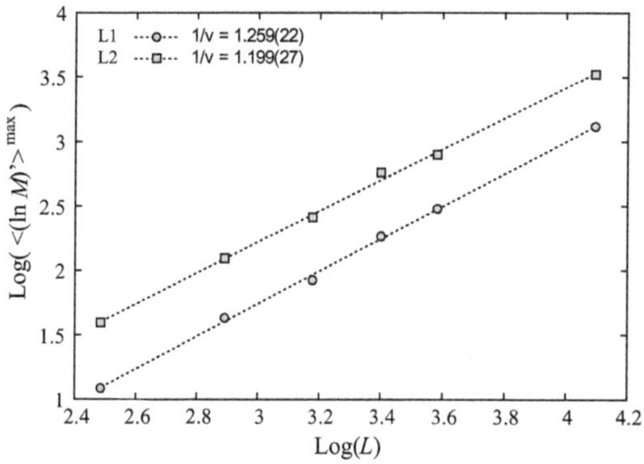

Figure 9.14 The maximum value of $\langle(\ln M)'\rangle = \langle E\rangle - \langle ME\rangle / \langle M\rangle$ versus L in the ln − ln scale for $J_s = -1$, where M is the sublattice order parameter. The slope of each line gives $1/\nu$. $L_k S_j$ denotes one sublattice magnetization of layer j.

reduction of degeneracy: Only collinear configurations survive by an entropy effect [82, 146, 349]. The infinite degeneracy is reduced to 6, i.e., the number of ways to put two AF spin pairs on a tetrahedron. The model is equivalent to 6-state Potts model. The first-order transition observed in the 3D case is in agreement with the Potts criterion according to which the transition in q-state Potts model is of first-order in 3D for $q \geq 3$.

In the case of a film with finite thickness studied here, it appears that the first-order character is lost.

A first possible cause is from the degeneracy. According to the results shown in the previous subsection, the GS degeneracy is 2 or 4 depending to J_s. If we compare to the Potts criterion according to which the transition is of first-order in 2D only when $q > 4$, then the transition in thin films should be of second order. That is indeed what we observed.

Another possible cause for the second-order transition observed here is from the role of the correlation in the film. For second-order transitions, some arguments, such as those from renormalization group, say that the correlation length in the direction perpendicular to the film is finite. Hence, it is irrelevant to the criticality. The film should have the 2D character. If a transition is of first order in 3D, i.e., the correlation length is finite at the transition temperature, then in thin films the thickness effect may be important: If the thickness is larger than the correlation length at the transition, than the first-order transition should remain. On the other hand, if the thickness is smaller than that correlation length, the spins then feel an "infinite" correlation length across the film thickness. As a consequence, two pictures can be thought of: (i) The whole system may be correlated and the first-order character is to become a second-order one; (ii) the correlation length is longer but still finite, and the transition remains of first order.

At this point, we would like to emphasize that in the case of simple surface conditions, i.e., no significant deviation of the surface parameters with respect to those of the bulk, the bulk behavior is observed when the thickness becomes larger than a few dozens of atomic layers [75, 78]: Surface effects are insignificant on some thermodynamic properties such as the value of the critical temperature, the mean value of magnetization at a given T, ... It should be, however, stressed that the criticality is very different. It depends on the correlation length compared to the thickness: For example, we have obtained in the case of simple cubic films with Ising model the critical exponents identical to those of 2D Ising universality class up to thickness of nine layers [270]. Due to the small thickness used here, we think that the 2D character should be assumed.

Now for the anisotropy, remember that in the case studied here, we do not deal with the discrete Ising model but rather an Ising-like Heisenberg model. The deviation from the 2D values may then result in part from a complex coupling between the Ising-like symmetry and the continuous nature of the classical Heisenberg spins. This deviation may be important if the anisotropy constant D is small as in the case studied here. From the renormalization group calculations, since anisotropy is a relevant parameter, one expects that any finite anisotropy will lead to Ising-like critical behavior, but with corrections due to the continuous nature of Heisenberg spins before one enters the linear regime around the Ising fixed point.

To conclude this paragraph, we believe, from physical arguments given above, that the critical exponents obtained above which do not belong to any known universality class may result from different physical mechanisms. This is a subject of future investigations. We will come back to the question of criticality in thin films in Chapter 16.

9.4 Green's Function Results

We can rewrite the full Hamiltonian (9.1) in the local framework as

$$\begin{aligned}\mathcal{H} = &- \sum_{<i,j>} J_{i,j} \Big\{ \frac{1}{4} \left(\cos\theta_{ij} - 1\right) \left(S_i^+ S_j^+ + S_i^- S_j^-\right) \\ &+ \frac{1}{4} \left(\cos\theta_{ij} + 1\right) \left(S_i^+ S_j^- + S_i^- S_j^+\right) \\ &+ \frac{1}{2} \sin\theta_{ij} \left(S_i^+ + S_i^-\right) S_j^z - \frac{1}{2} \sin\theta_{ij} S_i^z \left(S_j^+ + S_j^-\right) \\ &+ \cos\theta_{ij} S_i^z S_j^z \Big\} - \sum_{<i>} I_i (S_i^z)^2 \end{aligned} \tag{9.8}$$

where $\cos\left(\theta_{ij}\right)$ is the angle between two NN spins.

To study properties of quantum spins over a large region of temperatures, there are only a few methods which give relatively correct results. Among them, the GF method is known to recover the exact results at very low-T obtained from the spin wave theory. In addition, it is better than the spin wave theory at higher temperatures and can be used up to the transition temperature with of course less precision on the nature of the phase transition. We choose here this method to study quantum effects at low T and to obtain the phase diagram at high T.

The GF method can be used for non-collinear spin configurations [285]. In the case studied here, one has a collinear one because of the Ising-like anisotropy. In this case, we define two double-time GF by [340]

$$G_{ij}(t,t') = \ll S_i^+(t); S_j^-(t') \gg \tag{9.9}$$

$$F_{ij}(t,t') = \ll S_i^-(t); S_j^+(t') \gg \tag{9.10}$$

The equations of motion for $G_{ij}(t,t')$ and $F_{ij}(t,t')$ are written by

$$i\frac{d}{dt}G_{i,j}(t,t') = \left\langle \left[S_i^+(t), S_j^-(t')\right]\right\rangle \delta(t-t') - \left\langle\left\langle \left[\mathcal{H}, S_i^+(t)\right]; S_j^-(t')\right\rangle\right\rangle \tag{9.11}$$

$$i\frac{d}{dt}F_{i,j}(t,t') = \left\langle \left[S_i^-(t), S_j^-(t')\right]\right\rangle \delta(t-t') - \left\langle\left\langle \left[\mathcal{H}, S_i^-(t)\right]; S_j^-(t')\right\rangle\right\rangle \tag{9.12}$$

We shall neglect higher-order correlations by using the Tyablikov decoupling scheme [43] which is known to be valid for exchange terms [121]. Then, we introduce the following Fourier transforms:

$$G_{i,j}(t,t') = \frac{1}{\Delta}\iint d\mathbf{k}_{xy}\frac{1}{2\pi}\int_{-\infty}^{+\infty} d\omega e^{-i\omega(t-t')} g_{n,n'}(\omega,\mathbf{k}_{xy})\, e^{i\mathbf{k}_{xy}\cdot(\mathbf{R}_i-\mathbf{R}_j)} \tag{9.13}$$

$$F_{i,j}(t,t') = \frac{1}{\Delta}\iint d\mathbf{k}_{xy}\frac{1}{2\pi}\int_{-\infty}^{+\infty} d\omega e^{-i\omega(t-t')} f_{n,n'}(\omega,\mathbf{k}_{xy})\, e^{i\mathbf{k}_{xy}\cdot(\mathbf{R}_i-\mathbf{R}_j)} \tag{9.14}$$

where ω is the spin wave frequency, $\mathbf{k}_{xy}$ denotes the wave vector parallel to xy planes, $\mathbf{R}_i$ is the position of the spin at the site i, n and n' are, respectively, the indices of the layers where the sites i and j belong to. The integral over $\mathbf{k}_{xy}$ is performed in the first Brillouin zone whose surface is Δ in the xy reciprocal plane.

The Fourier transforms of the retarded GF satisfy a set of equations rewritten under the following matrix form:

$$\mathbf{M}(\omega)\,\mathbf{g} = \mathbf{u} \tag{9.15}$$

where $\mathbf{M}(\omega)$ is a square matrix $(2N_z \times 2N_z)$, $\mathbf{g}$ and $\mathbf{u}$ are the column matrices which are defined as follows:

$$\mathbf{g} = \begin{pmatrix} g_{1,n'} \\ f_{1,n'} \\ \vdots \\ g_{N_z,n'} \\ f_{N_z,n'} \end{pmatrix}, \mathbf{u} = \begin{pmatrix} 2\langle S_1^z\rangle \delta_{1,n'} \\ 0 \\ \vdots \\ 2\langle S_{N_z}^z\rangle \delta_{N_z,n'} \\ 0 \end{pmatrix} \tag{9.16}$$

and

$$\mathbf{M}(\omega) = \begin{pmatrix} A_1^+ & B_1 & D_1^+ & D_1^- & \cdots \\ -B_1 & A_1^- & -D_1^- & -D_1^+ & \vdots \\ \vdots & \cdots & \cdots & \cdots & \vdots \\ \vdots & C_{N_z}^+ & C_{N_z}^- & A_{N_z}^+ & B_{N_z} \\ \cdots & -C_{N_z}^- & -C_{N_z}^+ & -B_{N_z} & A_{N_z}^- \end{pmatrix} \tag{9.17}$$

where

$$\begin{aligned} A_n^{\pm} = \omega \pm \Big[&\frac{1}{2} J_n \langle S_n^z\rangle (Z\gamma)(\cos\theta_n + 1) \\ &- J_n \langle S_n^z\rangle Z\cos\theta_n - 2I_n \langle S_n^z\rangle \\ &- 2J_{n,n+1}\langle S_{n+1}^z\rangle \cos\theta_{n,n+1}^{(a)} \\ &- 2J_{n,n+1}\langle S_{n+1}^z\rangle \cos\theta_{n,n+1}^{(b)} \\ &- 2J_{n,n-1}\langle S_{n-1}^z\rangle \cos\theta_{n,n-1}^{(a)} \\ &- 2J_{n,n-1}\langle S_{n-1}^z\rangle \cos\theta_{n,n-1}^{(b)} \Big] \end{aligned} \tag{9.18}$$

$$B_n = \frac{1}{2} J_n \langle S_n^z\rangle (Z\gamma)(\cos\theta_n - 1) \tag{9.19}$$

$$\begin{aligned} C_n^{\pm} = &J_{n,n-1}\langle S_n^z\rangle \left(\cos\theta_{n,n-1}^{(a)} \pm 1\right) \\ &+ J_{n,n-1}\langle S_n^z\rangle \left(\cos\theta_{n,n-1}^{(b)} \pm 1\right) \end{aligned} \tag{9.20}$$

$$\begin{aligned} D_n^{\pm} = &J_{n,n+1}\langle S_n^z\rangle \left(\cos\theta_{n,n+1}^{(a)} \pm 1\right) \\ &+ J_{n,n+1}\langle S_n^z\rangle \left(\cos\theta_{n,n+1}^{(b)} \pm 1\right) \end{aligned} \tag{9.21}$$

in which, $Z = 4$ is the number of in-plane NN, $\theta_{n,n\pm1}^{(a)}$ the angle between two NN spins of sublattice 1 and 3 belonging to the layers n and $n \pm 1$ (see Fig. 9.2), $\theta_{n,n\pm1}^{(b)}$ the angle between two NN spins of sublattice 1 and 4, θ_n the angle between two in-plane NN spins in the layer n, and

$$\gamma = \frac{1}{Z}\left[4\cos\left(\frac{k_x a}{2}\right)\cos\left(\frac{k_y a}{2}\right)\right]$$

Here, for compactness we have used the following notations:

(i) J_n and D_n are the in-plane interactions. In the present model, J_n is equal to J_s for the two surface layers and equal to J for the interior layers. All D_n are set to be D.
(ii) $J_{n,n\pm1}$ are the interactions between a spin in the n-th layer and its neighbor in the $(n\pm1)$-th layer. Of course, $J_{n,n-1} = 0$ if $n = 1$, $J_{n,n+1} = 0$ if $n = N_z$.

Solving $\det|\mathbf{M}| = 0$, we obtain the spin wave spectrum ω of the present system. The solution for the GF $g_{n,n}$ is given by

$$g_{n,n} = \frac{|\mathbf{M}|_n}{|\mathbf{M}|} \tag{9.22}$$

with $|\mathbf{M}|_n$ being the determinant made by replacing the n-th column of $|\mathbf{M}|$ by $\mathbf{u}$ in (9.16). Writing now

$$|\mathbf{M}| = \prod_i \left(\omega - \omega_i\left(\mathbf{k}_{xy}\right)\right) \tag{9.23}$$

one sees that $\omega_i\left(\mathbf{k}_{xy}\right)$, $i = 1, \ldots, N_z$, are poles of the GF $g_{n,n}$. $\omega_i\left(\mathbf{k}_{xy}\right)$ can be obtained by solving $|\mathbf{M}| = 0$. In this case, $g_{n,n}$ can be expressed as

$$g_{n,n} = \sum_i \frac{f_n\left(\omega_i\left(\mathbf{k}_{xy}\right)\right)}{\left(\omega - \omega_i\left(\mathbf{k}_{xy}\right)\right)} \tag{9.24}$$

where $f_n\left(\omega_i\left(\mathbf{k}_{xy}\right)\right)$ is

$$f_n\left(\omega_i\left(\mathbf{k}_{xy}\right)\right) = \frac{|\mathbf{M}|_n\left(\omega_i\left(\mathbf{k}_{xy}\right)\right)}{\prod_{j\neq i}\left(\omega_j\left(\mathbf{k}_{xy}\right) - \omega_i\left(\mathbf{k}_{xy}\right)\right)} \tag{9.25}$$

Next, using the spectral theorem which relates the correlation function $\langle S_i^- S_j^+\rangle$ to the GF [383], one has

$$\left\langle S_i^- S_j^+\right\rangle = \lim_{\varepsilon\to 0}\frac{1}{\Delta}\iint d\mathbf{k}_{xy}\int_{-\infty}^{+\infty}\frac{i}{2\pi}\left(g_{n,n'}\left(\omega + i\varepsilon\right)\right.$$
$$\left. - g_{n,n'}\left(\omega - i\varepsilon\right)\right)\cdot\frac{d\omega}{e^{\beta\omega}-1}e^{i\mathbf{k}_{xy}\cdot\left(\mathbf{R}_i-\mathbf{R}_j\right)} \tag{9.26}$$

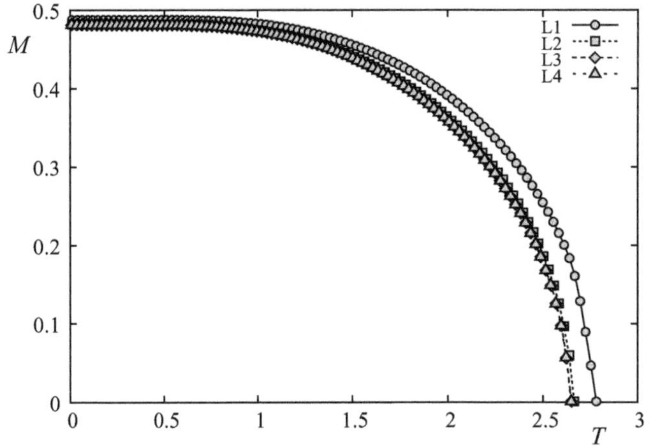

Figure 9.15 Layer magnetization of first four layers vs temperature for $J_s = -1.0$ and $D = 4$. L_j denotes the sublattice magnetization of layer j. Note that except the first layer (upper curve), all other layer magnetizations coincide in this figure scale.

where ϵ is an infinitesimal positive constant and $\beta = 1/k_B T$, k_B being the Boltzmann constant.

Using the GF presented above, we can calculate self-consistently various physical quantities as functions of temperature T. We start the self-consistent calculation from $T = 0$ with a small step for temperature: 5×10^{-3} at low T and 10^{-1} near T_c (in units of J/k_B). The convergence precision has been fixed at the fourth figure of the values obtained for the layer magnetizations. We know from the previous section that the spin configuration is collinear. Therefore in this section, we shall use a large value of Ising anisotropy D in order to get a rapid numerical convergence. For numerical calculation, we will use $D = 4$ and $J = -1$ and a size of 80^2 points in the first Brillouin zone.

Figure 9.15 shows the sublattice magnetizations of the first four layers. As seen, the first-layer one is larger than the other three just as in the case of the classical spins shown in Fig. 9.3. This difference in sublattice magnetization between layers vanishes at $J_s \simeq -0.8$ as seen in Fig. 9.16. Again here, one has a good agreement with the case of classical spins shown in Fig. 9.4.

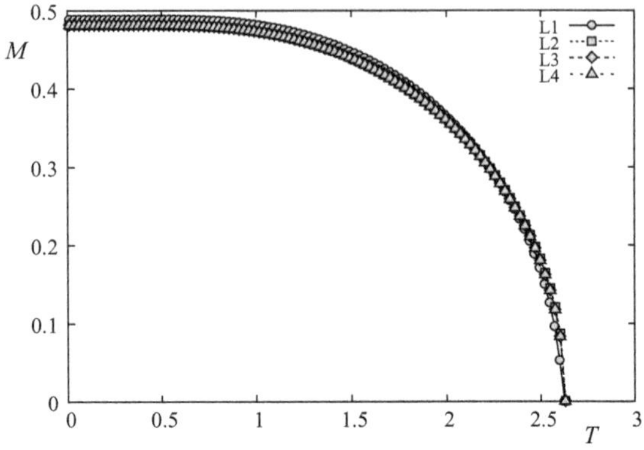

Figure 9.16 Layer magnetizations of first four layers vs temperature for $J_s = -0.8$ and $D = 4$. L_j denotes the sublattice magnetization of layer j. Note that except the first layer at low T (upper curve) all other layer magnetizations coincide in this figure scale.

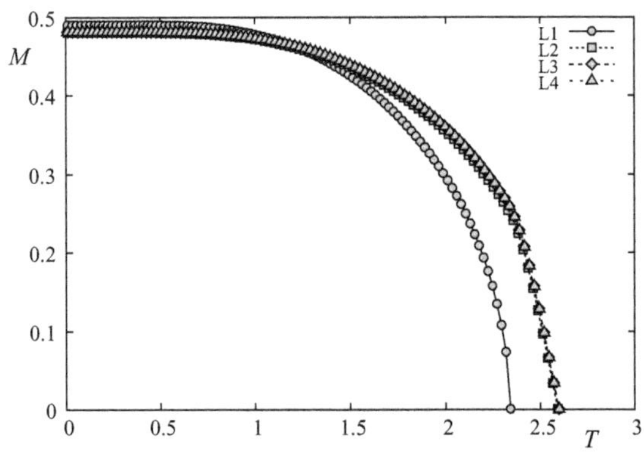

Figure 9.17 Layer magnetization of first four layers vs temperature for $J_s = -0.5$ and $D = 4$. L_j denotes the sublattice magnetization of layer j. Note that the first layer makes a crossover: It is higher at low T and smaller at high T than all other layer magnetizations which coincide in this figure scale. See text for comments on the crossover of surface magnetization.

For $J_s > -0.8$, the sublattice magnetization of the first layer is larger at low T and higher at high T as seen in Fig. 9.17 for $J_s = -0.5$. This crossover of sublattice magnetizations comes from the competition between quantum fluctuations and the strength of

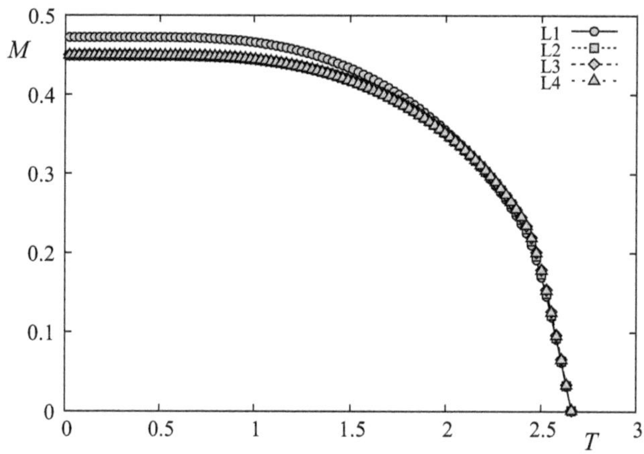

Figure 9.18 Layer magnetization of first four layers vs temperature for $J_s = -0.1$ and $D = 4$. L_j denotes the sublattice magnetization of layer j. Only at low T the surface magnetization is distinct (upper curve) from the other ones.

J_s: When $|J_s|$ is small, quantum fluctuations of the surface layer are small yielding a small zero-point spin contraction for surface spins at $T = 0$. So, surface magnetization is higher than the interior ones. At higher T, however, small $|J_s|$ gives rise to a small local field for surface spins which in turn yields a smaller surface magnetization at high T. This crossover has been found earlier in antiferromagnetic superlattices and films [79, 80].

For $J_s = -0.1$, there is no more crossover at low T as seen in Fig. 9.18. Moreover, there is only a single transition at $T_c \simeq 2.65$ for both surface and interior layers.

We summarize in Fig. 9.19 the phase diagram for the quantum spin case obtained with the GF method. The vertical discontinued line indicates the boundary between ordered phases of types I and II. Phase III is paramagnetic. Note the following interesting points:

(i) For $J_s < -0.4$ there is a surface transition distinct from that of interior layers.
(ii) For $J_s < -0.8$, surface transition occurs at a temperature higher than that of interior layers.

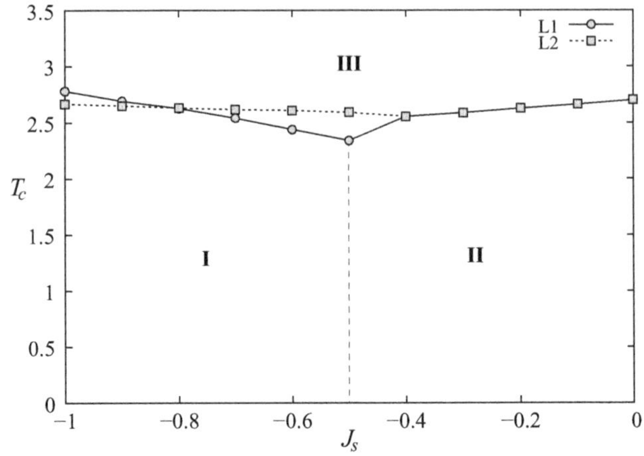

Figure 9.19 Phase diagram obtained by the Green's function method with $D = 4$. L_j denotes the transition temperature of the sublattice magnetization of layer j. Errors are smaller than symbol sizes. See text for comments.

(iii) There is a reentrance between $J_S = -0.4$ and $J_S = -0.5$. This is very similar to the phase diagram of the classical spins obtained by MC simulations shown in Fig. 9.7.

9.5 Concluding Remarks

We have shown, by means of a Green's function method and MC simulations, the results of the Heisenberg spin model with an Ising-like interaction anisotropy in thin films of stacked triangular lattices. The two surfaces of the film are frustrated. We found that surface spin configuration is non-collinear when surface antiferromagnetic interaction is smaller than a critical value J_S^c. In the non-collinear regime, the surface layer is disordered at a temperature lower than that for interior layers ("soft" surface). This can explain the so-called "magnetically dead surface" observed in some materials [36, 374]. The surface transition disappears for J_S larger than the critical value J_S^c. A phase diagram is established in the space (T, J_S). A good agreement between the Green's function method and the MC simulation is observed. This is due to the fact that at high temperatures where the transition takes place, the quantum nature of spins used in Green's function is lost so that we should find results of classical spins used in MC simulations. We have also studied by MC histogram technique the critical behavior of the phase transition using the finite-size effects. The result of the ratio of critical exponents γ/ν shows that the nature of the transition is complicated due to the influence of several physical mechanisms. The symmetry of the ground state alone cannot explain such a result. We have outlined a number of the most relevant mechanisms. Finally, we note that in surface magnetism the low surface magnetization experimentally observed [36, 374] has been generally attributed to the effects of the reduction of magnetic moments of surface atoms and/or the surface-localized low-lying magnon modes. The model considered in this chapter adds another origin for the low surface magnetization: surface frustration. It completes the list of possible explanations for experimental observations in thin films.

Chapter 10

Heisenberg Thin Films with Frustrated Surfaces

In this chapter, we show the results obtained by extensive Monte Carlo (MC) simulations and analytical Green's function (GF) method for a thin film made of stacked triangular layers of atoms using the Heisenberg spin model. We suppose that the in-plane surface interaction J_s can be antiferromagnetic or ferromagnetic while all other interactions are ferromagnetic. We suppose the film in addition an Ising-like interaction anisotropy. We show that the ground-state spin configuration is non-linear when J_s is lower than a critical value J_s^c. The film surfaces are then frustrated. In the frustrated case, there are two phase transitions related to the disordering of the surface and the interior layers. There is a good agreement between MC and GF results. In addition, we show from MC histogram calculation that the value of the ratio of critical exponents γ/ν of the observed transitions is deviated from the values of two- and three-dimensional Ising universality classes. The origin of this deviation is discussed using general physical arguments.

A part of the results shown here has been published in Ref. [248].

Physics of Magnetic Thin Films: Theory and Simulation
Hung T. Diep

ISBN 978-981-4877-42-8 (Hardcover), 978-1-003-12110-7 (eBook)
www.jennystanford.com

10.1 Introduction

The frustration is known to cause a great number of striking effects in various bulk spin systems. Its effects have been extensively studied during the last three decades theoretically, experimentally and numerically. Frustrated models serve not only as testing grounds for theories and approximations, but also to interpret experiments [85].

In the same period, surface physics and systems of nanoscales have been also enormously studied. This is due in particular to applications in magnetic recording, magnetic sensors, spin transport, ... let alone fundamental theoretical interests. Much is understood theoretically and experimentally in thin films where surfaces are 'clean,' i.e., no impurities, no steps, no islands, no defects of any kind [35, 75, 78, 79, 252]. Less is known at least theoretically on complicated thin films with special surface conditions such as defects [63, 320], arrays of dots and magnetization reversal phenomenon [132, 164, 189, 232, 298, 306, 307].

Section 10.2 is devoted to the description of our model. The ground state in the case of classical spins is determined as a function of the surface interaction. In Section 10.3, we consider the case of quantum spins and we apply the GF technique to determine the layer magnetizations and the transition temperature as a function of the surface interaction. The classical ground state determined in Section 10.2 is used here as starting (input)configuration for quantum spins. We are interested here in the effect of magnetic frustration on magnetic properties of thin films. A phase diagram is established showing interesting surface behaviors. Results from MC simulations for classical spins are shown in Section 10.4 and compared to those obtained by the GF method. We also calculate by the MC histogram technique the critical behavior of the phase transition observed here.

10.2 Model

It is known that many well-established theories failed to deal with frustrated spin systems [85]. Among the striking effects due to frustration, let us mention the high ground-state (GS) degeneracy associated often with new symmetries which give rise sometimes to new kinds of phase transition. One of the systems which are most studied is the antiferromagnetic triangular lattice. Due to its geometry, the spins are frustrated under nearest-neighbor (NN) antiferromagnetic interaction. In the case of Heisenberg or XY models, the frustration results in a 120° GS structure: The NN spins form a 120° angle alternately in the clockwise and counter-clockwise senses which are called left and right chiralities (see Fig. 18.6 in the solution to Problem 6 of Chapter 3). Another popular model is Villain's model where the frustration is caused by the competition between ferro- and antiferromagnetic interactions as seen in Problem 7 of Chapter 3 and its solution in Section 18.3).

10.2.1 Hamiltonian

In this chapter, we consider a thin film made up by stacking N_z planes of triangular lattice of $L \times L$ lattice sites.

The Hamiltonian is given by

$$\mathcal{H} = -\sum_{\langle i,j\rangle} J_{i,j}\mathbf{S}_i \cdot \mathbf{S}_j - \sum_{<i,j>} I_{i,j} S_i^z S_j^z \tag{10.1}$$

where $\mathbf{S}_i$ is the Heisenberg spin at the lattice site i, $\sum_{\langle i,j\rangle}$ indicates the sum over the NN spin pairs $\mathbf{S}_i$ and $\mathbf{S}_j$. The last term, which will be taken to be very small, is needed to make the film with a finite thickness to have a phase transition at a finite temperature in the case where all exchange interactions $J_{i,j}$ are ferromagnetic. This guarantees the existence of a phase transition at finite temperature, since it is known that a strictly two-dimensional system with an isotropic non-Ising spin model (XY or Heisenberg model) does not have long-range ordering at finite temperature [231].

Interaction between two NN surface spins is equal to J_s. Interaction between NN on adjacent layers and interaction between NN on an interior layer are supposed to be ferromagnetic and all equal to $J = 1$ for simplicity. The two surfaces of the film are frustrated if J_s is antiferromagnetic ($J_s < 0$).

10.2.2 Ground State

In this paragraph, we suppose that the spins are classical. The classical GS can be easily determined as shown below. Note that for antiferromagnetic systems, even for bulk materials, the quantum GS cannot be exactly determined (see Chapter 3). The classical GS is often used as starting configuration for quantum spins. We will follow the same line hereafter.

For $J_s > 0$ (ferromagnetic interaction), the GS is ferromagnetic. However, when J_s is negative the surface spins are frustrated. Therefore, there is a competition between the non-collinear surface ordering and the ferromagnetic ordering due to the ferromagnetic interaction from the spins of the beneath layer.

We first determine the GS configuration for $I = I_s = 0.1$ by using the steepest descent method: Starting from a random spin configuration, we calculate the magnetic local field at each site and align the spin of the site in its local field. In doing so for all spins and repeat until the convergence is reached, we obtain in general the GS configuration, without metastable states in the present model. The result shows that when J_s is smaller than a critical value J_s^c the magnetic GS is obtained from the planar $120°$ spin structure, supposed to be in the XY plane, by pulling them out of the spin XY plane by an angle β. The three spins on a triangle on the surface form thus an 'umbrella' with an angle α between them and an angle β between a surface spin and its beneath neighbor (see Fig. 10.1). This non-planar structure is due to the interaction of the spins on the beneath layer, just like an external applied field in the z direction. Of course, when J_s is larger than J_s^c one has the collinear ferromagnetic GS as expected: The frustration is not strong enough to resist the ferromagnetic interaction from the beneath layer.

We show in Fig. 10.2 $\cos(\alpha)$ and $\cos(\beta)$ as functions of J_s. The critical value J_s^c is found between -0.18 and -0.19. This value can be calculated analytically by assuming the 'umbrella structure'. For GS analysis, it suffices to consider just a cell shown in Fig. 10.1. This is justified by the numerical determination discussed above. Furthermore, we consider as a single solution all configurations obtained from each other by any global spin rotation.

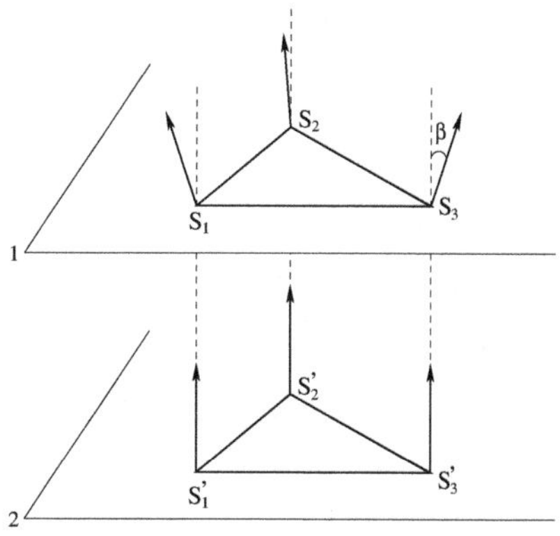

Figure 10.1 Non collinear surface spin configuration. Angles between spins on layer 1 are all equal (noted α), while angles between vertical spins are β.

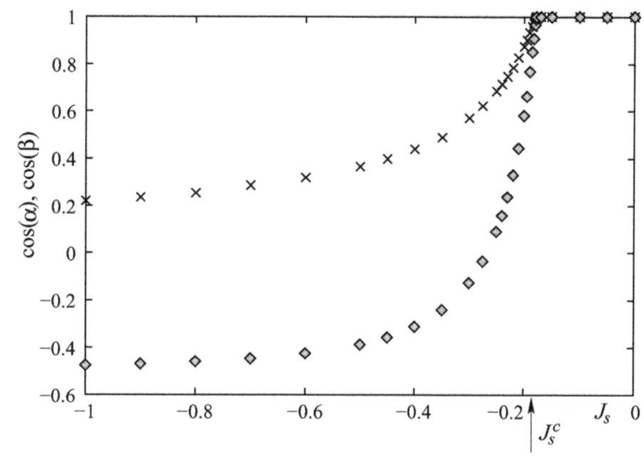

Figure 10.2 cos(α) (diamonds) and cos(β) (crosses) as functions of J_s. Critical value of J_s^c is shown by the arrow.

Let us consider the full Hamiltonian (10.1). For simplicity, the interaction inside the surface layer is set equal J_s ($-1 \leq J_s \leq 1$) and all others are set equal to $J > 0$. Also, we suppose that $I_{i,j} = I_s$ for spins on the surfaces with the same sign as J_s and all other $I_{i,j}$ are equal to $I > 0$ for the inside spins including interaction between a surface spin and the spin on the beneath layer.

The spins are numbered as in Fig. 10.1: S_1, S_2 and S_3 are the spins in the surface layer (first layer), S'_1, S'_2 and S'_3 are the spins in the internal layer (second layer). The Hamiltonian for the cell is written as

$$\begin{aligned}
H_p = &-6\,[J_s\left(\mathbf{S}_1\cdot\mathbf{S}_2+\mathbf{S}_2\cdot\mathbf{S}_3+\mathbf{S}_3\cdot\mathbf{S}_1\right)\\
&+I_s\left(S_1^zS_2^z+S_2^zS_3^z+S_3^zS_1^z\right)\\
&+J\left(\mathbf{S}'_1\cdot\mathbf{S}'_2+\mathbf{S}'_2\cdot\mathbf{S}'_3+\mathbf{S}'_3\cdot\mathbf{S}'_1\right)\\
&+I\left(S_1'^zS_2'^z+S_2'^zS_3'^z+S_3'^zS_1'^z\right)]\\
&-2J\left(\mathbf{S}_1\cdot\mathbf{S}'_1+\mathbf{S}_2\cdot\mathbf{S}'_2+\mathbf{S}_3\cdot\mathbf{S}'_3\right)\\
&-2I\left(S_1^zS_1'^z+S_2^zS_2'^z+S_3^zS_3'^z\right)
\end{aligned}\tag{10.2}$$

Let us decompose each spin into two components: an xy component, which is a vector, and a z component $\mathbf{S}_i = (\mathbf{S}_i^{\parallel}, S_i^z)$. Only surface spins have xy vector components. The angle between these xy components of NN surface spins is $\gamma_{i,j}$ which is chosen by ($\gamma_{i,j}$ is in fact the projection of α defined above on the xy plane)

$$\gamma_{1,2}=0,\ \gamma_{2,3}=\frac{2\pi}{3},\ \gamma_{3,1}=\frac{4\pi}{3}\tag{10.3}$$

The angles β_i and β'_i of the spin $\mathbf{S}_i$ and $\mathbf{S}'_i$ with the z axis are by symmetry

$$\begin{cases}\beta_1=\beta_2=\beta_3=\beta\\ \beta'_1=\beta'_2=\beta'_3=0\end{cases}$$

The total energy of the cell (10.2), with $S_i = S'_i = \frac{1}{2}$, can be rewritten as

$$H_p=-\frac{9(J+I)}{2}-\frac{3(J+I)}{2}\cos\beta-\frac{9(J_s+I_s)}{2}\cos^2\beta+\frac{9J_s}{4}\sin^2\beta\tag{10.4}$$

By a variational method, the minimum of the cell energy corresponds to

$$\frac{\partial H_p}{\partial\beta}=\left(\frac{27}{2}J_s+9I_s\right)\cos\beta\sin\beta+\frac{3}{2}(J+I)\sin\beta=0\tag{10.5}$$

We have

$$\cos\beta=-\frac{J+I}{9J_s+6I_s}\tag{10.6}$$

For given values of I_s and I, we see that the solution (10.6) exists for $J_s \leq J_s^c$ where the critical value J_s^c is determined by $-1 \leq \cos\beta \leq 1$. For $I = -I_s = 0.1$, $J_s^c \approx -0.1889J$ in excellent agreement with the numerical result.

The classical GS determined here will be used as input GS configuration for quantum spins considered in the next section.

10.3 Green's Function Method

Let us consider the quantum spin case. For a given value of J_s, we shall use the GF method to calculate the layer magnetizations as functions of temperature. The details of the method in the case of non-collinear spin configuration have been given in Ref. [285]. We briefly recall it here and show the application to the present model.

10.3.1 Formalism

We can rewrite the full Hamiltonian (10.1) in the local framework of the classical GS configuration as

$$\begin{aligned}\mathcal{H} = &-\sum_{<i,j>} J_{i,j}\Bigg\{\frac{1}{4}\left(\cos\theta_{ij}-1\right)\left(S_i^+S_j^+ + S_i^-S_j^-\right)\\&+\frac{1}{4}\left(\cos\theta_{ij}+1\right)\left(S_i^+S_j^- + S_i^-S_j^+\right)\\&+\frac{1}{2}\sin\theta_{ij}\left(S_i^+ + S_i^-\right)S_j^z - \frac{1}{2}\sin\theta_{ij}S_i^z\left(S_j^+ + S_j^-\right)\\&+\cos\theta_{ij}S_i^zS_j^z\Bigg\} - \sum_{<i,j>} I_{i,j}S_i^zS_j^z \qquad (10.7)\end{aligned}$$

where $\cos(\theta_{ij})$ is the angle between two NN spins determined classically in the previous section.

Following Tahir-Kheli and ter Haar [340], we define two double-time GFs by

$$G_{ij}(t,t') = \ll S_i^+(t); S_j^-(t') \gg \qquad (10.8)$$

$$F_{ij}(t,t') = \ll S_i^-(t); S_j^+(t') \gg \qquad (10.9)$$

The equations of motion for $G_{ij}(t, t')$ and $F_{ij}(t, t')$ read

$$i\frac{d}{dt}G_{i,j}\left(t,t'\right) = \left\langle\left[S_i^+\left(t\right), S_j^-\left(t'\right)\right]\right\rangle\delta\left(t-t'\right) - \left\langle\left\langle\left[\mathcal{H}, S_i^+\left(t\right)\right]; S_j^-\left(t'\right)\right\rangle\right\rangle \quad (10.10)$$

$$i\frac{d}{dt}F_{i,j}\left(t,t'\right) = \left\langle\left[S_i^-\left(t\right), S_j^-\left(t'\right)\right]\right\rangle\delta\left(t-t'\right) - \left\langle\left\langle\left[\mathcal{H}, S_i^-\left(t\right)\right]; S_j^-\left(t'\right)\right\rangle\right\rangle \quad (10.11)$$

We will follow the same method as that used in Chapter 9: using the Tyablikov decoupling scheme [43] and the Fourier transforms of the retarded GFs, we have a set of equations rewritten under a matrix form

$$\mathbf{M}\left(\omega\right)\mathbf{g} = \mathbf{u} \quad (10.12)$$

where $\mathbf{M}\left(\omega\right)$ is a square matrix $(2N_z \times 2N_z)$, $\mathbf{g}$ and $\mathbf{u}$ are the column matrices which are defined as follows:

$$\mathbf{g} = \begin{pmatrix} g_{1,n'} \\ f_{1,n'} \\ \vdots \\ g_{N_z,n'} \\ f_{N_z,n'} \end{pmatrix}, \qquad \mathbf{u} = \begin{pmatrix} 2\left\langle S_1^z\right\rangle\delta_{1,n'} \\ 0 \\ \vdots \\ 2\left\langle S_{N_z}^z\right\rangle\delta_{N_z,n'} \\ 0 \end{pmatrix} \quad (10.13)$$

and

$$\mathbf{M}\left(\omega\right) = \begin{pmatrix} A_1^+ & B_1 & D_1^+ & D_1^- & \cdots \\ -B_1 & A_1^- & -D_1^- & -D_1^+ & \vdots \\ \vdots & \cdots & \cdots & \cdots & \vdots \\ \vdots & C_{N_z}^+ & C_{N_z}^- & A_{N_z}^+ & B_{N_z} \\ \cdots & -C_{N_z}^- & -C_{N_z}^+ & -B_{N_z} & A_{N_z}^- \end{pmatrix} \quad (10.14)$$

where

$$A_n^\pm = \omega \pm \Big[\frac{1}{2}J_n\left\langle S_n^z\right\rangle(Z\gamma)\left(\cos\theta_n + 1\right) - J_n\left\langle S_n^z\right\rangle Z\cos\theta_n - J_{n,n+1}\left\langle S_{n+1}^z\right\rangle\cos\theta_{n,n+1} - J_{n,n-1}\left\langle S_{n-1}^z\right\rangle\cos\theta_{n,n-1} - ZI_n\left\langle S_n^z\right\rangle - I_{n,n+1}\left\langle S_{n+1}^z\right\rangle - I_{n,n-1}\left\langle S_{n-1}^z\right\rangle\Big] \quad (10.15)$$

$$B_n = \frac{1}{2} J_n \left\langle S_n^z \right\rangle (\cos\theta_n - 1)(Z\gamma) \tag{10.16}$$

$$C_n^{\pm} = \frac{1}{2} J_{n,n-1} \left\langle S_n^z \right\rangle (\cos\theta_{n,n-1} \pm 1) \tag{10.17}$$

$$D_n^{\pm} = \frac{1}{2} J_{n,n+1} \left\langle S_n^z \right\rangle (\cos\theta_{n,n+1} \pm 1) \tag{10.18}$$

in which, $Z = 6$ is the number of in-plane NN, $\theta_{n,n\pm1}$ the angle between two NN spins belonging to the layers n and $n \pm 1$, θ_n the angle between two in-plane NN in the layer n, and

$$\gamma = \left[2\cos(k_x a) + 4\cos(k_y a/2)\cos\left(k_y a\sqrt{3}/2\right)\right]/Z$$

Here, for compactness we have used the following notations:

(i) J_n and I_n are the in-plane interactions. In the present model J_n is equal to J_s for the two surface layers and equal to J for the interior layers. All I_n are set to be I.

(ii) $J_{n,n\pm1}$ and $I_{n,n\pm1}$ are the interactions between a spin in the nth layer and its neighbor in the $(n \pm 1)^{th}$ layer. Of course, $J_{n,n-1} = I_{n,n-1} = 0$ if $n = 1$, $J_{n,n+1} = I_{n,n+1} = 0$ if $n = N_z$.

Solving $\det|\mathbf{M}| = 0$, we obtain the spin wave spectrum ω of the present system. We follow the same method in Chapter 9, we arrive at

$$\left\langle S_i^- S_j^+ \right\rangle = \lim_{\varepsilon\to 0} \frac{1}{\Delta} \iint d\mathbf{k}_{xy} \int_{-\infty}^{+\infty} \frac{i}{2\pi} \left(g_{n,n'}(\omega + i\varepsilon) - g_{n,n'}(\omega - i\varepsilon)\right) \frac{d\omega}{e^{\beta\omega} - 1} e^{i\mathbf{k}_{xy}\cdot(\mathbf{R}_i - \mathbf{R}_j)} \tag{10.19}$$

where ϵ is an infinitesimal positive constant and $\beta = 1/k_B T$, k_B being the Boltzmann constant. For spin $S = 1/2$, the thermal average of the z component of the i-th spin belonging to the n-th layer is given by

$$\left\langle S_i^z \right\rangle = \frac{1}{2} - \left\langle S_i^- S_i^+ \right\rangle \tag{10.20}$$

In the following, we shall use the case of spin one-half. Note that for the case of general S, the expression for $\left\langle S_i^z \right\rangle$ is more complicated since it involves higher quantities such as $\left\langle (S_i^z)^2 \right\rangle$.

Using the GF presented above, we can calculate self-consistently various physical quantities as functions of temperature T. The first important quantity is the temperature dependence of the angle between each spin pair. This can be calculated in a self-consistent manner at any temperature by minimizing the free energy at each temperature to get the correct value of the angle as it has been done for a frustrated bulk spin systems [305]. In the following, we limit ourselves to the self-consistent calculation of the layer magnetizations which allows us to establish the phase diagram as seen in the following.

For numerical calculation, we used $I = 0.1J$ with $J = 1$. For positive J_s, we take $I_s = 0.1$ and for negative J_s, we use $I_s = -0.1$. A size of 80^2 points in the first Brillouin zone is used for numerical integration. We start the self-consistent calculation from $T = 0$ with a small step for temperature 5×10^{-3} or 10^{-1} (in units of J/k_B). The convergence precision has been fixed at the fourth figure of the values obtained for the layer magnetizations.

10.3.2 Phase Transition and Phase Diagram of the Quantum Case

First we show an example where $J_s = -0.5$ in Fig. 10.3. As seen, the surface-layer magnetization is much smaller than the second-layer one. In addition there is a strong spin contraction at $T = 0$

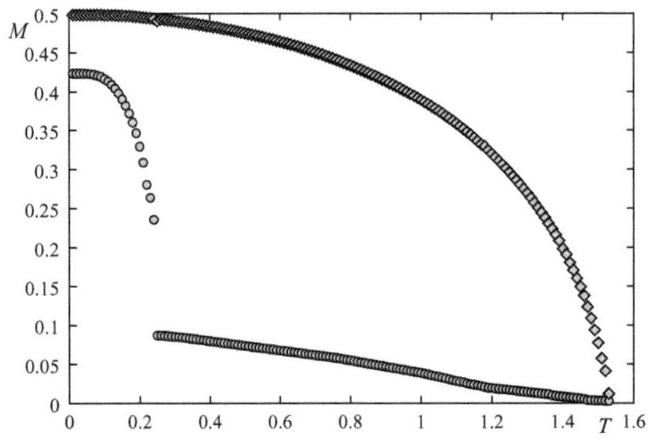

Figure 10.3 First two layer-magnetizations obtained by the GF technique vs. T for $J_s = -0.5$ with $I = -I_s = 0.1$. The surface-layer magnetization (lower curve) is much smaller than the second-layer one. See text for comments.

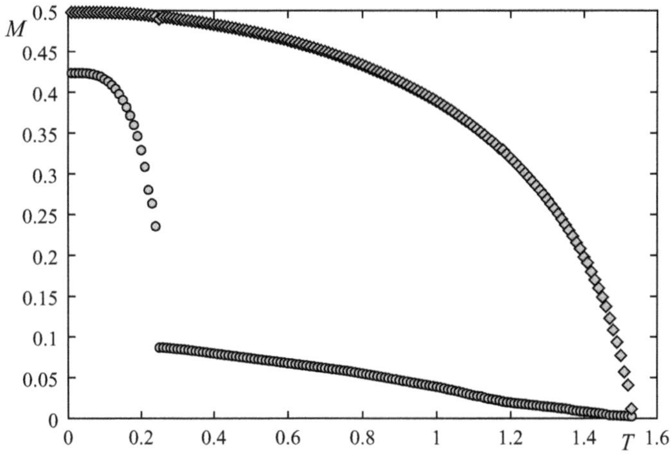

Figure 10.4 First two layer-magnetizations obtained by the GF technique vs. T for $J_s = 0.5$ with $I = I_s = 0.1$.

for the surface layer. This is due to the antiferromagnetic nature of the in-plane surface interaction J_s. One sees that the surface becomes disordered at a temperature $T_1 \simeq 0.2557$, while the second layer remains ordered up to $T_2 \simeq 1.522$. Therefore, the system is partially disordered for temperatures between T_1 and T_2. This result is very interesting because it confirms again the existence of the partial disorder in quantum spin systems observed earlier in bulk frustrated quantum spin systems [285, 305]. Note that between T_1 and T_2, the ordering of the second layer acts as an external field on the first layer, inducing therefore a small value of its magnetization. A further evidence of the existence of the surface transition will be provided with the surface susceptibility in the MC results shown below.

Figure 10.4 shows the non-frustrated case where $J_s = 0.5$, with $I = I_s = 0.1$. As seen, the first-layer magnetization is smaller than the second-layer one. There is only one transition temperature. Note the difficulty for numerical convergency when the magnetizations come close to zero.

We show in Fig. 10.5 the phase diagram in the space (J_s, T). Phase I denotes the ordered phase with surface non collinear spin configuration, phase II indicates the collinear ordered state, and

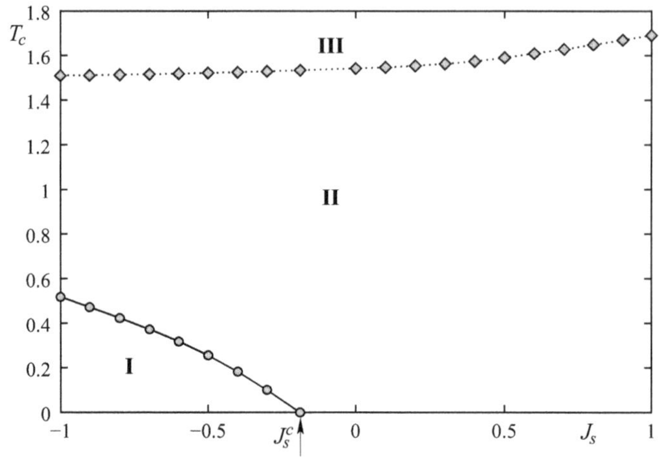

Figure 10.5 Phase diagram in the space (J_s, T) for the quantum Heisenberg model with $N_z = 4, I = |I_s| = 0.1$. See text for the description of phases I to III.

phase III is the paramagnetic phase. Note that the surface transition does not exist for $J_s \geq J_s^c$.

10.4 Monte Carlo Results

It is known that methods for quantum spins, such as the spin wave theory or the GF method presented above, are not satisfactory at high temperatures. The spin wave theory, even with magnon–magnon interactions taken into account, cannot go to temperatures close to T_c. The GF method on the other hand can go up to T_c but due to the decoupling scheme, it cannot give correct critical behavior at T_c. Fortunately, we know that quantum spin systems behave as their classical counterparts at high T. So, to see if the phase diagram obtained in the previous section for the quantum model is correct or not, we can consider its classical version and use MC simulations to obtain the phase diagram for comparison. MC simulations are excellent means to overcome approximations used in analytic calculations for the high T region as discussed above.

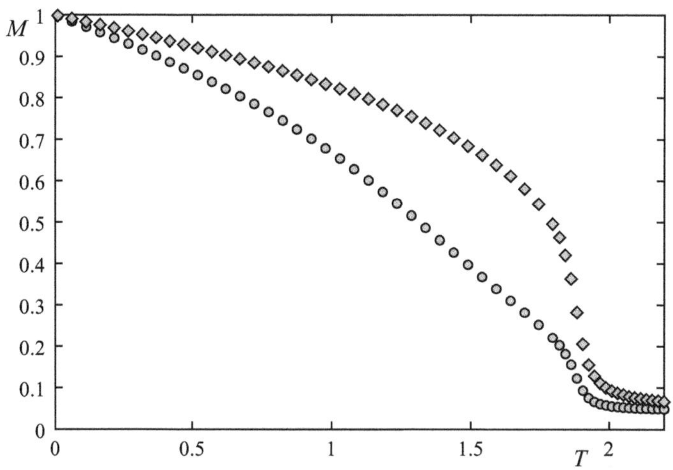

Figure 10.6 Magnetizations of layer 1 (circles) and layer 2 (diamonds) versus temperature T in unit of J/k_B for $J_s = 0.5$ with $I = I_s = 0.1$, $L = 36$.

In this paragraph, we show the results obtained by MC simulations with the Hamiltonian (10.1) but the spins are the classical Heisenberg model of magnitude $S = 1$.

The film sizes are $L \times L \times N_z$ where $N_z = 4$ is the number of layers (film thickness) taken as in the quantum case presented above. We use here $L = 24, 36, 48, 60$ to study finite-size effects as shown below. Periodic boundary conditions are used in the XY planes. The equilibrating time is about 10^6 MC steps per spin and the averaging time is 2×10^6 MC steps per spin. $J = 1$ is taken as unit of energy in the following.

Let us show in Fig. 10.6 the layer magnetization of the first two layers as a function of T, in the case $J_s = 0.5$ with $N_z = 4$ (the third and fourth layers are symmetric). Although we observe a smaller magnetization for the surface layer, there is clearly no surface transition. This is very similar to the quantum case.

In Fig. 10.7, we show a frustrated case where $J_s = -0.5$. The surface layer in this case becomes disordered at a temperature much lower than that for the second layer. Note that the surface magnetization is not saturated to 1 at $T = 0$. This is because the

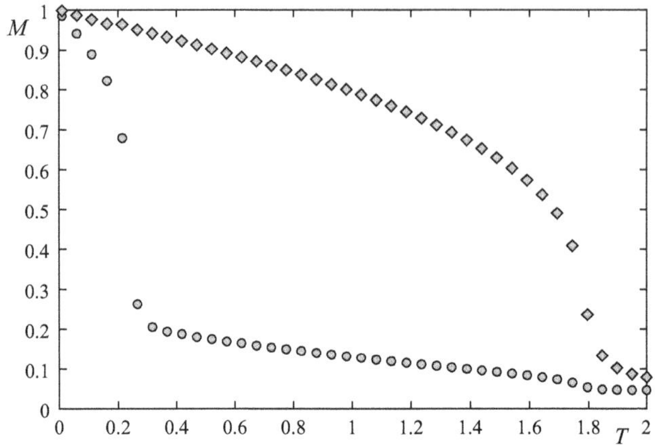

Figure 10.7 Magnetizations of layer 1 (circles) and layer 2 (diamonds) versus temperature T in unit of J/k_B for $J_s = -0.5$ with $I = -I_s = 0.1$, $L = 36$.

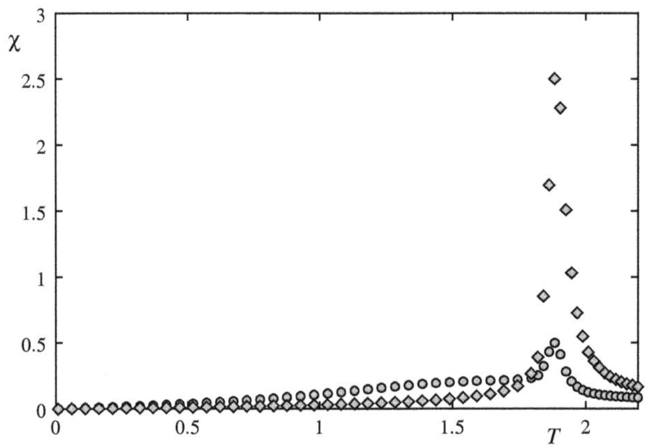

Figure 10.8 Susceptibilities of layer 1 (circles) and layer 2 (diamonds) versus temperature T in unit of J/k_B for $J_s = 0.5$ with $I = I_s = 0.1$, $L = 36$.

surface spins make an angle with the z axis so their z component is less than 1 in the GS.

Figure 10.8 shows the susceptibilities of the first and second layers in the case where $J_s = 0.5$ with $I = I_s = 0.1$ where one

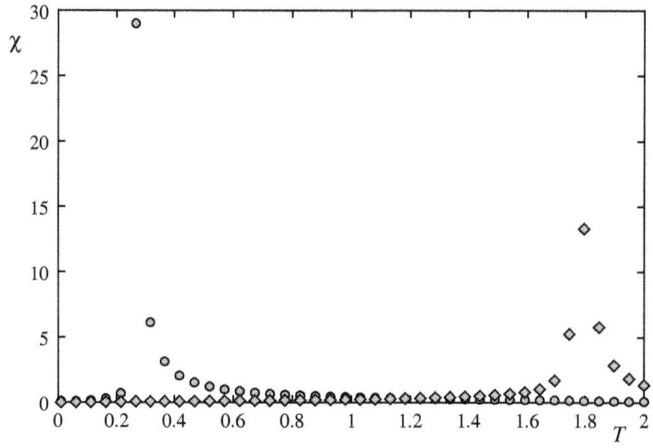

Figure 10.9 Susceptibility of layer 1 (circles) and layer 2 (diamonds) versus temperature T in unit of J/k_B for $J_s = -0.5$ with $I = -I_s = 0.1$, $L = 36$. Note that for clarity, the susceptibility of the layer 2 has been multiplied by a factor 5.

observes the peaks at the same temperature indicating a single transition in contrast to the frustrated case shown in Fig. 10.9. These results confirm the above results of layer magnetizations.

To establish the phase diagram, the transition temperatures are taken at the change of curvature of the layer magnetizations, i.e., at the maxima of layer susceptibilities shown before. Figure 10.10 shows the phase diagram obtained in the space (J_s, T). It is interesting to see that this phase diagram resembles remarkably that obtained for the quantum counterpart model shown in Fig. 10.5.

Let us study the finite-size effect of the phase transitions shown in Fig. 10.10. To this end, we use the histogram technique, which has been proved so far to be excellent for the calculation of critical exponents [110–112]. The principle is as follows (see details in Chapter 6). Using the Metropolis algorithm to determine approximately the critical temperature region, then choosing a temperature T_0 as close as possible to the presupposed transition temperature. We then make a very long run at T_0 to establish an energy histogram. From formulas established using the statistical canonical distribution, we can calculate physical quantities in a continuous manner for temperatures around T_0 [110–112]. We can

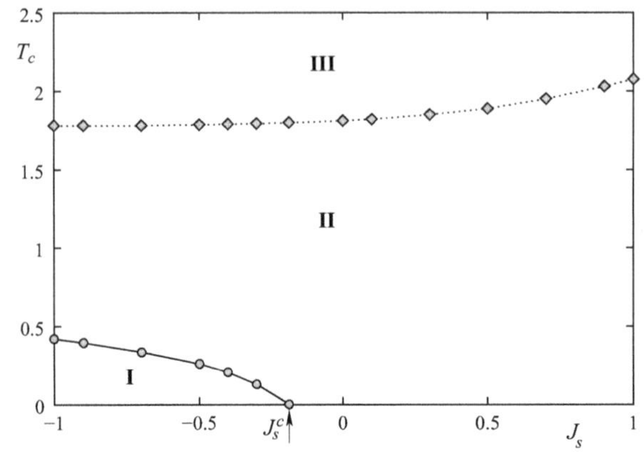

Figure 10.10 Phase diagram in the space (J_s, T) for the classical Heisenberg model with $N_z = 4$, $I = |I_s| = 0.1$. Phases I to III have the same meanings as those in Fig. 10.5.

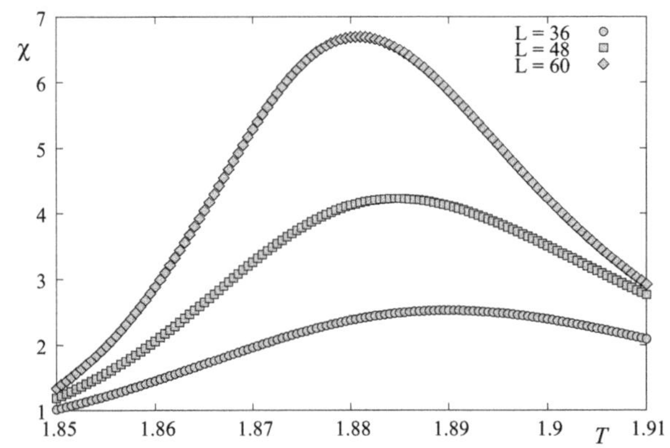

Figure 10.11 Susceptibility versus T for $L = 36, 48, 60$ with $J_s = 0.5$ and $I = I_s = 0.1$.

then identify with precision the transition temperature as well as the maximal values of fluctuation quantities such as specific heat and susceptibility.

Figure 10.11 shows the susceptibility versus T for $L = 36, 48, 60$ in the case of $J_s = 0.5$. For presentation convenience, the size $L = 24$ has been removed since the peak for this case is rather flat in the

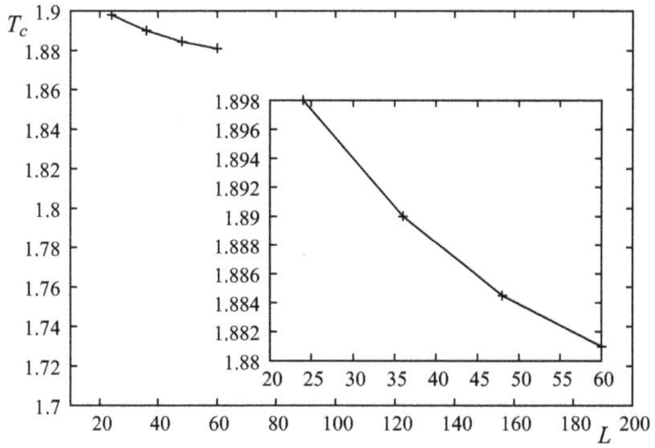

Figure 10.12 Transition temperature versus L with $J_s = 0.5$ and $I = I_s = 0.1$. The inset shows the enlarged scale.

scale of the figure. However, it shall be used to calculate the critical exponent γ for the transition. As seen, the maximum $\chi^{\max}$ of the susceptibility increases with increasing L.

For completeness, we show in Fig. 10.12 the transition temperature as a function of L. A rough extrapolation to infinity gives $T_c^{\infty} \simeq 1.86 \pm 0.02$.

In the frustrated case, i.e., $J_s < J_s^c$, we perform the same calculation to estimate the finite-size effect. Note that in this case there are two phase transitions. We show in Fig. 10.13 the layer susceptibilities as functions of T for different L. As seen, both surface and second-layer transitions have a strong size dependence.

We show in Figs. 10.14 and 10.15 the size dependence of the transition temperatures T_1 (surface transition) and T_2 (second-layer transition).

The size dependence of the maxima observed above allows us to estimate the ratio γ/ν (see Chapter 6 for details on the finite-size scaling laws). We show now $\ln \chi^{\max}$ as a function of $\ln L$ for the different cases studied above. Figures 10.16a and 10.16b correspond, respectively, to the transitions of surface and second layer occurring in the frustrated case with $J_s = -0.5$, while Fig. 10.16c corresponds to the unique transition occurring in the non frustrated case with $J_s = 0.5$. The slopes of these straight lines give $\gamma/\nu \simeq 1.864 \pm 0.034$ (a), 1.878 ± 0.027 (b), 1.801 ± 0.027 (c). The errors were estimated from the mean least-square fitting

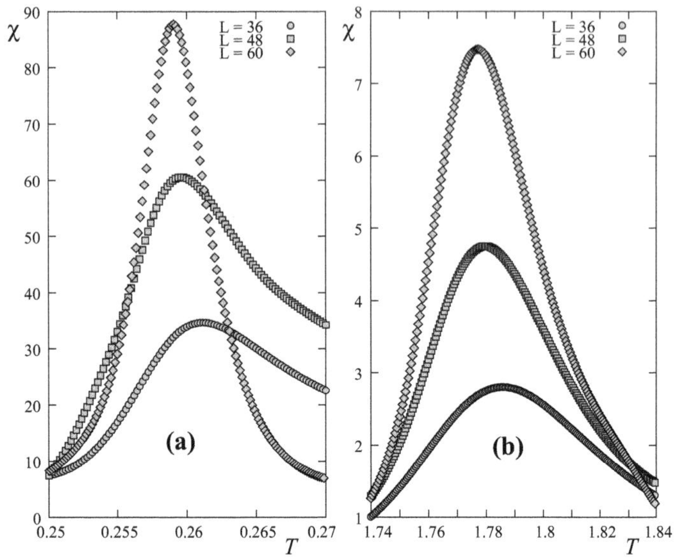

Figure 10.13 Layer susceptibilities versus T for $L = 36, 48, 60$ with $J_s = -0.5$ and $I = -I_s = 0.1$. Left (right) figure corresponds to the first (second) layer susceptibility.

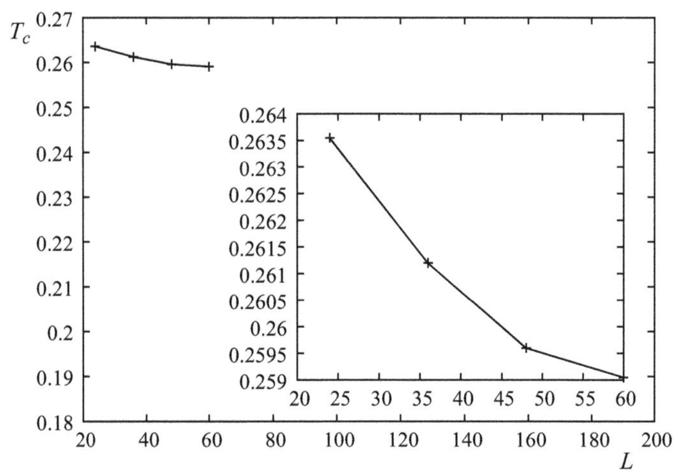

Figure 10.14 Transition temperature versus L for the surface layer in the case $J_s = -0.5$ with $I = -I_s = 0.1$. The inset shows the enlarged scale.

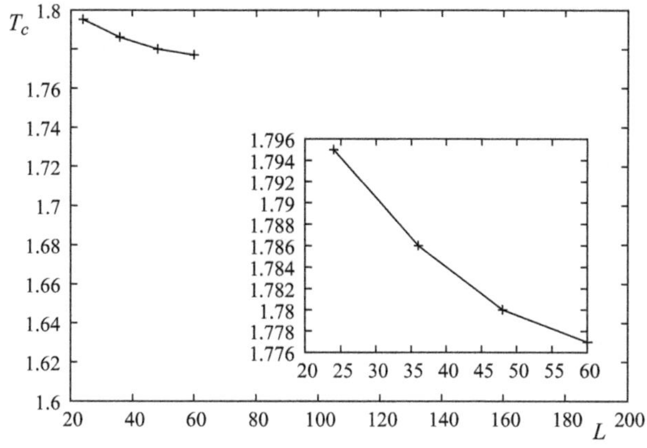

Figure 10.15 Transition temperature versus L for the second layer in the case $J_s = -0.5$ with $I = -I_s = 0.1$. The inset shows the enlarged scale.

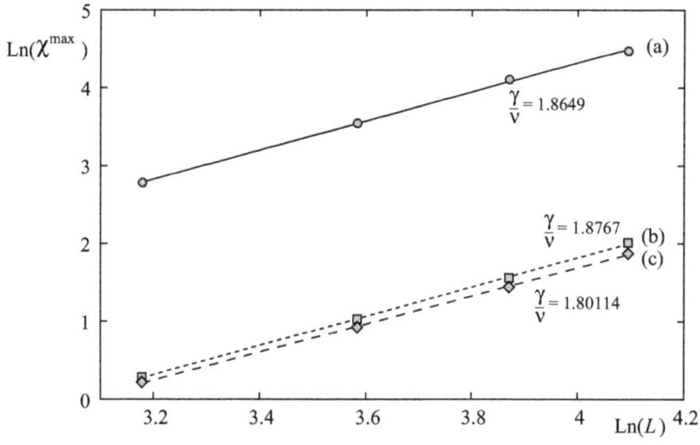

Figure 10.16 Maximum of surface-layer susceptibility versus L for L = 24, 36, 48, 60 with $J_s = -0.5$ (a,b), $J_s = 0.5$ (c) and $I = |I_s| = 0.1$, in the ln − ln scale. The slope gives γ/ν indicated in the figure for each case. See text for comments.

and errors on the peak values obtained with different values of T_0 (multiple-histogram technique). Within errors, the first two values, which correspond to the frustrated case, can be considered as identical, while the last one corresponding to the non-frustrated case is different. We will return to this point later.

At this stage, we would like to emphasize the following points. First, we observe that these values of γ/ν are found to be between those of the two-dimensional (2D) Ising model ($\gamma/\nu = 1.75$) and the three-dimensional (3D) one ($\gamma/\nu = 1.241/0.63 \simeq 1.97$). A question which naturally arises on the roles of the Ising-like anisotropy term, of the two-fold chiral symmetry and of the film thickness. The roles of the anisotropy term and the chiral symmetry are obvious: the Ising character should be observed in the result (we return to this point below). It is, however, not clear for the effect of the thickness. Some arguments, such as those from the renormalization group, say that the correlation length in the direction perpendicular to the film is finite; hence, it is irrelevant to the criticality; the 2D character, therefore, should be theoretically preserved. We think that such arguments are not always true because it is difficult to conceive that when the film thickness becomes larger and larger the 2D universality should remain. Instead, we believe that that the finite thickness of the film affects the 2D universality in one way or another, giving rise to "effective" critical exponents with values deviated from the 2D ones. The larger the thickness is, the stronger this deviation becomes. The observed values of γ/ν shown above may contain an effect of a 2D-3D crossover. At this point, we would like to emphasize that, in the case of simple surface conditions, i.e., no significant deviation of the surface parameters with respect to those of the bulk, the bulk behavior is observed when the thickness becomes larger than a dozen of atomic layers [75, 78]: surface effects are insignificant on thermodynamic properties of the film. There are, therefore, reasons to believe that there should be a crossover from 2D to 3D at some film thickness. Of course, this is an important issue which needs to be theoretically clarified in the future. We return now to the effect of Ising anisotropy and chiral symmetry. The deviation from the 2D values may result in part from a complex coupling between the Ising symmetry, due to anisotropy and chirality, and the continuous nature of the classical Heisenberg spins studied here. This deviation may be important if the anisotropy constant I is small. For the effect chiral symmetry, it is a complex matter. To show the complexity in determining the critical universality with chiral symmetry, let us discuss about a simpler case with similar chiral symmetry: the XY model on the fully frustrated Villain's lattice (see model described in Problem 7 of Chapter 3 and its solution in Section 18.3). There has been a large number of investigations on the nature of the transition observed in this 2D case in the context of the frustration effects [45, 85, 264]. In this model, though the chirality symmetry argument says that the transition should be of 2D Ising universality class, many investigators found a deviation of critical exponents from those of the 2D case (see review in Ref. [45]). For example, the following values are found for the critical exponent ν: $\nu = 0.889$ [287] and $\nu = 0.813$ [203]. These values are close to that obtained for the single transition in a mixed XY-Ising model which is 0.85 [204, 257]. It is now believed that the XY character of the spins affects the Ising chiral symmetry giving rise to those deviated critical exponents. Similarly, in the case of thin film studied here, we do not deal with the discrete Ising model but rather an Ising-like Heisenberg model. The Ising character due to chiral symmetry of the transition at the surface is believed to be perturbed by the continuous nature of Heisenberg spins. The transition of interior layer shown in Fig. 10.16b suffers similar but not the same effects because of the absence of chiral symmetry on this layer. So the value γ/ν is a little different. In the non frustrated case shown in Fig. 10.16c, the deviation from the 2D Ising universality class is less important because of the absence of the chiral symmetry. This small deviation is believed to stem mainly from the continuous nature of Heisenberg spins.

To conclude this paragraph, we believe, from physical arguments given above, that the deviation from 2D Ising universality class of the transitions observed here is due to the following causes: the effect of the coupling between the continuous degree of freedom of Heisenberg spin to the chiral symmetry, the small Ising-like anisotropy and the film thickness.

10.5 Concluding Remarks

We have used the GF method and MC simulations to study the Heisenberg spin model with an Ising-like interaction anisotropy in thin films of stacked triangular lattices. The two surfaces of the film are frustrated. We found that surface spin configuration is non-collinear when the surface antiferromagnetic interaction is smaller than a critical value J_s^c. In the non-collinear regime, the surface layer is disordered at a temperature lower than that for interior layers ("soft" surface) giving rise to the so-called "magnetically dead surface" observed in some materials [36, 374]. The surface transition disappears for J_s larger than the critical value J_s^c. Using the GF method we have established a phase diagram in the space (T, J_s) which is in a good agreement with the MC result. This is due to the fact that at high temperatures where the transition takes place, the quantum nature of spins used in the GF is lost so that we should find results of classical spins obtained by MC simulations. We have also studied by MC histogram technique the critical behavior of the phase transition using the finite-size effects. The result of the ratio of critical exponents γ/ν shows that the nature of the transition is complicated due to the influence of several physical mechanisms. The symmetry of the ground state alone cannot explain such a result. We have outlined a number of the most relevant mechanisms. We will return to the criticality of thin films in Chapter 16.

Chapter 11

Phase Transition in Helimagnetic Thin Films

In this chapter, we show the main properties of a helimagnetic thin film with quantum Heisenberg spin model by using the Green's function (GF) method. Surface spin configuration is calculated by minimizing the spin interaction energy. It is shown that the angles between spins near the surface are strongly modified with respect to the bulk configuration. Taking into account this surface spin reconstruction, we calculate self-consistently the spin wave spectrum and the layer magnetizations as functions of temperature up to the disordered phase. The spin wave spectrum shows the existence of a surface-localized branch which causes a low surface magnetization. We show that quantum fluctuations give rise to a crossover between the surface magnetization and interior-layer magnetizations at low temperatures. We calculate the transition temperature and show that it depends strongly on the helical angle. Results are in agreement with existing experimental observations on the stability of helical structure in thin films and on the insensitivity of the transition temperature with the film thickness. We also study effects of various parameters such as surface exchange and anisotropy interactions. Monte Carlo simulations for the classical spin model are also carried out for comparison with the quantum theoretical result.

Physics of Magnetic Thin Films: Theory and Simulation
Hung T. Diep

ISBN 978-981-4877-42-8 (Hardcover), 978-1-003-12110-7 (eBook)
www.jennystanford.com

The results shown in this chapter are taken from Ref. [89] for the body-centered cubic (BCC) lattice, and from Ref. [102] for the simple cubic (SC) case.

11.1 Introduction

Helimagnets were discovered a long time ago by Yoshimori [369] and Villain [350]. In the simplest model, the helimagnetic ordering is non-collinear due to a competition between nearest-neighbor (NN) and next-nearest-neighbor (NNN) interactions: For example, a spin in a chain turns an angle θ with respect to its previous neighbor. Low-temperature properties in helimagnets such as spin waves [91, 139, 286, 289] and heat capacity [269] have been extensively investigated. Helimagnets belong to a class of frustrated vector-spin systems. In spite of their long history, the nature of the phase transition in bulk helimagnets as well as in other non-collinear magnets such as stacked triangular XY and Heisenberg antiferromagnets has been elucidated only recently [90, 187, 250, 251]. For reviews on many aspects of frustrated spin systems, the reader is referred to Ref. [85].

In this chapter, we study a helimagnetic thin film with the quantum Heisenberg spin model. Surface effects in thin films have been widely studied theoretically, experimentally and numerically, during the past three decades [36, 374]. Nevertheless, surface effects in helimagnets have only been recently studied: surface spin structures [229], Monte Carlo (MC) simulations [64], magnetic field effects on the phase diagram in Ho [291] and a few experiments [176, 177]. We will compare our work to these in the conclusion.

Helical magnets present potential applications in spintronics with predictions of spin-dependent electron transport in these magnetic materials [150, 166, 357]. We shall use the GF method to study a quantum spin model on a helimagnetic thin film of body-centered cubic (BCC) lattice. The GF method has been initiated by Zubarev [383] for collinear bulk magnets (ferromagnets and antiferromagnets) and by Diep et al. for thin films of collinear spin configurations [78]. For non-collinear magnets, the GF method has also been developed for bulk helimagnets [286] and for frustrated films in Refs. [248, 249] with results presented in the previous chapters. Note that surface effects in thin films of stacked triangular antiferromagnets with non-collinear 120° spin configuration have been investigated by the method of equation of motion [230]. However, the model in these works did not possess a surface spin reconstruction and do not belong to the family of helical structures as our model described below.

In helimagnets, the presence of a surface modifies the competing forces acting on surface spins. As a consequence, as will be shown below, the angles between neighboring spins become non-uniform, making calculations harder. This explains why there was no microscopic calculation for helimagnetic films before Refs. [89, 102].

Note that for illustration, we use below the BCC lattice structure, but the results shown below are valid for different lattices, not restricted to the BCC crystal, provided modifications on the coordination number and therefore on the value of the critical value $(J_2/J_1)_c$ (J_1: NNN interaction, J_2: NNN interaction). For example, the BCC case has $(J_2/J_1)_c = 1$, while the simple cubic lattice has $(J_2/J_1)_c = 1/4$.

In Section 11.2, the model is presented and classical ground state (GS) of the helimagnetic film is determined. In Section 11.3, the general GF method for non-uniform spin configurations is shown in details. The GF results are shown in Section 11.4 where the spin wave spectrum, the layer magnetizations and the transition temperature are shown. Effects of surface interaction parameters and the film thickness are discussed. The case of helimagnetic films with simple cubic (SC) lattice is shown in Section 11.5.

11.2 Model and Classical Ground State

Let us recall that bulk helical structures are due to the competition of various kinds of interaction [19, 224, 276, 350, 369]. We consider hereafter the simplest model for a film: The helical ordering is along one direction, namely the c-axis perpendicular to the film surface.

We consider a thin film of BCC lattice of N_z layers, with two symmetrical surfaces perpendicular to the c-axis, for simplicity. The exchange Hamiltonian reads

$$\mathcal{H} = \mathcal{H}_e + \mathcal{H}_a \tag{11.1}$$

where the isotropic exchange part is given by

$$\mathcal{H}_e = -\sum_{\langle i,j \rangle} J_{i,j} \mathbf{S}_i \cdot \mathbf{S}_j \tag{11.2}$$

$J_{i,j}$ being the interaction between two quantum Heisenberg spins $\mathbf{S}_i$ and $\mathbf{S}_j$ occupying the lattice sites i and j. The anisotropic part is chosen as

$$\mathcal{H}_a = -\sum_{<i,j>} I_{i,j} S_i^z S_j^z \cos\theta_{ij} \tag{11.3}$$

where $I_{i,j}$ is the anisotropic interaction along the in-plane local spin-quantization axes z of $\mathbf{S}_i$ and $\mathbf{S}_j$, supposed to be positive, small compared to J_1, and limited to NN on the c-axis. Let us mention that according to the theorem of Mermin and Wagner [231] continuous isotropic spin models such as XY and Heisenberg spins do not have long-range ordering at finite temperatures in two dimensions. Since we are dealing with the Heisenberg model in a thin film, it is useful to add an anisotropic interaction to create a long-range ordering and a phase transition at finite temperatures.

To generate a bulk helimagnetic structure, the simplest way is to take a ferromagnetic interaction between NNs, say J_1 (> 0), and an antiferromagnetic interaction between NNNs, $J_2 < 0$. It is obvious that if $|J_2|$ is smaller than a critical value $|J_2^c|$, the classical GS spin configuration is ferromagnetic [91, 139, 289]. Since our purpose is to investigate the helimagnetic structure near the surface and surface effects, let us consider the case of a helimagnetic structure only in the c-direction perpendicular to the film surface. In such a case, we assume a non-zero J_2 only on the c-axis (see Section 3.4). This assumption simplifies formulas but does not change the physics of the problem since including the uniform helical angles in two other directions parallel to the surface will not introduce additional surface effects. Note that the bulk case of the above quantum spin model have been studied by the GF method [286].

Let us recall that the helical structure in the bulk is planar: Spins lie in planes perpendicular to the c-axis: the angle between two NNs in the adjacent planes is a constant and is given by $\cos\alpha = -J_1/J_2$ for a BCC lattice (see method of calculation for the bulk configuration in Section 3.4). The helical structure exists, therefore, if $|J_2| \geq J_1$, namely $|J_2^c|(\text{bulk}) = J_1$ (see Fig. 11.1, top). To simplify the presentation, we take a zero anisotropy $I_{i,j} = 0$. The effect of $I_{i,j}$ on the GS will be shown at the end of this section.

To calculate the classical GS surface spin configuration, we write down the expression of the energy of spins along the c-axis, starting

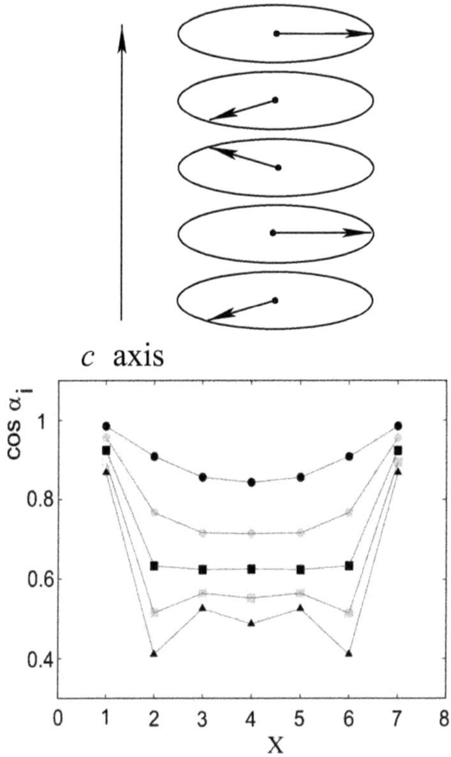

Figure 11.1 Top: Bulk helical structure along the c-axis, in the case $\alpha = 2\pi/3$, namely $J_2/J_1 = -2$. Bottom: (color online) Cosinus of $\alpha_1 = \theta_1 - \theta_2, \cdots, \alpha_7 = \theta_7 - \theta_8$ across the film for $J_2/J_1 = -1.2, -1.4, -1.6, -1.8, -2$ (from top) with $N_z = 8$: a_i stands for $\theta_i - \theta_{i+1}$ and x indicates the film layer i where the angle a_i with the layer $(i+1)$ is shown. The values of the angles are given in Table 11.1: a strong rearrangement of spins near the surface is observed.

from the surface:

$$
\begin{aligned}
E = & -Z_1 J_1 \cos(\theta_1 - \theta_2) - Z_1 J_1[\cos(\theta_2 - \theta_1) + \cos(\theta_2 - \theta_3)] + \cdots \\
& -J_2 \cos(\theta_1 - \theta_3) - J_2 \cos(\theta_2 - \theta_4) \\
& -J_2[\cos(\theta_3 - \theta_1) + \cos(\theta_3 - \theta_5)] + \cdots
\end{aligned} \tag{11.4}
$$

where $Z_1 = 4$ is the number of NNs in a neighboring layer, θ_i denotes the angle of a spin in the i-th layer made with the Cartesian x axis of the layer. The interaction energy between two NN spins in the two adjacent layers i and j depends only on the difference $\alpha_i \equiv \theta_i - \theta_{i+1}$. The GS configuration corresponds to the minimum of E. We have to solve the set of equations:

$$\frac{\partial E}{\partial \alpha_i} = 0, \quad \text{for } i = 1, N_z - 1 \tag{11.5}$$

Explicitly, we have

$$\frac{\partial E}{\partial \alpha_1} = 8J_1 \sin\alpha_1 + 2J_2 \sin(\alpha_1 + \alpha_2) = 0 \tag{11.6}$$

$$\frac{\partial E}{\partial \alpha_2} = 8J_1 \sin\alpha_2 + 2J_2 \sin(\alpha_1 + \alpha_2) + 2J_2 \sin(\alpha_2 + \alpha_3) = 0 \tag{11.7}$$

$$\frac{\partial E}{\partial \alpha_3} = 8J_1 \sin\alpha_3 + 2J_2 \sin(\alpha_2 + \alpha_3) + 2J_2 \sin(\alpha_3 + \alpha_4) = 0 \tag{11.8}$$

$$\frac{\partial E}{\partial \alpha_4} = \cdots$$

where we have expressed the angle between two NNNs as follows: $\theta_1 - \theta_3 = \theta_1 - \theta_2 + \theta_2 - \theta_3 = \alpha_1 + \alpha_2$, etc. In the bulk case, putting all angles α_i in Eq. 11.7 equal to α we get $\cos\alpha = -J_1/J_2$ as expected. For the spin configuration near the surface, let us consider in the first step only three parameters α_1 (between the surface and the second layer), α_2 and α_3. We take $\alpha_n = \alpha$ from $n = 4$ inward up to $n = N_z/2$, the other half being symmetric. Solving the first two equations, we obtain

$$\tan\alpha_2 = -\frac{2J_2(\sin\alpha_3 + \sin\alpha_1)}{8J_1 + 2J_2(\cos\alpha_3 + \cos\alpha_1)} \tag{11.9}$$

The iterative numerical procedure is as follows: (i) replacing α_3 by $\alpha = \arccos(-J_1/J_2)$ and solving (11.6) and (11.9) to obtain α_1 and α_2, (ii) replacing these values into (11.8) to calculate α_3, (iii) using this value of α_3 to solve again (11.6) and (11.9) to obtain new values of α_1 and α_2, (iv) repeating steps (ii) and (iii) until the convergence is reached within a desired precision. In the second step, we use α_1, α_2 and α_3 to calculate by iteration α_4, assuming a bulk value for α_5. In the third step, we use α_i ($i = 1 - 4$) to calculate α_5 and so on. The results calculated for various J_2/J_1 are shown in Fig. 11.1 (bottom) for a film of $N_z = 8$ layers. The values obtained are shown in Table 11.1. Results of $N_z = 16$ will be shown later.

Some remarks are in order: (i) result shown is obtained by iteration with errors less than $10^{-4\circ}$, (ii) strong angle variations

Table 11.1 Values of $\cos\theta_{n,n+1} = \alpha_n$ between two adjacent layers are shown for various values of J_2/J_1

J_2/J_1	$\cos_{1,2}$	$\cos_{2,3}$	$\cos_{3,4}$	$\cos_{4,5}$	α(bulk)
−1.2	0.985 (9.79°)	0.908 (24.73°)	0.855 (31.15°)	0.843 (32.54°)	33.56°
−1.4	0.955 (17.07°)	0.767 (39.92°)	0.716 (44.28°)	0.714 (44.41°)	44.42°
−1.6	0.924 (22.52°)	0.633 (50.73°)	0.624 (51.38°)	0.625 (51.30°)	51.32°
−1.8	0.894 (26.66°)	0.514 (59.04°)	0.564 (55.66°)	0.552 (56.48°)	56.25°
−2.0	0.867 (29.84°)	0.411 (65.76°)	0.525 (58.31°)	0.487 (60.85°)	60°

Only angles of the first half of the 8-layer film are shown: Other angles are, by symmetry, $\cos\theta_{7,8} = \cos\theta_{1,2}$, $\cos\theta_{6,7} = \cos\theta_{2,3}$, $\cos\theta_{5,6} = \cos\theta_{3,4}$. The values in parentheses are angles in degrees. The last column shows the value of the angle in the bulk case (infinite thickness). For presentation, angles are shown with two digits.

are observed near the surface with oscillation for strong J_2, (iii) the angles at the film center are close to the bulk value α (last column), meaning that the surface reconstruction affects just a few atomic layers, (iv) the bulk helical order is stable just a few atomic layers away from the surface even for films thicker that $N_z = 8$ (see below). This helical stability has been experimentally observed in holmium films [206].

Note that using the numerical steepest descent method described in Ref. [248] we find the same result.

Let us discuss now the effect of the anisotropy on the GS configuration. The form of Eq. (11.3) simplifies a lot: Since the interaction is limited to NN, it suffices to replace in Eq. (11.4) the parameter J_1 by $J_1' = J_1 + I_1$ where $I_1 = I_{i,j}$ for any NN pairs (i, j). The GS calculation is done exactly in the same manner. That is the reason why we choose the form of Eq. (11.3). The GS configuration is slightly modified but the method and the general aspects of the results described above remain valid. Of course, the calculation of the spin wave spectrum and layer magnetizations presented below take into account the GS modification at each value of I_1. Choosing another form of anisotropy, for example, taking a standard single-ion anisotropy $-I(S_i^z)^2$ along the spin local axis, will add just a constant in Eq. (11.4) [because $(S_i^z)^2 = 1$ in the GS]. So, it will not affect the GS configuration.

In the following, using the spin configuration obtained at each J_2/J_1 we calculate the spin wave excitation and properties of the film such as the zero-point spin contraction, the layer magnetizations and the critical temperature.

11.3 Green's Function Method

Let us define the local spin coordinates as follows: the quantization axis of spin $\mathbf{S}_i$ is on its ζ_i axis which lies in the plane, the η_i axis of $\mathbf{S}_i$ is along the c-axis, and the ξ_i axis forms with η_i and ζ_i axes a direct trihedron. Since the spin configuration is planar, all spins have the same η axis. Furthermore, all spins in a given layer are parallel. Let $\hat{\xi}_i$, $\hat{\eta}_i$ and $\hat{\zeta}_i$ be the unit vectors on the local (ξ_i, η_i, ζ_i) axes. We use the following local transformation which has been used for the first time in Ref. [139] and described in Section 3.4:

$$\mathbf{S}_i = S_i^x \hat{\xi}_i + S_i^y \hat{\eta}_i + S_i^z \hat{\zeta}_i \tag{11.10}$$

$$\mathbf{S}_j = S_j^x \hat{\xi}_j + S_j^y \hat{\eta}_j + S_j^z \hat{\zeta}_j \tag{11.11}$$

We have (see Fig. 11.2)

$$\hat{\xi}_j = \cos\theta_{ij}\hat{\zeta}_i + \sin\theta_{ij}\hat{\xi}_i$$

$$\hat{\zeta}_j = -\sin\theta_{ij}\hat{\zeta}_i + \cos\theta_{ij}\hat{\xi}_i$$

$$\hat{\eta}_j = \hat{\eta}_i$$

where $\cos\theta_{ij} = \cos(\theta_i - \theta_j)$ is the angle between two spins i and j.

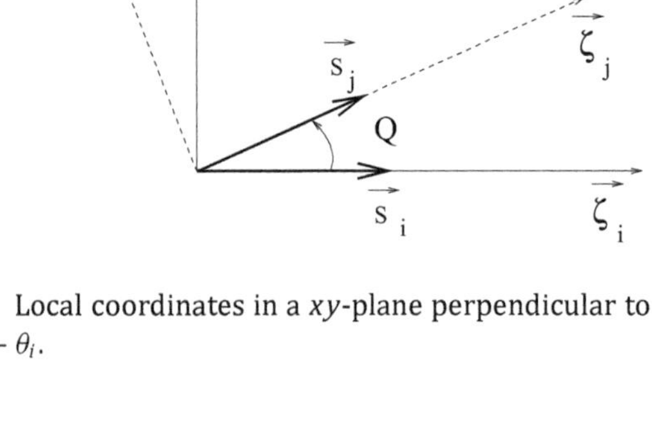

Figure 11.2 Local coordinates in a xy-plane perpendicular to the c-axis. Q denotes $\theta_j - \theta_i$.

Note that in the laboratory coordinate system, namely in the film coordinates, the z direction coincides with the c-direction or the $\hat{\eta}$ axis perpendicular to the film surface, while the x and y directions are taken to be the BCC crystal axes in the film plane.

Replacing these into Eq. (11.11) to express $\mathbf{S}_j$ in the $(\hat{\xi}_i, \hat{\eta}_i, \hat{\zeta}_i)$ coordinates, then calculating $\mathbf{S}_i \cdot \mathbf{S}_j$, we obtain the following exchange Hamiltonian from (11.2):

$$\begin{aligned}\mathcal{H}_e = -\sum_{<i,j>} J_{i,j} \Bigg\{ & \frac{1}{4}\left(\cos\theta_{ij} - 1\right)\left(S_i^+ S_j^+ + S_i^- S_j^-\right) \\ & + \frac{1}{4}\left(\cos\theta_{ij} + 1\right)\left(S_i^+ S_j^- + S_i^- S_j^+\right) \\ & + \frac{1}{2}\sin\theta_{ij}\left(S_i^+ + S_i^-\right) S_j^z \\ & - \frac{1}{2}\sin\theta_{ij} S_i^z \left(S_j^+ + S_j^-\right) + \cos\theta_{ij} S_i^z S_j^z \Bigg\} \end{aligned} \quad (11.12)$$

11.3.1 General Formulation for Non-Collinear Magnets

We define the following two double-time Green's functions in the real space:

$$\begin{aligned} G_{i,j}(t,t') &= \ll S_i^+(t); S_j^-(t') \gg \\ &= -i\theta(t-t') < \left[S_i^+(t), S_j^-(t')\right] > \end{aligned} \quad (11.13)$$

$$\begin{aligned} F_{i,j}(t,t') &= \ll S_i^-(t); S_j^-(t') \gg \\ &= -i\theta(t-t') < \left[S_i^-(t), S_j^-(t')\right] > \end{aligned} \quad (11.14)$$

We need these two functions because the equation of motion of the first function generates functions of the second type, and vice versa. These equations of motion are

$$\begin{aligned} i\hbar\frac{d}{dt}G_{i,j}(t,t') &= \left\langle\left[S_i^+(t), S_j^-(t')\right]\right\rangle \delta(t-t') \\ &\quad - \left\langle\left\langle\left[\mathcal{H}, S_i^+(t)\right]; S_j^-(t')\right\rangle\right\rangle \end{aligned} \quad (11.15)$$

$$\begin{aligned} i\hbar\frac{d}{dt}F_{i,j}(t,t') &= \left\langle\left[S_i^-(t), S_j^-(t')\right]\right\rangle \delta(t-t') \\ &\quad - \left\langle\left\langle\left[\mathcal{H}, S_i^-(t)\right]; S_j^-(t')\right\rangle\right\rangle \end{aligned} \quad (11.16)$$

where the spin operators and their commutation relations are given by

$$S_j^{\pm} = S_j^x \hat{\xi}_j \pm i S_j^y \hat{\eta}_j$$

$$\left[S_j^+, S_l^-\right] = 2S_j^z \delta_{j,l}$$

$$\left[S_j^z, S_l^{\pm}\right] = \pm S_j^{\pm} \delta_{j,l}$$

Expanding the commutators in Eqs. (11.15)–(11.16), and using the Tyablikov decoupling scheme [346] for higher-order functions, for example, $\langle\langle S_{i'}^z S_i^+(t); S_j^-(t')\rangle\rangle \simeq < S_{i'}^z > \langle\langle S_i^+(t); S_j^-(t')\rangle\rangle$, etc., we obtain the following general equations for non-collinear magnets:

$$\begin{aligned}
i\hbar \frac{dG_{i,j}(t,t')}{dt} &= 2 < S_i^z > \delta_{i,j}\delta(t-t') \\
&\quad - \sum_{i'} J_{i,i'}[< S_i^z > (\cos\theta_{i,i'} - 1) \times F_{i',j}(t,t') \\
&\quad + < S_i^z > (\cos\theta_{i,i'} + 1) G_{i',j}(t,t') \\
&\quad - 2 < S_{i'}^z > \cos\theta_{i,i'} G_{i,j}(t,t')] \\
&\quad + 2\sum_{i'} I_{i,i'} < S_{i'}^z > \cos\theta_{i,i'} G_{i,j}(t,t')
\end{aligned} \tag{11.17}$$

$$\begin{aligned}
i\hbar \frac{dF_{i,j}(t,t')}{dt} &= \sum_{i'} J_{i,i'}[< S_i^z > (\cos\theta_{i,i'} - 1) \times G_{i',j}(t,t') \\
&\quad + < S_i^z > (\cos\theta_{i,i'} + 1) F_{i',j}(t,t') \\
&\quad - 2 < S_{i'}^z > \cos\theta_{i,i'} F_{i,j}(t,t')] \\
&\quad - 2\sum_{i'} I_{i,i'} < S_{i'}^z > \cos\theta_{i,i'} F_{i,j}(t,t')
\end{aligned} \tag{11.18}$$

11.3.2 BCC Helimagnetic Films

In the case of a BCC thin film with a (001) surface, the above equations yield a closed system of coupled equations within the Tyablikov decoupling scheme [346]. For clarity, we separate the sums on NN interactions and NNN interactions as follows:

$$
\begin{aligned}
i\hbar\frac{dG_{i,j}(t,t')}{dt} &= 2 < S_i^z > \delta_{i,j}\delta(t-t') \\
&- \sum_{k'\in NN} J_{i,k'}[< S_i^z > (\cos\theta_{i,k'} - 1) \times F_{k',j}(t,t') \\
&+ < S_i^z > (\cos\theta_{i,k'} + 1)G_{k',j}(t,t') \\
&- 2 < S_{k'}^z > \cos\theta_{i,k'}G_{i,j}(t,t')] \\
&+ 2\sum_{k'\in NN} I_{i,k'} < S_{k'}^z > \cos\theta_{i,k'}G_{i,j}(t,t') \\
&- \sum_{i'\in NNN} J_{i,i'}[< S_i^z > (\cos\theta_{i,i'} - 1) \times F_{i',j}(t,t') \\
&+ < S_i^z > (\cos\theta_{i,i'} + 1)G_{i',j}(t,t') \\
&- 2 < S_{i'}^z > \cos\theta_{i,i'}G_{i,j}(t,t')] \qquad (11.19)
\end{aligned}
$$

$$
\begin{aligned}
i\hbar\frac{dF_{k,j}(t,t')}{dt} &= \sum_{i'\in NN} J_{k,i'}[< S_k^z > (\cos\theta_{k,i'} - 1) \times G_{i',j}(t,t') \\
&+ < S_k^z > (\cos\theta_{k,i'} + 1)F_{i',j}(t,t') \\
&- 2 < S_{i'}^z > \cos\theta_{k,i'}F_{k,j}(t,t')] \\
&- 2\sum_{i'\in NN} I_{k,i'} < S_{i'}^z > \cos\theta_{k,i'}F_{k,j}(t,t') \\
&+ \sum_{k'\in NNN} J_{k,k'}[< S_k^z > (\cos\theta_{k,k'} - 1) \times G_{k',j}(t,t') \\
&+ < S_k^z > (\cos\theta_{k,k'} + 1)F_{k',j}(t,t') \\
&- 2 < S_{k'}^z > \cos\theta_{k,k'}F_{k,j}(t,t')] \qquad (11.20)
\end{aligned}
$$

For simplicity, except otherwise stated, all NN interactions $(J_{k,k'}, I_{k,k'})$ are taken equal to (J_1, I_1) and all NNN interactions are taken equal to J_2 in the following. Furthermore, let us denote, in the film coordinates defined above, the Cartesian components of the spin position $\mathbf{R}_i$ by (ℓ_i, m_i, n_i).

We now introduce the following in-plane Fourier transforms:

$$
\begin{aligned}
G_{i,j}(t,t') &= \frac{1}{\Delta}\iint_{BZ} d\mathbf{k}_{xy}\frac{1}{2\pi}\int_{-\infty}^{+\infty} d\omega e^{-i\omega(t-t')} \\
&\times g_{n_i,n_j}(\omega,\mathbf{k}_{xy})\, e^{i\mathbf{k}_{xy}\cdot(\mathbf{R}_i-\mathbf{R}_j)} \qquad (11.21)
\end{aligned}
$$

$$
\begin{aligned}
F_{k,j}(t,t') &= \frac{1}{\Delta}\iint_{BZ} d\mathbf{k}_{xy}\frac{1}{2\pi}\int_{-\infty}^{+\infty} d\omega e^{-i\omega(t-t')} \\
&\times f_{n_k,n_j}(\omega,\mathbf{k}_{xy})\, e^{i\mathbf{k}_{xy}\cdot(\mathbf{R}_k-\mathbf{R}_j)} \qquad (11.22)
\end{aligned}
$$

where ω is the spin wave frequency, $\mathbf{k}_{xy}$ denotes the wave-vector parallel to xy planes and $\mathbf{R}_i$ is the position of the spin at the site i. n_i, n_j and n_k are, respectively, the z-component indices of the layers where the sites $\mathbf{R}_i$, $\mathbf{R}_j$ and $\mathbf{R}_k$ belong to. The integral over $\mathbf{k}_{xy}$ is performed in the first Brillouin zone (BZ) whose surface is Δ in the xy reciprocal plane. For convenience, we denote $n_i = 1$ for all sites on the surface layer, $n_i = 2$ for all sites of the second layer and so on.

Note that for a three-dimensional case, making a 3D Fourier transformation of Eqs. (11.19)–(11.20) we obtain the spin wave dispersion relation in the absence of anisotropy:

$$\hbar\omega = \pm\sqrt{A^2 - B^2} \tag{11.23}$$

where

$$\begin{aligned} A &= J_1 \left\langle S^z \right\rangle [\cos\theta + 1] Z\gamma + 2ZJ_1 \left\langle S^z \right\rangle \cos\theta \\ &\quad + J_2 \left\langle S^z \right\rangle [\cos(2\theta) + 1] Z_c \cos(k_z a) + 2Z_c J_2 \left\langle S^z \right\rangle \cos(2\theta) \\ B &= J_1 \left\langle S^z \right\rangle (\cos\theta - 1) Z\gamma + J_2 \left\langle S^z \right\rangle [\cos(2\theta) - 1] Z_c \cos(k_z a) \end{aligned}$$

where $Z = 8$ (NN number), $Z_c = 2$ (NNN number on the c-axis), $\gamma = \cos(k_x a/2)\cos(k_y a/2)\cos(k_z a/2)$ (a: lattice constant). We see that $\hbar\omega$ is zero when $A = \pm B$, namely at $k_x = k_y = k_z = 0$ ($\gamma = 1$) and at $k_z = 2\theta$ along the helical axis. The case of ferromagnets (antiferromagnets) with NN interaction only is recovered by putting $\cos\theta = 1\ (-1)$ [78].

Let us return to the film case. We make the in-plane Fourier transforms Eqs. (11.21)–(11.22) for Eqs. (11.19)–(11.20). We obtain the following matrix equation:

$$\mathbf{M}(\omega)\,\mathbf{h} = \mathbf{u} \tag{11.24}$$

where $\mathbf{M}(\omega)$ is a square matrix of dimension $(2N_z \times 2N_z)$, $\mathbf{h}$ and $\mathbf{u}$ are the column matrices which are defined as follows:

$$\mathbf{h} = \begin{pmatrix} g_{1,n'} \\ f_{1,n'} \\ \vdots \\ g_{n,n'} \\ f_{n,n'} \\ \vdots \\ g_{N_z,n'} \\ f_{N_z,n'} \end{pmatrix}, \quad \mathbf{u} = \begin{pmatrix} 2\left\langle S_1^z \right\rangle \delta_{1,n'} \\ 0 \\ \vdots \\ 2\left\langle S_{N_z}^z \right\rangle \delta_{N_z,n'} \\ 0 \end{pmatrix} \tag{11.25}$$

where, taking $\hbar = 1$ hereafter,

$$\mathbf{M}(\omega) = \begin{pmatrix} \omega + A_1 & 0 & B_1^+ & C_1^+ & D_1^+ & E_1^+ & 0 & 0 & 0 & 0 & 0 & 0 \\ 0 & \omega - A_1 & -C_1^+ & -B_1^+ & -E_1^+ & -D_1^+ & 0 & 0 & 0 & 0 & 0 & 0 \\ \cdots & \cdots & \cdots & \cdots & \cdots & \cdots & \cdots & \cdots & \cdots & \cdots & \cdots & \cdots \\ \cdots & D_n^- & E_n^- & B_n^- & C_n^- & \omega + A_n & 0 & B_n^+ & C_n^+ & D_n^+ & E_n^+ & \cdots \\ \cdots & -E_n^- & -D_n^- & -C_n^- & -B_n^- & 0 & \omega - A_n & -C_n^+ & -B_n^+ & -E_n^+ & -D_n^+ & \cdots \\ \cdots & \cdots & \cdots & \cdots & \cdots & \cdots & \cdots & \cdots & \cdots & \cdots & \cdots & \cdots \\ 0 & 0 & 0 & 0 & 0 & 0 & D_{N_z}^- & E_{N_z}^- & B_{N_z}^- & C_{N_z}^- & \omega + A_{N_z} & 0 \\ 0 & 0 & 0 & 0 & 0 & 0 & -E_{N_z}^- & -D_{N_z}^- & -C_{N_z}^- & -B_{N_z}^- & 0 & \omega - A_{N_z} \end{pmatrix} \quad (11.26)$$

where

$$A_n = -8J_1(1+d)\left[\left\langle S_{n+1}^z \right\rangle \cos\theta_{n,n+1} + \left\langle S_{n-1}^z \right\rangle \cos\theta_{n,n-1}\right] - 2J_2\left[\left\langle S_{n+2}^z \right\rangle \cos\theta_{n,n+2} + \left\langle S_{n-2}^z \right\rangle \cos\theta_{n,n-2}\right]$$

where $n = 1, 2, \cdots, N_z$, $d = I_1/J_1$, and

$$B_n^{\pm} = 4J_1 \left\langle S_n^z \right\rangle (\cos\theta_{n,n\pm 1} + 1)\gamma$$
$$C_n^{\pm} = 4J_1 \left\langle S_n^z \right\rangle (\cos\theta_{n,n\pm 1} - 1)\gamma$$
$$E_n^{\pm} = J_2 \left\langle S_n^z \right\rangle (\cos\theta_{n,n\pm 2} - 1)$$
$$D_n^{\pm} = J_2 \left\langle S_n^z \right\rangle (\cos\theta_{n,n\pm 2} + 1)$$

In the above expressions, $\theta_{n,n\pm 1}$ the angle between a spin in the layer n and its NN spins in layers $n \pm 1$, etc., and $\gamma = \cos\left(\frac{k_x a}{2}\right)\cos\left(\frac{k_y a}{2}\right)$.

Solving $\det|\mathbf{M}| = 0$, we obtain the spin wave spectrum ω of the present system: For each value (k_x, k_y), there are $2N_z$ eigenvalues of ω corresponding to two opposite spin precessions as in antiferromagnets (the dimension of $\det|\mathbf{M}|$ is $2N_z \times 2N_z$). Note that the above equation depends on the values of $< S_n^z >$ $(n = 1, \ldots, N_z)$. Even at temperature $T = 0$, these z-components are not equal to 1/2 because we are dealing with an antiferromagnetic system where fluctuations at $T = 0$ give rise to the so-called zero-point spin contraction [87]. Worse, in our system with the existence of the film surfaces, the spin contractions are not spatially uniform as will be seen below. So the solution of $\det|\mathbf{M}| = 0$ should be found by iteration. This will be explicitly shown hereafter.

The solution for $g_{n,n}$ is given by

$$g_{n,n}(\omega) = \frac{|\mathbf{M}|_{2n-1}}{|\mathbf{M}|} \qquad (11.27)$$

where $|\mathbf{M}|_{2n-1}$ is the determinant made by replacing the $(2n-1)$-th column of $|\mathbf{M}|$ by $\mathbf{u}$ given by Eq. (11.25) [note that $g_{n,n}$ occupies the $(2n-1)$-th line of the matrix $\mathbf{h}$]. Writing now

$$|\mathbf{M}| = \prod_i \left[\omega - \omega_i\left(\mathbf{k}_{xy}\right)\right] \qquad (11.28)$$

we see that $\omega_i\left(\mathbf{k}_{xy}\right)$, $i = 1, \ldots, 2N_z$, are poles of $g_{n,n}$. $\omega_i\left(\mathbf{k}_{xy}\right)$ can be obtained by solving $|\mathbf{M}| = 0$. In this case, $g_{n,n}$ can be expressed as

$$g_{n,n}(\omega) = \sum_i \frac{D_{2n-1}\left(\omega_i\left(\mathbf{k}_{xy}\right)\right)}{\left[\omega - \omega_i\left(\mathbf{k}_{xy}\right)\right]} \qquad (11.29)$$

where $D_{2n-1}\left(\omega_i\left(\mathbf{k}_{xy}\right)\right)$ is

$$D_{2n-1}\left(\omega_i\left(\mathbf{k}_{xy}\right)\right) = \frac{|\mathbf{M}|_{2n-1}\left(\omega_i\left(\mathbf{k}_{xy}\right)\right)}{\prod_{j\neq i}\left[\omega_j\left(\mathbf{k}_{xy}\right) - \omega_i\left(\mathbf{k}_{xy}\right)\right]} \qquad (11.30)$$

Next, using the spectral theorem which relates the correlation function $\langle S_i^- S_j^+ \rangle$ to Green's function [383], we have

$$\langle S_i^- S_j^+ \rangle = \lim_{\varepsilon \to 0} \frac{1}{\Delta} \iint d\mathbf{k}_{xy} \int_{-\infty}^{+\infty} \frac{i}{2\pi} \big(g_{n,n'}(\omega + i\varepsilon) - g_{n,n'}(\omega - i\varepsilon) \big) \frac{d\omega}{e^{\beta\omega} - 1} e^{i\mathbf{k}_{xy} \cdot (\mathbf{R}_i - \mathbf{R}_j)}, \quad (11.31)$$

where ϵ is an infinitesimal positive constant and $\beta = (k_B T)^{-1}$, k_B being the Boltzmann constant.

Using Green's function presented above, we can calculate self-consistently various physical quantities as functions of temperature T. The magnetization $\langle S_n^z \rangle$ of the n-th layer is given by

$$\begin{aligned} \langle S_n^z \rangle &= \frac{1}{2} - \langle S_n^- S_n^+ \rangle \\ &= \frac{1}{2} - \lim_{\epsilon \to 0} \frac{1}{\Delta} \iint d\mathbf{k_{xy}} \int_{-\infty}^{+\infty} \frac{i}{2\pi} [g_{n,n}(\omega + i\epsilon) \\ &\quad - g_{n,n}(\omega - i\epsilon)] \frac{d\omega}{e^{\beta\omega} - 1} \end{aligned} \quad (11.32)$$

Replacing Eq. (11.29) in Eq. (11.32) and making use of the following identity:

$$\frac{1}{x - i\eta} - \frac{1}{x + i\eta} = 2\pi i \delta(x) \quad (11.33)$$

we obtain

$$\langle S_n^z \rangle = \frac{1}{2} - \frac{1}{\Delta} \iint dk_x dk_y \sum_{i=1}^{2N_z} \frac{D_{2n-1}(\omega_i)}{e^{\beta\omega_i} - 1} \quad (11.34)$$

where $n = 1, \ldots, N_z$. As $< S_n^z >$ depends on the magnetizations of the neighboring layers via $\omega_i (i = 1, \cdots, 2N_z)$, we should solve by iteration the equations (11.34) written for all layers, namely for $n = 1, \ldots, N_z$, to obtain the magnetizations of layers 1, 2, 3, ..., N_z at a given temperature T. Note that by symmetry, $< S_1^z > = < S_{N_z}^z >$, $< S_2^z > = < S_{N_z-1}^z >$, $< S_3^z > = < S_{N_z-2}^z >$, and so on. Thus, only $N_z/2$ self-consistent layer magnetizations are to be calculated.

The value of the spin in the layer n at $T = 0$ is calculated by

$$\langle S_n^z \rangle (T = 0) = \frac{1}{2} + \frac{1}{\Delta} \iint dk_x dk_y \sum_{i=1}^{N_z} D_{2n-1}(\omega_i) \quad (11.35)$$

where the sum is performed over N_z negative values of ω_i (for positive values the Bose–Einstein factor is equal to 0 at $T = 0$).

The transition temperature T_c can be calculated in a self-consistent manner by iteration, letting all $< S_n^z >$ tend to zero, namely $\omega_i \to 0$. Expanding $e^{\beta\omega_i} - 1 \to \beta_c\omega_i$ on the right-hand side of Eq. (11.34) where $\beta_c = (k_B T_c)^{-1}$, we have by putting $\langle S_n^z\rangle = 0$ on the left-hand side,

$$\beta_c = 2\frac{1}{\Delta}\iint dk_x dk_y \sum_{i=1}^{2N_z} \frac{D_{2n-1}(\omega_i)}{\omega_i} \tag{11.36}$$

There are N_z such equations using Eq. (11.34) with $n = 1, \ldots, N_z$. Since the layer magnetizations tend to zero at the transition temperature from different values, it is obvious that we have to look for a convergence of the solutions of the equations Eq. (11.36) to a single value of T_c. The method to do this will be shown below.

11.4 Spin Waves: Results from the Green's Function Method

Let us take $J_1 = 1$, namely ferromagnetic interaction between NN. We consider the helimagnetic case where the NNN interaction J_2 is negative and $|J_2| > J_1$. The non-uniform GS spin configuration across the film has been determined in Section 11.2 for each value of $p = J_2/J_1$. Using the values of $\theta_{n,n\pm1}$ and $\theta_{n,n\pm2}$ to calculate the matrix elements of $|\mathbf{M}|$, then solving $\det|\mathbf{M}| = 0$ we find the eigenvalues $\omega_i (i = 1, \ldots, 2N_z)$ for each $\mathbf{k}_{xy}$ with an input set of $\langle S_n^z\rangle(n = 1, \ldots, N_z)$ at a given T. Using Eq. (11.34) for $n = 1, \ldots, N_z$ we calculate the output $\langle S_n^z\rangle(n = 1, \ldots, N_z)$. Using this output set as input, we calculate again $\langle S_n^z\rangle(n = 1, \ldots, N_z)$ until the input and output are identical within a desired precision P. Numerically, we use a Brillouin zone of 100^2 wave-vector values, and we use the obtained values $\langle S_n^z\rangle$ at a given T as input for a neighboring T. At low T and up to $\sim \frac{4}{5}T_c$, only a few iterations suffice to get $P \leq 1\%$. Near T_c, several dozens of iteration are needed to get the convergence. We show below our results.

11.4.1 Spectrum

We calculated the spin wave spectrum as described above for each a given J_2/J_1. The spin wave spectrum depends on the temperature via the temperature-dependence of layer magnetizations. Let us show in Fig. 11.3 the spin wave frequency ω versus $k_x = k_y$ in the case of an 8-layer film where $J_2/J_1 = -1.4$ at two temperatures $T = 0.1$ and $T = 1.02$ (in units of $J_1/k_B = 1$). Some remarks are in order:

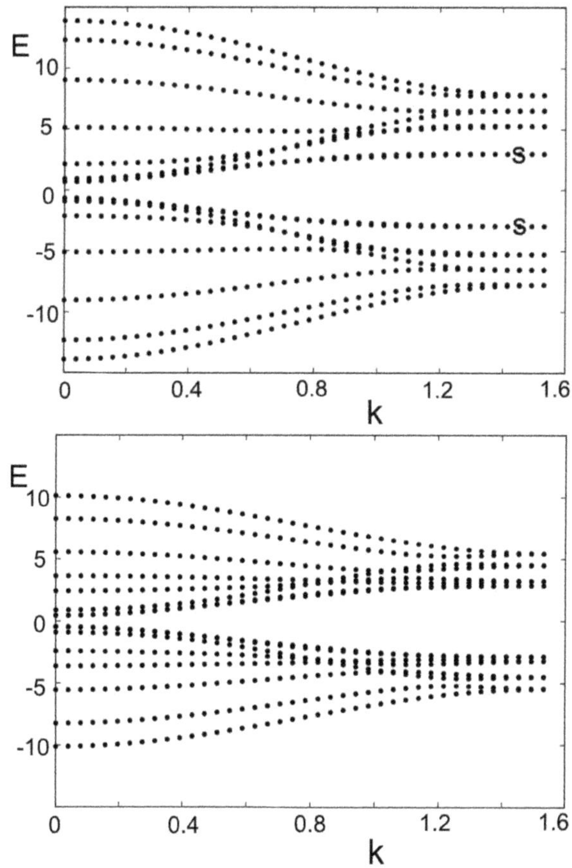

Figure 11.3 Spectrum $E = \hbar\omega$ versus $k \equiv k_x = k_y$ for $J_2/J_1 = -1.4$ at $T = 0.1$ (top) and $T = 1.02$ (bottom) for $N_z = 8$ and $d = 0.1$. The surface branches are indicated by s.

(i) There are 8 positive and 8 negative modes corresponding two opposite spin precessions. Unlike ferromagnets, spin waves in antiferromagnets and non-collinear spin structures have opposite spin precessions which describe the opposite circular motion of each sublattice spins [87]. The negative sign does not mean spin wave negative energy, but it indicates just the precession contrary to the trigonometric sense.

(ii) Note that there are two degenerate acoustic surface branches lying at low energy on each side. This degeneracy comes from the two symmetrical surfaces of the film. These surface modes propagate parallel to the film surface but are damped from the surface inward.

(iii) As T increases, layer magnetizations decrease (see below), reducing, therefore, the spin wave energy as seen in Fig. 11.3 (bottom).

(iv) If the spin magnitude $S \neq 1/2$, then the spectrum is shifted toward higher frequency since it is proportional to S.

(v) Surface spin wave spectrum (and bulk spin waves) can be experimentally observed by inelastic neutron scattering in ferromagnetic and antiferromagnetic films [36, 374]. To our knowledge, such experiments have not been performed for helimagnets.

11.4.2 Spin Contraction at $T = 0$ and Transition Temperature

It is known that in antiferromagnets, quantum fluctuations give rise to a contraction of the spin length at zero temperature (see Chapter 3 and Ref. [87]). We will see here that a spin under a stronger antiferromagnetic interaction has a stronger zero-point spin contraction. The spins near the surface serve for such a test. In the case of the film considered above, spins in the first and in the second layers have only one antiferromagnetic NNN while interior spins have two NNN, so the contraction at a given J_2/J_1 is expected to be stronger for interior spins. This is verified with the results shown in Fig. 11.4. When $|J_2|/J_1$ increases, namely the antiferromagnetic interaction becomes stronger, we observe stronger contractions. Note that the contraction tends to zero when

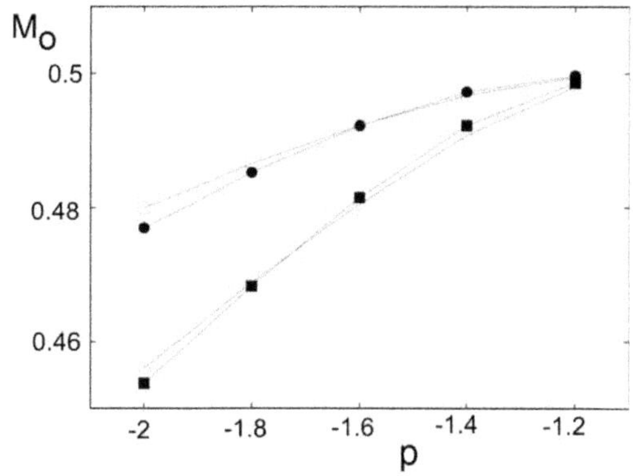

Figure 11.4 (Color online) Spin lengths of the first four layers at $T = 0$ for several values of $p = J_2/J_1$ with $d = 0.1$, $N_z = 8$. Black circles, void circles, black squares and void squares are for first, second, third and fourth layers, respectively. See text for comments.

the spin configuration becomes ferromagnetic, namely J_2 tends to -1.

11.4.3 Layer Magnetizations

Let us show two examples of the magnetization, layer by layer, from the film surface in Figs. 11.5 and 11.6, for the case where $J_2/J_1 = -1.4$ and -2 in a $N_z = 8$ film. Let us comment on the case $J_2/J_1 = -1.4$:

(i) the shown result is obtained with a convergence of 1%. For temperatures closer to the transition temperature T_c, we have to lower the precision to a few percents which reduces the clarity because of their close values (not shown).

(ii) the surface magnetization, which has a large value at $T = 0$ as seen in Fig. 11.4, crosses the interior layer magnetizations at $T \simeq 0.42$ to become much smaller than interior magnetizations at higher temperatures. This crossover phenomenon is due to the competition between quantum fluctuations, which dominate low-T behavior, and the low-lying surface spin wave modes which strongly diminish the surface magnetization at higher T. Note that the second-layer magnetization makes also

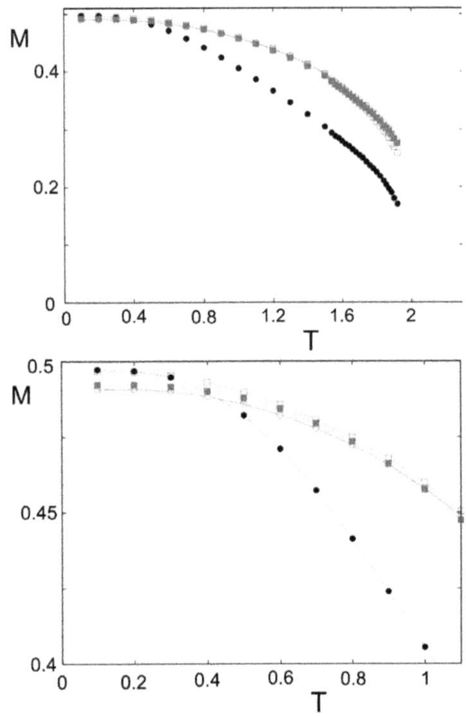

Figure 11.5 (Color online) Layer magnetizations as functions of T for $J_2/J_1 = -1.4$ with $d = 0.1$, $N_z = 8$ (top). Zoom of the region at low T to show crossover (bottom). Black circles, blue void squares, magenta squares and red void circles are for first, second, third and fourth layers, respectively. See text.

a crossover at $T \simeq 1.3$. Similar crossovers have been observed in quantum antiferromagnetic films [79] and quantum superlattices [80].

Similar remarks can be also made for the case $J_2/J_1 = -2$.

Note that although the layer magnetizations are different at low temperatures, they will tend to zero at a unique transition temperature as seen below. The reason is that as long as an interior layer magnetization is not zero, it will act on the surface spins as an external field, preventing them to become zero.

The temperature where layer magnetizations tend to zero is calculated by Eq. (11.36). Since all layer magnetizations tend to

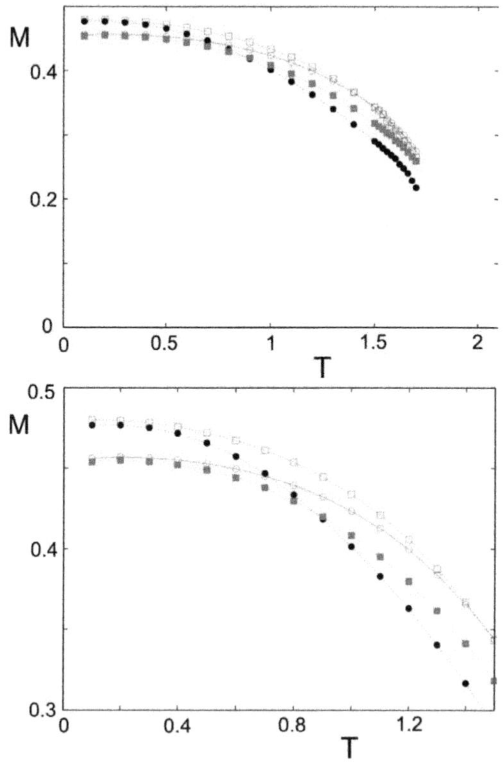

Figure 11.6 (Color online) Layer magnetizations as functions of T for $J_2/J_1 = -2$ with $d = 0.1$, $N_z = 8$ (top). Zoom of the region at low T to show crossover (bottom). Black circles, blue void squares, magenta squares and red void circles are for first, second, third and fourth layers, respectively. See text.

zero from different values, we have to solve self-consistently N_z equations (11.36) to obtain the transition temperature T_c. One way to do it is to use the self-consistent layer magnetizations obtained as described above at a temperature as close as possible to T_c as input for Eq. (11.36). As long as the T is far from T_c the convergence is not reached: We have four 'pseudo-transition temperatures' T_{cs} as seen in Fig. 11.7, one for each layer. The convergence of these T_{cs} can be obtained by a short extrapolation from temperatures when they are rather close to each other. T_c is thus obtained with a very small extrapolation error as seen in Fig. 11.7 for $p = J_2/J_1 = -1.4$:

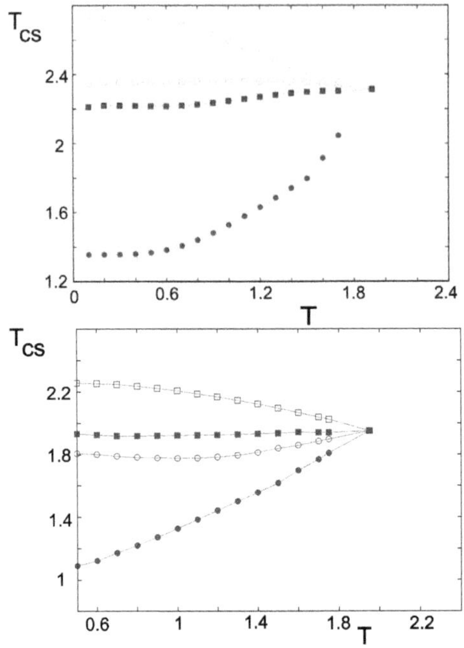

Figure 11.7 (Color online) Top: Transition temperature is calculated at $p = J_2/J_1 = -1.4$ for $d = 0.1$, $N_z = 8$: At each temperature, using the self-consistent values of layer magnetizations at $T < T_c$, Eq. (11.36) is solved to obtain T_{cs} for each layer. The convergence is reached when T_{cs} tend to a single value T_c. One has $T_c \simeq 2.313 \pm 0.010$. Red circles, black void circles, blue squares and blue void squares are T_{cs} obtained from Eq. (11.36) for first, second, third and fourth layers, respectively, at different temperatures. Bottom: Extrapolation by lines to obtain T_c is shown for surface parameter $J_1^s/J_1 = 0.7$. The precision for self-consistent convergence is 1% for layer magnetizations. See text for comments.

$T_c \simeq 2.313 \pm 0.010$. The results for several $p = J_2/J_1$ are shown in Fig. 11.8.

11.4.4 Effect of Anisotropy and Surface Parameters

The results shown above have been calculated with an in-plane anisotropy interaction $d = 0.1$. Let us show now the effect of d. Stronger d will enhance all the layer magnetizations and increase T_c. Figure 11.9 shows the surface magnetization versus T for several

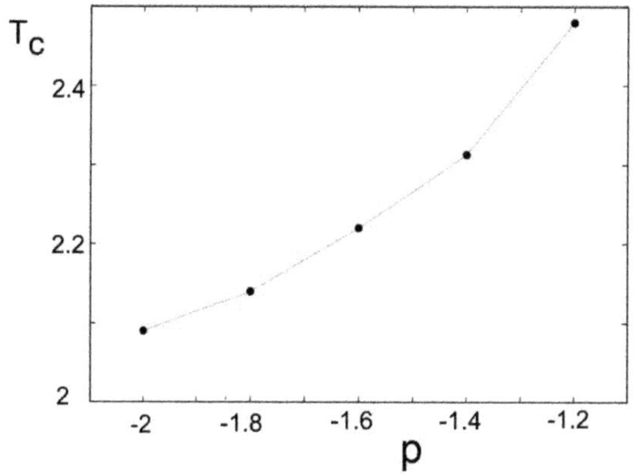

Figure 11.8 (Color online) Transition temperature versus $p = J_2/J_1$ for an 8-layer film with $d = 0.1$. See text for comments.

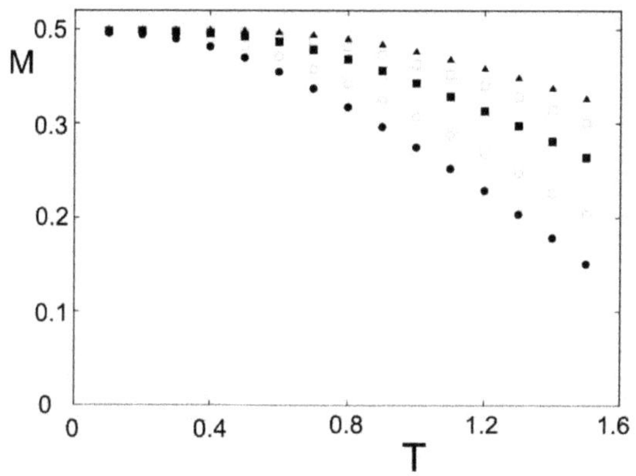

Figure 11.9 Surface magnetization versus T for $d = 0.05$ (circles), 0.1 (void circles), 0.2 (squares), 0.3 (void squares) and 0.4 (triangles), with $J_2/J_1 = -1.4$, $N_z = 16$.

values of d (other layer magnetizations are not shown to preserve the figure clarity). The transition temperatures are 2.091 ± 0.010, 2.313 ± 0.010, 2.752 ± 0.010, 3.195 ± 0.010 and 3.630 ± 0.010 for $d = 0.05$, 0.1, 0.2, 0.3 and 0.4, respectively. These values versus d lie on a remarkable straight line.

Let us examine the effects of the surface anisotropy and exchange parameters d_s and J_1^s. As seen above, even in the case where the surface interaction parameters are the same as those in the bulk the surface spin wave modes exist in the spectrum. These localized modes cause a low surface magnetization observed in Figs. 11.5 and 11.6. Here, we show that with a weaker NN exchange interaction between surface spins and the second-layer ones, namely $J_1^s < J_1$, the surface magnetization becomes even much smaller with respect to the magnetizations of interior layers. This is shown in Fig. 11.10 for several values of J_1^s. We observe again here the crossover of layer magnetizations at low T due to quantum fluctuations as discussed earlier. The transition temperature strongly decreases with J_1^s: We have $T_c = 2.103 \pm 0.010, 1.951 \pm 0.010, 1.880 \pm 0.010$ and 1.841 ± 0.010 for $J_1^s = 1, 0.7, 0.5$ and 0.3, respectively ($N_z = 16, J_2/J_1 = -2$, $d = d_s = 0.1$). Note that the value $J_1^s = 0.5$ is a very particular value: the GS configuration is a uniform configuration with all angles equal

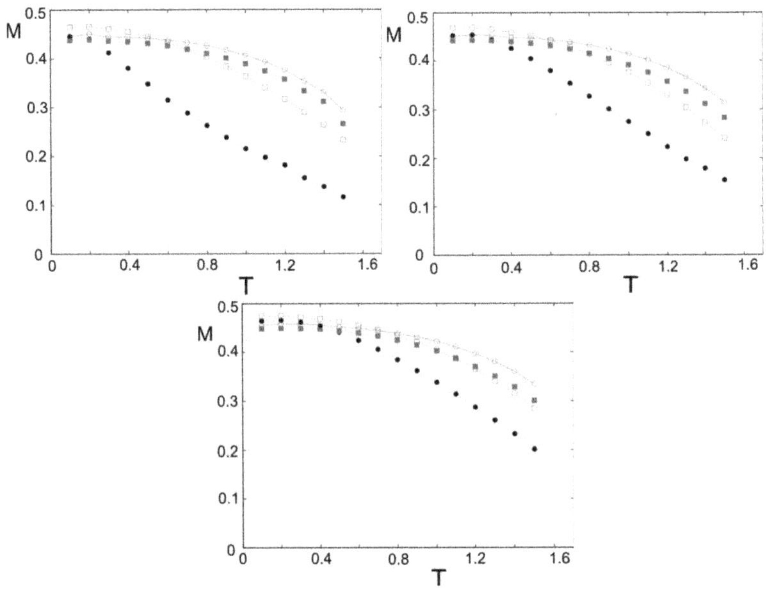

Figure 11.10 (Color online) Layer magnetizations as functions of T for the surface interaction $J_1^s = 0.3$ (top, left), 0.5 (top, right) and 0.7 (bottom) with $J_2/J_1 = -2$, $d = 0.1$ and $N_z = 16$. Black circles, blue void squares, magenta squares and red void circles are for first, second, third and fourth layers, respectively.

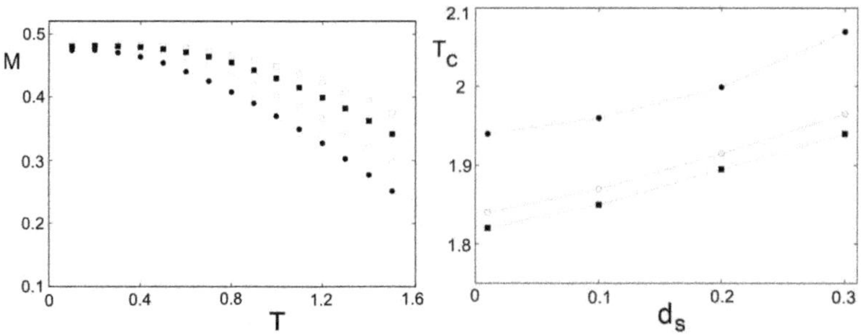

Figure 11.11 Left: Surface magnetization versus T for $d_s = 0.01$ (circles), 0.1 (void circles), 0.2 (squares) and 0.3 (void squares), with $J_1^s = 1, J_2/J_1 = -2$ and $N_z = 16$. Right: Transition temperature versus d_s for J_1^s=0.7, 0.5 and 0.3 (curves from up to down), with $J_2/J_1 = -2, d = 0.1$ and $N_z = 16$.

to 60°, namely there is no surface spin rearrangement. This can be explained if we look at the local field acting on the surface spins: the lack of neighbors is compensated by this weak positive value of J_1^s so that their local field is equal to that of a bulk spin. There is thus no surface reconstruction. Nevertheless, as T increases, thermal effects will strongly diminish the surface magnetization as seen in Fig. 11.10 (middle). As for the surface anisotropy parameter d_s, it affects strongly the layer magnetizations and the transition temperature: We show in Fig. 11.11 the surface magnetizations and the transition temperature for several values of d_s.

11.4.5 Effect of the Film Thickness

We have performed calculations for $N_z = 8$, 12 and 16. The results show that the effect of the thickness at these values is not significant: The difference lies within convergence errors. Note that the classical ground state of the first four layers is almost the same: For example, here are the values of cosinus of the angles of the film first half for $N_z = 16$ which are to be compared with the values for $N_z = 8$ given in Table 11.1, for $p = J_2/J_1 = -2$ (in parentheses are angles in degree):

0.86737967 (29.844446), 0.41125694 (65.716179), 0.52374715 (58.416061), 0.49363765 (60.420044), 0.50170541 (59.887100),

0.49954081 (60.030373), 0.50013113 (59.991325), 0.49993441 (60.004330).

From the 4th layer, the angle is almost equal to the bulk value (60°).

At $p = J_2/J_1 = -2$, the transition temperature is 2.090 ± 0.010 for $N_z = 8$, 2.093 ± 0.010 for $N_z = 12$ and 2.103 ± 0.010 for $N_z = 16$. These are the same within errors. At smaller thicknesses, the difference can be seen. However, for helimagnets in the z direction, thicknesses smaller than 8 do not allow to see fully the surface helical reconstruction which covers the first four layers (see Section 11.2).

At this stage, it is interesting to note that our result is in excellent agreement with experiments: It has been experimentally observed that the transition temperature does not vary significantly in MnSi films in the thickness range of 11–40 nm [176]. One possible explanation is that the helical structure is very stable as seen above: The surface perturbs the bulk helical configuration only at the first four layers, so the bulk 'rigidity' dominates the transition. This has been experimentally seen in holmium films [206].

11.4.6 Classical Helimagnetic Films: Monte Carlo Simulation

To appreciate quantum effects causing crossovers of layer magnetizations presented above at low temperatures, we show here some results of the classical counterpart model: Spins are classical XY spins of amplitude $S = 1$. We take the XY spins rather than the Heisenberg spins for comparison with the quantum case because in the latter case we have used an in-plane Ising-like anisotropy interaction d. Monte Carlo simulations have been carried out over film samples of $100 \times 100 \times 16$. Periodic boundary conditions are applied in the xy plane. One million of MC steps are discarded to equilibrate the system and another million of MC steps are used for averaging. The layer magnetizations versus T are shown in Fig. 11.12 for the case where surface interaction $J_1^s = 1$ (top) and 0.3 (bottom) with $J_2/J_1 = -2$ and $N_z = 16$. One sees that (i) by extrapolation there is no spin contraction at $T = 0$ and

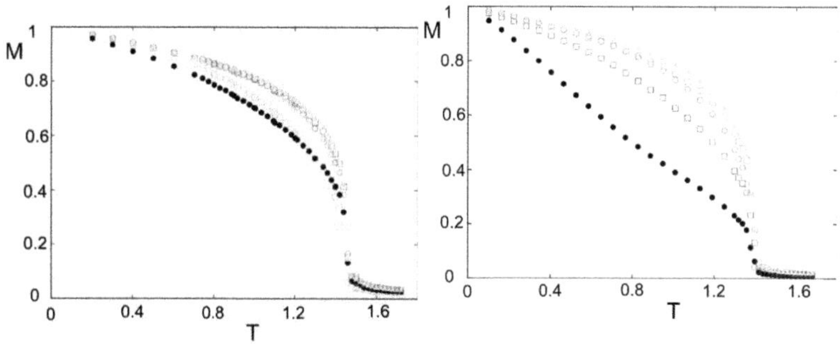

Figure 11.12 (Color online) Monte Carlo results: Layer magnetizations as functions of T for the surface interaction $J_1^s = 1$ (left) and 0.3 (right) with $J_2/J_1 = -2$ and $N_z = 16$. Black circles, blue void squares, cyan squares and red void circles are for first, second, third and fourth layers, respectively.

there is no crossover of layer magnetizations at low temperatures, (ii) from the intermediate temperature region up to the transition the relative values of layer magnetizations are not always the same as in the quantum case: For example, at $T = 1.2$, one has $M_1 < M_3 < M_4 < M_2$ in Fig. 11.12 (top) and $M_1 < M_2 < M_4 < M_3$ in Fig. 11.12 (bottom) which are not the same as in the quantum case shown in Fig. 11.6 (top) and Fig. 11.10 (top). Our conclusion is that even at temperatures close to the transition, helimagnets may have slightly different behaviors according to their quantum or classical nature. Extensive MC simulations with size effects and detection of the order of the phase transition do not fall within the scope of this present chapter.

11.5 Simple Cubic Helimagnetic Films

In the above section, we have presented the work performed for the case of a helimagnetic film with the BCC lattice structure [89]. A similar study has been carried out for the case of a helimagnetic film with the simple cubic lattice [102].

We consider a thin film of simple cubic lattice of N_z layers described by the Hamiltonian

$$\mathcal{H} = -\sum_{\langle i,j \rangle} J_{i,j} \mathbf{S}_i \cdot \mathbf{S}_j - \sum_{<i,j>} I_{i,j} S_i^z S_j^z \cos\theta_{ij} \tag{11.37}$$

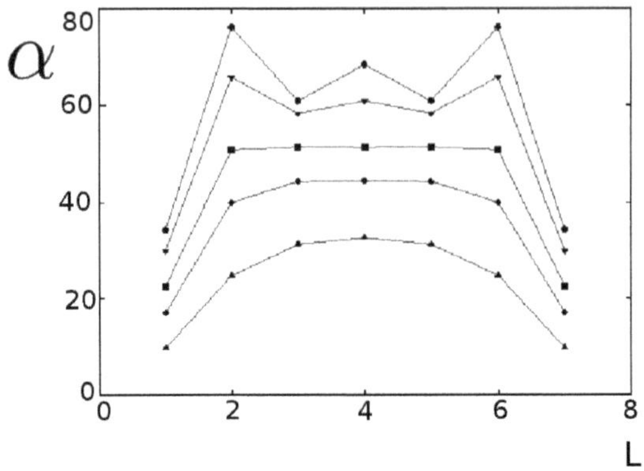

Figure 11.13 Angle $\alpha_1 \ldots \alpha_7$ across the film for $J_2/J_1 = -0.6, -0.5, -0.4, -0.35, -0.3$ (from top) with $N_z = 8$.

where $\mathbf{S}_i$ is the Heisenberg spin at the lattice site i, J_{ij} is the exchange interaction between two NN spins. As before, we have added an anisotropic term $I_{i,j}$ along the spin-quantization axes z taken to be very small. This anisotropy is necessary to have a phase transition at a finite temperature in a thin film (according to the theorem of Mermin and Wagner[231], 2D systems of continuous spins with a short-range interaction do not have long-range ordering at finite T).

The helimagnetic order results from the competition between NN ferromagnetic interaction ($J_1 > 0$) and the NNN aniferromagnetic exchange interaction ($J_2 < 0$) in the z. direction. Let α_i be the angle of a spin in the i-th layer made with the spin in the next layer. In the bulk case the helimagnetic structure is possible only for $|J_2| > J_1/4$ (see Section 3.4 and Ref. [102]).

In a film with a thickness, we have used the steepest descent method [248] to calculate the turn angle between spins of adjacent layers. The result is shown in Fig. 11.13 where we see the spin configurations near the two surfaces are strongly modified with respect to the bulk turn angle.

Using the GF method as described above we have calculated the spin wave frequency ω versus $k_x = k_y$ for various surface exchange interactions J_s in the case of an 8-layer film with an anisotropy

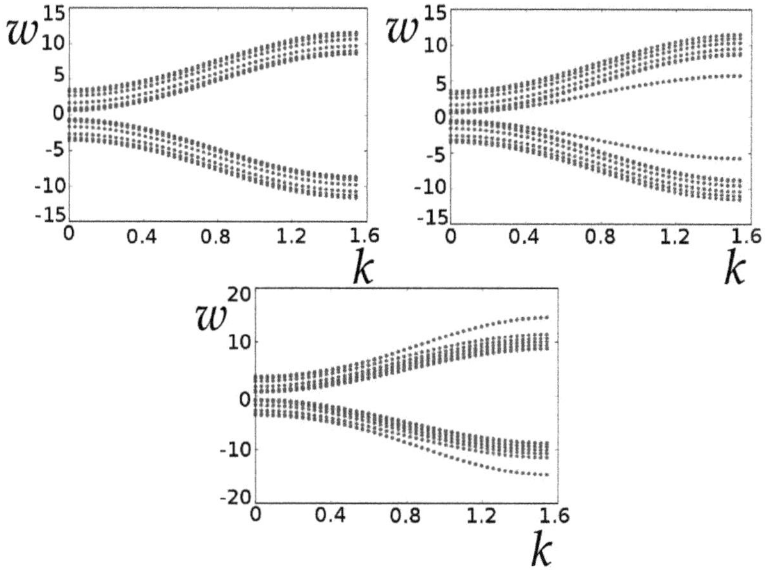

Figure 11.14 Spin wave frequency spectrum versus $k \equiv k_x = k_y$ in the case where $N_z = 8, J_2 = -1$ and $d = 0.1$ for $J_s/J_1 = 1$(top, left), $J_s/J_1 = 0.6$ (top, right) and $J_s/J_1 = 1.6$ (bottom).

$d/J_1 = 0.1$. We note the existence of acoustic surface modes which lie in the low-energy region for $J_s/J_1 = 0.6$ (see Fig. 11.14, middle) and optical surface modes which lie in the high-energy region for $J_s = 1.6/J_1$ (see Fig. 11.14, bottom). Note that no such modes exist in the case $J_s/J_1 = 1$ (see Fig. 11.14, top).

The surface magnetization, which has a large value at $T = 0$ as seen in Fig. 11.15, crosses the interior layer magnetizations at $T = 0.6$ to become smaller than interior magnetizations at higher temperatures. This crossover phenomenon is due to the competition between quantum fluctuations, which dominate the low-T behavior, and the low-lying surface spin wave modes which strongly diminish the surface magnetization at higher T. Note that the second-layer magnetization makes also a crossover at $T = 0.6$. Similar crossovers have been observed in quantum antiferromagnetic films [79], quantum superlattices [80] and the BCC case presented above.

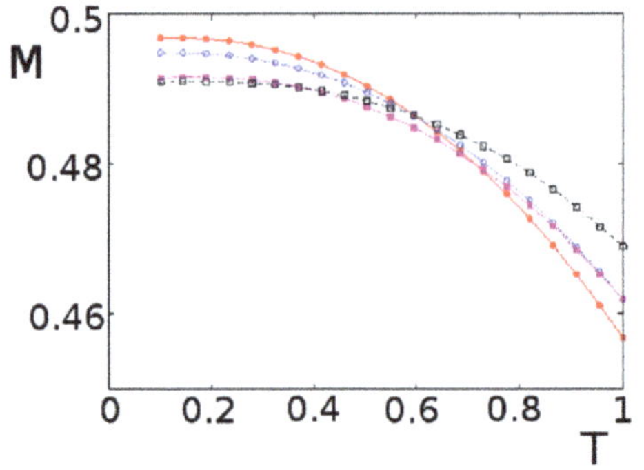

Figure 11.15 (Color online) Layer magnetization as a function of T for $J_2/J_1 = -0.7$ with $d/J_1 = d_s/J_1 = 0.1$, $N_z = 8$: Red circles, blue void circles, magenta squares and black void squares are magnetizations of the first, second, third and fourth layers, respectively.

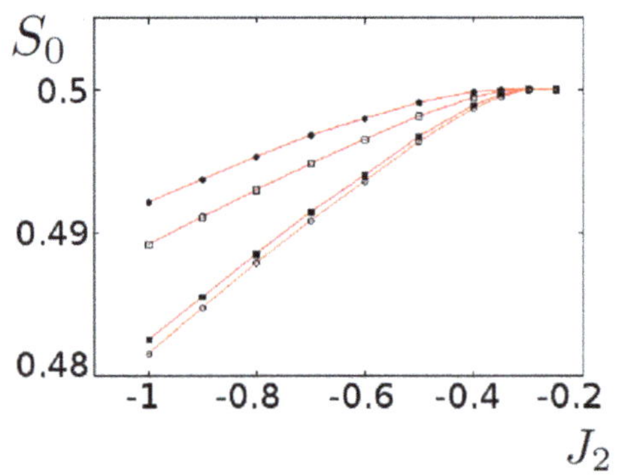

Figure 11.16 Spin length at $T = 0$ versus J_2, with $d = d_s = 0.1$, $N_z = 8$ ($J_1 = 1$): Black circles, void squares, black squares and void circles are data for spins in first, second, third and fourth layers, respectively.

We show the zero-point contraction for various values of J_2 ($J_1 = 1$) in Fig. 11.16. When $|J_2|$ increases, namely the antiferromagnetic interaction becomes stronger, we observe stronger contractions. Note that the contraction tends to zero when the spin configuration becomes ferromagnetic, namely J_2 tends to -0.25.

Let us show the effect of the surface anisotropy in Fig. 11.17: Stronger d_s makes larger surface magnetization and compensates the lack of a neighbor for surface spins.

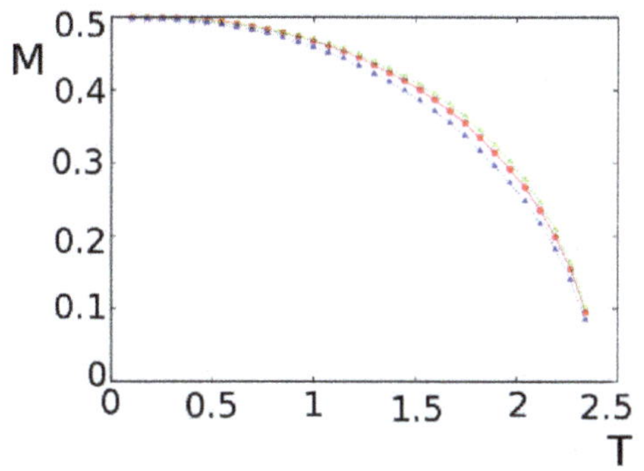

Figure 11.17 Layer magnetization as function of T, effect of d_s: $d_s/J_1 = 0.2$, with $J_2/J_1 = -0.5$, $d/J_1 = 0.1$, $N_z = 8$. Red circles, green void triangles, blue triangles and magenta circles are magnetizations of the first, second, third and fourth layers, respectively.

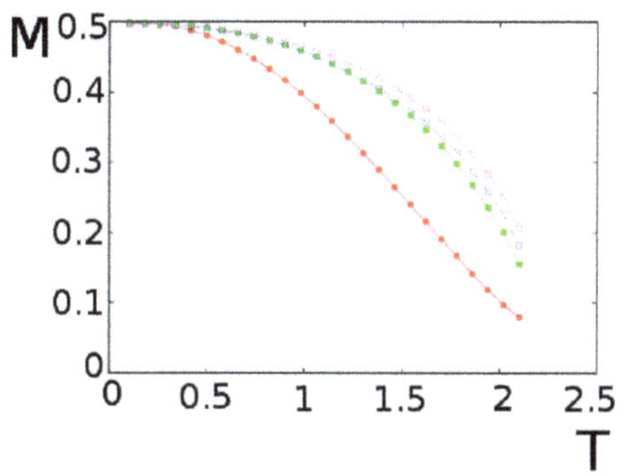

Figure 11.18 Layer magnetizations as functions of T, effect of J_s: $J_s = 0.4$. with $J_2/J_1 = -0.5$, $d/J_1 = 0.1$, $N_z = 8$. Red circles, green squares, blue void squares, and magenta void squares are magnetizations of the first, second, third and fourth layers, respectively.

We show now the effect of the surface interaction J_s. Figure 11.18 shows the case of weak J_s where a low surface magnetization is observed. This "soft surface" is due to the effect of the acoustic surface modes such as those seen in Fig. 11.14 (top, right). On the contrary, when the surface interaction is strong, we have the case of "hard" surface as seen in Fig. 11.19 for a strong J_s [the optical

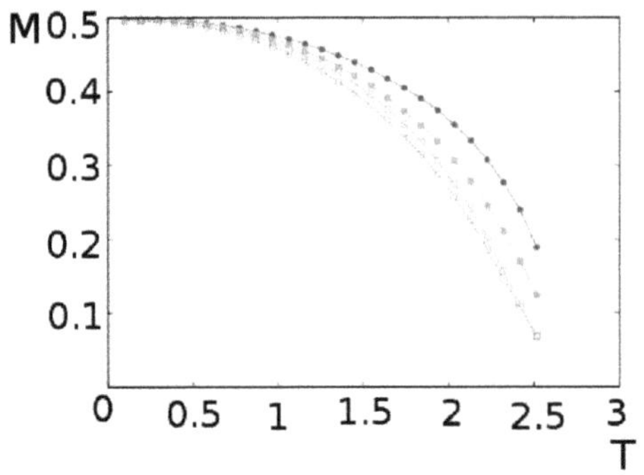

Figure 11.19 Layer magnetizations versus T, effect of J_s: $J_s/J_1 = 1.6$. $J_2/J_1 = -0.5$, $d/J_1 = 0.1$, $N_z = 8$. The color code is given in the previous figure.

surface modes seen in Fig. 11.14 (bottom) are responsible for the hard surface].

11.6 Conclusion

We have studied in this chapter surface effects in a helimagnetic film of BCC and SC lattices with quantum Heisenberg spins. Note that the method presented above can be applied to any lattice structure, and the results found here are valid for general helimagnetic structures. Note also that the results have been shown for the case of spin $S = 1/2$ where quantum fluctuations are strong at low temperatures. Rare-earth elements Ho and Dy with helical structures along the c-axis and ferromagnetic in the basal planes which are very similar to the present model are expected to bear the same features as what has been shown above. However, at low temperatures, due to their larger spin amplitudes, $S = 7/2$ and $5/2$ for Ho and Dy, quantum fluctuations are certainly weaker and the crossover may occur with smaller difference. Numerical applications of our formalism should be performed to get precise values for these cases.

In this chapter, the classical bulk ground-state spin configuration in a thin film is exactly calculated and is found to be strongly modified near the film surface. The surface spin rearrangement is found at the first four layers in our model, regardless of the bulk angle, namely the NNN interaction strength J_2. The spin wave excitation is calculated using a general GF technique for non-collinear spin configurations. The layer magnetization as a function of temperature and the transition temperature are shown for various interaction parameters. Among the striking features presented in the present chapter, let us mention

(i) the crossover of layer magnetizations at low temperatures due to the competition between quantum fluctuations and thermal effects,
(ii) the existence of low-lying surface spin wave modes which cause a low surface magnetization,
(iii) a strong effect of the surface exchange interaction (J_1^s) which drastically modifies the surface spin configuration and gives rise to a very low surface magnetization,
(iv) the transition temperature varies strongly with the helical angle but it is insensitive to the film thickness in agreement with experiments performed on MnSi films [176] and holmium [206],
(v) the classical spin model counterpart gives features slightly different from those of the quantum model, both at low and high temperatures.

Let us make some comments on works of similar models. The work by Mello et al. [229] have treated almost the same model as ours using a hexagonal anisotropy which corresponds to the case of Dy in which the helical angle is $\simeq 60°$. However, the authors of this work studied only the classical spin configuration at $T = 0$. Rodrigues et al. [291] have used exactly the same model as Mello et al. but with application to the Ho case. They have used the mean-field estimation to establish the phase diagram in the space (T, H) (H: magnetic field) and shown that surface effects affect the phase diagram. The MC work of Cinti et al. [64] was based on a classical spin model with a Hamiltonian, very different from ours, including a 6-constant interaction (a kind of dipolar interaction) in the c-direction.

To conclude, let us emphasize that the general theoretical method proposed here allows us to study at a microscopic level surface spin waves and their physical consequences at finite temperatures in helimagnetic films with non-collinear spin configurations. It can be used in more complicated situations such as helimagnets with Dzyaloshinskii–Moriya interactions [177]. We will show this case in Chapters 13–15.

Chapter 12

Partial Phase Transition in Helimagnetic Thin Films in a Field

We study the phase transition in a helimagnetic film with Heisenberg spins under an applied magnetic field in the c direction perpendicular to the film. The helical structure is due to the antiferromagnetic interaction between next-nearest neighbors in the c direction. Helimagnetic films in zero field are known to have a strong modification of the in-plane helical angle near the film surfaces. We show that spins react to a moderate applied magnetic field by creating a particular spin configuration along the c axis. With increasing temperature (T), using Monte Carlo simulations we show that the system undergoes a phase transition triggered by the destruction of the ordering of a number of layers. This partial phase transition is shown to be intimately related to the ground-state spin structure. We show why some layers undergo a phase transition while others do not. The Green's function method for non-collinear magnets is also carried out to investigate effects of quantum fluctuations. Non-uniform zero-point spin contractions and a crossover of layer magnetizations at low T are shown and discussed.

The results shown in this chapter have been published in Ref. [105].

Physics of Magnetic Thin Films: Theory and Simulation
Hung T. Diep

ISBN 978-981-4877-42-8 (Hardcover), 978-1-003-12110-7 (eBook)
www.jennystanford.com

12.1 Introduction

Helimagnets have been subject of intensive investigations over the past four decades since the discovery of its ordering [350, 369]: In the bulk, a spin in a space direction turns an angle θ with respect to the orientation of its previous nearest neighbor (see Fig. 12.1). This helical structure can take place in several directions simultaneously with different helical angles. The helical structure shown in Fig. 12.1 is due to the competition between the interaction between nearest neighbors (NN) and the antiferromagnetic interaction between next-nearest neighbors (NNN). Other helimagnetic structures have also been very early investigated [28, 29, 169]. Spin wave properties in bulk helimagnets have been investigated by spin wave theories [91, 139, 289] and Green's function method [286]. Heat capacity in bulk MnSi has been experimentally investigated [269].

We confine ourselves to the case of a Heisenberg helical film in an applied magnetic field. Helimagnets are special cases of a large family of periodic non-collinear spin structures called frustrated systems of XY and Heisenberg spins. The frustration has several

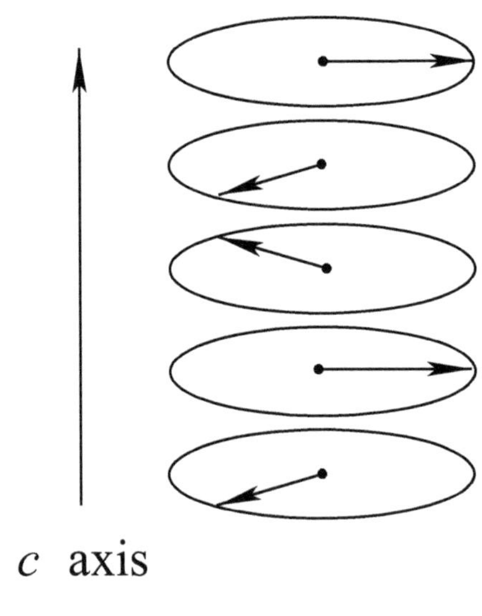

Figure 12.1 Spin configuration along the c direction in the bulk case where $J_2/J_1 = -1$, $H/J_1 = 0$.

origins: (i) It can be due to the geometry of the lattice such as the triangular lattice, the face-centered cubic (FCC) and hexagonal-close-packed (HCP) lattices, with antiferromagnetic NN interaction [19, 82, 83, 224, 276]. (ii) It can be due to competing interactions between NN and NNN such as the case of helimagnets [350, 369] shown in Fig. 12.1. (iii) It can be due to the competition between the exchange interaction which favors collinear spin configurations and the Dzyaloshinskii–Moriya (DM) interaction which favors perpendicular spin arrangements.

Effects of the frustration have been extensively studied in various systems during the past 30 years. The reader is referred to recent reviews on bulk frustrated systems given in Ref. [85]. When frustration effects are coupled with surface effects, the situation is often complicated. Let us mention our previous works on a frustrated surface [248] and on a frustrated FCC antiferromagnetic film [249] where surface spin rearrangements and surface phase transitions have been found (see Chapters 9 and 10). We have also recently shown results in zero field of thin films of body-centered cubic (BCC) and simple cubic (SC) structures in the previous chapter. The helical angle along the c axis perpendicular to the film surface was found to strongly vary in the vicinity of the surface. The phase transition and quantum fluctuations have been presented.

In this chapter, we are interested in the effect of an external magnetic field applied along the c axis perpendicular to the film surface of a helimagnet with both classical and quantum Heisenberg spins. Note that without an applied field, the spins lie in the xy planes: Spins in the same plane are parallel while two NN in the adjacent planes form an angle α which varies with the position of the planes [102] (see the previous chapter), unlike in the bulk. As will be seen below, the applied magnetic field gives a very complex spin configuration across the film thickness. We determine this ground state (GS) by the numerical steepest descent method. We will show by Monte Carlo (MC) simulation that the phase transition in the field is due to the disordering of a number of layers inside the film. We identify the condition under which a layer becomes disordered. This partial phase transition is not usual in thin films where one observes more often the disordering of the surface layer, not an interior layer. At low temperatures, we investigate effects of quantum fluctuations using a Green's function (GF) method for non-collinear spin configurations.

Section 12.2 is devoted to the description of the model and the determination of the classical GS. The structure of the GS spin configuration is shown as a function of the applied field. Section 12.3 is used to show the MC results at finite temperatures where a partial phase transition is observed. Effects of the magnetic field strength and the film thickness are shown. The GF method is described in Section 12.4 and its results on the layer magnetizations at low temperatures are displayed and discussed in terms of quantum fluctuations.

12.2 Model: Determination of the Classical Ground State

We consider a thin film of SC lattice of N_z layers stacked in the c direction. Each lattice site is occupied by a Heisenberg spin. For the GS determination, the spins are supposed to be classical spins in this section. The Hamiltonian is given by

$$\mathcal{H} = -\sum_{\langle i,j \rangle} J_{i,j} \mathbf{S}_i \cdot \mathbf{S}_j - \sum_i \mathbf{H} \cdot \mathbf{S}_i \tag{12.1}$$

where $J_{i,j}$ is the interaction between two spins $\mathbf{S}_i$ and $\mathbf{S}_j$ occupying the lattice sites i and j and $\mathbf{H}$ denotes an external magnetic field applied along the c axis. To generate helical angles in the c direction, we suppose an antiferromagnetic interaction J_2 between NNN in the c direction in addition to the ferromagnetic interaction J_1 between NN in all directions. For simplicity, we suppose that J_1 is the same everywhere. For this section, we shall suppose J_2 is the same everywhere for the presentation clarity. Note that in the bulk in zero field, the helical angle along the c axis is given by $\cos\alpha = -\frac{J_1}{4J_2}$ for a SC lattice [87] with $|J_2| > 0.25J_1$. Below this value, the ferromagnetic ordering is stable (see Section 3.4).

In this chapter, we will study physical properties as functions of J_2/J_1, H/J_1 and k_BT/J_1. Hereafter, for notation simplicity we will take $J_1 = 1$ and $k_B = 1$. The temperature is thus in unit of J_1/k_B, the field and the energy are in unit of J_1.

In a film, the angles between NN in adjacent planes are not uniform across the film: A strong variation is observed near the surfaces. An exact determination can be done by energy minimization [89] or by numerical steepest descent method [248, 249]. The latter is particularly efficient for complex situations such as the present case where the spins are no longer in the xy planes in an applied field: A spin in the i-th layer is determined by two parameters which are the angle with its NN in the adjacent plane, say $\alpha_{i,i+1}$, and the azimuthal angle β_i formed with the c axis. Since there is no competing interaction in the xy planes, spins in each plane are parallel. We shall use here the steepest descent method which consists in calculating the local field at each site and aligning the spin in its local field to minimize its energy. The reader is referred to Ref. [248] for a detailed description. In so doing for all sites and repeating many times until a convergence to the lowest energy is obtained with a desired precision (usually at the sixth digit, namely at $\simeq 10^{-6}$ per cents), one obtains the GS configuration. Note that we have used several thousands of different initial conditions to check the convergence to a single GS for each set of parameters.

Figures 12.2a and 12.2b show the spin components S^z, S^y and S^x for all layers. The spin lengths in the xy planes are shown in Fig. 12.2c. Since the spin structure in a field is complicated and plays an important role in the partial phase transition shown in the next section, let us describe it in details and explain the physical reason lying behind:

(1) Several planes have negative z spin components. This can be understood by examining the competition between the magnetic field which tends to align spins in the c direction, and the antiferromagnetic interaction J_2 which tries to preserve the antiferromagnetic ordering. This is very similar to the case of collinear antiferromagnets: in a weak magnetic field, the spins remain antiparallel, and in a moderate field, the so-called "spin flop" occurs: The neighboring spins stay antiparallel with each other but turn themselves perpendicular to the field direction to reduce the field effect [87].
(2) Due to the symmetry of the two surfaces, one observes the following symmetry with respect to the middle of the film:

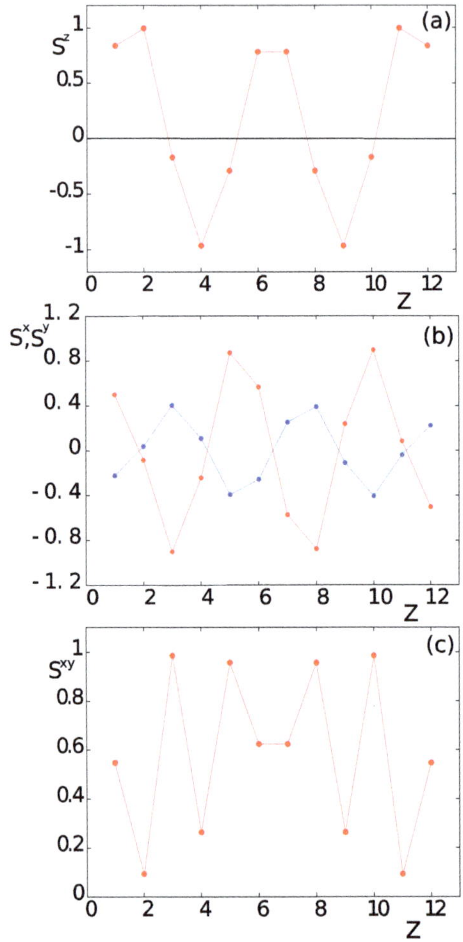

Figure 12.2 Spin components across the film in the case where $H = 0.2$. The horizontal axis Z represents plane Z ($Z = 1$ is the first plane etc.): (a) S^z; (b) S^x (red) and S^y (blue); (c) Modulus S^{xy} of the projection of the spins on the xy plane. See text for comments.

(i) $S_1^z = S_{N_z}^z$, $S_2^z = S_{N_z-1}^z$, $S_3^z = S_{N_z-2}^z$ etc.
(ii) $S_1^y = -S_{N_z}^y$, $S_2^y = -S_{N_z-1}^y$, $S_3^y = -S_{N_z-2}^y$ etc.
(iii) $S_1^x = -S_{N_z}^x$, $S_2^x = -S_{N_z-1}^x$, $S_3^x = -S_{N_z-2}^x$ etc.

Note that while the z components are equal, the x and y components are antiparallel (Fig. 12.2b): The spins preserve their

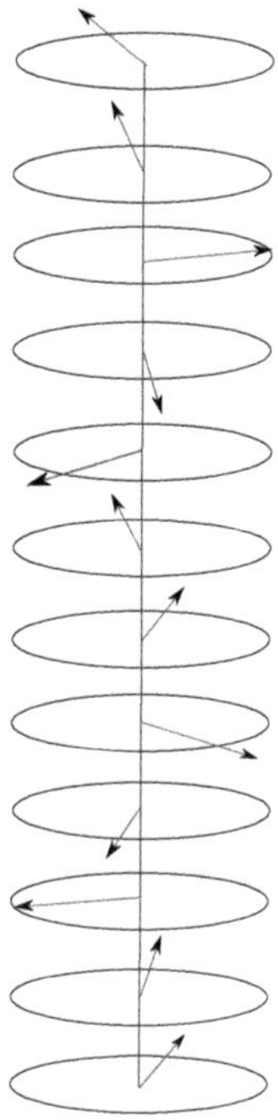

Figure 12.3 Spin configuration in the case where $H = 0.4$, $J_2 = -1$, $N_z = 12$. The circles in the xy planes with radius equal to 1 are plotted to help identify the orientation of each spin. The spins when viewed along the c axis are shown in Fig. 12.4d.

antiferromagnetic interaction for the transverse components. This is similar to the case of spin flop in the bulk (see p. 86 of Ref. [87]). Only at a very strong field that all spins turn into the field direction.

(3) The GS spin configuration depends on the film thickness. An example will be shown in the next section.

A full view of the "chain" of N_z spins along the c axis between the two surfaces is shown in Fig. 12.3.

Note that the angle in the xy plane is determined by the NNN interaction J_2. Without field, the symmetry is about the c axis, so x and y spin components are equivalent (see Fig. 12.1). Under the field, due to the surface effect, the spins make different angles with the c axis giving rise to different z components for the layers across the film as shown in Fig. 12.2a. Of course, the symmetry axis is still the c axis, so all S^x and S^y are invariant under a rotation around the c axis. Figure 12.2b shows the symmetry of S^x as that of S^y across the film as outlined in remark (iii). Figure 12.2b is thus an instantaneous configuration between S^x and S^y for each layer across the film. As the simulation time is going on these components rotate about the c axis but their symmetry outlined in remark (iii) is valid at any time. The xy spin modulus S^{xy} shown in Fig. 12.2c, on the other hand, is time-invariant. The phase transition occurring for layers with large S^{xy} (xy disordering) is shown in the next section.

The GS spin configuration depends on the field magnitude H. If H increases, we observe an interesting phenomenon: Fig. 12.4 shows the spin configurations projected on the xy plane (top view) for increasing magnetic field. We see that the spins of each chain tend progressively to lie in a same plane perpendicular to the xy planes (Figs. 12.4a–c). The "planar zone" observed in Fig. 12.4c occurs between $H \simeq 0.35$ and 0.5. For stronger fields they are no more planar (Fig. 12.4d-f). Note that the larger the xy component is, the smaller the z component becomes: For example, in Fig. 12.4a, the

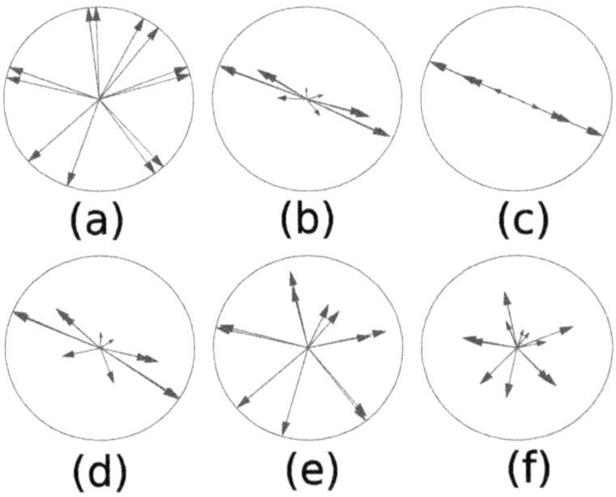

Figure 12.4 Top view of S^{xy} (projection of spins on xy plane) across the film for several values of H: (a) 0, (b) 0.03, (c) 0.2, (d) 0.4, (e) 0.7, (f) 1.7. The radius of the circle, equal to 1, is the spin full length: For high fields, spins are strongly aligned along the c axis, S^{xy} is therefore much smaller than 1.

spins are in the xy plane without field ($H = 0$) and in Fig. 12.4f they are almost parallel to the c axis because of a high field.

12.3 Phase Transition

We recall that for bulk materials, in spite of their long history, the nature of the phase transition in non-collinear magnets such as stacked triangular XY and Heisenberg antiferromagnets has been elucidated only recently [90, 187, 250, 251]. On the other hand, surface effects in thin films have been intensively studied during the past three decades [36, 87, 374]. Most of theoretical studies were limited to collinear magnetic orderings. Phase transitions in thin films with non-collinear ground states have been only recently studied [89, 102, 229, 248, 249]. MC simulations of a helimagnetic thin film [64] and a few experiments in helimagnets [176, 177] have also been carried out. These investigations were motivated by the fact that helical magnets present a great potential of applications in spintronics with spin-dependent electron transport [150, 166, 357].

As described in the previous section, the planar helical spin configuration in zero field becomes non-planar in a perpendicular field. In order to interpret the phase transition shown below, let us mention that a layer having a large z spin-component parallel to the field cannot have a phase transition because its magnetization will never become zero. This is similar to a ferromagnet in a field. However, layers having large negative z spin-components (antiparallel to the field) can undergo a transition due to the magnetization reversal at a higher temperature similarly to an antiferromagnet in a field. In addition, the xy spin-components whose xy fluctuations are not affected by the perpendicular field can make a transition. Having mentioned these, we expect that some layers will undergo a phase transition, while others will not. This is indeed what we observed in MC simulations shown in the following.

For MC simulations, we use the Metropolis algorithm (see Chapter 6) and a sample size $N \times N \times N_z$ with $N = 20, 40, 60, 100$ for detecting lateral-size effects and $N_z = 8, 12, 16$ for thickness effects. The equilibrium time is 10^5 MC steps/spin and the thermal average is performed with the following 10^5 MC steps/spin.

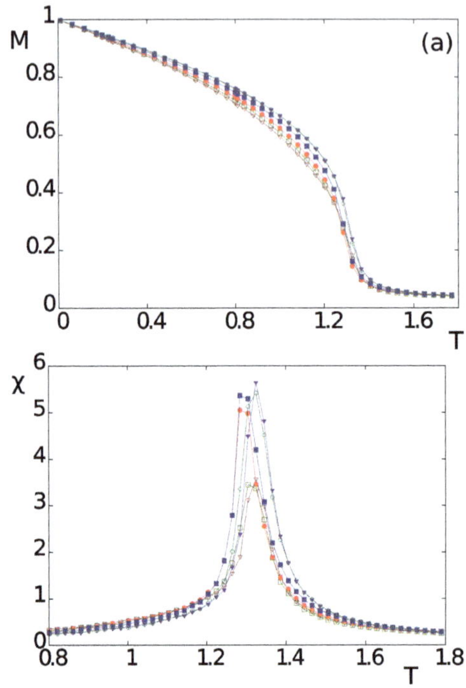

Figure 12.5 (a) Layer magnetization and (b) layer magnetic susceptibility versus T for $H = 0$, $J_2 = -1$, $N_z = 12$. Dark olive green void squares for the first layer, maroon void triangles for the second, red circles for the third, indigo triangles for the fourth, dark blue squares for the fifth, dark green void circles for the sixth layer.

12.3.1 Results of 12-Layer Film

In order to appreciate the effect of the applied field, let us show first the case where $H = 0$ in Fig. 12.5. We see there that all layers undergo a phase transition within a narrow region of T.

In an applied field, as seen earlier, in the GS all layers do not have the same characteristics so one expects different behaviors. Figure 12.6 shows the layer magnetizations and the layer susceptibilities as functions of T for $H = 0.2$ with $J_2 = -1$, $N_z = 12$ (only the first six layers are shown, the other six are symmetric). Several remarks are listed below:

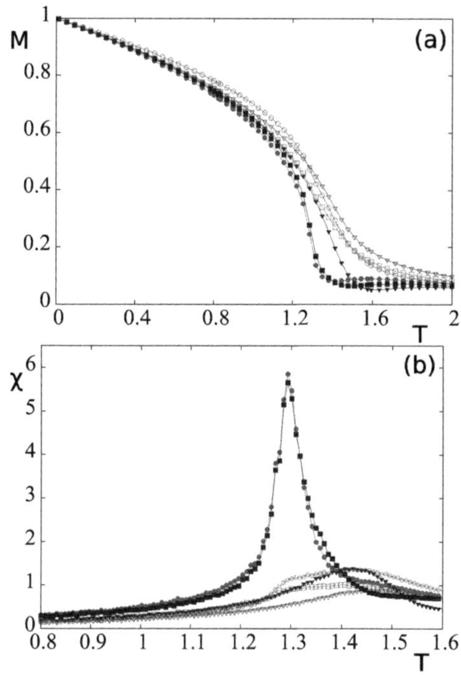

Figure 12.6 (a) Layer magnetization and (b) layer magnetic susceptibility versus T for $H = 0.2, J_2 = -1, N_z = 12$. Dark olive green void squares for the first layer, maroon void triangles for the second, red circles for the third, indigo triangles for the fourth, dark blue squares for the fifth, dark green void circles for the sixth layer.

(1) Only layer 3 and layer 5 have a phase transition: Their magnetizations strongly fall down at the transition temperature. This can be understood from what we have anticipated above: These layers have the largest xy components (see Fig. 12.2c). Since the correlation between xy components do not depend on the applied field, the temperature destroys the in-plane ferromagnetic ordering causing the transition. It is not the case for the z components which are kept non-zero by the field. Of course, symmetric layers 8 and 10 have the same transition (not shown).

(2) Layers with small amplitudes of xy components do not have a strong transverse ordering at finite T: The absence of

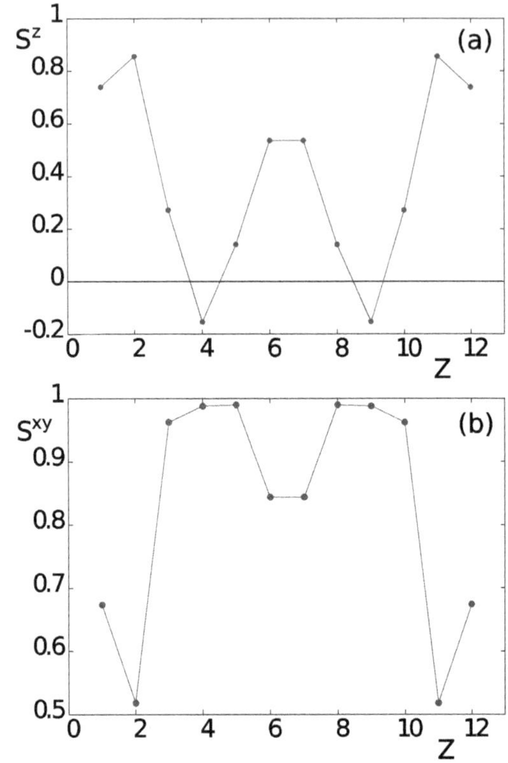

Figure 12.7 (a) S^z and (b) S^{xy} across the film with $H = 0.7$.

pronounced peaks in the susceptibility indicates that they do not make a transition (see Fig. 12.6).

(3) Note that the xy spin components of layers 3 and 5 are disordered at $T_c \simeq 1.275$ indicated by pronounced peaks of the susceptibility.

What we learn from the example shown above is that under an applied magnetic field the film can have a partial transition: Some layers with large xy spin components undergo a phase transition (destruction of their transverse xy correlation). This picture is confirmed by several simulations for various field strengths. Another example is shown in the case of a strong field $H = 0.7$: The GS is shown in Fig. 12.7, where we observe large xy spin components

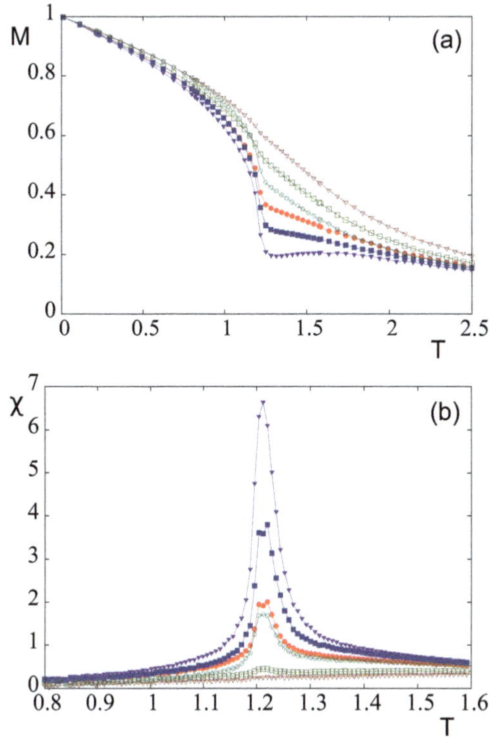

Figure 12.8 (a) Layer magnetization and (b) layer magnetic susceptibility versus T for $H = 0.7, J_2 = -1, N_z = 12$. Dark olive green void squares for the first layer, maroon void triangles for the second, red circles for the third, indigo triangles for the fourth, dark blue squares for the fifth, dark green void circles for the sixth layer.

of layers 3, 4, and 5 (and symmetric layers 7, 8 and 9). We should expect a transition for each of these layers. This is indeed the case: We show these transitions in Fig. 12.8, where sharp peaks of the susceptibilities of these layers are observed.

We close this section by showing some size effects. Figure 12.9 shows the effect of lateral size (xy planes) on the layer susceptibility. As expected in a continuous transition, the peaks of the susceptibilities of the layers undergoing a transition grow strongly with the layer lattice size.

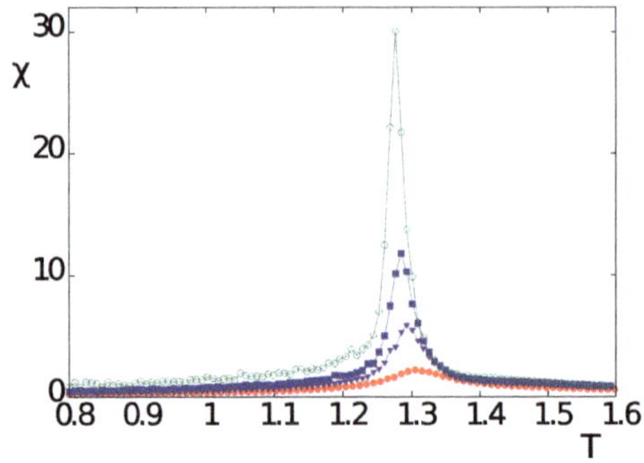

Figure 12.9 Magnetic susceptibility of the third layer versus T for $H = 0.2$, $J_2 = -1$, $N_z = 12$. Dark green void circles, dark blue squares, indigo triangles, red circles are susceptibilities for layer lattice sizes 100×100, 60×60, 40×40 and 20×20, respectively.

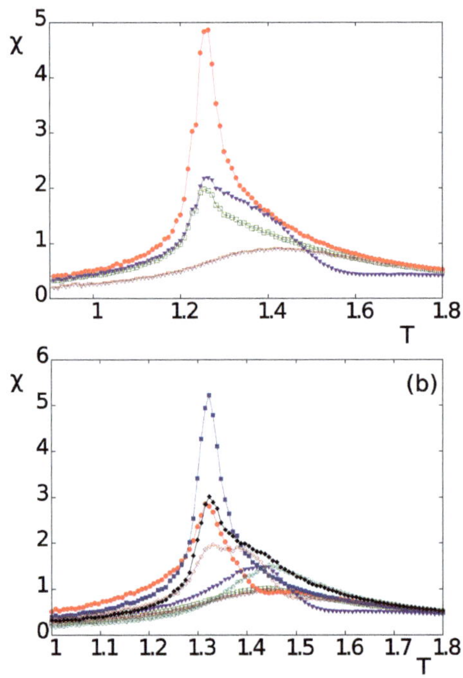

Figure 12.10 Magnetic susceptibility versus T for two thicknesses with $H = 0.2$, $J_2 = -1$: (a) $N_z = 8$, (b) $N_z = 16$. Dark olive green void squares are for the first layer, maroon void triangles for the second, red circles for the third, indigo triangles for the fourth, dark blue squares for the fifth, dark green void circles for the sixth, black diamonds for the seventh, dark brown void diamonds for the eighth. See text for comments.

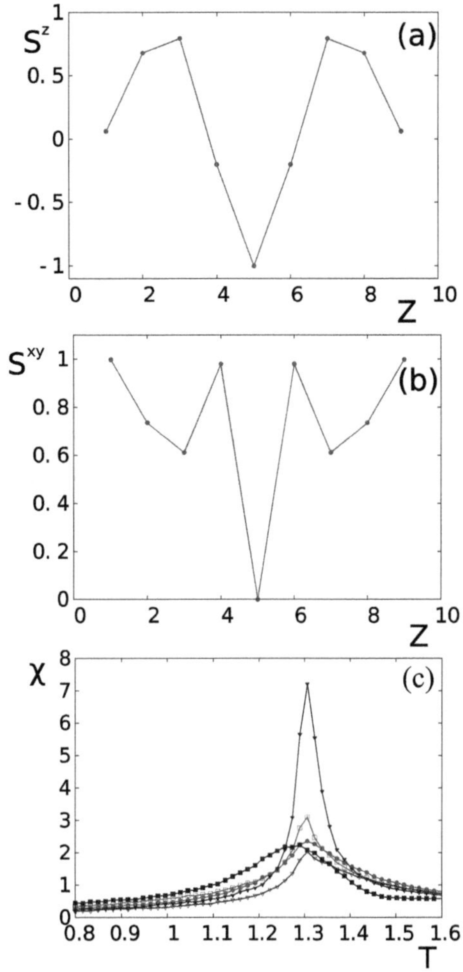

Figure 12.11 Nine-layer film: Spin components across the film in the case where $H = 0.2$. The horizontal axis Z represents plane Z ($Z = 1$ is the first plane etc.): (a) S^z; (b) modulus S^{xy} of the projection of the spins on the xy plane; (c) layer susceptibilities versus T: Dark olive green void squares are for the first layer, maroon void triangles for the second, red circles for the third, indigo triangles for the fourth, dark blue squares for the fifth layer, respectively. Only layers 4 and 1 (also layers 9 and 6, not shown) undergo a transition. See text for comments.

12.3.2 Effects of the Film Thickness

As for the thickness effects, we note that changing the thickness (odd or even number of layers) will change the GS spin configuration so that the layers with largest xy components are not the same. As a consequence, the layers which undergo the transition are not the same for different thicknesses. We show in Fig. 12.10 the layer susceptibilities for $N_z = 8$ and 16. For $N_z = 8$, the layers which undergo a transition are the first, third and fourth layers with pronounced peaks, while for $N = 16$, the layers which undergo a transition are the third, fifth, seventh and eighth layers.

Let us show the case of an odd number of layers. Figure 12.11 shows the results for $N_z = 9$ with $H = 0.2$, $J_2 = -1$. Due to the odd layer number, the center of symmetry is the middle layer (5th layer). As seen, the layers 1 and 4 and their symmetric counterparts (layers 9 and 6) have largest xy spin modulus (Fig. 12.11b). The transition argument shown above predicts that these layers have a transversal phase transition in these xy planes. This is indeed seen in Fig. 12.11c, where the susceptibility of layer 4 has a strong peak at the transition. The first layer, due to the lack of neighbors, has a weaker peak. The other layers do not undergo a transition. They show only a rounded maximum.

12.4 Quantum Fluctuations, Layer Magnetizations and Spin Wave Spectrum

We shall extend here the method used in Chapter 11 and Ref. [89] for zero field to the case where an applied magnetic field is present. The method remains essentially the same except the fact that each spin is defined not only by its angles with the NN in the adjacent layers but also by its azimuthal angle formed with the c axis as seen in Section 12.2.

We use in the following the Hamiltonian (12.1) but with quantum Heisenberg spins $\mathbf{S}_i$ of magnitude 1/2. In addition, it is known that in two dimensions there is no long-range order at finite temperature for isotropic spin models [231] with short-range interaction. Since our films have small thickness, it is useful to add an anisotropic interaction to stabilize the long-range ordering at finite temperatures. Let us use the following anisotropy between $\mathbf{S}_i$ and $\mathbf{S}_j$ which stabilizes the angle between their local quantization axes S_i^z and S_j^z:

$$\mathcal{H}_a = -I_1 \sum_{<i,j>} S_i^z S_j^z \cos\theta_{ij} \tag{12.2}$$

where I_1 is supposed to be positive, small compared to J_1, and limited to NN.

The general method has been recently described in details in Refs. [89, 102] and in Chapter 11. To save space, let us give the results for the simple cubic helimagnetic film in a field. We define the following two double-time Green's functions in the real space:

$$\begin{aligned} G_{i,j}(t,t') &= \ll S_i^+(t); S_j^-(t') \gg \\ &= -i\theta(t-t') < \left[S_i^+(t), S_j^-(t')\right] > \end{aligned} \tag{12.3}$$

$$\begin{aligned} F_{i,j}(t,t') &= \ll S_i^-(t); S_j^-(t') \gg \\ &= -i\theta(t-t') < \left[S_i^-(t), S_j^-(t')\right] > \end{aligned} \tag{12.4}$$

Writing the equations of motion of these functions and using the Tyablikov decoupling scheme to reduce the higher-order functions, we obtain the general equations for non-collinear magnets [89]. We next introduce the following in-plane Fourier transforms $g_{n,n'}$ and $f_{n,n'}$ of the G and F Green's functions, we finally obtain the following coupled equations:

$$\begin{aligned} & D_n^- g_{n-2,n'} + E_n^- f_{n-2,n'} + B_n^- g_{n-1,n'} + C_n^- f_{n-1,n'} \\ & +(\omega + A_n) g_{n,n'} + B_n^+ g_{n+1,n'} + C_n^+ f_{n+1,n'} + D_n^+ g_{n+2,n'} \\ & +E_n^+ f_{n+2,n'} = 2\left\langle S_n^z \right\rangle \delta_{n,n'} \end{aligned} \tag{12.5}$$

$$\begin{aligned} & -E_n^- g_{n-2,n'} - D_n^- f_{n-2,n'} - C_n^- g_{n-1,n'} - B_n^- f_{n-1,n'} \\ & +(\omega - A_n) f_{n,n'} - C_n^+ g_{n+1,n'} - B_n^+ f_{n+1,n'} \\ & -E_n^+ g_{n+2,n'} - D_n^+ f_{n+2,n'} = 0 \end{aligned} \tag{12.6}$$

where $n = 1, 2, \ldots, N_z$, $d_n = I_1/J_1^{\parallel}$, $\gamma = (\cos k_x a + \cos k_y a)/2$. The coefficients are given by

$$\begin{aligned} A_n = & -8J_1^{\parallel} < S_n^z > (1 + d_n - \gamma) \\ & -2 < S_{n+1}^z > \cos\theta_{n,n+1}(d_n + J_1^{\perp}) \\ & -2 < S_{n-1}^z > \cos\theta_{n,n-1}(d_n + J_1^{\perp}) \\ & -2J_2 < S_{n+2}^z > \cos\theta_{n,n+2} \\ & -2J_2 < S_{n-2}^z > \cos\theta_{n,n-2} - H\cos\zeta_n \end{aligned}$$

$$B_n^{\pm} = 2J_1^{\perp}\left\langle S_n^z\right\rangle(\cos\theta_{n,n\pm1}+1)$$
$$C_n^{\pm} = 2J_1^{\perp}\left\langle S_n^z\right\rangle(\cos\theta_{n,n\pm1}-1)$$
$$E_n^{\pm} = J_2\left\langle S_n^z\right\rangle(\cos\theta_{n,n\pm2}-1)$$
$$D_n^{\pm} = J_2\left\langle S_n^z\right\rangle(\cos\theta_{n,n\pm2}+1)$$

ω is the spin wave frequency, k_x and k_y denote the wave vector components in the xy planes, n is the index of the layer along the c axis with $n = 1$ being the surface layer, $n = 2$ the second layer and so on. The angle ζ_n is the azimuthal angle formed by a spin in the layer n with the c axis. Note that (i) if $n = 1$ then there are no $n-1$ and $n-2$ terms in the matrix coefficients, (ii) if $n = 2$ then there are no $n-2$ terms, (iii) if $n = N_z$ then there are no $n+1$ and $n+2$ terms, (iv) if $n = N_z - 1$ then there are no $n+2$ terms. Besides, we have distinguished the in-plane NN interaction $J_1^{\parallel}$ from the inter-plane NN one $J_1^{\perp}$. If we write all equations explicitly for $n = 1, \ldots, N_z$ we can put these equations under a matrix of dimension $2N_z \times 2N_z$. Solving this matrix equation, one gets the spin wave frequencies ω at a given wave vector and a given T.

The layer magnetizations can be calculated at finite temperatures self-consistently. The numerical method to carry out this task has been described in details in Ref. [89]. It is noted that in bulk antiferromagnets and helimagnets the spin length is contracted at $T = 0$ due to quantum fluctuations [87]. Therefore, we also calculate the layer magnetization at $T = 0$ [78, 89]. It is interesting to note that due to the difference of the local field acting on a spin near the surface, the spin contraction is expected to be different for different layers.

We show in Fig. 12.12 the spin length of different layers at $T = 0$ for $N = 12$ and $J_2 = -1$ as functions of H. All spin contractions are not sensitive for H lower than 0.4, but rapidly become smaller for further increasing H. They spin lengths are all saturated at the same value for $H > 2$. Figure 12.13 shows the spin length as a function of J_2. When $J_2 \geq -0.4$, the spin configuration becomes ferromagnetic, and as a consequence the contraction tends to 0. Note that in zero field, the critical value of J_2 is -0.25. In both Figs. 12.12 and 12.13, the surface layer and the third layer have smaller contractions than the other layers. This can be understood

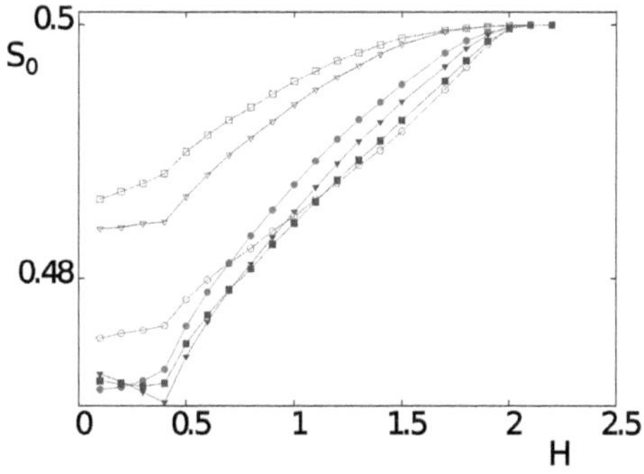

Figure 12.12 Spin lengths at $T = 0$ versus applied magnetic field H. Dark olive green void squares correspond to the spin length of the first layer, maroon void triangles to that of the second, red circles to the third, indigo triangles to the fourth, dark blue squares to the fifth, dark green void circles to the sixth layer.

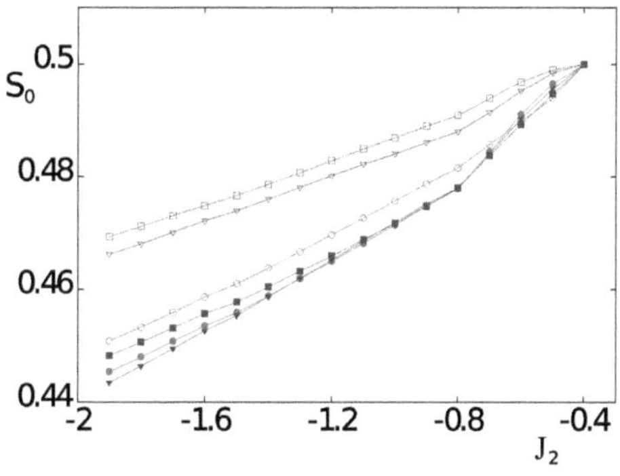

Figure 12.13 Spin lengths at $T = 0$ versus J_2. Dark olive green void squares correspond to the spin length of the first layer, maroon void triangles to that of the second, red circles to the third, indigo triangles to the fourth, dark blue squares to the fifth, dark green void circles to the sixth layer.

by examining the antiferromagnetic contribution to the GS energy of a spin in these layers: They are smaller than those of the other layers.

We show in Fig. 12.14 the layer magnetizations versus T for the case where $J_2 = -1$ and $N_z = 12$ (top figure). The low-T region is enlarged in the inset where one observes a crossover between the magnetizations of layers 1, 3 and 6 at $T \simeq 0.8$: Below this temperature, $M_1 > M_3 > M_6$, and above they become $M_1 < M_3 < M_6$.

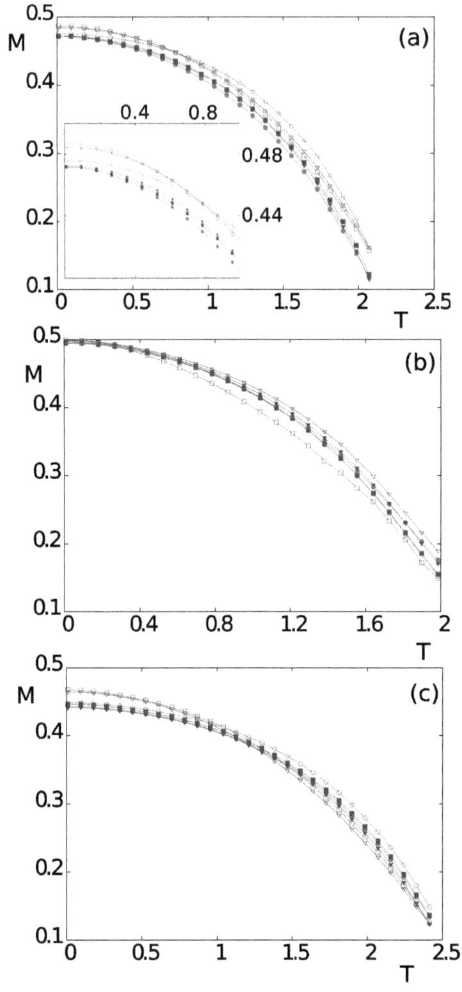

Figure 12.14 Layer magnetizations versus T for several values of J_2 with $H = 0.2$, and $N_z = 12$: (a) $J_2 = -1$, (b) $J_2 = -0.5$, (c) $J_2 = -2$. Dark olive green void squares correspond to the magnetization of the first layer, maroon void triangles to the second, red circles to the third, indigo triangles to the fourth, dark blue squares to the fifth, dark green void circles to the sixth layer. The inset in the top figure shows an enlarged region at low T. See text for comments.

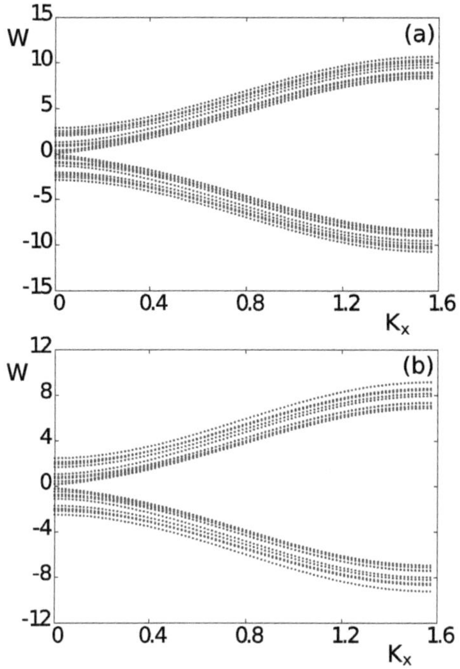

Figure 12.15 Spin wave spectrum versus $k_x = k_y$ where W stands for spin wave frequency ω in Eqs. (12.5)–(12.6), at (a) $T = 0.353$ and (b) $T = 1.212$, with $H = 0.2$.

This crossover is due to the competition between several complex factors: For example, quantum fluctuations have less effect on the surface magnetization making it larger than magnetizations of interior planes at low T as explained above (see Fig. 12.12), while the missing of neighbors for surface spins tends to diminish the surface magnetization at high T [79, 89]. The middle figure shows the case where $J_2 = -0.5$ closer to the ferromagnetic limit. The spin length at $T = 0$ is almost 0.5 (very small contraction) and there is no visible crossover observed in the top figure. The bottom figure shows the case $J_2 = -2$, which is the case of a strong helical angle. We observe then a crossover at a higher T ($\simeq$1.2) which is in agreement with the physical picture given above on the competition between quantum and thermal fluctuations. Note that we did not attempt to get closer to the transition temperature, namely $M < 0.1$ because the convergence of the self-consistency then becomes bad.

Before closing this section, let us discuss about the spin wave spectrum. Let us remind that to solve self-consistently Eqs. (12.5)–(12.6) at each T, we use as inputs $< S_1^z >, < S_2^z >, \ldots, < S_{N_z}^z >$ to search for the eigenvalues ω for each vector (k_x, k_y) and then calculate the outputs $< S_1^z >, < S_2^z >, \ldots, < S_{N_z}^z >$. The self-consistent solution is obtained when the outputs are equal to the inputs at a desired convergence precision fixed at the fifth digit (see other details in Ref. [89]). Figure 12.15 shows the spin wave spectrum in the direction $k_x = k_y$ of the Brillouin zone at $T = 0.353$ and $T = 1.212$ for comparison. As seen, as T increases the spin wave frequency decreases. Near the transition (not shown), it tends to zero. Figure 12.16 shows the spin wave spectrum at $T = 0.353$ for

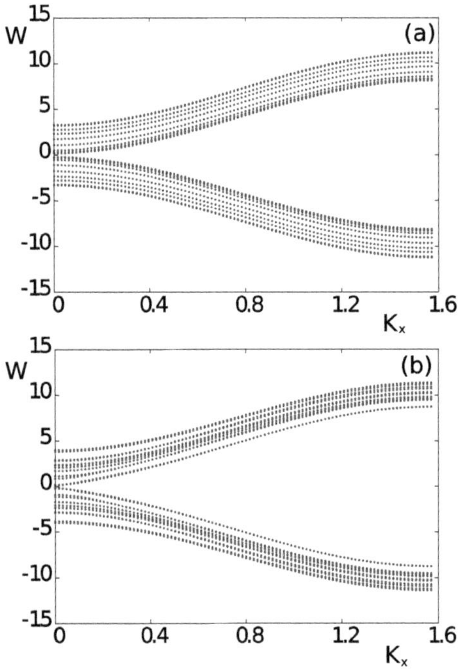

Figure 12.16 Spin wave spectrum versus $k_x = k_y$ at $T = 0.353$ where W stands for spin wave frequency ω in Eqs. (12.5)–(12.6), for (a) $J_2 = -0.5$ and (b) $J_2 = -2$, with $H = 0.2$.

$J_2 = 0.5$ and $J_2 = -1$, for comparison. Examining them closely, we see that the distribution of the spin wave modes (positions of the branches in the spectrum) is quite different for the two cases. When summed up for calculating the layer magnetizations, they give rise to the difference observed for the two cases shown in Fig. 12.14.

12.5 Conclusion

In this chapter, we have shown (i) the GS spin configuration of a Heisenberg helimagnetic thin film in a magnetic field applied along the c axis perpendicular to the film, (ii) the phase transition occurring in the film at a finite temperature, (iii) quantum effects at low T and the temperature dependence of the layer magnetizations as well as the spin wave spectrum.

We emphasize that under the applied magnetic field, the spin configurations of the layers in the GS are different with each other across the film. When the temperature increases, the layers with large xy spin-components undergo a phase transition where the transverse (in-plane) xy ordering is destroyed. This "transverse" transition is possible because the xy spin-components are perpendicular to the field. Other layers with small xy spin-components, namely large z components, do not make a transition because the ordering in S^z is maintained by the applied field. The transition of a number of layers with large xy spin-components, not all layers, is a new phenomenon discovered here with our present model.

We have also investigated the quantum version of the model by using the Green's function method. The results show that the zero-point spin contraction is different from layer to layer. We also find a crossover of layer magnetizations which depends on J_2, namely on the magnitude of helical angles.

Experiments are often performed on materials with helical structures more complicated than the model considered above. However, the clear physical pictures given in our present analysis are believed to be useful in the search for the interpretation of experimental data.

Chapter 13

Spin Waves in Systems with Dzyaloshinskii–Moriya Interaction

In this chapter, we study the magnetic properties of a system of quantum Heisenberg spins interacting with each other via a ferromagnetic exchange interaction J and an in-plane Dzyaloshinskii–Moriya interaction D. The non-collinear ground state due to the competition between J and D is determined. We employ a self-consistent Green's function theory to calculate the spin wave spectrum and the layer magnetizations at finite T in two and three dimensions as well as in a thin film with surface effects. Analytical details and the validity of the method are shown and discussed. Numerical solutions are shown for realistic physical interaction parameters. Discussion on possible experimental verifications is given.

The main results of this chapter have been published in Refs. [93, 104].

13.1 Introduction

The Dzyaloshinskii–Moriya (DM) interaction was proposed to explain the weak ferromagnetism which was observed in antiferro-

Physics of Magnetic Thin Films: Theory and Simulation
Hung T. Diep

ISBN 978-981-4877-42-8 (Hardcover), 978-1-003-12110-7 (eBook)
www.jennystanford.com

magnetic Mn compounds. The phenomenological Landau–Ginzburg model introduced by I. Dzyaloshinskii [99] was microscopically derived by T. Moriya [240]. The interaction between two spins $\mathbf{S}_i$ and $\mathbf{S}_j$ is written as

$$\mathbf{D}_{i,j} \cdot \mathbf{S}_i \wedge \mathbf{S}_j \tag{13.1}$$

where $\mathbf{D}_{i,j}$ is a vector which results from the displacement of non-magnetic ions located between $\mathbf{S}_i$ and $\mathbf{S}_j$, for example, in Mn–O–Mn bonds. The direction of $\mathbf{D}_{i,j}$ depends on the symmetry of the displacement [240]. The definition of $\mathbf{D}_{i,j}$ is given below (see also Section 15.1).

There have been a large number of investigations on the effect of the DM interaction in various materials, both experimentally and theoretically for weak ferromagnetism in perovskite compounds (see the references cited in Refs. [100, 316], for example). However, the interest in the DM interaction goes beyond the weak ferromagnetism: For example, it has been recently shown in various works that the DM interaction is at the origin of topological skyrmions [2, 37, 150, 166, 213, 224, 242, 296, 313, 357, 370, 371] and new kinds of magnetic domain walls [144, 292]. The increasing interest in skyrmions results from the fact that skyrmions may play an important role in the electronic transport which is at the heart of technological application devices [113].

In this chapter, we are interested in the spin wave (SW) properties of a system of spins interacting with each other via a DM interaction in addition to the symmetric isotropic Heisenberg exchange interaction. The competition between these interactions gives rise to a non-collinear spin configuration in the ground state (GS). Unlike helimagnets where the helical GS spin configuration results from the competition between the ferromagnetic nearest-neighbor (NN) and antiferromagnetic next-nearest neighbor (NNN) interactions [350, 369], the DM interaction favors the perpendicular spin configuration. This gives rise to a non-trivial SW behavior as will be seen below. Note that there has been a number of early works dealing with some aspects of the SW properties in DM systems [237, 282, 328, 353, 373].

Section 13.2 is devoted to the description of the model and the determination of the GS. Section 13.3 shows the formulation of our self-consistent Green's function (GF) method. Section 13.4 shows results on the SW spectrum and the magnetization in two dimensions (2D) and three dimensions (3D). The case of thin films with free surfaces is shown in Section 13.5 where layer magnetizations at finite temperature (T) and the thickness effect are presented. Discussion and experimental suggestion are made in Section 13.6.

13.2 Model and Ground State

We consider a thin film of simple cubic (SC) lattice of N layers stacked in the y direction perpendicular to the film surface. For the reason which is shown below, we choose the film surface as a xz plane. The Hamiltonian is given by

$$\mathcal{H} = \mathcal{H}_e + \mathcal{H}_{DM} \tag{13.2}$$

$$\mathcal{H}_e = -\sum_{\langle i,j\rangle} J_{i,j}\mathbf{S}_i \cdot \mathbf{S}_j \tag{13.3}$$

$$\mathcal{H}_{DM} = \sum_{\langle i,j\rangle} \mathbf{D}_{i,j} \cdot \mathbf{S}_i \wedge \mathbf{S}_j \tag{13.4}$$

where $J_{i,j}$ and $\mathbf{D}_{i,j}$ are the exchange and DM interactions, respectively, between two Heisenberg spins $\mathbf{S}_i$ and $\mathbf{S}_j$ of magnitude $S = 1/2$ occupying the lattice sites i and j.

The SC lattice can support the DM interaction in the absence of the inversion symmetry [224, 240, 242] as in MnSi. The absence of inversion symmetry can be also achieved by the positions of non-magnetic ions between magnetic ions. The vector $\mathbf{D}$ between two magnetic ions is defined as $\mathbf{D}_{i,j} = A\mathbf{r}_i \wedge \mathbf{r}_j$ where $\mathbf{r}_i$ is the vector connecting the non-magnetic ion to the spin $\mathbf{S}_i$ and $\mathbf{r}_j$ is that to the spin $\mathbf{S}_j$, A being a constant. One sees that $\mathbf{D}_{i,j}$ is perpendicular to the plane formed by $\mathbf{r}_i$ and $\mathbf{r}_j$.

For simplicity, let us consider the case where the in-plane and inter-plane exchange interactions between NN are both ferromagnetic and denoted by $J_{\parallel}$ and $J_{\perp}$, respectively. The vector $\mathbf{D}_{i,j}$ is chosen by supposing the situation where non-magnetic ions are positioned in the plane xz. With this choice, $\mathbf{D}_{i,j}$ is perpendicular to the xz plane and our model gives rise to a planar spin configuration, namely the spins stay in xz planes. There are no situations where the spins are out of plane. This simplifies our calculation. The DM interaction is supposed to be between NN in the plane with a constant D.

Due to the competition between the exchange J term which favors the collinear configuration, and the DM term which favors the perpendicular one, we expect that the spin $\mathbf{S}_i$ makes an angle $\theta_{i,j}$ with its neighbor $\mathbf{S}_j$. Therefore, the quantization axis of $\mathbf{S}_i$ is not the same as that of $\mathbf{S}_j$. Let us call $\hat{\zeta}_i$ the quantization axis of $\mathbf{S}_i$ and $\hat{\xi}_i$ its perpendicular axis in the xz plane. The third axis $\hat{\eta}_i$, perpendicular to the film surface, is chosen in such a way to make $(\hat{\xi}_i, \hat{\eta}_i, \hat{\zeta}_i)$ an orthogonal direct frame. Writing $\mathbf{S}_i$ and $\mathbf{S}_j$ in their respective local coordinates, one has

$$\mathbf{S}_i = S_i^x \hat{\xi}_i + S_i^y \hat{\eta}_i + S_i^z \hat{\zeta}_i \tag{13.5}$$

$$\mathbf{S}_j = S_j^x \hat{\xi}_j + S_j^y \hat{\eta}_j + S_j^z \hat{\zeta}_j \tag{13.6}$$

We choose the vector $\mathbf{D}_{i,j}$ perpendicular to the xz plane, namely

$$\mathbf{D}_{i,j} = De_{i,j}\hat{\eta}_i \tag{13.7}$$

where $e_{i,j} = +1(-1)$ if $j > i$ $(j < i)$ for NN on the $\hat{\xi}_i$ or $\hat{\zeta}_i$ axis. Note that $e_{j,i} = -e_{i,j}$.

To determine the GS, the easiest way is to use the steepest descent method: We calculate the local field acting on each spin from its neighbors and we align the spin in its local-field direction to minimize its energy. Repeating this for all spins and iterating many times until the convergence is reached with a desired precision (usually at the sixth digit, namely at $\simeq 10^{-6}$ per cents), we obtain the lowest energy state of the system (see Ref. [248]). Note that we have used several thousands of different initial conditions to check the convergence to a single GS for each set of parameters. Choosing $\mathbf{D}_{i,j}$ lying perpendicular to the spin plane (i.e., xz plane) as indicated in Eq. (13.7), we determine the GS as a function of D by the steepest descent method. An example is shown in Fig. 13.1 for $\theta = \pi/6$ $(D = -0.577)$ with $J_{\parallel} = J_{\perp} = 1$. We see that each spin has the same angle with its four NN in the plane (angle between NN in adjacent planes is zero).

In the present model, the DM interaction is supposed between the NN in the plane xz, so we see in Fig. 13.1 that in the GS the angle

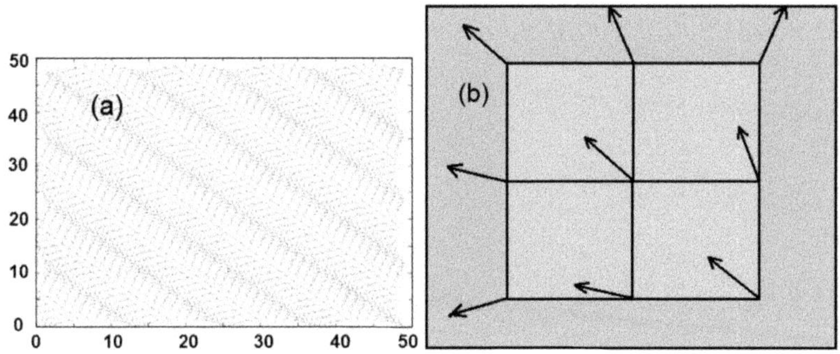

Figure 13.1 (a) The ground state is a planar configuration on the xz plane. The figure shows the case where $\theta = \pi/6$ ($D = -0.577$) along the x and z axes, with $J_{\shortparallel} = J_{\perp} = 1$, obtained by using the steepest descent method; (b) a zoom is shown around a spin with its nearest neighbors in the xz plane.

between in-plane NN is not zero. The spin configuration is planar, meaning that the turn angle is in the plane: Each spin turns an angle θ with respect to its NN, and it is in both directions in the plane (see Fig. 13.1b). There is no helicoidal configuration. Note that the spin structure is different from that of MnSi where the turn angle is in a plane perpendicular to the screw axis. This case has been studied in our previous papers [89, 102]. Note also that the helical-wave vector in the MnSi case is along the screw axis, while in our case, it lies along each of the two axes x and z of the xz plane (see Fig. 13.1b).

Let us show the relation between θ, D and $J_{\shortparallel}$: The energy of the spin $\mathbf{S}_i$ is written as

$$E_i = -4J_{\shortparallel}S^2\cos\theta - 2J_{\perp}S^2 + 4DS^2\sin\theta \qquad (13.8)$$

where $\theta = |\theta_{i,j}|$ and care has been taken on the signs of $\sin\theta_{i,j}$ and $e_{i,j}$ when counting NN, namely two opposite NN have opposite signs. The minimization of E_i yields

$$\frac{dE_i}{d\theta} = 0 \Rightarrow -\frac{D}{J_{\shortparallel}} = \tan\theta \Rightarrow \theta = \arctan\left(-\frac{D}{J_{\shortparallel}}\right) \qquad (13.9)$$

The value of θ for a given $\frac{D}{J_{\shortparallel}}$ is precisely what obtained by the steepest descent method.

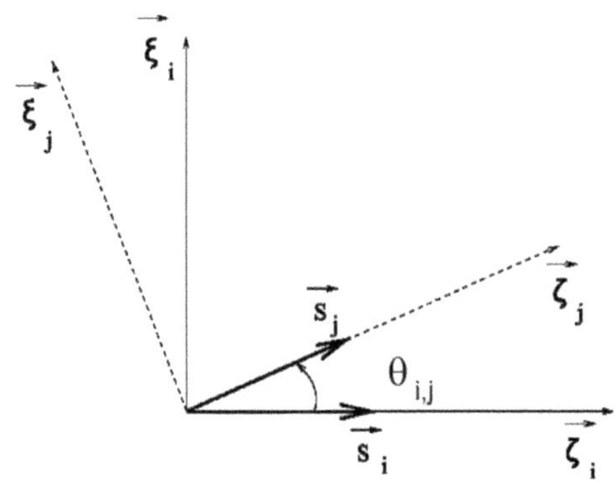

Figure 13.2 Local coordinates in the xz plane. The spin quantization axes of $\mathbf{S}_i$ and $\mathbf{S}_j$ are $\hat{\zeta}_i$ and $\hat{\zeta}_j$, respectively.

Note that the perpendicular axes $\hat{\eta}_i$ and $\hat{\eta}_j$ coincide. Now, expressing the local frame of $\mathbf{S}_j$ in the local frame of $\mathbf{S}_i$, we have

$$\hat{\zeta}_j = \cos\theta_{i,j}\hat{\zeta}_i + \sin\theta_{i,j}\hat{\xi}_i \quad (13.10)$$

$$\hat{\xi}_j = -\sin\theta_{i,j}\hat{\zeta}_i + \cos\theta_{i,j}\hat{\xi}_i \quad (13.11)$$

$$\hat{\eta}_j = \hat{\eta}_i \quad (13.12)$$

so that

$$\mathbf{S}_j = S_j^x(\cos\theta_{i,j}\hat{\zeta}_i - \sin\theta_{i,j}\hat{\xi}_i) + S_j^y\hat{\eta}_i + S_j^z(\cos\theta_{i,j}\hat{\zeta}_i + \sin\theta_{i,j}\hat{\xi}_i) \quad (13.13)$$

The DM term of Eq. (13.4) can be rewritten as

$$\begin{aligned}\mathbf{S}_i \wedge \mathbf{S}_j &= (-S_i^z S_j^y - S_i^y S_j^x \sin\theta_{i,j} + S_i^y S_j^z \cos\theta_{i,j})\hat{\xi}_i \\ &+(S_i^x S_j^x \sin\theta_{i,j} + S_i^z S_j^z \sin\theta_{i,j})\hat{\eta}_i \\ &+(S_i^x S_j^y - S_i^y S_j^z \sin\theta_{i,j} - S_i^y S_j^x \cos\theta_{i,j})\hat{\zeta}_i \quad (13.14)\end{aligned}$$

Using Eq. (13.7), we have

$$\begin{aligned}\mathcal{H}_{DM} &= \sum_{\langle i,j\rangle} \mathbf{D}_{i,j} \cdot \mathbf{S}_i \wedge \mathbf{S}_j \\ &= D\sum_{\langle i,j\rangle}(S_i^x S_j^x e_{i,j}\sin\theta_{i,j} + S_i^z S_j^z e_{i,j}\sin\theta_{i,j}) \\ &= \frac{D}{4}\sum_{\langle i,j\rangle}[(S_i^+ + S_i^-)(S_j^+ + S_j^-)e_{i,j}\sin\theta_{i,j} \\ &+4S_i^z S_j^z e_{i,j}\sin\theta_{i,j}] \quad (13.15)\end{aligned}$$

where we have replaced $S^x = (S^+ + S^-)/2$. Note that $e_{i,j} \sin\theta_{i,j}$ is always positive since for a NN on the positive axis direction, $e_{i,j} = 1$ and $\sin\theta_{i,j} = \sin\theta$ where θ is positively defined, while for a NN on the negative axis direction, $e_{i,j} = -1$ and $\sin\theta_{i,j} = \sin(-\theta) = -\sin\theta$.

Note that for non-collinear spin configurations, the local spin coordinates allow one to use the commutation relations between spin operators of a spin which are valid only when the z spin component is defined on its quantification axis. This method has been applied for helimagnets [91, 139, 286].

13.3 Self-Consistent Green's Function Method: Formulation

The GF method has been developed for collinear [78, 79] and non-collinear surface spin configurations in thin films [89, 102, 248, 249] and in magneto-ferroelectric superlattices [317]. Let us briefly recall here the principal steps of calculation and give the results for the present model. In the following, we consider the case of spin one-half $S = 1/2$. Expressing the Hamiltonian in the local coordinates, we obtain

$$\begin{aligned}\mathcal{H} = &-\sum_{<i,j>} J_{i,j}\Bigg\{\frac{1}{4}\left(\cos\theta_{i,j}-1\right)\left(S_i^+S_j^+ + S_i^-S_j^-\right)\\ &+\frac{1}{4}\left(\cos\theta_{i,j}+1\right)\left(S_i^+S_j^- + S_i^-S_j^+\right)\\ &+\frac{1}{2}\sin\theta_{i,j}\left(S_i^+ + S_i^-\right)S_j^z - \frac{1}{2}\sin\theta_{i,j}S_i^z\left(S_j^+ + S_j^-\right)\\ &+\cos\theta_{i,j}S_i^zS_j^z\Bigg\}\\ &+\frac{D}{4}\sum_{\langle i,j\rangle}[(S_i^+ + S_i^-)(S_j^+ + S_j^-)e_{i,j}\sin\theta_{i,j} + 4S_i^zS_j^ze_{i,j}\sin\theta_{i,j}]\end{aligned} \tag{13.16}$$

As said in the previous section, the spins lie in the xz planes, each on its quantization local z axis (Fig. 13.2).

Note that unlike the sinus term of the DM Hamiltonian, Eq. (13.15), the sinus terms of $\mathcal{H}_e$, the third line of Eq. (13.16), are zero when summed up on opposite NN (no $e_{i,j}$ to compensate). The third line disappears, therefore, in the following.

At this stage, it is very important to note that the standard commutation relations between spin operators S^z and $S^{\pm}$ are defined with z as the spin quantization axis. In non-collinear spin configurations, calculations of SW spectrum using commutation relations without paying attention to this are wrong.

It is known that in two dimensions (2D) there is no long-range order at finite temperature (T) for isotropic spin models with short-range interaction [231]. Thin films have small thickness; therefore, to stabilize the ordering at finite T, it is useful to add an anisotropic interaction. We use the following anisotropy between $\mathbf{S}_i$ and $\mathbf{S}_j$ which stabilizes the angle determined above between their local quantization axes S_i^z and S_j^z:

$$\mathcal{H}_a = -\sum_{<i,j>} I_{i,j} S_i^z S_j^z \cos\theta_{i,j} \tag{13.17}$$

where $I_{i,j}$ is supposed to be positive, small compared to $J_{\parallel}$, and limited to NN. Hereafter we take $I_{i,j} = I_1$ for NN pair in the xz plane, for simplicity. As it turns out, this anisotropy helps stabilize the ordering at finite T in 2D as discussed. It helps also stabilize the SW spectrum at $T = 0$ in the case of thin films but it is not necessary for 2D and 3D at $T = 0$. The total Hamiltonian is finally given by

$$\mathcal{H} = \mathcal{H}_e + \mathcal{H}_{DM} + \mathcal{H}_a \tag{13.18}$$

We define the following two double-time GF's in the real space:

$$\begin{aligned} G_{i,j}(t,t') &= \ll S_i^+(t); S_j^-(t') \gg \\ &= -i\theta(t-t') < \left[S_i^+(t), S_j^-(t')\right] > \end{aligned} \tag{13.19}$$

$$\begin{aligned} F_{i,j}(t,t') &= \ll S_i^-(t); S_j^-(t') \gg \\ &= -i\theta(t-t') < \left[S_i^-(t), S_j^-(t')\right] > \end{aligned} \tag{13.20}$$

The equations of motion of these functions read

$$\begin{aligned} i\hbar\frac{dG_{i,j}(t,t')}{dt} &= < \left[S_i^+(t), S_j^-(t')\right] > \delta(t-t') \\ &\quad - \ll \left[\mathcal{H}, S_i^+\right]; S_j^- \gg \end{aligned} \tag{13.21}$$

$$\begin{aligned} i\hbar\frac{dF_{i,j}(t,t')}{dt} &= < \left[S_i^-(t), S_j^-(t')\right] > \delta(t-t') \\ &\quad - \ll \left[\mathcal{H}, S_i^-\right]; S_j^- \gg \end{aligned} \tag{13.22}$$

For the $\mathcal{H}_e$ and $\mathcal{H}_a$ parts, the above equations of motion generate terms such as $\ll S_l^z S_i^{\pm}; S_j^- \gg$ and $\ll S_l^{\pm} S_i^{\pm}; S_j^- \gg$. These functions can be approximated by using the Tyablikov decoupling to reduce to the above-defined G and F functions:

$$\ll S_l^z S_i^{\pm}; S_j^- \gg \simeq < S_l^z > \ll S_i^{\pm}; S_j^- \gg \tag{13.23}$$

$$\ll S_l^{\pm} S_i^{\pm}; S_j^- \gg \simeq < S_l^{\pm} > \ll S_i^{\pm}; S_j^- \gg \simeq 0 \tag{13.24}$$

The last expression is due to the fact that transverse SW motions $< S_l^{\pm} >$ are zero with time. For the DM term, the commutation relations $[\mathcal{H}, S_i^{\pm}]$ give rise to the following term:

$$D \sum_l \sin\theta [\mp S_i^z (S_l^+ + S_l^-) \pm 2 S_i^{\pm} S_l^z] \tag{13.25}$$

which leads to the following type of GFs:

$$\ll S_i^z S_l^{\pm}; S_j^- \gg \simeq < S_i^z > \ll S_l^{\pm}; S_j^- \gg \tag{13.26}$$

Note that we have replaced $e_{i,j} \sin\theta_{i,j}$ by $\sin\theta$ where θ is positive. The above equation is related to G and F functions [see Eq. (13.24)]. The Tyablikov decoupling scheme [43, 346] neglects higher-order functions.

We now introduce the following in-plane Fourier transforms $g_{n,n'}$ and $f_{n,n'}$ of the G and F Green's functions:

$$G_{i,j}(t, t', \omega) = \frac{1}{\Delta} \int_{BZ} d\mathbf{k}_{xz} e^{-i\omega(t-t')} \times g_{n,n'}(\omega, \mathbf{k}_{xz}) e^{i\mathbf{k}_{xz}\cdot(\mathbf{R}_i - \mathbf{R}_j)} \tag{13.27}$$

$$F_{i,j}(t, t', \omega) = \frac{1}{\Delta} \int_{BZ} d\mathbf{k}_{xz} e^{-i\omega(t-t')} \times f_{n,n'}(\omega, \mathbf{k}_{xz}) e^{i\mathbf{k}_{xz}\cdot(\mathbf{R}_i - \mathbf{R}_j)} \tag{13.28}$$

where the integral is performed in the first xz Brillouin zone (BZ) of surface Δ, ω is the spin wave frequency, n and n' are the indices of the layers along the c axis to which $\mathbf{R}_i$ and $\mathbf{R}_j$ belong ($n = 1$ being the surface layer, $n = 2$ the second layer and so on). We finally obtain the following matrix equation:

$$\mathbf{M}(E)\,\mathbf{h} = \mathbf{u} \tag{13.29}$$

where $\mathbf{M}(E)$ is a square matrix of dimension $(2N \times 2N)$, $\mathbf{h}$ and $\mathbf{u}$ are the column matrices which are defined as follows:

$$\mathbf{h} = \begin{pmatrix} g_{1,n'} \\ f_{1,n'} \\ \vdots \\ g_{n,n'} \\ f_{n,n'} \\ \vdots \\ g_{N,n'} \\ f_{N,n'} \end{pmatrix}, \qquad \mathbf{u} = \begin{pmatrix} 2\left\langle S_1^z \right\rangle \delta_{1,n'} \\ 0 \\ \vdots \\ 2\left\langle S_N^z \right\rangle \delta_{N,n'} \\ 0 \end{pmatrix} \tag{13.30}$$

where $E = \hbar\omega$ and $\mathbf{M}(E)$ is given by

$$\begin{pmatrix} E+A_1 & B_1 & C_1 & 0 & 0 & 0 & 0 & 0 & 0 \\ -B_1 & E-A_1 & 0 & -C_1 & 0 & 0 & 0 & 0 & 0 \\ \cdots & \cdots & \cdots & \cdots & \cdots & \cdots & \cdots & \cdots & \cdots \\ \cdots & 0 & C_n & 0 & E+A_n & B_n & C_n & 0 & 0 \\ \cdots & 0 & 0 & -C_n & -B_n & E-A_n & 0 & -C_n & 0 \\ \cdots & \cdots & \cdots & \cdots & \cdots & \cdots & \cdots & \cdots & \cdots \\ 0 & 0 & 0 & 0 & 0 & C_N & 0 & E+A_N & B_N \\ 0 & 0 & 0 & 0 & 0 & 0 & -C_N & -B_N & E-A_N \end{pmatrix} \tag{13.31}$$

with

$$\begin{aligned} A_n = & -J_{\shortparallel}[8 < S_n^z > \cos\theta(1+d_n) - 4 < S_n^z > \gamma(\cos\theta + 1)] \\ & -2J_{\perp}(< S_{n-1}^z > + < S_{n+1}^z >) \\ & -4D\sin\theta < S_n^z > \gamma + 8D\sin\theta < S_n^z > \end{aligned} \tag{13.32}$$

$$B_n = 4J_{\shortparallel} < S_n^z > \gamma(\cos\theta - 1) - 4D\sin\theta < S_n^z > \gamma \tag{13.33}$$

$$C_n = 2J_{\perp} < S_n^z > \tag{13.34}$$

where $n = 1, 2, \ldots, N$, $d_n = I_1/J_{\shortparallel}$, $\gamma = (\cos k_x a + \cos k_z a)/2$, k_x and k_z denote the wave-vector components in the xz planes, a the lattice constant. Note that (i) if $n = 1$ (surface layer) then there are no $n-1$ terms in the matrix coefficients, (ii) if $n = N$ then there are no $n+1$ terms. Besides, we have distinguished the in-plane NN interaction $J_{\shortparallel}$ from the inter-plane NN one $J_{\perp}$.

In the case of a thin film, the SW eigenvalues at a given wave vector $\mathbf{k} = (k_x, k_z)$ are calculated by diagonalizing the matrix 13.31.

Taking $S = 1/2$, the layer magnetization of the layer n is given by (see technical details in Section 8.4 of Chapter 8 and in Ref. [87]):

$$\langle S_n^z \rangle = \frac{1}{2} - \frac{1}{\Delta} \iint dk_x dk_z \sum_{i=1}^{2N} \frac{Q_{2n-1}(E_i)}{e^{E_i/k_B T} - 1} \tag{13.35}$$

where $n = 1, \ldots, N$, and $Q_{2n-1}(E_i)$ is the determinant obtained by replacing the $(2n-1)$-th column of $\mathbf{M}$ by $\mathbf{u}$ at E_i.

The layer magnetizations can be calculated at finite temperatures self-consistently using the above formula. The numerical method to carry out this task has been described in details in Refs. [89]. One can summarize here: (i) Using a set of trial values (inputs) for $\langle S_n^z \rangle$ ($n = 1, \ldots, N$), one diagonalizes the matrix to find spin wave energies E_i which are used to calculate the outputs $\langle S_n^z \rangle$ ($n = 1, \ldots, N$) by using Eq. (13.35); (ii) using the outputs as inputs to iterate the equations; (iii) if the output values are the same as the inputs within a precision (usually at 0.001%), the iteration is stopped. The method is thus self-consistent.

The value of the spin in the layer n at $T = 0$ is calculated by (see Section 8.4 and [87, 89]):

$$\langle S_n^z \rangle (T = 0) = \frac{1}{2} + \frac{1}{\Delta} \iint dk_x dk_z \sum_{i=1}^{N} Q_{2n-1}(E_i) \tag{13.36}$$

where the sum is performed over N negative values of E_i (for positive values, the Bose–Einstein factor in Eq. (13.35) is equal to 0 at $T = 0$).

The transition temperature T_c can be calculated by letting $\langle S_n^z \rangle$ on the left-hand side of Eq. (13.35) to go to zero. The energy E_i tends then to zero, so that we can make an expansion of the exponential at $T = T_c$. We have

$$\left[\frac{1}{k_B T_c}\right] = \frac{2}{\Delta} \iint dk_x dk_z \sum_{i=1}^{2N} \frac{Q_{2n-1}(E_i)}{E_i} \tag{13.37}$$

13.4 Two and Three Dimensions: Spin Wave Spectrum and Magnetization

Consider just one single xz plane. The above matrix is reduced to two coupled equations:

$$(E + A_n)g_{n,n'} + B_n f_{n,n'} = 2 < S_n^z > \delta(n, n')$$
$$-B_n g_{n,n'} + (E - A_n) f_{n,n'} = 0 \quad (13.38)$$

where A_n is given by (13.32) but without $J_\perp$ term for the 2D case considered here. Coefficients B_n and C_n are given by (13.33) and (13.34) with $C_n = 0$. The poles of the GF are the eigenvalues of the SW spectrum which are given by the secular equation

$$(E + A_n)(E - A_n) + B_n^2 = 0$$
$$[E + A_n][E - A_n] + B_n^2 = 0$$
$$E^2 - A_n^2 + B_n^2 = 0$$
$$E = \pm\sqrt{(A_n + B_n)(A_n - B_n)} \quad (13.39)$$

where $\pm$ indicate the left and right SW precessions. Several remarks are in order:

(i) If $\theta = 0$, we have $B_n = 0$ and the last three terms of A_n are zero. We recover then the ferromagnetic SW dispersion relation

$$E = 2ZJ_{\parallel} < S_n^z > (1 - \gamma) \quad (13.40)$$

where $Z = 4$ is the coordination number of the square lattice (taking $d_n = 0$).

(ii) If $\theta = \pi$, we have $A_n = 8J_{\parallel} < S_n^z >$, $B_n = -8J_{\parallel} < S_n^z > \gamma$. We recover then the antiferromagnetic SW dispersion relation

$$E = 2ZJ_{\parallel} < S_n^z > \sqrt{1 - \gamma^2} \quad (13.41)$$

(iii) in the presence of a DM interaction, we have $0 < \cos\theta < 1$ $(0 < \theta < \pi/2)$. If $d_n = 0$, the quantity in the square root of Eq. (13.39) is always ≥ 0 for any θ. It is zero at $\gamma = 1$. The SW spectrum is, therefore, stable at the long-wavelength limit. The anisotropy d_n gives a gap at $\gamma = 1$.

As said earlier, the necessity to include an anisotropy has a double purpose: It permits a gap and stabilizes a long-range ordering at finite T in 2D systems.

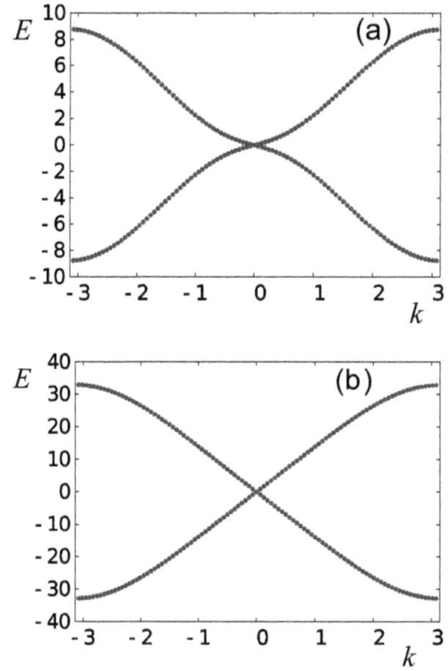

Figure 13.3 Spin wave spectrum $E(k)$ versus $k \equiv k_x = k_z$ for (a) $\theta =$ 0.524 radian and (b) $\theta = 1.393$ in two dimensions at $T = 0.1$. Positive and negative branches correspond to right and left precessions. A small d ($= 0.001$) has been used to stabilized the ordering at finite T in 2D. See text for comments.

Figure 13.3 shows the SW spectrum calculated from Eq. (13.39) for $\theta = 30°$ ($\pi/6$ radian) and $80°$ (1.396 radian). The spectrum is symmetric for positive and negative wave vectors and for left and right precessions. Note that for small θ (i.e., small D) E is proportional to k^2 at low k (cf. Fig. 13.3a), as in ferromagnets. However, as θ increases, we observe that E becomes linear in k as seen in Fig. 13.3b. This is similar to antiferromagnets. The change of behavior is progressive with increasing θ, we do not observe a sudden transition from k^2 to k behavior. This feature is also observed in three dimensions (3D) and in thin films as seen below.

It is noted that, thanks to the existence of the anisotropy d, we avoid the logarithmic divergence at $k = 0$ so that we can observe a long-range ordering at finite T in 2D. We show in Fig. 13.4 the magnetization M ($\equiv < S^z >$) calculated by Eq. (13.35) for one layer using $d = 0.001$. It is interesting to observe that M depends strongly

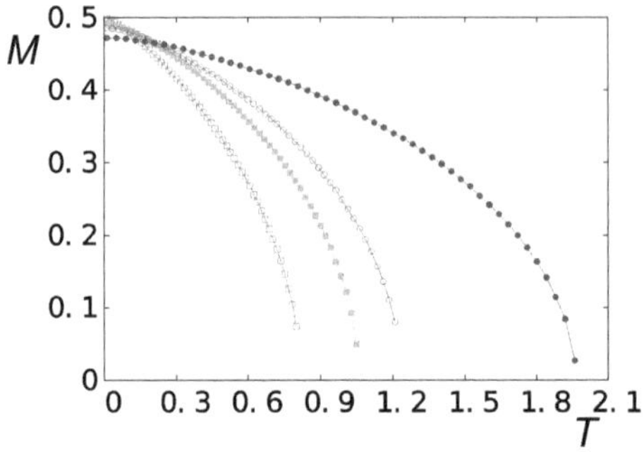

Figure 13.4 Magnetizations M versus temperature T for a monolayer (2D) $\theta = 0.175$ (radian), $\theta = 0.524$, $\theta = 0.698$, $\theta = 1.047$ (void magenta squares, green filled squares, blue void circles and filled red circles, respectively). A small d ($= 0.001$) has been used to stabilized the ordering at finite T in 2D. See text for comments.

on θ: At high T, larger θ yields stronger M. However, at $T = 0$ the spin length is smaller for larger θ due to the so-called spin contraction [87] calculated by Eq. (13.36). As a consequence, there is a crossover of magnetizations obtained with different θ at low T as shown in Fig. 13.4.

Let us study the 3D case. The crystal is periodic in three directions. We can use the Fourier transformation in the y direction, namely $g_{n\pm1} = g_n e^{\pm ik_y a}$ and $f_{n\pm1} = f_n e^{\pm ik_y a}$. The matrix (13.30) is reduced to two coupled equations of g and f functions, omitting index n,

$$\begin{aligned}(E + A')g + Bf &= 2 < S^z > \\ -Bg + (E - A')f &= 0\end{aligned} \quad (13.42)$$

where

$$\begin{aligned}A' = -J_{\parallel}[8 &< S^z > \cos\theta(1 + d) \\ &-4 < S^z > \gamma(\cos\theta + 1)] \\ &+4J_{\perp} < S^z > \cos(k_y a) \\ &-4D\sin\theta < S^z > \gamma \\ &+8D\sin\theta < S^z >\end{aligned} \quad (13.43)$$

$$\begin{aligned}B = 4J_{\parallel} &< S^z > \gamma(\cos\theta - 1) \\ &-4D\sin\theta < S^z > \gamma\end{aligned} \quad (13.44)$$

The spectrum is given by

$$E = \pm\sqrt{(A' + B)(A' - B)} \tag{13.45}$$

If $\cos\theta = 1$ (ferromagnetic), one has $B = 0$. By regrouping the Fourier transforms in three directions, one obtains the 3D ferromagnetic dispersion relation $E = 2Z < S^z > (1 - \gamma^2)$ where $\gamma = [\cos(k_x a) + \cos(k_y a) + \cos(k_z a)]/3$ and $Z = 6$, coordination number of the simple cubic lattice. Unlike the 2D case where the angle is inside the plane so that the antiferromagnetic case can be recovered by setting $\cos\theta = -1$ as seen above, one cannot use the above formula to find the antiferromagnetic case because in the 3D formulation it was supposed a ferromagnetic coupling between planes, namely there is no angle between adjacent planes in the above formulation.

The same consideration as in the 2D case treated above shows that for $d = 0$ the spectrum $E \geq 0$ for positive precession and $E \leq 0$ for negative precession, for any θ. The limit $E = 0$ is at $\gamma = 1$ ($\mathbf{k} = 0$). Thus there is no instability due to the DM interaction. Using Eq. (13.45), we have calculated the 3D spectrum. This is shown in Fig. 13.5 for a small and a large value of θ. As in the 2D case, we observe $E \propto k$ when $k \to 0$ for large θ. Main properties of the system are dominated by the in-plane DM behavior.

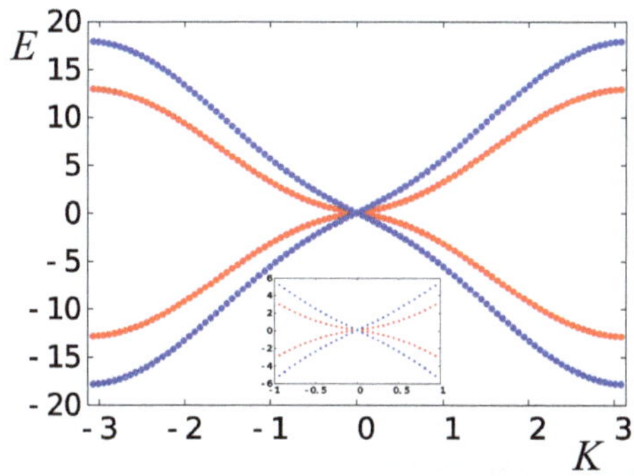

Figure 13.5 Spin wave spectrum $E(k)$ versus $k \equiv k_x = k_z$ for $\theta = \pi/6$ (red circles) and $\theta = \pi/3$ (blue circles) in three dimensions at $T = 0.1$, with $d = 0$. Note the linear-k behavior at low k for the large value of θ (inset). See text for comments.

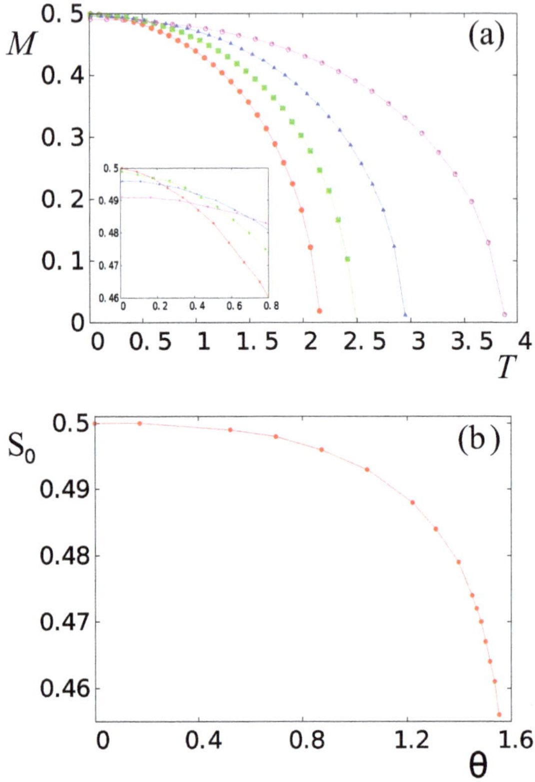

Figure 13.6 (a) Magnetization M versus temperature T for a 3D crystal $\theta = 0.175$ (radian), $\theta = 0.524, \theta = 0.785, \theta = 1.047$ (red circles, green squares, blue triangles and void magenta circles, respectively), with $d = 0$. Inset: Zoom showing the crossover of magnetizations at low T for different θ, (b) The spin length S_0 at $T = 0$ versus θ. See text for comments.

Figure 13.6a displays the magnetization M versus T for several values of θ. As in the 2D case, when θ is not zero, the spins have a contraction at $T = 0$: A stronger θ yields a stronger contraction. This generates a magnetization crossover at low T shown in the inset of Fig. 13.6a. The spin length at $T = 0$ versus θ is displayed in Fig. 13.6b. Note that the spin contraction in 3D is smaller than that in 2D. This is expected since quantum fluctuations are stronger at lower dimensions.

13.5 The Case of a Thin Film: Spin Wave Spectrum, Layer Magnetizations

In the 2D and 3D cases shown above, there is no need at $T = 0$ to use a small anisotropy d. However, in the case of thin films shown below, due to the lack of neighbors at the surface, the introduction of a DM interaction destabilizes the spectrum at long wavelength $\mathbf{k} = 0$. Depending on θ, we have to use a value for d_n larger or equal to a "critical value" d_c to avoid imaginary SW energies at $\mathbf{k} = 0$. The critical value d_c is shown in Fig. 13.7 for a four-layer film. Note that at the perpendicular configuration $\theta = \pi/2$, no SW excitation is possible: SW cannot propagate in a perpendicular spin configuration since the wave-vectors cannot be defined.

We show now a SW spectrum at a given thickness N. There are $2N$ energy values half of them are positive and the other half negative (left and right precessions): E_i $(i = 1, \ldots, 2N)$. Figure 13.8 shows the case of a film of 8 layers with $J_{\parallel} = J_{\perp} = 1$ for a weak and a strong value of D (small and large θ). As in the 2D and 2D cases, for strong D, E is proportional to k at small k (cf. Fig. 13.8b). It is noted that this behavior concerns only the first mode. The upper modes remain in the k^2 behavior.

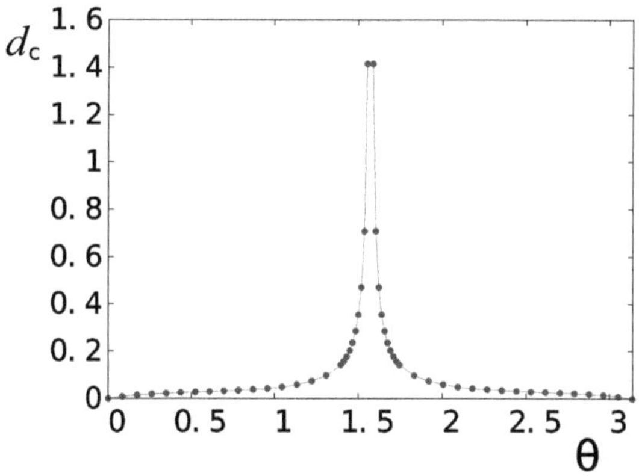

Figure 13.7 Value d_c above which the SW energy $E(\mathbf{k} = 0)$ is real as a function of θ (in radian), for a four-layer film. Note that no spin wave excitations are possible near the perpendicular configuration $\theta = \pi/2$. See text for comments.

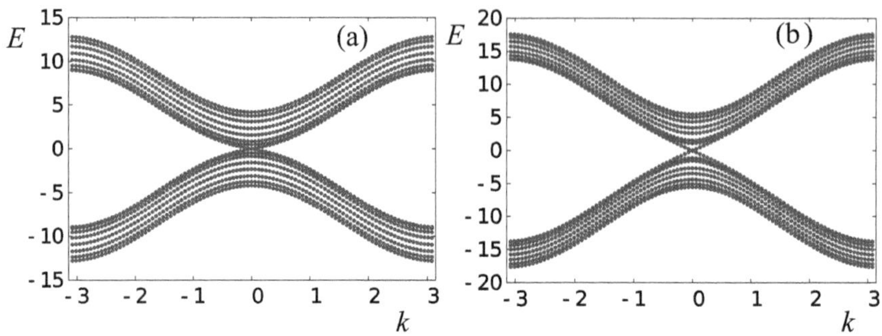

Figure 13.8 Spin wave spectrum $E(k)$ versus $k \equiv k_x = k_z$ for a thin film of 8 layers: (a) $\theta = \pi/6$ (in radian) (b) $\theta = \pi/3$, using $d = d_c$ for each case ($d_c = 0.012$ and 0.021, respectively). Positive and negative branches correspond to right and left precessions. Note the linear-k behavior at low k for the large θ case. See text for comments.

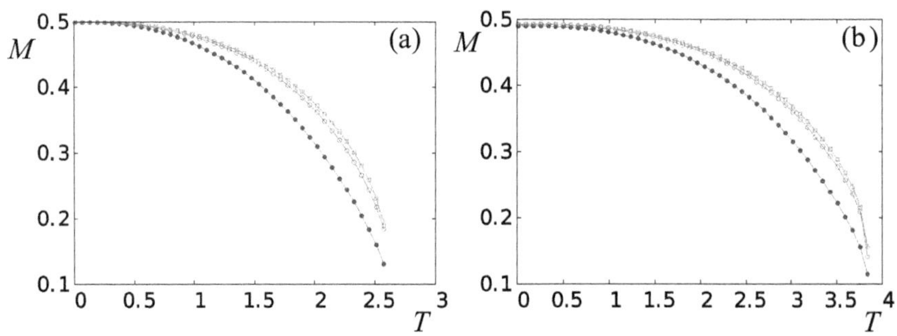

Figure 13.9 Eight-layer film: layer magnetizations M versus temperature T for (a) $\theta = \pi/6$ (radian), (b) $\theta = \pi/3$, with $d = 0.1$. Red circles, blue void circles, green void triangles and magenta squares correspond, respectively, to the first, second, third and fourth layer.

Figure 13.9 shows the layer magnetizations of the first four layers in an eight-layer film (the other half is symmetric) for several values of θ. In each case, we see that the surface layer magnetization is smallest. This is a general effect of the lack of neighbors for surface spins even when there is no surface-localized SW as in the present simple-cubic lattice case [78, 87].

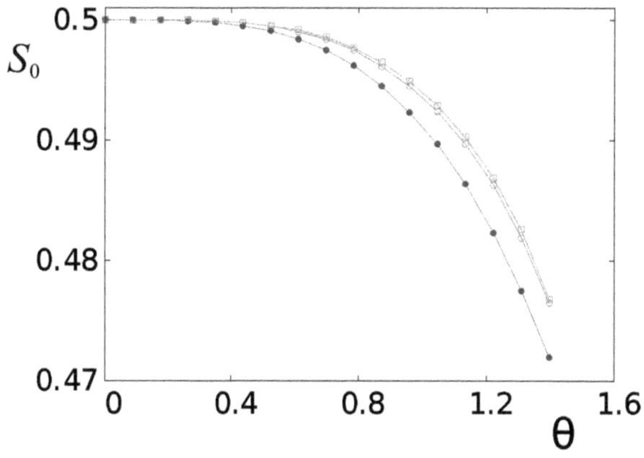

Figure 13.10 Spin length S_0 at $T = 0$ of the first 4 layers as a function of θ, for $N = 8$, $d = 0.1$. Red circles, blue void circles, green void triangles and magenta squares correspond, respectively, to the first, second, third and fourth layer.

The spin length at $T = 0$ for a eight-layer film is shown in Fig. 13.10 as a function of θ. One observes that the spins are strongly contracted with large θ.

Let us touch upon the surface effect in the present model. We know that for the simple cubic lattice, if the interactions are the same everywhere in the film, then there is no surface localized modes, and this is true with DM interaction (see spectrum in Fig. 13.8) and without DM interaction (see Ref. [78]). In order to create surface modes, we have to take the surface exchange interactions different from the bulk ones. Low-lying branches of surface modes which are "detached" from the bulk spectrum are seen in the SW spectrum shown in Fig. 13.11a with $J_{\parallel}^{s} = 0.5$, $J_{\perp}^{s} = 0.5$. These surface modes strongly affect the surface magnetization as observed in Fig. 13.11b: The surface magnetization is strongly diminished with increasing T. The role of surface-localized modes on the strong decrease of the surface magnetization as T increases has already been analyzed more than 40 years ago [78].

We show now the effect of the film thickness in the present model. The case of thickness $N = 12$ is shown in Fig. 13.12a with $\theta = \pi/6$ where the layer magnetizations versus T are shown in details. The gap at $k = 0$ due to d is shown in Fig. 13.12b as a

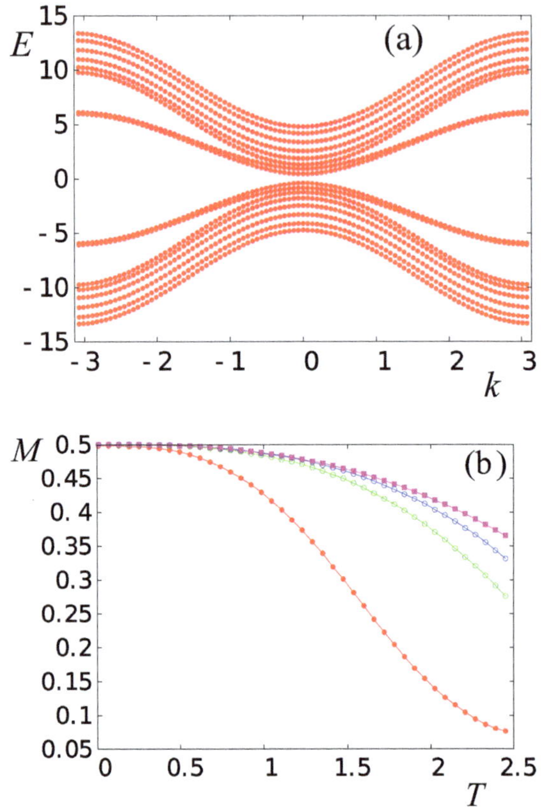

Figure 13.11 Surface effect: (a) spin wave spectrum $E(k)$ versus $k = k_x = k_z$ for a thin film of 8 layers: $\theta = \pi/6$, $d = 0.2$, $J^s_{\parallel} = 0.5$, $J^s_{\perp} = 0.5$, the gap at $k = 0$ is due to d. The surface-mode branches are detached from the bulk spectrum. (b) Layer magnetizations versus T for the first, second, third and fourth layer (red circles, green void circles, blue void circles and magenta filled squares, respectively). See text for comments.

function of the film thickness N for $d = 0.1$ and $\theta = \pi/6$, at $T = 0$. We see that the gap depends not only on d but also on the value of the surface magnetization which is larger for thicker films. The transition temperature T_c versus the thickness N is shown in Fig. 13.12c where one observes that T_c tends rapidly to the bulk value (3D) which is $\simeq 2.82$ for $d = 0.1$.

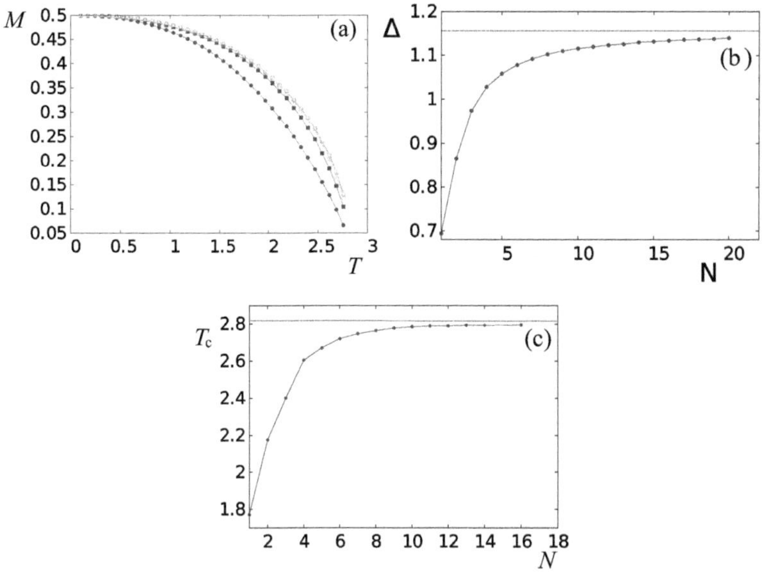

Figure 13.12 Twelve-layer film: (a) Layer magnetizations versus T for $\theta = \pi/6$ and anisotropy $d = 0.1$. Red circles, blue squares, green void squares magenta circles, void turquoise triangles and brown triangles correspond, respectively, to first, second, third, fourth, fifth and sixth layer, (b) Gap at $k = 0$ as a function of film thickness N for $\theta = \pi/6, d = 0.1$, at $T = 0.1$, (c) Critical temperature T_c versus the film thickness N calculated with $\theta = \pi/6$ and $d = 0.1$ using Eq. (13.37). Note that for infinite thickness (namely 3D), $T_c \simeq 2.8$ for $d = 0.1$.

13.6 Discussion and Experimental Suggestion

Several results found above can be experimentally verified. Let us confine ourselves in the case of thin films in this section, although the following discussion applies also in the 2D and 3D crystals.

The first striking aspect is the very particular effect of the DM interaction in the SW spectrum: We have seen in Fig. 13.8b that the first mode is linear in k for small k, but not the upper modes. This is not the case of ferromagnetic and antiferromagnetic interactions in thin films: For a ferromagnetic interaction, all modes have the k^2 behavior at small k [87], and for antiferromagnetic interactions all modes have the linear-k behavior. Thus, the DM interaction affects only the first mode, and the J-term in the Hamiltonian, Eq. (13.3), maintains the ferromagnetic behavior for higher-energy modes in the SW spectrum. Experiments which measure spin wave spectra such as neutron scattering [216] and spin wave resonance can verify if the first mode in the SW spectrum shown in Fig. 13.8b is linear in k for large DM interactions. For SW higher-energy modes, there has been an experimental breakthrough with the spin polarized electron energy loss spectroscopy (SPEELS): This technique allows us to detect very high-energy surface magnons, up to 240 meV. It has been proved to be very efficient to probe the dispersion of magnons in the ultrathin ferromagnetic films [275]. We believe that using such high-efficient experimental means, the effect of the DM interaction shown in this chapter can be verified.

The second striking feature is the surface effect on the SW spectrum and the surface magnetization shown, respectively, in Figs. 13.11a and 13.11b. These can be experimentally verified with the above-mentioned techniques. Note that the slope of the first SW mode allows us to deduce the interaction parameter; therefore, the DM interaction strength D. Note that neutron scattering can be used for a complete determination of the magnetic properties of a system at finite temperatures [216].

The third interesting result is the dependence of the critical temperature T_c on the film thickness shown in Fig. 13.12c. This can be completely determined by neutron scattering and SPEELS which measure the magnetization as a function of temperature of films as thin as a few atomic layers [216, 275].

We believe that, in spite of cumbersome mathematical details shown in Section 13.3, numerical results coming out from the formulation shown in Sections 13.4 and 13.5 are very physically meaningful. We expect, therefore, experimental verifications on materials showing a DM interaction such as those mentioned in Refs. [100, 113, 144, 316].

13.7 Concluding Remarks

By a self-consistent Green's function theory, we obtain the expression of the spin wave dispersion relation in 2D and 3D as well as in a thin film. Due to the competition between ferromagnetic interaction J and the perpendicular DM interaction D, the GS is non-linear with an angle θ which is shown to explicitly depend on the ratio D/J. The spectrum is shown to depend on θ and the layer magnetization is calculated self-consistently as a function of temperature up to the critical temperature T_c.

We have obtained new and interesting results. In particular we have showed that (i) the spin wave excitation in 2D and 3D crystals is stable at $T = 0$ with the non-collinear spin configuration induced by the DM interaction D without the need of an anisotropy, (ii) in the case of thin films, we need a small anisotropy d to stabilize the spin wave excitations because of the lack of neighbors at the surface, (iii) the spin wave energy E depends on D, namely on θ: At the long wavelength limit, E is proportional to k^2 for small D but E is linear in k for strong D, in 2D and 3D as well as in a thin film, (iv) quantum fluctuations are inhomogeneous for layer magnetizations near the surface, (v) unlike in some previous works, spin waves in systems with asymmetric DM interactions are found to be symmetric with respect to opposite propagation directions.

We have suggested some experiments to verify the above results in Section 13.6.

Chapter 14

Skyrmions in Thin Films

We generate a crystal of skyrmions in two dimensions (2D) using a Heisenberg Hamiltonian including the ferromagnetic interaction J, the Dzyaloshinskii–Moriya interaction D, and an applied magnetic field H. The ground state (GS) is determined by minimizing the interaction energy. We show that the GS is a skyrmion crystal in a region of (D, H). The stability of this skyrmion crystalline phase at finite temperatures is shown by a study of the time-dependence of the order parameter using Monte Carlo simulations. We observe that the relaxation is very slow and follows a stretched exponential law. The skyrmion crystal phase is shown to undergo a transition to the paramagnetic state at a finite temperature.

We also study a 2D skyrmion system on fluctuating surfaces with periodic boundary conditions using a Monte Carlo simulation technique. In this model, not only spins but also lattice vertices are integrated into the partition function of the model. From the obtained results, we conclude that skyrmions are stable even on fluctuating surfaces, contrary to initial expectations.

The main results of this chapter have been recently published in Refs. [103] and [106]. The reader is referred to those papers for more details.

Physics of Magnetic Thin Films: Theory and Simulation
Hung T. Diep

ISBN 978-981-4877-42-8 (Hardcover), 978-1-003-12110-7 (eBook)
www.jennystanford.com

14.1 Introduction: Magnetic Field Effect, Excitations of Skyrmions

Skyrmions have been extensively investigated in condensed matter physics [37, 39, 113, 207] since its theoretical formulation by Skyrme [325] in the context of nuclear matter.

There are several mechanisms and interactions leading to the appearance of skyrmions in various kinds of matter. The most popular one is certainly the Dzyaloshinskii–Moriya (DM) interaction which was initially proposed to explain the weak ferromagnetism observed in antiferromagnetic Mn compounds. The phenomenological Landau–Ginzburg model introduced by I. Dzyaloshinskii [99] was microscopically derived by T. Moriya [240]. This demonstration shows that the DM interaction comes from the second-order perturbation of the exchange interaction between two spins which is not zero only under some geometrical conditions of non-magnetic atoms found between them. The order of magnitude of DM interaction, D, is therefore, perturbation theory obliges, small. The explicit form of the DM interaction will be given in the next section. However, we can think that the demonstration of Moriya [240] is a special case and the general Hamiltonian may have the same form but with different microscopic origin.

The DM interaction has been shown to generate skyrmions in various kinds of crystals. For example, it can generate a crystal of skyrmions in which skyrmions arrange themselves in a periodic structure [51, 195, 196, 294]. Skyrmions have been shown to exist in crystal liquids [1, 39, 207]. A single skyrmion has also been observed [371]. Existence of skyrmion crystals have been found in thin films [109, 370]. Direct observation of the skyrmion Hall effect has been realized [165]. Artificial skyrmion lattices have been devised for room temperatures [126]. Experimental observations of skyrmion lattices have been realized in MnSi in 2009 [24, 242] and in doped semiconductors in 2010 [371]. At this stage, it should be noted that skyrmion crystals can also be created by competing exchange interactions without DM interactions [141, 263]. So, mechanisms for creating skyrmions are multiple.

We note that spin wave excitations in systems with a DM interaction in the helical phase without skyrmions have been investigated by many authors [104, 237, 282, 328, 353, 373].

Applications of skyrmions in spintronics have been largely discussed and their advantages compared to early magnetic devices such as magnetic bubbles have been pointed out in a recent review by W. Kang et al. [171]. Among the most important applications of skyrmions, let us mention skyrmion-based racetrack memory [266], skyrmion-based logic gates [377, 379], skyrmion-based transistor [183, 323, 378] and skyrmion-based artificial synapse and neuron devices [158, 210].

In this chapter, we study a skyrmion crystal created by the competition between the nearest-neighbor (NN) ferromagnetic interaction J and the DM interaction of magnitude D under an applied magnetic field **H**. We show by Monte Carlo (MC) simulation that the skyrmion crystal is stable at finite temperatures up to a transition temperature T_c where the topological structure of each skyrmion and the periodic structure of skyrmions are destroyed.

The chapter is organized as follows. Section 14.2 is devoted to the description of the model and the method to determine the ground state (GS). It is shown that our model generates a skyrmion crystal with a perfect periodicity at temperature $T = 0$. The GS phase diagram in the space (D, H) is presented. The phase transition of the skyrmion crystal is studied in Section 14.3. Results showing the stability of the skyrmion crystal at finite T obtained from MC simulations are shown in Section 14.4. We show in this section that the relaxation of the skyrmions is very slow and follows a stretched exponential law. The stability of the skyrmion phase is destroyed at a phase transition to the paramagnetic state. In Section 14.5, we study the effect of the fluctuating lattice, namely the elastic effect, on the stability of the skyrmion crystal. We find that the skyrmion crystal is stable against the lattice deformation. Concluding remarks are given in Section 14.6.

14.2 Model and Ground State

The DM interaction energy between two spins $\mathbf{S}_i$ and $\mathbf{S}_j$ is written as

$$\mathbf{D}_{i,j} \cdot \mathbf{S}_i \wedge \mathbf{S}_j \tag{14.1}$$

where $\mathbf{D}_{i,j}$ is a vector which results from the displacements of non-magnetic ions located between $\mathbf{S}_i$ and $\mathbf{S}_j$, for example, in Mn-O-Mn bonds in the historical papers [99, 240]. The direction of $\mathbf{D}_{i,j}$ depends on the symmetry of the displacements [240].

Theoretical and experimental investigations on the effect of the DM interaction in various materials have been extensively carried out in the context of weak ferromagnetism observed in perovskite compounds (see references cited in Refs. [100, 316]). As said in the Introduction, the interest in the DM interaction goes beyond the weak ferromagnetism. It has been shown that the DM interaction is at the origin of topological skyrmions [2, 37, 103, 106, 150, 166, 213, 224, 242, 296, 313, 357, 370, 371] and new kinds of magnetic domain walls [144, 292]. The increasing interest in skyrmions results from the fact that skyrmions may play an important role in technological application devices [113, 377].

In this chapter, we consider for simplicity the two-dimensional (2D) case where the spins are on a square lattice in the xy plane. We are interested in the stability of the skyrmion crystal generated in a system of spins interacting with each other via a DM interaction and a symmetric isotropic Heisenberg exchange interaction in an applied field perpendicular to the xy plane. All interactions are limited to NN. The full Hamiltonian is given by

$$\mathcal{H} = -J \sum_{\langle ij \rangle} \mathbf{S_i} . \mathbf{S_j} + D \sum_i \vec{S_i} \wedge (\vec{S}_{i+x} + \vec{S}_{i+y})$$
$$-H \sum_i S_i^z \tag{14.2}$$

where the DM interaction and the exchange interaction are taken between NN on both x and y directions. Rewriting it in a convenient form, we have

$$\mathcal{H} = -J\sum_{\langle ij\rangle}\mathbf{S_i}\cdot\mathbf{S_j} + D\sum_i[S_i^yS_{i+x}^z - S_i^zS_{i+x}^y - S_i^xS_{i+y}^z$$
$$+S_i^zS_{i+y}^x] - H\sum_i S_i^z$$
$$= -J\sum_{\langle ij\rangle}\mathbf{S_i}\cdot\mathbf{S_j} + D\sum_i[S_i^y(S_{i+x}^z - S_{i-x}^z)$$
$$-S_i^z(S_{i+x}^y - S_{i-x}^y) - S_i^x(S_{i+y}^z - S_{i-y}^z)$$
$$+S_i^z(S_{i+y}^x - S_{i-y}^x)] - H\sum_i S_i^z \quad (14.3)$$

For the i-th spin, one has

$$\mathcal{H}_i = -S_i^xH_i^x - S_i^yH_i^y - S_i^zH_i^zy \quad (14.4)$$

where the local-field components are given by

$$H_i^x = J\sum_{NN}S_j^x + D(S_{i+y}^z - S_{i-y}^z) \quad (14.5)$$

$$H_i^y = J\sum_{NN}S_j^y - D(S_{i+x}^z - S_{i-x}^z) \quad (14.6)$$

$$H_i^zz = J\sum_{NN}S_j^z + D(S_{i+x}^y - S_{i-x}^y) - D(S_{i+y}^x - S_{i-y}^x) + H \quad (14.7)$$

To determine the ground state (GS), we minimize the energy of each spin, one after another. This can be numerically achieved as the following. At each spin, we calculate its local-field components acting on it from its NN using the above equations. Next we align the spin in its local field, i.e., taking $S_i^x = H_i^x/\sqrt{H_i^x **2 + H_i^y **2 + H_i^z **2}$ etc. The denominator is the modulus of the local field. In doing so, the spin modulus is normalized to be 1. As seen from Eq. (14.4), the energy of the spin $\mathbf{S}_i$ is minimum. We take another spin and repeat the same procedure until all spins are visited. This achieves one iteration. We have to do a sufficient number of iterations until the system energy converges. For the skyrmion case, it takes about 1000 iterations to have the fifth-digit convergence. We have used random initial configurations and we observed that a number of them lead to metastable states. This is seen by comparing the energies of several thousands of initial configurations and examining the snapshots. A good GS has always a perfect hexagonal structure as shown below.

Note that the GS configuration does not depend on the system size if it is large enough. This size is from $N = 50$ for the value of $D = 1$ and $H = 0.5$ used in the example shown below.

An example of GS are displayed in Fig. 14.1 for the crystal size 50×50: A crystal of skyrmions is seen using $D = 1$ and $H = 0.5$ (in unit of $J = 1$). Note that we use here an example of large D for clarity, but we will show later in the phase diagram that the skyrmion crystal exists in a much smaller region for small D.

Let us give some comments on Fig. 14.1:

(i) The periodic boundary conditions have been used, leading to periodic GS spin configurations.

(ii) The size of each skyrmion depends on the value of D for a given H. In the case presented in this figure the diameter of each skyrmion is 10 lattice spacings. For larger D, the diameter is reduced, there are more skyrmions for a given lattice size. For small D, the system tends to a single large skyrmion centered at the middle of the lattice.

(iii) The skyrmions form a triangular lattice (see the top figure of Fig. 14.1). Note that the underlined lattice is a square lattice where each site is occupied by a spin. There is thus a six-fold degeneracy of the skyrmion lattice due to the global rotation of the spin by $2\pi/6$.

(iv) All the skyrmions have the same chirality. There is thus, in addition to the six-fold degeneracy mentioned above, a two-fold chirality degeneracy of the configuration shown in Fig. 14.1. Note that two neighboring skyrmions are separated by spins perfectly aligned along the field (the spin at the center of each skyrmion is a perfectly aligned antiparallel to the field).

In Fig. 14.2a we show a GS at $H = 0$ where domains of long and round islands of up spins separated by labyrinths of down spins are mixed. When H is increased, vortices begin to appear. The GS is a mixing of long islands of up spins and vortices as seen in Fig. 14.2b obtained with $D = 1$ and $H = 0.25$. This phase can be called "labyrinth phase" or "stripe phase."

It is interesting to note that skyrmion crystals with texture similar to those shown in Figs. 14.1 and 14.2 have been experimentally observed in various materials [126, 242, 370, 372],

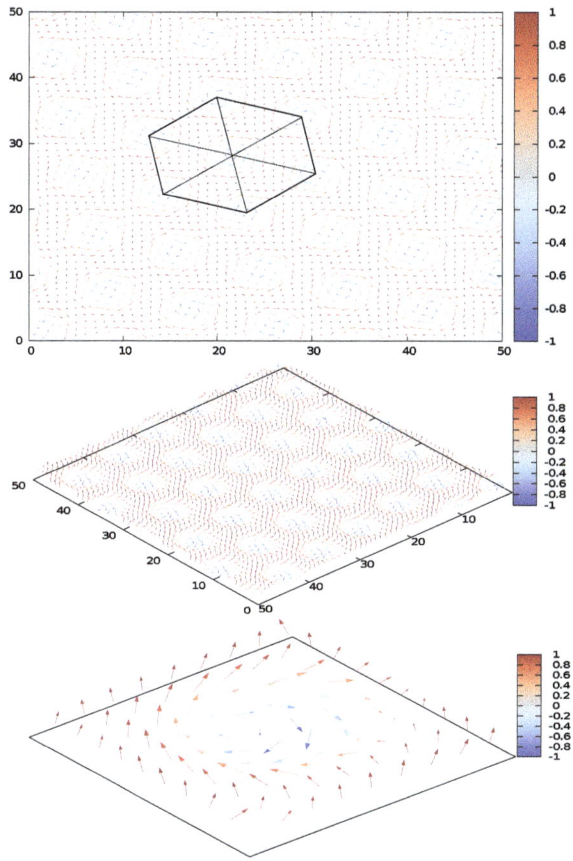

Figure 14.1 Ground state for $D/J = 1$ and $H/J = 0.5$, a crystal of skyrmions is observed. Top: Skyrmion crystal viewed in the xy plane. Middle: a 3D view. Bottom: zoom of the structure of a single vortex. The value of S_z is indicated on the color scale. See text for comments.

but the most similar skyrmion crystal was observed in two-dimensional $Fe_{0.5}Co_{0.5}Si$ by Yu et al. using Lorentz transmission electron microscopy [371].

We have performed the GS calculation taking many values in the plane (D, H). The phase diagram is established in Fig. 14.3. Above the blue line is the field-induced ferromagnetic phase. Below the red line is the labyrinth phase with a mixing of skyrmions and rectangular domains. The skyrmion crystal phase is found in a

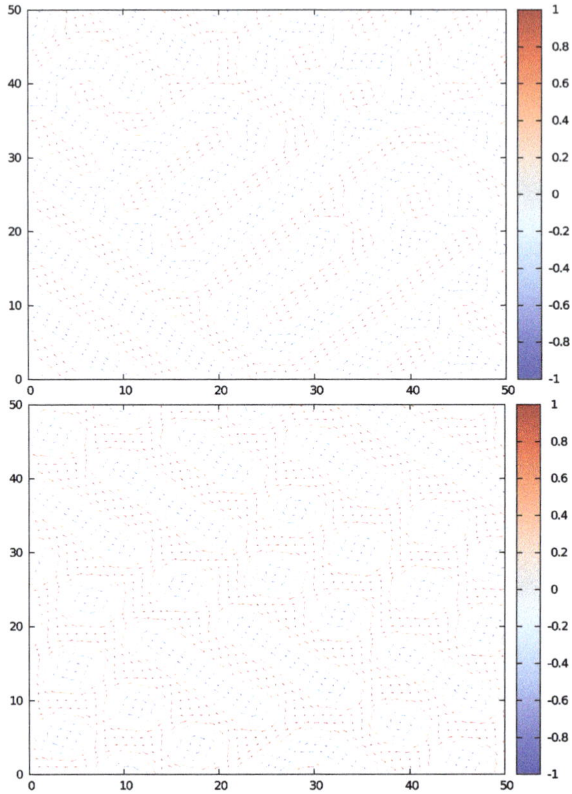

Figure 14.2 (Top) Ground state for $D/J = 1$ and $H/J = 0$, a mixing of domains of long and round islands; (bottom) ground state for $D/J = 1$ and $H/J = 0.25$, a mixing of domains of long islands and vortices. We call these structures the "labyrinth phase."

narrow region between these two lines, down to infinitesimal D. In order to enlarge the stability region of skyrmions, we may need to include other kinds of interaction such as dipole-dipole interaction, anisotropies and/or other competing interactions. However, this is out of the scope of this work.

Although the effects of the temperature (T) will be shown in the next section, we anticipate here the phase diagram shown in Fig. 14.4 in the space (T, H) for an overview before the detailed

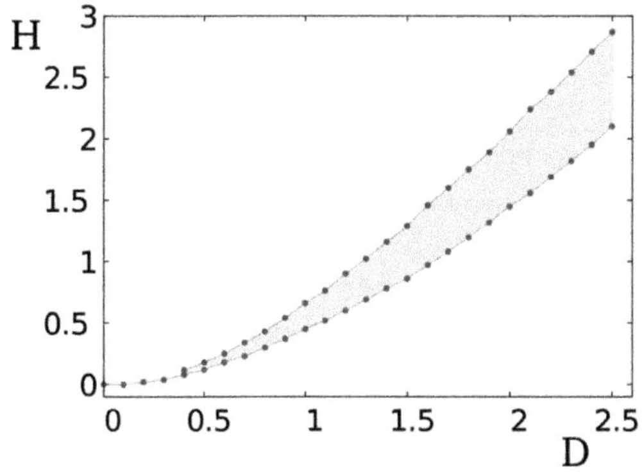

Figure 14.3 Phase diagram in the (D, H) plane for size $N = 100$.

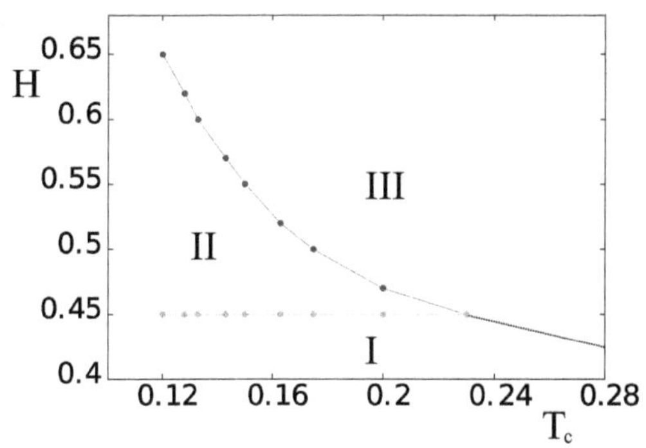

Figure 14.4 Phase diagram in the (T, H) plane for size $N = 100$. Phases I, II and III indicate the labyrinth phase, the skyrmion crystal phase and the field-induced ferromagnetic phase. The discontinued line separating phase I and III at high T is not a transition line.

data presentation given below. The definitions of phases I, II and III are given in the caption. Note that the discontinued line is not a transition line it schematically indicates a region where spins turned progressively to the field at high T.

In the following, we are interested in the phase transition of the skyrmion crystal. The stability of the skyrmion crystal phase at finite temperatures T is studied by calculating the relaxation time.

14.3 Skyrmion Crystal: Phase Transition

In this section, we show results obtained from MC simulations on a sheet of square lattice of size $N \times N$ with periodic boundary conditions. The first step is to determine the GS spin configuration by minimizing the spin energy by iteration as described above. Using this GS configuration, we heat the system from $T = 0$ to a temperature T during an equilibrating time t_0 before averaging physical quantities over the next 10^6 MC steps per spin. The time t_0 is the "waiting time" during which the system relaxes before we perform averaging during the next t_a.

The definition of an order parameter for a skyrmion crystal is not obvious. Taking advantage of the fact that we know the GS, we define the order parameter as the projection of an actual spin configuration at a given T on its GS and we take the time average. This order parameter is thus defined as

$$M(T) = \frac{1}{N^2(t_a - t_0)} \sum_i |\sum_{t=t_0}^{t_a} \mathbf{S}_i(T, t) \cdot S_i^0(T = 0)| \qquad (14.8)$$

where $\mathbf{S}_i(T, t)$ is the i-th spin at the time t, at temperature T, and $\mathbf{S}_i(T = 0)$ is its state in the GS. The order parameter $M(T)$ is close to 1 at very low T where each spin is only weakly deviated from its state in the GS. $M(T)$ is zero when every spin strongly fluctuates in the paramagnetic state. The above definition of $M(T)$ is similar to the Edward-Anderson order parameter used to measure the degree of freezing in spin glasses [234]: We follow each spin during the time evolution and take the spatial average at the end.

We show in Fig. 14.5 the order parameter M versus T (red data points) as well as the average z spin component (blue data points) calculated by the projection procedure for the total time $t = 10^5 + 10^6$ MC steps per spin. As seen, both two curves indicate a phase transition at $T_c \simeq 0.26J/k_B$. The fact that M does not vanish above T_c is due to the effect of the applied field. Note that each skyrmion has a center with spins of negative z components (the most negative at the center), the spins turn progressively to positive z components while going away from the center.

We can also define another order parameter: Since the field acts on the z direction, in the GS and in the skyrmion crystalline phase we have both positive and negative S_z. In the paramagnetic state,

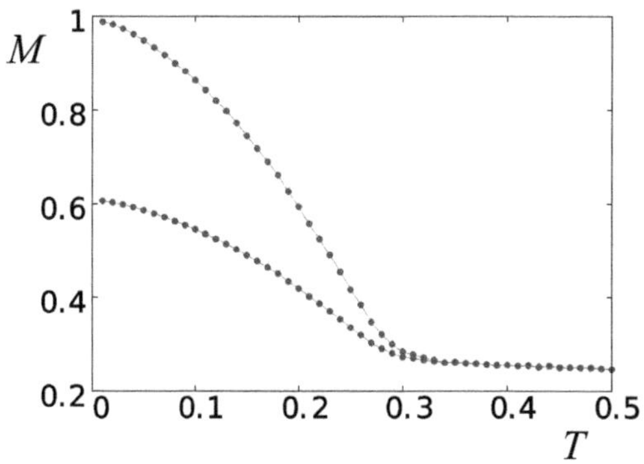

Figure 14.5 Red circles: Order parameter defined in Eq. (14.8) versus T, for $H = 0.5$ and $N = 1800$, averaged during $t_a = 10^5$ MC steps per spin after an equilibrating time $t_0 = 10^5$ MC steps. Blue crosses: the projection of the S_z on S_z^0 of the ground state as defined in Eq. (14.8) but for the z components only. See text for comments.

the negative S_z will turn to the field direction. We define thus the following parameters using the z spin-components

$$Q_+(T) = \frac{1}{N^2(t_a - t_0)} \sum_{S_i^z>0} \sum_{t_0}^{t_a} S_i^z(T, t) \tag{14.9}$$

$$Q_-(T) = \frac{1}{N^2(t_a - t_0)} \sum_{S_i^z<0} \sum_{t_0}^{t_a} S_i^z(T, t) \tag{14.10}$$

Figure 14.6 shows Q_+ and Q_- versus T. As seen, at the transition Q_+ undergoes a change of curvature and Q_- becomes zero. All spins have positive S_z after the transition due to spin reversal by the field.

The results obtained at the end of the simulation may depend on the overall time $t = t_0 + t_a$. In simple systems, the choices of t_0 and t_a can be guided by testing the time-dependence of physical quantities, and the values of t_0 and t_a are chosen when physical quantities do not depend on these run times. However, in disordered systems such as spin glasses and in complicated systems such as frustrated systems, the relaxation time is very long and out of the reach of simulation time. In such cases, we have to recourse to some scaling relations in order to deduce the values of physical quantities at equilibrium [261, 273]. We show below how to obtain the value of an order parameter at the infinite time.

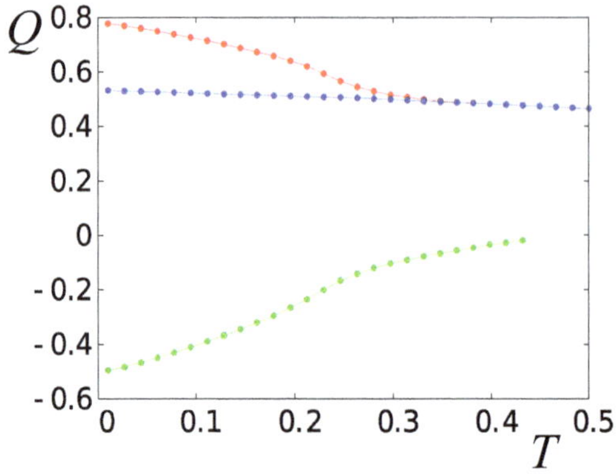

Figure 14.6 Order parameters defined in Eqs. (14.9)–(14.10) versus T, for $H = 0.5$ and $N = 800$, $t_0 = 10^5$, $t_a = 10^6$. Red circles: total positive z component ($Q_+(T)$). Green circles: total negative z component $Q_-(T)$. Blue circles: total z component. See text for comments.

14.4 Stability of Skyrmion Crystal at Finite Temperatures

In order to detect the dependence of $M(T)$ on the total MC time $t = t_0 + t_a$, we calculate the average of $M(T)$ over 10^6 MC steps per spin, after a waiting time t_0 as said above. We record the values of $M(T)$ in different runs with t_0 varying from 10^4 to 10^6 MC steps per spin. We plot these results as a function of different total time t in Fig. 14.7 for three temperatures. As said above, to find the value extrapolated at the infinite time, we use the stretched exponential relaxation defined by

$$M(T, t) = A \exp\left[-(t/\tau)^{\alpha}\right] + c \tag{14.11}$$

where t is the total simulation time, α is the stretched exponent, A a temperature-dependent constant, and τ the relaxation time. Note that this definition, without the constant c, has been used by many previous authors in the context of spin glasses [9, 32, 94, 225, 335]. We have introduced c which is the infinite-time limit of $M(T)$. We have taken t from 10^4 to 10^6 MC steps per spin in the simulation. At the infinite-time limit, c is zero for $T \gg T_c$, and $c \neq 0$ for $T < T_c$. Figure 14.7 shows $M(T, t)$ as a function of time t in unit of 10^3 MC

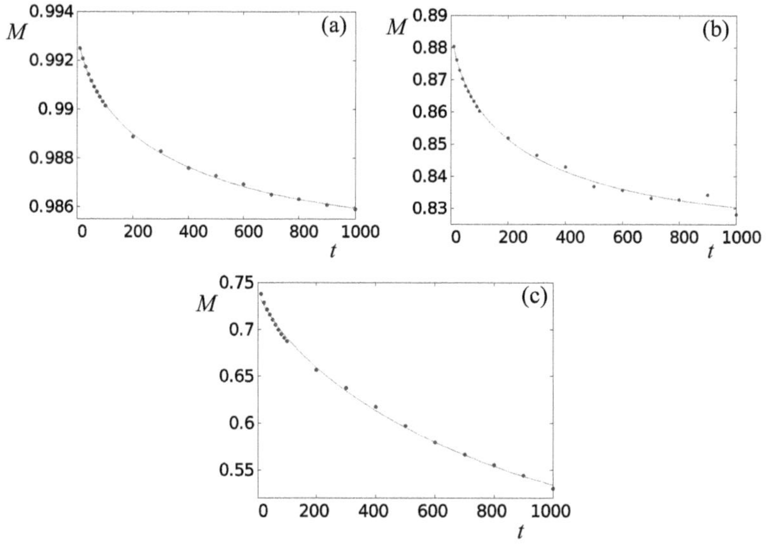

Figure 14.7 The order parameter M defined by Eq. (14.8) versus MC time t in unit of 1000 MC steps per spin, for $H = 0.5$ and $N = 800$ (a) $T = 0.01$, values of fitting parameters $\alpha = 0.6$, $A = 0.008 \pm 0.00001$, $\tau = (364 \pm 19)10^3$, $c = 0.984505 \pm 0.00012$; (b) $T = 0.094$, $\alpha = 0.6$, $A = 0.0653 \pm 0.0013$, $\tau = (277 \pm 36)10^3$, $c = 0.822 \pm 0.018$); (c) $T = 0.17$, $\alpha = 0.8$, $A = 0.31 \pm 0.01$, $\tau = 891 \pm 100)10^3$, $c = 0.43 \pm 0.017$.

steps per spin, for three temperatures $T = 0.01, 0.094$ and 0.17. As seen, the fit with Eq. (14.11) presented by the continuous line is very good for the whole range of t.

Several remarks are in order:

(i) The precision of all parameters are between 1% to 5% depending on the parameter.
(ii) The value of α can vary a little bit according to the choice and the precision of the other parameters in the fitting but this variation is within a very small window of values around the value given above. For example, at $T = 0.17$, α can only be in the interval $[0.8 \pm 0.02]$. The value of α can vary with temperature as seen here: At low T, $\alpha = 0.6$, and at a higher T, we have $\alpha = 0.8$. This variation has been seen in other systems, in particular in spin glasses [256].

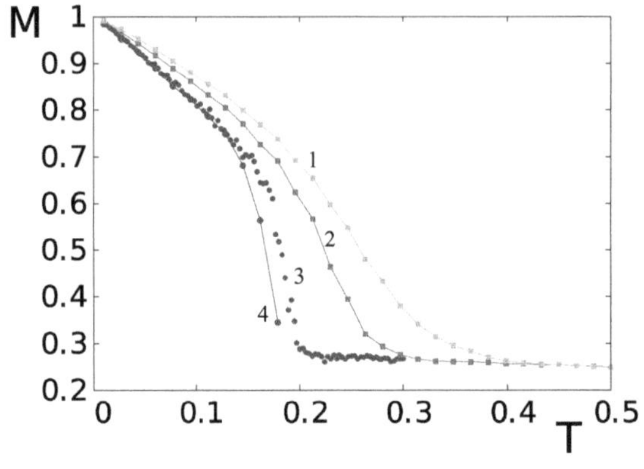

Figure 14.8 The order parameter $M(T)$ versus T for several waiting times t, for $H = 0.5$ and $N = 800$: curves 1, 2, 3 and 4 are, respectively, for $t = 10^5, 2 \times 10^5, 10^6, \infty$ (by fitting with Eq. (14.11).

(iii) The stretched exponential behavior is believed to be due to the non-uniform spin configuration since collinear spin states (ferromagnetic an antiferromagnetic orderings) follow a non-stretched exponential law [273]. The stretched exponential law was found in disordered phases such as spin glasses [9, 32, 94, 225, 335]. Of course if D is small, the configuration tends to a ferromagnetic one, the relaxation tends to a non-stretched law.

(iv) The relaxation time, within statistical errors, is approximately constant at low T, but it increases rapidly when T tends to T_c as seen in the value of τ at $T = 0.17$. This increase is a consequence of the so-called "critical slowing-down" when the system enters the critical region.

Let us show $M(T)$ as a function of T in Fig. 14.8 using the results of different run times from $t = 10^4 + 10^6$ MC steps per spin to $t = 10^6 + 10^6$. The values at infinite time for each T deduced from Eq. (14.11) is also shown. We see that while the total time $10^5 + 10^6$ MC steps per spin is sufficient at low T, it is not enough at higher T. That was the reason why we should use Eq. (14.11) to find the value of $M(T)$ at infinite time to be sure that the skyrmion crystal is stable at finite temperatures.

We have studied finite-size effects on the phase transition at T_c and we have seen that from $N = 800$, all curves coincide: There is no observable finite-size effects for $N \geq 800$. The present 2D skyrmion lattice phase, therefore, remains stable for the infinite dimension, unlike the ferromagnetic Heisenberg model in 2D [231]. Note that the nature of the ordering at low T and the phase transition observed here: In spite of the vortex nature of skyrmions, the transition is not due to the Kosterlitz–Thouless vortex mechanism observed in ferromagnetic XY spin systems [191, 192] since our "vortices" are stable without changing their position up to the transition temperature.

14.5 Skyrmion Crystal: Effect of Lattice Elasticity

In this section, we would like to see whether the skyrmion phase observed in the preceding sections is stable under large lattice deformations [106]. For this purpose, it is interesting to assume the lattice degrees of freedom as a dynamical variable. Here we should note that "dynamical" variable refers to the variable that is integrated in the partition function in the context of statistical mechanics. Therefore, a fluctuating lattice is considered to be a direct tool for this purpose, where the lattice fluctuation is allowed not only into the in-plane direction but also into the out-of-plane direction [107, 127, 145, 174, 246, 260, 279, 360]. Skyrmions on fluctuating lattice are interesting also from the viewpoint of liquid crystal skyrmions [39, 207].

It is widely accepted that skyrmion configurations, which are observed on materials such as FeGe, MnSi, or Cu_2OSeO_3, are promising candidates for future computer memories because of their dramatic energy-saving property in electric transportation. Due to the topological nature of the spin configuration, its stability against external stimuli such as thermal fluctuations is one of its most interesting properties [37, 40, 99, 240, 296, 325]. For these reasons, many experimental and theoretical studies have been conducted by assuming competing interactions, such as ferromagnetic (FM) interaction, Dzyaloshinskii–Moriya (DM), and Zeeman energy under constant magnetic fields [51, 96, 163, 242,

243, 379]. In particular, skyrmions in 2D systems have attracted a great deal of attention due to their potential applications in memory devices [20, 38, 122, 135]. However, the creation/annihilation of skyrmions and the stability of the skyrmion phase are yet to be studied in the context of such applications.

Along this research direction, several investigations have been conducted on the responses to mechanical stresses, which have a non-trivial effect on the spin configuration [245, 290]. In Ref. [321], Shi and Wang numerically find that skyrmions are stabilized due to the magnetoelastic coupling even without external magnetic field by introducing an explicit magnetoelastic interaction energy. This confirms that skyrmions may be generated by various mechanisms. We come back to this point later in the next section. Nii et al. recently reported that MnSi is an anisotropic elastic object [258], and they also reported the experimental results of a stress-induced skyrmion-to-conical phase transition in MnSi, where the applied uniaxial stresses create/annihilate skyrmions under suitable external magnetic fields [259]. It is also reported by Seki et al. that skyrmions in Cu_2OSeO_3 can be stabilized by a uniaxial tensile strain [313, 314], and Levatic et al. reported that mechanical pressure significantly increases the skyrmion pocket in Cu_2OSeO_3 [208] and in MnSi [57]. From these experiments, it was clarified that uniaxial mechanical stimuli stabilize/destabilize the skyrmion phase, or, in other words, a small lattice deformation enhances the stability/instability of the skyrmion phase. This indicates that the interplay between two different degrees of freedom spin and lattice, which can be rephrased by spin and orbit, plays an important role in mechanical responses, which has been extensively studied via DM interaction energy. Chen et al. numerically studied skyrmions in MnSi by using a 3D model and suggested that anisotropies in DM and FM interactions play an important role in their stability under uniaxial stress in a wide range of temperatures including low temperature region [60]. These anisotropies are naturally expected when crystalline lattices are considerably deformed. Thus, it is interesting to study the skyrmion stability under lattice deformations at finite temperatures.

The DM interaction on the continuous 2D plane depends on both spins and coordinates, and hence, we are able to calculate mechanical properties such as surface tension under the presence of skyrmions if the coordinate, or equivalently the surface, is treated as a dynamical variable [107, 127, 145, 174, 246, 260, 279, 360]. Moreover, the DM interaction is invariant under an arbitrary rotation around the axis perpendicular to the surface. However, its discrete version, which will be described in the next section, has only a discrete symmetry on 2D rigid lattices because the rigid lattices have only discrete symmetries. In contrast, the fluctuating lattices, where the edge direction of the lattice becomes almost random, are not regarded as a rigid lattice by any rotation. For this reason, the discrete DM interaction is expected to be influenced by the surface fluctuations.

Therefore, it is worthwhile studying skyrmions to see the interplay between spins and surface elasticity on fluctuating surfaces. In this section, we calculate the surface tension τ by Monte Carlo (MC) simulations to study the influences of skyrmion configurations on τ. Three different spin configurations are expected on the surface: stripe, skyrmion, and FM phases, as in the case of rigid lattices [103, 196, 371, 372]. Phase diagrams are also obtained in the space temperature versus external magnetic field. In this calculation, mechanical stress can be applied to the elastic lattices only in the form of strain, which is given by fixing the boundary to a suitable size or shape. Such an effective stress becomes isotropic in general, because the lattice structure changes relatively freely due to the fact that the vertex position is a dynamical variable. Nevertheless, the uniaxial stress condition is simulated by modifying the boundary lengths of the rectangular lattice, for example, from its isotropic lengths (L_1, L_2) to $(L_1/\xi, \xi L_2)$, $(\xi \neq 1)$. As a result of this geometry modification, we have to check whether the stripe domain axis aligns along the uniaxial stress direction. We also examine whether this uniaxial stress condition influences the stability of skyrmion phase under an external magnetic field.

14.5.1 Triangulated Lattices and Skyrmion Model

To study skyrmions on the fluctuating lattices, we have to combine the skyrmion model and the elastic surface model [107, 127, 174, 246, 260, 360]. If the two Hamiltonians are merged, the total

Hamiltonian becomes slightly lengthy, but the definition itself is straightforward. Indeed, the Hamiltonian S is given by a linear combination of several terms such as

$$\begin{aligned}
S(\mathbf{r},\sigma) &= S_1 + \kappa S_2 + \lambda S_F + \delta S_{\rm DM} + S_B \\
S_1 &= \sum_{ij} \ell_{ij}^2, \quad S_2 = \sum_{ij} (1 - \mathbf{n}_i \cdot \mathbf{n}_j) \\
S_F &= \sum_{ij} (1 - \sigma_i \cdot \sigma_j), \quad S_{\rm DM} = \sum_{ij} \mathbf{e}_{ij} \cdot \sigma_i \times \sigma_j \\
S_B &= \sum_{i} \sigma_i \cdot \mathbf{B}, \quad \mathbf{B} = (0, 0, B)
\end{aligned} \tag{14.12}$$

The first and second terms are the discrete Hamiltonians of the surface model of Helfrich and Polyakov [145, 279]. These S_1 and S_2 are called Gaussian bond potential and bending energy, respectively, and $\ell_{ij}(= |\mathbf{r}_j - \mathbf{r}_i|)$ is the length of bond ij, and $\mathbf{n}_i$ and $\mathbf{n}_j$ in S_2 are the unit normal vectors of the triangles that share the bond ij. The parameter κ is the bending rigidity of the surface [107, 127, 174, 246, 360].

The third term λS_F is the energy for describing the FM interaction between two neighboring spins with the interaction strength λ. The fourth term $\delta S_{\rm DM}$ describes the DM interaction with the interaction coefficient δ, where $\mathbf{e}_{ij}(= (\mathbf{r}_j - \mathbf{r}_i)/|\mathbf{r}_j - \mathbf{r}_i|)$ is the unit tangential vector from $\mathbf{r}_i$ to $\mathbf{r}_j$. This vector $\mathbf{e}_{ij}$ corresponds to $\partial \mathbf{r}$ in the continuous $S_{\rm DM}$ which varies on the fluctuating lattices. It is also possible to use $\ell_{ij}(= \mathbf{r}_j - \mathbf{r}_i)$ for $\partial \mathbf{r}$; however, here we choose to use the unit length vector $\mathbf{e}_{ij}$ for $\partial \mathbf{r}$. The continuous expression $\sigma \times \partial_a \sigma$ can also be reduced to $\sigma_j \times \sigma_i$ in the discrete $S_{\rm DM}$, because the differential $\partial_a \sigma$ is replaced by the difference $\sigma_j - \sigma_i$. Thus, the term $S_{\rm DM}$ plays a role in the interaction between σ and $\mathbf{r}$, the spins and the surface shape. The final term S_B is the energy for the external magnetic field $\mathbf{B} = (0, 0, B)$.

We note that the effects of surface fluctuations on spins are taken into consideration only via the DM interaction energy $S_{\rm DM}$. The tangential vector $\mathbf{e}_{ij}$ in $S_{\rm DM}$ is deformed by the surface fluctuations as mentioned above. On the other hand, the FM interaction can also reflect surface fluctuations, because the magnetoelastic coupling is not negligible on the fluctuating surfaces due to the spin–orbit coupling [303]. However, in the case of skyrmion deformations in a strained crystal, it is expected that the deformation of DM interaction plays an essential role, which is precisely checked in Ref. [322]. So, in this section, we simply use the ordinary undeformed S_F in Eq. (14.12). In other words, we neglect the effect of magnetoelastic coupling in the FM interaction and we study only the effects of surface elasticity on spins via the DM interaction deformation.

The partition function is given by

$$Z(\beta, \kappa, \lambda, \delta, B) = \sum_{\sigma} \int \prod_{i=1}^{N-N_{\rm BD}} d\mathbf{r}_i \int \prod_{i=1}^{N_{\rm BD}} d\mathbf{r}_i C(A_p) \exp\left[-\beta S(\mathbf{r}, \sigma)\right] \tag{14.13}$$

where $\beta = 1/T$, ($\Leftrightarrow$ $k_B = 1$). $\sum_{\sigma}$ denotes the sum of all possible values of σ. The symbol β is the inverse temperature, and the simulation unit is defined by the relation $k_B = 1$, and, hence, the temperature T is the reduced temperature. The symbols $\int \prod_i d\mathbf{r}_{i=1}^{N-N_{\rm BD}}$, $\int \prod_i d\mathbf{r}_{i=1}^{N_{\rm BD}}$ denote the three-dimensional and two-dimensional multiple integrations of the vertices inside and on the boundary, respectively, where $N_{\rm BD}$ is the total number of vertices on the boundary. The symbol $C(A_p)$ denotes a constraint on the integration of the boundary vertices so that the projected area remains fixed to A_p. This constraint is a function of $\mathbf{r}_1, \cdots, \mathbf{r}_{N_{\rm BD}}$. However, this function $C(\mathbf{r}_1, \cdots, \mathbf{r}_{N_{\rm BD}}; A_p)$ is too complex to write down with A_p, and for this reason we use $C(A_p)$ for the short notation [358].

14.5.2 Results

We studied a 2D skyrmion system on fluctuating surfaces with periodic boundary conditions using a Monte Carlo simulation technique. Note that in this model, not only spins but also lattice vertices are integrated into the partition function of the model. From the obtained results, we conclude that skyrmions are stable even on fluctuating surfaces, contrary to initial expectations. These results have been recently published in Ref. [106]. The reader is referred to this paper for details.

The Dzyaloshinskii–Moriya (DM) interaction is assumed together with the ferromagnetic interaction and Zeeman energy under an external magnetic field. We assume that the lattice fluctuations are only reflected in the DM interaction, and the magnetoelastic coupling in the ferromagnetic interaction is not taken into consideration. Note that in our model the competition of S_F and S_{DM} suffices to create a skyrmion crystal. To keep the model as simple as possible in order to analyze the elasticity effect, we did not include long-range magnetic dipole-dipole interaction and anisotropy in the model. These ingredients will yield another kind of skyrmions which give rise to another behavior under elasticity and uniaxial strains. In addition to those skyrmion Hamiltonians, the Helfrich–Polyakov Hamiltonian for membranes is assumed. A surface governed by these Hamiltonians spans a rectangular frame of a fixed projected area, and the surface (or frame) tension is calculated. In this lattice calculation, the effect of stress is introduced by strains, which are imposed on the lattices by fixing the edge lengths of the boundary constant with periodic boundary conditions. In this sense, this effective stress is automatically introduced by the boundary frame. No mechanical stress can be imposed on the lattice as an input, but it is obtained as an output of the simulations.

We obtain numerical results that are consistent with the experimentally observed fact that a lateral pressure imposed on a 3D material creates skyrmions under a longitudinal magnetic field. Our result indicates that skyrmions are stable on 2D fluctuating surfaces, where there is no crystalline structure in contrast to the rigid lattices including real crystalline materials.

We further examine how uniaxial strains influence spin configurations. To impose a uniaxial strain on the lattice, we assume an anisotropic condition for the edge length of the rectangular boundary. As a result of this calculation, we find that the effective pressure or stress increment makes the stripe domain align to the stress direction under the same magnetic field and temperature as in the case without the uniaxial strain. As mentioned in Ref. [106], the responses of stripe domains under uniaxial strains depend on the physical mechanism which generate skyrmions. Our model is not consistent with the reported experimental results of Ref. [71], but the fact that our stripes are parallel to uniaxial strains may be found in other materials and may have interesting transport applications. Of course, a more general magnetoelastic coupling included not only in S_{DM} but also in S_F should be considered in our future work [322, 355]. In contrast, almost no difference is observed in the skyrmion phase, and this also confirms the stability of the skyrmion phase on fluctuating surfaces.

We would like to comment on the studies of skyrmion stability on fluctuating lattices from the viewpoint of Thiele equation [342]. In Ref. [190], an interaction between skyrmion pairs on two separated surfaces is studied by Thiele equation. The authors in Ref. [190] assume that the bilayer lattice is flat, and hence, the effects of surface fluctuations are not taken into consideration. Therefore, to clarify the influence of the surface fluctuations on the skyrmion stability, it is interesting to study their model on a bilayer which consists of fluctuating and triangulated lattices. One possible technique is simply to assume that the fluctuating lattices are frozen and only spin variables are time dependent.

Finally, we also comment on a possibility of experimental studies. Skyrmions can only be observed on the materials such as FeGe, MnSi, or Cu_2OSeO_3 as mentioned in Ref. [106], and we have no possibility at present to observe and study skyrmions on elastic surfaces such as ferroelectric polymers under the condition similar to those described in this. However, it seems possible to measure the effects of local stresses even on the above-mentioned materials FeGe, etc. by using the technique developed for individual manipulation of superconducting magnetic vortex [194]. It is interesting to study the stability of skyrmions under local stress or strains. This may be the first step toward the understanding of the role of surface fluctuations in the skyrmion stability.

14.6 Conclusion

In this chapter, we have shown that the competition between a ferromagnetic interaction J and a Dzyaloshinskii–Moriya interaction D under an applied magnetic field H in two dimensions generate a skyrmion crystal in a region of the phase space (D, H). The spin model is the classical Heisenberg model. We have numerically determined the ground state by minimizing the local energy, spin by spin, using an iteration procedure. The skyrmion lattice is then heated to a finite temperature by the use of Monte Carlo simulations. We have shown that the skyrmion lattice is stable up to a finite temperature T_c beyond which the system becomes disordered. We have also shown that the relaxation follows a stretched exponential law. We believe that such a stability can be experimentally observed in real systems.

Perhaps, our model is the simplest model able to generate a skyrmion crystal without the need to include more complicated interactions such as long-range dipolar interactions, easy and uniaxial anisotropies [196, 213, 372]. Note that other authors have shown that skyrmion crystals can also be generated only with frustrated short-range interactions without even the DM interaction [141, 263]. So, we have to keep in mind that there are many interaction mechanisms of different nature which can generate skyrmion crystals. The origin of a skyrmion crystal determines its structure, its stability, its dynamics and in short its properties (see a discussion on this point in [372]). Experiments can be used, therefore, to determine the interaction mechanism which is at the origin of the formation of a skyrmion crystal in the material under investigation.

We have also studied the effect of the lattice elasticity on the skyrmion stability in two dimensions. The Helfrich–Polyakov Hamiltonian for membranes has been assumed in addition to other skyrmion terms in the previous sections. A surface governed by these terms spans a rectangular frame of a fixed projected area, and the surface (or frame) tension is calculated. We found that the fluctuating surface does not destroy the skyrmion structure. The effect of stress has been calculated. We obtained numerical results consistent with the experimentally observed fact. Our result indicates that skyrmions are stable on 2D fluctuating surfaces where there is no crystalline structure. We further examined how uniaxial strains influence the spin configuration. We find that the effective pressure or stress increment makes the stripe domain align to the stress direction under the same magnetic field and temperature as in the case without the uniaxial strain. Details are given in Ref. [106].

Chapter 15

Skyrmions in Superlattices

In this chapter, we study the effects of Dzyaloshinskii–Moriya (DM) magnetoelectric coupling between ferroelectric and magnetic layers in a superlattice formed by alternate magnetic and ferroelectric films. Magnetic films are films of simple cubic lattice with Heisenberg spins interacting with each other via an exchange J and a DM interaction with the ferroelectric interface. Electrical polarizations of ± 1 are assigned at simple cubic lattice sites in the ferroelectric films. We determine the ground-state (GS) spin configuration in the magnetic film. In zero field, the GS is periodically non-collinear and in an applied field $\mathbf{H}$ perpendicular to the layers, it shows the existence of skyrmions at the interface. Using the Green's function (GF) method we study the spin waves (SW) excited in a monolayer and also in a bilayer sandwiched between ferroelectric films, in zero field. We show that the DM interaction strongly affects the long-wavelength SW mode. We calculate also the magnetization at low temperature T. We use next Monte Carlo simulations to calculate various physical quantities at finite temperatures such as the critical temperature, the layer magnetization and the layer polarization, as functions of the interface magnetoelectric DM coupling and the applied magnetic field. Phase transition to the disordered phase is studied in detail.

Physics of Magnetic Thin Films: Theory and Simulation
Hung T. Diep

ISBN 978-981-4877-42-8 (Hardcover), 978-1-003-12110-7 (eBook)
www.jennystanford.com

The main results of this chapter have been published in Refs. [317, 318].

15.1 Introduction

Non-uniform spin structures, which are quite interesting by themselves, became the subject of close attention after the discovery of electrical polarization in some of them [97]. The existence of polarization is possible due to the inhomogeneous magnetoelectric effect, namely that electrical polarization can occur in the region of magnetic inhomogeneity. It is known that the electric polarization vector is transformed in the same way as the combination of the magnetization vector and the gradient of the magnetization vector, meaning that these values can be related by the proportionality relation. In Ref. [241], it was found that in a crystal with cubic symmetry the relationship between electrical polarization and inhomogeneous distribution of the magnetization vector has the following form:

$$\mathbf{P} = \gamma\chi_e(\mathbf{M}\cdot(\nabla\cdot\mathbf{M}) - (\mathbf{M}\cdot\nabla)\cdot\mathbf{M}) \tag{15.1}$$

here γ is the magnetoelectric coefficient, and χ_e the permittivity. In non-collinear structures, the microscopic mechanism of the coupling of polarization and the relative orientation of the magnetization vectors is based on the interaction of Dzyaloshinskii–Moriya [61, 180, 316]. The corresponding term in the Hamiltonian is

$$H_{DM} = \mathbf{D}_{i,j}\cdot\mathbf{S_i}\times\mathbf{S_j} \tag{15.2}$$

where $\mathbf{S_i}$ is the spin of the i-th magnetic ion, and $\mathbf{D}_{i,j}$ is the Dzyaloshinskii–Moriya vector. The vector $\mathbf{D}_{i,j}$ is proportional to the vector product $\mathbf{R}\times\mathbf{r_{i,j}}$ of the vector $\mathbf{R}$ which specifies the displacement of the ligand (for example, oxygen) and the unit vector $\mathbf{r_{i,j}}$ along the axis connecting the magnetic ions i and j (see Fig. 15.1a). We write

$$\mathbf{D}_{i,j}\propto\mathbf{R}\times\mathbf{r_{i,j}} \tag{15.3}$$

Thus, the Dzyaloshinskii–Moriya interaction connects the angle between the spins and the magnitude of the displacement of

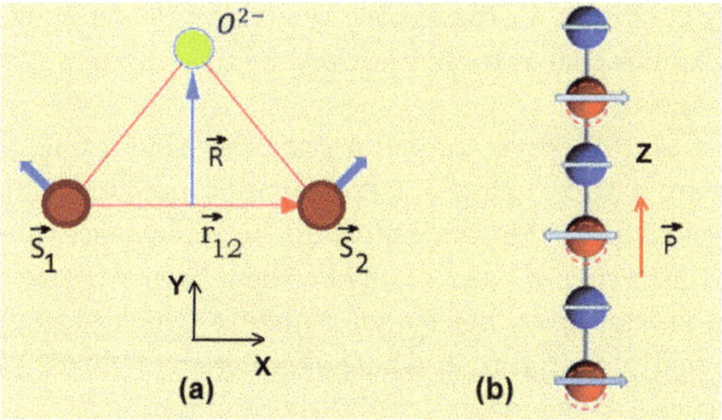

Figure 15.1 (a) Schema of Dzyaloshinskii–Moriya interaction, spins are in the *xy* plane; (b) microscopic mechanisms of creation of electric spontaneous polarization **P** due to displacements of atoms (red) in the same direction *z*.

non-magnetic ions. In some micromagnetic structures all ligands are shifted in one direction, which leads to the appearance of macroscopic electrical polarization (see Fig. 15.1b). By nature, this interaction is a relativistic amendment to the indirect exchange interaction, and is relatively weak [284]. In the case of magnetically ordered matter, the contribution of the Dzyaloshinskii–Moriya interaction to the free energy can be represented as Lifshitz antisymmetric invariants containing spatial derivatives of the magnetization vector. In analogy, the vortex magnetic configuration can be stable via Skyrme mechanism [37].

Skyrmions were theoretically predicted more than twenty years ago as stable micromagnetic structures [41]. The idea came from nuclear physics, where the elementary particles were represented as vortex configurations of continuous fields. The stability of such configurations was provided by the "Skyrme mechanism"—the components in Lagrangians containing antisymmetric combinations of spatial derivatives of field components [325]. For a long time, skyrmions have been the subject only of theoretical studies. In particular, it was shown that such structures can exist in antiferromagnets [42] and in magnetic metals [296]. In the latter case, the model included the possibility of changing the magnitude of the magnetization vector and spontaneous emergence of the skyrmion lattice without the application of external magnetic field. A necessary condition for the existence of skyrmions in bulk samples was the absence of an inverse transformation in the crystal magnetic symmetry group.

Diep et al. [92] have studied a crystal of skyrmions generated on a square lattice using a ferromagnetic exchange interaction and a Dzyaloshinskii–Moriya interaction between nearest-neighbors under an external magnetic field. They have shown that the skyrmion crystal has a hexagonal structure which is shown to be stable up to a temperature T_c where a transition to the paramagnetic phase occurs and the dynamics of the skyrmions at $T < T_c$ follows a stretched exponential law. In Ref. [296] it was shown that the most extensive class of candidates for the detection of skyrmions includes the surfaces and interfaces of magnetic materials, where the geometry of the material breaks the central symmetry and, therefore, can lead to the appearance of chiral interactions similar to the Dzyaloshinskii–Moriya interaction. In addition, skyrmions are two-dimensional solitons, the stability of which is provided by the local competition of short-range interactions exchange and Dzyaloshinskii–Moriya interactions [92, 184].

The idea of using skyrmions in memory devices nowadays is reduced to the information encoding using the presence or absence of a skyrmion in certain area of the material. A numerical simulation of the creation and displacement of skyrmions in thin films was carried out in Ref. [302] using a spin-polarized current. The advantage of skyrmions with respect to the domain boundaries in such magnetic memory circuits (e.g., racetrack memory, see Ref. [266]) is the relatively low magnitude of the currents required to move the skyrmions along the "track." For the first time, skyrmions were experimentally detected in the MnSi helimagnet [242]. Below the Curie temperature in MnSi spins are aligned in helicoidal or conical structure (the field was applied along the [100] axis), depending on the magnitude of the applied magnetic field. Similar experimental results were obtained for the compound $Fe_{1-x}Co_xSi$, $x = 0.2$ [243]. The investigation of $Fe_{0.5}Co_{0.5}Si$ made it possible to take the next important step in the study of skyrmions—to directly observe them using Lorentz electron microscopy [371].

The dependence of the stability of the skyrmion lattice on the sample thickness of FeGe was studied in more detail in Ref. [370].

Studies have confirmed that the thinner was the film, the greater was the "stability region" of skyrmions.

Skyrmions as the most compact isolated micromagnetic objects are of great practical interest as memory elements [184]. The stability of skyrmions [92] can make the memory on their basis non-volatile, and low control currents will reduce the cost of rewriting compared to similar technologies based on domain boundaries. In Refs. [313] and [315], magnetic and electrical properties of the skyrmion lattice were studied in the multiferroic Cu_2OSeO_3. It has been shown that that energy consumption can be minimized by using the electric field to control the micromagnetic structures. It is worth noting that the multiferroics $BaFe_{12-x-0.05}Sc_xMg_{0.05}O_{19}$ may also have a skyrmion structure [295, 372]. The manipulations with skyrmions were first demonstrated in the diatomic PdFe layer on the iridium substrate, and the importance of this achievement for the technology of information storing is difficult to overestimate: It makes possible to write and read the individual skyrmions using a spin-polarized tunneling current [293]. In Ref. [283], the possibility of the nucleation of skyrmions by the electric field by means of an inhomogeneous magnetoelectric effect was established.

Recent studies are focused on the interface-induced skyrmions. Therefore, the superstructures naturally lead to the interaction of skyrmions on different interfaces, which has unique dynamics compared to the interaction of the same-interface skyrmions. In Ref. [190], a theoretical study of two skyrmions on two-layer systems was carried using micromagnetic modeling, as well as an analysis based on the Thiele equation, which revealed a reaction between them, such as the collision and a bound state formation. The dynamics sensitively depends on the sign of DM interaction, i.e., the helicity, and the skyrmion numbers of two skyrmions, which are well described by the Thiele equation. In addition, the colossal spin-transfer-torque effect of bound skyrmion pair on antiferromagnetically coupled bilayer systems was discovered. In Ref. [226] the study of the Thiele equation was carried for current-induced motion in a skyrmion lattice through two soluble models of the pinning potential.

We consider in this chapter a superlattice composed of alternate magnetic films and ferroelectric films. The aim is to propose a new model for the coupling between the magnetic film and the ferroelectric film by introducing a DM-like interaction. It turns out that this interface coupling gives rise to non-collinear spin configurations in zero applied magnetic field and to skyrmions in a field **H** applied perpendicularly to the films. Using the Green's function method, we study spin wave excitations in zero field of a monolayer and a bilayer. We find that the DM interaction affects strongly the long-wavelength mode. Monte Carlo simulations are carried out to study the phase transition of the superlattice as functions of the interface coupling strength.

Section 15.2 is devoted to the description of our model and the determination of the ground-state spin configuration with and without applied magnetic field. Section 15.3 shows the results obtained by Monte Carlo simulations for the phase transition in the system as a function of the interface DM coupling. In Section 15.4, we show the results of spin wave spectrum obtained by the Green's function technique in zero field for a monolayer and a bilayer. Section 15.5 shows the effect of the frustration on the skyrmion structure. The frustration is introduced by adding the competing interactions J_1 and J_2 between nearest neighbors (NN) and next-nearest neighbors (NNN).

15.2 Model and Ground State

15.2.1 Model

Consider a superlattice composed of alternate magnetic and ferroelectric films (see Fig. 15.2a). Both have the structure of simple cubic lattice of the same lattice constant, for simplicity. The Hamiltonian of this multiferroic superlattice is expressed as:

$$\mathcal{H} = H_m + H_f + H_{mf} \tag{15.4}$$

where H_m and H_f are the Hamiltonians of the ferromagnetic and ferroelectric subsystems, respectively, while H_{mf} is the Hamiltonian of magnetoelectric interaction at the interface between two adjacent films.

We describe the Hamiltonian of the magnetic film with the Heisenberg spin model on a simple cubic lattice:

$$H_m = -\sum_{i,j} J_{ij}^m \mathbf{S}_i \cdot \mathbf{S}_j - \sum_i \mathbf{H} \cdot \mathbf{S}_i \tag{15.5}$$

where $\mathbf{S}_i$ is the spin on the i-th site, $\mathbf{H}$ is the external magnetic field, $J_{ij}^m > 0$ the ferromagnetic interaction parameter between a spin and its nearest neighbors (NN) and the sum is taken over NN spin pairs. We consider $J_{ij}^m > 0$ to be the same, namely J^m, for spins everywhere in the magnetic film. The external magnetic field $\mathbf{H}$ is applied along the z-axis which is perpendicular to the plane of the layers. The interaction of the spins at the interface will be given below.

For the ferroelectric film, we suppose for simplicity that electric polarizations are Ising-like vectors of magnitude 1, pointing in the $\pm z$ direction. The Hamiltonian is given by

$$H_f = -\sum_{i,j} J_{ij}^f \mathbf{P}_i \cdot \mathbf{P}_j - \sum_i E^z P_i^z \tag{15.6}$$

where $\mathbf{P}_i$ is the polarization on the i-th lattice site, $J_{ij}^f > 0$ the interaction parameter between NN and the sum is taken over NN sites. Similar to the ferromagnetic subsystem we will take the same $J_{ij}^f = J^f$ for all ferroelectric sites. We apply the external electric field $\mathbf{E}$ along the z-axis.

We suppose the following Hamiltonian for the magnetoelectric interaction at the interface:

$$H_{mf} = \sum_{i,j,k} J_{ijk}^{mf} \, \mathbf{D}_{i,j} \cdot \left[\mathbf{S_i} \times \mathbf{S_j}\right] \tag{15.7}$$

In this expression $J_{ijk}^{mf} \, \mathbf{D}_{i,j}$ plays the role of the DM vector which is perpendicular to the xy plane. Using Eqs. (15.2)–(15.3), one has

$$\begin{aligned} \mathbf{D}_{i,j} &= \mathbf{R} \times \mathbf{r}_{i,j} \\ \mathbf{D}_{j,i} &= \mathbf{R} \times \mathbf{r}_{j,i} = -\mathbf{D}_{i,j} \end{aligned} \tag{15.8}$$

Now, let us define for our model

$$J_{ijk}^{mf} = J_{i,j}^{mf} P_k \tag{15.9}$$

which is the DM interaction parameter between the electric polarization $\mathbf{P}_k$ at the interface ferroelectric layer and the two NN spins $\mathbf{S_i}$ and $\mathbf{S_j}$ belonging to the interface ferromagnetic layer.

Hereafter, we suppose $J_{i,j}^{mf} = J^{mf}$ independent of (i, j). Selecting $\mathbf{R}$ in the xy plane perpendicular to $\mathbf{r}_{i,j}$ (see Fig. 15.1) we can write $\mathbf{R} \times \mathbf{r}_{i,j} = a\mathbf{z}\, e_{i,j}$ where $e_{i,j} = -e_{j,i} = 1$, a is a constant and $\mathbf{z}$ the unit vector on the z axis.

It is worth at this stage to specify the nature of the DM interaction to avoid a confusion often seen in the literature. The term $\left[\mathbf{S}_i \times \mathbf{S}_j\right]$ changes its sign with the permutation of i and j, but the whole DM interaction defined in Eq. (15.2) does not change its sign because $D_{i,j}$ changes its sign with the permutation as seen in Eq. (15.3). Note that if the whole DM interaction is antisymmetric then when we perform the lattice sum, nothing of the DM interaction remains in the Hamiltonian. This explains why we need the coefficient $e_{i,j}$ introduced above and present in Eq. (15.10) below.

We collect all these definitions we write H_{mf} in a simple form

$$
\begin{aligned}
H_{mf} &= \sum_{i,j,k} J^{mf}\, P_k\, (\mathbf{R} \times \mathbf{r}_{i,j}) \cdot \left[\mathbf{S_i} \times \mathbf{S_j}\right] \\
&= \sum_{i,j,k} J^{mf}\, P_k\, e_{i,j} \mathbf{z} \cdot \left[\mathbf{S_i} \times \mathbf{S_j}\right] \\
&= \sum_{i,j,k} J^{mf}\, e_{i,j}\, \mathbf{P}_k \cdot \left[\mathbf{S_i} \times \mathbf{S_j}\right]
\end{aligned}
\tag{15.10}
$$

where the constant a is absorbed in J^{mf}.

As seen in Eq. (15.10), the coefficient of the interface coupling is proportional to $<P_k>$ which depends on T. If $<P_k>$ becomes zero before the loss of skyrmion texture, we will not see the latter. Therefore, we have chosen the polarization of the Ising type with ferroelectric interaction parameter J^f in a way that its transition temperature is higher than that of the magnetic part. Note that the magnetic transition is driven by the competition between T and the magnetic texture (skyrmions) which is a result of the competition between J, the DM interaction (namely $<P_k>$) and field H.

We note that the DM interaction is taken only between NN spin. If we choose the DM vector $\mathbf{D}$ perpendicular to the xy plane then the DM interaction energy is minimum when the spins are in the xy plane because $\mathbf{D}$ is parallel to $[\mathbf{S}_i \times \mathbf{S}_j]$. One can choose any orientation for $\mathbf{D}$ but in that case to have the minimum energy the plane containing $\mathbf{S}_i$ and $\mathbf{S}_j$ should be perpendicular to $\mathbf{D}$: The spins are not in the xy plane, making the spin configuration analysis difficult.

The superlattice and the interface interaction are shown in Fig. 15.2. A polarization at the interface interact with 5 spins on the magnetic layer according to Eq. (15.10), for example (see Fig. 15.2b):

$$J^{mf}\mathbf{P}_1 \cdot [e_{1,2}(\mathbf{S_1} \times \mathbf{S_2}) + e_{1,3}(\mathbf{S_1} \times \mathbf{S_3}) + e_{1,4}(\mathbf{S_1} \times \mathbf{S_4}) + e_{1,5}(\mathbf{S_1} \times \mathbf{S_5})] \quad (15.11)$$

Since we suppose $\mathbf{P}_k$ is a vector of magnitude 1 pointing along the z axis, namely its z component is $P_k^z = \pm 1$, we will use hereafter P_k^z for electric polarization instead of $\mathbf{P}_k$.

From Eq. (15.10), we see that the magnetoelectric interaction J^{mf} favors a canted spin structure. It competes with the exchange interaction J of H_m which favors collinear spin configurations. Usually the magnetic or ferroelectric exchange interaction is the leading term in the Hamiltonian, so that in many situations the magnetoelectric effect is negligible. However, in nanofilms of superlattices the magnetoelectric interaction is crucial for the creation of non-collinear long-range spin order.

Note that the hypothesis that $\mathbf{P}_k$ is in the z direction in order to have the polarization proportional to the DM vector [Eqs. (15.7)–(15.10)]. The DM vector is taken in the z direction in order to have spins in the magnetic layers lying in the xy plane, in the absence of an applied field (see Sections 15.2.2.1 and 15.4). The polarization is in addition supposed of the Ising type since in this chapter, this assumption allows us to have the DM vector in a fixed direction z. The assumption is justified by the fact that in ferroelectric materials, if atoms are displaced in the same direction it gives rise to a spontaneous polarization in that direction as illustrated in Fig. 15.1b.

15.2.2 Ground State

15.2.2.1 Ground state in zero magnetic field

Let us analyze the structure of the ground state (GS) in zero magnetic field. Since the polarizations are along the z axis, the interface DM interaction is minimum when $\mathbf{S_i}$ and $\mathbf{S_j}$ lie in the xy interface plane and perpendicular to each other. However, the ferromagnetic

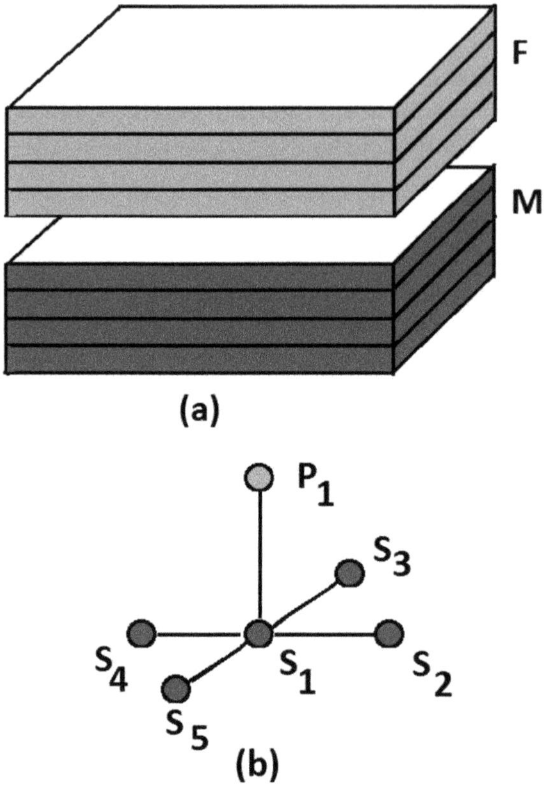

Figure 15.2 (a) The superlattice composed of alternately a ferroelectric layer indicated by F and a magnetic layer indicated by M; (b) A polarization P_1 at the interface interacts with 5 spins in the magnetic layer. See text for expression.

exchange interaction among the spins will compete with the DM perpendicular configuration. The resulting configuration is non-collinear. We will determine it below, but at this stage, we note that the ferroelectric film has always polarizations along the z axis even when interface interaction is turned on.

Let us determine the GS spin configurations in magnetic layers in zero field. If the magnetic film has only one monolayer, the minimization of H_{mf} in zero magnetic field is done as follows.

By symmetry, each spin has the same angle θ with its four NN in the xy plane. The energy of the spin $\mathbf{S}_i$ gives the relation between θ and J^m

$$E_i = -4J^m S^2 \cos\theta + 8J^{mf} P^z S^2 \sin\theta \quad (15.12)$$

where $\theta = |\theta_{i,j}|$ and care has been taken on the signs of $\sin\theta_{i,j}$ when counting NN, namely two opposite NN have opposite signs, and the opposite coefficient e_{ij}, as given in Eq. (15.11). Note that the coefficient 4 of the first term is the number of in-plane NN pairs, and the coefficient 8 of the second term is due to the fact that each spin has 4 coupling DM pairs with the NN polarization in the upper ferroelectric plane, and 4 with the NN polarization of the lower ferroelectric plane (we are in the case of a magnetic monolayer). The minimization of E_i yields, taking $P^z = 1$ in the GS and $S = 1$,

$$\frac{dE_i}{d\theta} = 0 \Rightarrow \frac{-2J^{mf}}{J^m} = \tan\theta \Rightarrow \theta = \arctan(-\frac{2J^{mf}}{J^m}) \quad (15.13)$$

The value of θ for a given $\frac{-2J^{mf}}{J^m}$ is precisely what obtained by the numerical minimization of the energy. We see that when $J^{mf} \to 0$, one has $\theta \to 0$, and when $J^{mf} \to -\infty$, one has $J^{mf} \to \pi/2$ as it should be. Note that we will consider here $J^{mf} < 0$ so as to have $\theta > 0$.

The above relation between the angle and J^{mf} will be used in the last section to calculate the spin waves in the case of a magnetic monolayer sandwiched between ferroelectric films.

In the case when the magnetic film has a thickness, the angle between NN spins in each magnetic layer is different from that of the neighboring layer. It is more convenient using the numerical minimization method called "steepest descent method" to obtain the GS spin configuration. This method consists in minimizing the energy of each spin by aligning it parallel to the local field acting on it from its NN. This is done as follows. We generate a random initial spin configuration, then we take one spin and calculate the interaction field from its NN. We align it in the direction of this field, and take another spin and repeat the procedure until all spins are considered. We go again for another sweep until the total energy converges to a minimum. In principle, with this iteration procedure the system can be stuck in a meta-stable state when there is a strong interaction disorder such as in spin-glasses. But for uniform, translational interactions, we have never encountered such a problem in many systems studied so far.

We use a sample size $N \times N \times L$. For most calculations, we select $N = 40$ and $L = 8$ using the periodic boundary conditions in the xy plane. For simplicity, when we investigate the effect of the exchange couplings on the magnetic and ferroelectric properties, we take the same thickness for the magnetic and ferroelectric films, namely $L_a = L_b = 4 = L/2$. Exchange parameters between spins and polarizations are taken as $J^m = J^f = 1$ for the simulation. For simplicity we will consider the case where the in-plane and inter-plane exchange magnetic and ferroelectric interactions between nearest neighbors are both positive. All the results are obtained with $J^m = J^f = 1$ for different values of the interaction parameter J^{mf}.

We investigated the following range of values for the interaction parameters J^{mf}: From $J^{mf} = -0.05$ to $J^{mf} = -6.0$ with different values of the external magnetic and electric fields. We note that the steepest descent method calculates the real ground state with the minimum energy to the value $J^{mf} = -1.25$. After larger values, the angle θ tends to $\pi/2$ so that all magnetic exchange terms (scalar products) will be close to zero, the minimum energy corresponds to the DM energy. Figure 15.3 shows the GS configurations of the magnetic interface layer for small values of J^{mf}: -0.1, -0.125, -0.15. Such small values yield small values of angles between spins so that the GS configurations have ferromagnetic and non-collinear domains. Note that angles in magnetic interior layers are different but the GS configurations are of the same texture (not shown).

For larger values of J^{mf}, the GS spin configurations have periodic structures with no more mixed domains. We show in Fig. 15.4 examples where $J^{mf} = -0.45$ and -1.2. Several remarks are in order:

(i) Each spin has the same turning angle θ with its NN in both x and y direction. The schematic zoom in Fig. 15.4c shows that the spins on the same diagonal (spins 1 and 2, spins 3 and 4) are parallel. This explains the structures shown in Figs. 15.4a and 15.4b;

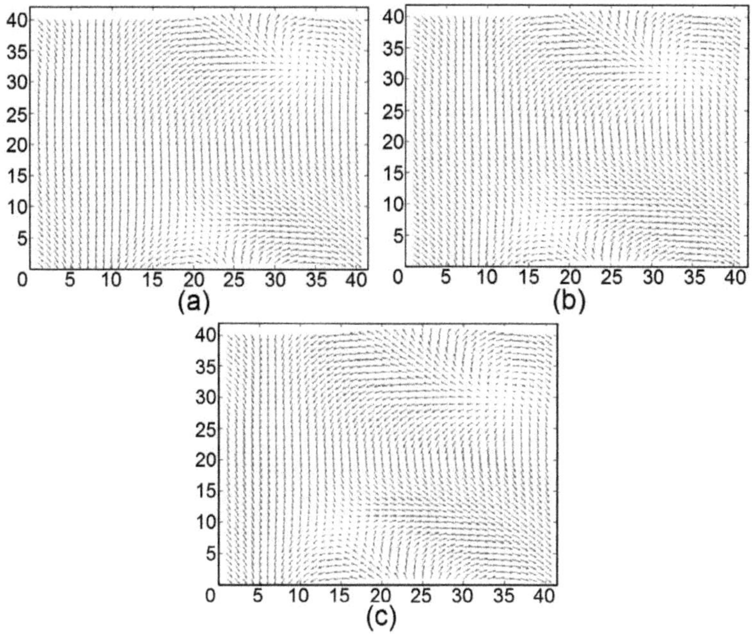

Figure 15.3 GS spin configuration for weak couplings: $J^{mf} = -0.1$ (a), -0.125 (b), -0.15 (c), with $H = 0$.

(ii) The periodicity of the diagonal parallel lines depends on the value of θ (comparing Fig. 15.4a and Fig. 15.4b). With a large size of N, the periodic conditions have no significant effects.

15.2.2.2 Ground state in applied magnetic field

We apply a magnetic field perpendicular to the xy plane. As we know, in systems where some spin orientations are incompatible with the field such as in antiferromagnets, the down spins cannot be turned into the field direction without losing its interaction energy with the up spins. To preserve this interaction, the spins turn into the direction almost perpendicular to the field while staying almost parallel with each other. This phenomenon is called "spin flop" [87]. In more complicated systems such as helimagnets in a field, more complicated reaction of spins to the field was observed,

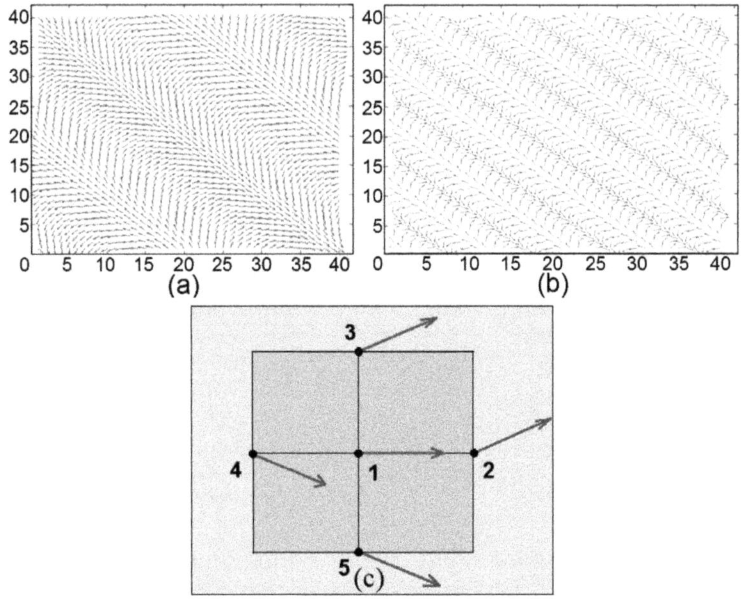

Figure 15.4 GS spin configurations for $J^{mf} = -0.45$ (a), -1.2 (b), with $H = 0$. Angles between NN are schematically zoomed (c). See text for comments.

leading to striking phenomena such as partial phase transition in thin helimagnetic films [105]. In the present system, there is a competition between the applied field which wants to align the spins along the z direction, and the DM interaction which wants the spins to be perpendicular which each other in the x plane. As a consequence, spins find a compromise which is the structure of skyrmions as shown below.

Figure 15.5a shows the ground state configuration for $J^{mf} = -1.1$ for first (surface) magnetic layer, with external magnetic layer $H = 0.1$. Figure 15.5b shows the 3D view. We can observe the beginning of the birth of skyrmions at the interface and in the interior magnetic layer.

Figure 15.6a shows the ground state configuration for $J^{mf} = -1.1$ for first (surface) magnetic layer, with external magnetic layer $H = 0.2$. Figure 15.6b shows the 3D view. We can observe the skyrmions for the surface and interior magnetic layer.

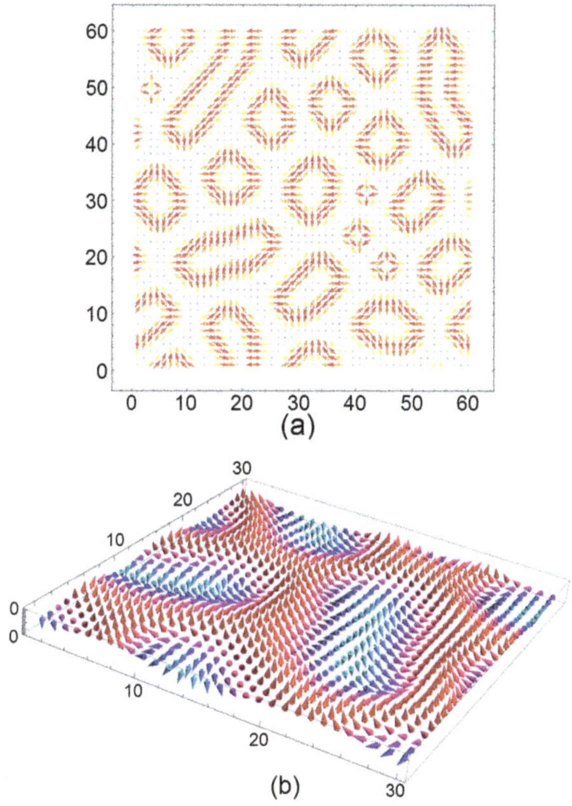

Figure 15.5 GS configuration of the surface magnetic layer for (a) $J^{mf} = -1.1$ and $H = 0.1$, (b) 3D view of the surface GS configuration.

Note that the skyrmions are found here in a range of sufficiently strong interface coupling and the applied field. The skyrmions are distributed in 3D space (not on a plane) in the magnetic layer. Figure 15.6 shows a cut in xy plane so that the projected sizes are not uniform. We have made a single magnetic layer: In that case, skyrmions are uniform on a 2D sheet (not shown). We note that the skyrmion and anti-skyrmion textures are not degenerate due to the DM asymmetry [see Eq. (15.10)]: Choosing the direction of **P** will fix the skyrmion turning direction, i.e., $[\mathbf{S}_i \times \mathbf{S}_j]$. Changing **P** will change skyrmions into antiskyrmions or vice versa.

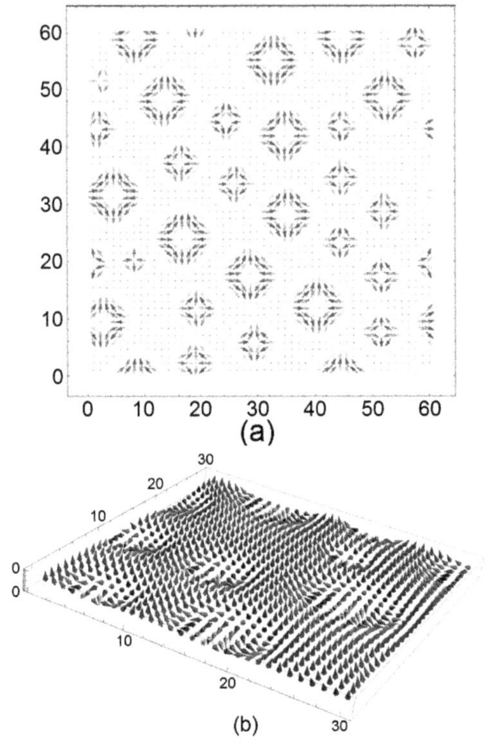

Figure 15.6 (a) GS configuration for the surface magnetic layer for $J^{mf} = -1.1$ and $H = 0.2$, (b) 3D view.

Figure 15.7 shows the GS configuration of the interface magnetic layer (top) for $J^{mf} = -1.1$, with external magnetic layer $H = 0.33$. The bottom figure shows the configurations of the second (interior) magnetic layer. We can observe skyrmions on both the interface and the interior magnetic layers.

Figure 15.8 shows the 3D view of the GS configuration for $J^{mf} = -1.1$, with $H = 0.33$ for the first (interface) magnetic layer and the second (interior) magnetic layer. We can observe skyrmions very pronounced for the surface layer but less contrast for the interior magnetic layer. For fields stronger than $H = 0.33$, skyrmions disappear in interior layers. At strong fields, all spins are parallel to the field, thus no skyrmions anywhere.

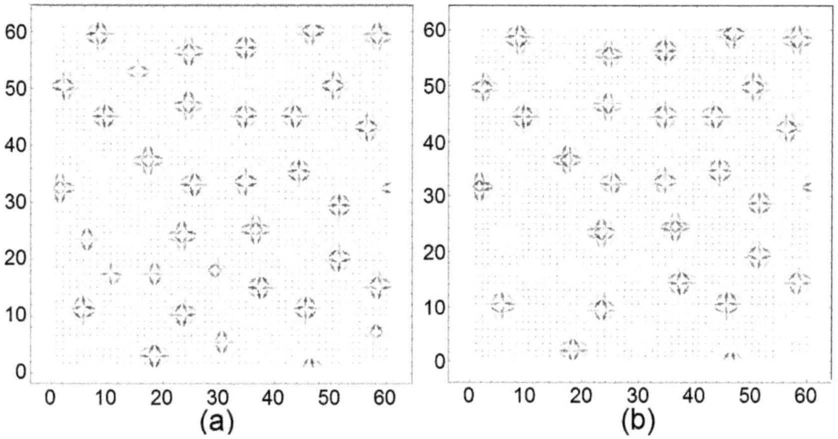

Figure 15.7 (a) GS configuration for the interface magnetic layer for $J^{mf} = -1.1$ and $H = 0.33$, (b) GS configurations for the second and third magnetic layers (they are identical). See text for comments.

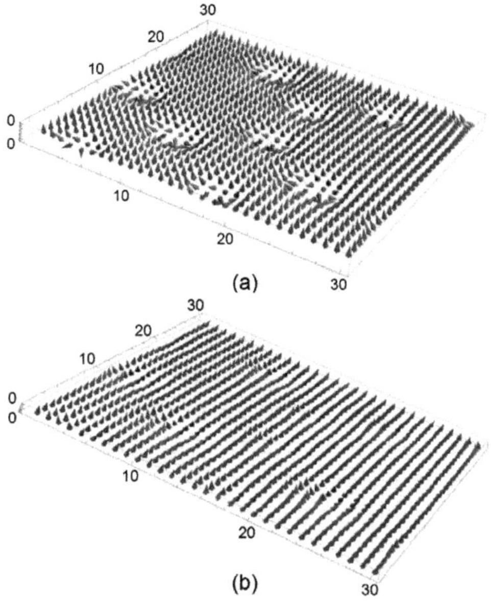

Figure 15.8 (a) 3D view of the GS configuration of the interface, (b) 3D view of the GS configuration of the second and third magnetic layers, for J^{mf} and $H = 0.33$.

15.3 Skyrmion Phase Transition: Monte Carlo Results

We use the Metropolis algorithm [48, 199] to calculate physical quantities of the system at finite temperatures T. As said above, we use mostly the size $N \times N \times L$ with $N = 40$ and thickness $L = L_m + L_f = 8$ (4 magnetic layers, 4 ferroelectric layers). Simulation times are 10^5 Monte Carlo steps (MCS) per spin for equilibrating the system and 10^5 MCS/spin for averaging. We calculate the internal energy and the layer order parameters of the magnetic (M_m) and ferroelectric (M_f) films.

The order parameter $M_f(n)$ of layer n is defined as

$$M_f(n) = \frac{1}{N^2} \langle | \sum_{i \in n} P_i^z | \rangle \tag{15.14}$$

where $\langle ... \rangle$ denotes the time average.

The definition of an order parameter for a skyrmion crystal is not obvious. Taking advantage of the fact that we know the GS, we define the order parameter as the projection of an actual spin configuration at a given T on its GS and we take the time average. This order parameter of layer n is thus defined as

$$M_m(n) = \frac{1}{N^2(t_a - t_0)} \sum_{i \in n} | \sum_{t=t_0}^{t_a} \mathbf{S}_i(T, t) \cdot \mathbf{S}_i^0(T = 0) | \tag{15.15}$$

where $\mathbf{S}_i(T, t)$ is the i-th spin at the time t, at temperature T, and $\mathbf{S}_i(T = 0)$ is its state in the GS. The order parameter $M_m(n)$ is close to 1 at very low T where each spin is only weakly deviated from its state in the GS. $M_m(n)$ is zero when every spin strongly fluctuates in the paramagnetic state. The above definition of $M_m(n)$ is similar to the Edward–Anderson (EA) order parameter used to measure the degree of freezing in spin glasses [234]. The EA order parameter, by definition, is calculated as follows. We follow each spin during the time. If it is frozen, then its time average is not zero. If it strongly fluctuates with time evolution, then its time average is zero. To calculate the overall degree of freezing, it suffices to add the square of each spin's time average. In doing so, we see that the EA order parameter does not express the nature of ordering, but only the degree of freezing.

In general, when the GS has several degenerate configurations such as the all-up and the all-down spin configurations in a ferromagnetic system of Ising spins, the system chooses one of the two when T tends to 0. The coexistence of several phases is not tolerated in such a case because the resulting energy is higher than that of a pure one (due to walls). However, in frustrated systems where one can construct a ground state by random stacking of frustrated units, one does not have a long-range ordering. Here, we wish to follow the evolution of the system ordering from $T = 0$, so we have to compare the configuration at temperature T at the time t with the GS we have selected to do the slow heating. That was what we did: We compare the actual configuration obtained by slowly heating the selected GS by projecting it on the selected GS, see Eq. (15.15). There are several possibilities: (i) If the spin structure is not stable when $T \neq 0$, $M_m(n)$ goes to zero with time, this is the case of the Kosterlitz–Thouless (XY spins in 2D with NN interaction); (ii) if the spin structure is frozen or ordered (SG, ferromagnets, ...), $M_m(n)$ is not zero at low T. In our case of skyrmion structure, we have observed the second case, namely the GS is stable up to a finite T.

If the system makes a global rotation during the simulation, then $\sum_{t=t_0}^{t_a} \mathbf{S}_i(T,t) \cdot \mathbf{S}_i^0(T=0) = 0$ for a long-time average. But the length of this run-time depends on the nature of ordering and the size of the system used in simulations. For large disordered systems such as SG and complicated non-collinear extended skyrmion structures, the global rotation may be forbidden or the time to realize it is out of reach in MC simulations. To see if a global rotation is realized or not, we have to make a finite-time scaling to deduce properties at the infinite time. This is very similar in spirit with the finite-size scaling used to deduce properties at the infinite crystal size. We have previously performed a finite-time scaling for the 2D skyrmion crystal [103]. In that work, we have used the same order parameter as Eq. (15.15). We have seen that skyrmions need much more than 10^6 MC steps per spin to relax to equilibrium. The order parameter follows a stretched exponential law as in SG and stabilized at non-zero values for $T < T_c$ at the infinite time. If there is a global rotation, we would not have non-zero values of $M_m(n)$ for $T < T_c$ at the infinite time. We note that in the present work, as in Ref. [103],

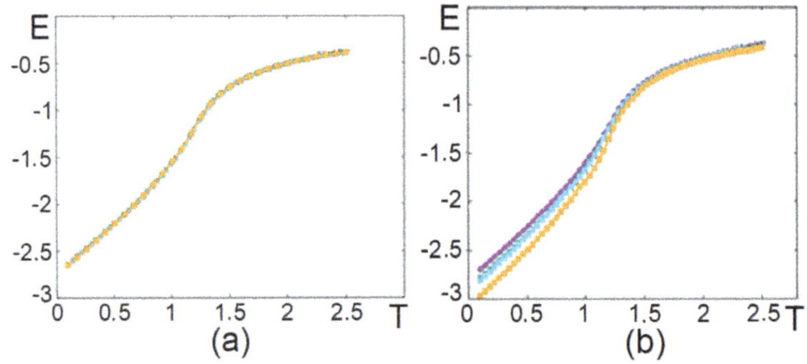

Figure 15.9 Energy of the magnetic film versus temperature T for (a) $J^{mf} = -0.1, J^{mf} = -0.125, J^{mf} = -0.15, J^{mf} = -0.2$ (all the lines are the same, see text for comments); (b) $J^{mf} = -0.45$ (purple line), $J^{mf} = -0.75$ (green line), $J^{mf} = -0.85$ (blue line) and $J^{mf} = -1.2$ (gold line), without an external magnetic field.

we have made a very slow heating of a selected GS and we did not observe a global rotation.

Note that the counting of topological charges around each skyrmion is numerically possible. In that case, the charge number evolves with T and goes to zero at the phase transition. The procedure is equivalent to projecting the skyrmion spin texture on its GS. We have chosen the projection one.

The total order parameters M_m and M_f are the sum of the layer order parameters, namely $M_m = \sum_n M_m(n)$ and $M_f = \sum_n M_f(n)$.

In Fig. 15.9, we show the dependence of energy of the magnetic film versus temperature, without an external magnetic field, for various values of the interface magnetoelectric interaction: In Fig. 15.9a for weak values $J^{mf} = -0.1$, $J^{mf} = -0.125$, $J^{mf} = -0.15$, $J^{mf} = -0.2$, and in Fig. 15.9b for stronger values $J^{mf} = -0.45, J^{mf} = -0.75, J^{mf} = -0.85, J^{mf} = -1.2$.

As said in the GS determination, when J^{mf} is weak, the GS is composed with large ferromagnetic domains at the interface (see Fig. 15.3). Interior layers are still ferromagnetic. The energy, therefore, does not vary with weak values of J^{mf} as seen in Fig. 15.9a. The phase transition occurs at the curvature change, namely maximum of the derivative or maximum of the specific heat,

$T_c^m \simeq 1.25$. Note that the energy at $T = 0$ is equal to -2.75 by extrapolating the curves in Fig. 15.9a to $T = 0$. This value is just the sum of energies of the spins across the layers: 2 interior spins with 6 NN, 2 interface spins with 5 NN. The energy per spin is thus (in ferromagnetic state): $E = -(2 \times 6 + 2 \times 5)/(4 \times 2) = -2.75$ (the factor 2 in the denominator is to remove the bond double counting in a crystal).

For stronger values of J^{mf}, the curves shown in Fig. 15.9b indicate a deviation of the ferromagnetic state due to the non-collinear interface structure. Nevertheless, we observe the magnetic transition at almost the same temperature, namely $T_c^m \simeq 1.25$. It means that spins in interior layers dominate the ordering.

We show in Fig. 15.10 the total order parameters of the magnetic film M_m and the ferroelectric film M_f versus T for various values of the parameter of the magnetoelectric interaction $J^{mf} = -0.1$, -0.125, -0.15, -0.2 and for $J^{mf} = -0.45$, -0.75, -0.85, -1.2, without an external magnetic field. Several remarks are in order:

(i) For the magnetic film, M_m shows strong fluctuations but we still see that all curves fall to zero at $T_c^m \simeq 1.25$. These fluctuations come from non-uniform spin configurations and also from the nature of the Heisenberg spins in low dimensions [231].

(ii) For the ferroelectric film, M_f behaves very well with no fluctuations. This is due to the Ising nature of electric polarizations supposed in the present model. The ferroelectric film undergoes a phase transition at $T_c^f \simeq 1.50$.

(iii) There are thus two transitions, one magnetic and one ferroelectric, separately. The magnetic transition occurs at a lower temperature. We know that in bulk crystals the transition temperature is approximately proportional to $1/n$ where n is the component number of the spin: $n = 3$ for the Heisenberg spin, $n = 1$ for the Ising spin [10, 380]. The fact that the ferroelectric transition occurs at a higher temperature observed in Fig. 15.10b is understood. The weak coupling with the magnetic film makes the two transition temperatures separately. Between T_c^m and T_c^f the superlattice is partially disordered: The magnetic part is disordered while the ferroelectric part is ordered. The partial disorder has been

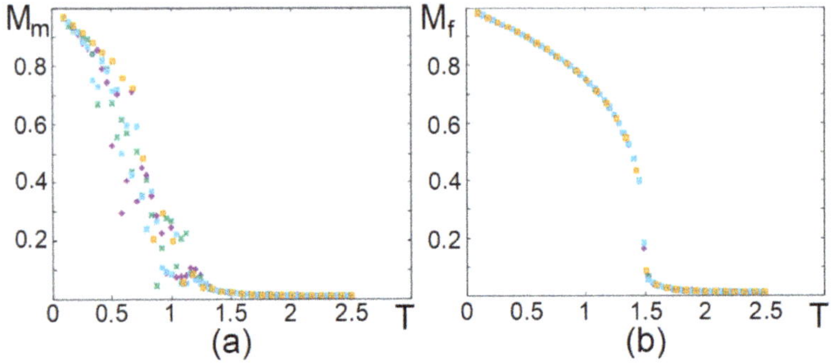

Figure 15.10 (a) Order parameter of the magnetic film M_m versus T; (b) Order parameter of the ferroelectric film M_f versus T, for $J^{mf} = -0.1$ (purple dots), $J^{mf} = -0.125$ (green dots), $J^{mf} = -0.15$ (blue dots), $J^{mf} = -0.2$ (gold dots), without an external magnetic field.

observed in many systems, for example, the surface layer of a thin film can become disordered at a low temperature while the bulk is still ordered [248]. One can also mention the partial phase transition in helimagnets in a field [105].

We show in Fig. 15.11 the order parameters of the magnetic and ferroelectric films at strong values of J^{mf} as functions of T, in zero field. We observe that the stronger $|J^{mf}|$ is, the lower T_c^m becomes. This is understood by the discussion given below Eq. 15.13 for a monolayer: The stronger $|J^{mf}|$ makes the larger angle θ. In the case of many magnetic layers shown in Fig. 15.11, the larger angle causes a stronger competition with the collinear ferromagnetic interaction of the interior layers. This enhanced competition gives rise to the destruction of the ordering at a lower temperature.

We examine the field effects now. Figure 15.12 shows the order parameter and the energy of the magnetic film versus T, for various values of the external magnetic field. The interface magnetoelectric interaction is $J^{mf} = -1.2$. Depending on the magnetic field, the non-collinear spin configuration survives up to a temperature between 0.5 and 1 (for $H = 0$). After the transition, spins align themselves in the field direction, giving a large value of the order parameter (Fig. 15.12a). The energy shows a sharp curvature change only for

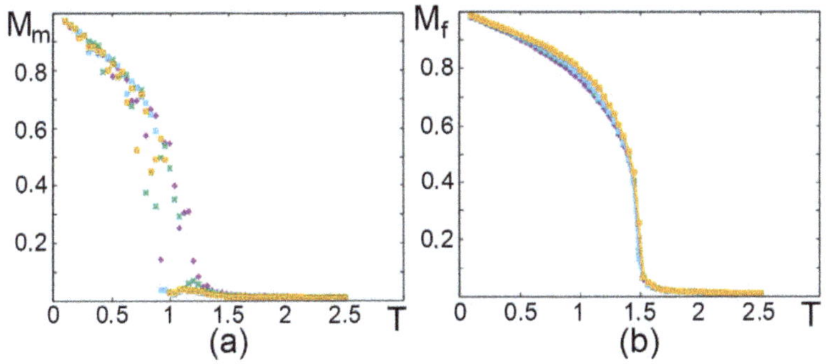

Figure 15.11 (a) Order parameter of the magnetic film versus T; (b) Order parameter of the ferroelectric film versus T for $J^{mf} = -0.45$ (purple dots), $J^{mf} = -0.75$ (green dots), $J^{mf} = -0.85$ (blue dots) and $J^{mf} = -1.2$ (gold dots), without an external magnetic field.

$H = 0$, meaning that the specific heat is sharp only for $H = 0$ and broadened more and more with increasing H.

We consider now the case of very strong interface couplings.

Figure 15.13a shows the magnetic order parameter versus T. The purple and green lines correspond to M for $J^{mf} = -2.5$ with $H^z = 1.0$ and $H^z = 1.5$, respectively; the blue and gold lines correspond to M for $J^{mf} = m - 6$ with $H^z = 0$ and $H^z = 1$. These curves indicate first-order phase transitions at $T_c^m = 1.05$ for ($J^{mf} = -2.5$, $H^z = 1$) (purple), at $T_c^m = 1.12$ for ($J^{mf} = -2.5$, $H^z = 1.5$) (green) and at $T_c^m = 1.25$ for ($J^{mf} = -6$, $H^z = 1$) (blue). In the case of zero field, namely ($J^{mf} = -6$, $H^z = 0$) (gold), one has a first-order phase transitions occurring at $T_c = 2.30$.

Let us discuss the nature of the transition in shown in Fig. 15.13a. When $H \neq 0$, the first transition at low temperature ($T \simeq 1.05 - 1.25$) is due to the destruction of the skyrmion structure. After this transition, the z spin components being not zero under an applied field come close to zero only at high T ($\simeq 2.3$). This is not a phase transition because the z components will never be zero in a field if J^{mf} is not so strong. When J^{mf} is very strong ($J^{mf} = -6$, blue data points), the DM interaction is so strong that the spins will lie in the xy plane in spite of H: We see that the z spin components are zero after the loss of the ferroelectric ordering at $T \simeq 2.3$. Note

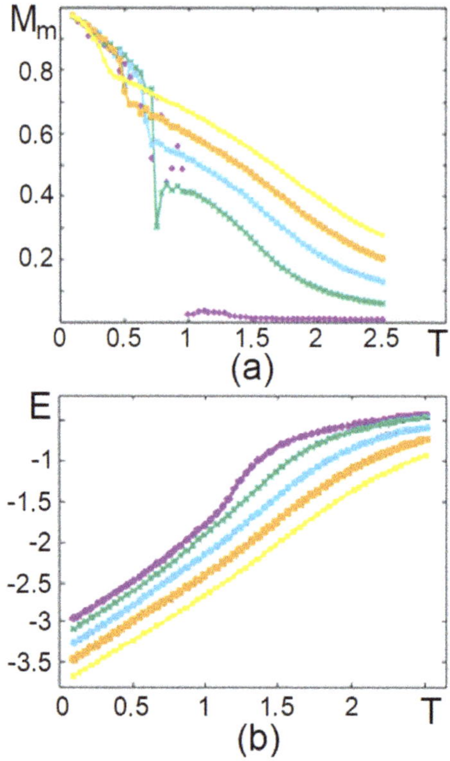

Figure 15.12 (a) Temperature dependence of (a) the magnetic order parameter; (b) the magnetic energy for $H = 0$ (purple dots), $H = 0.25$ (green line), $H = 0.5$ (blue line), $H = 0.75$ (gold line), $H = 1$ (yellow line). The interface magnetoelectric interaction is $J^{mf} = -1.2$.

that when $H = 0$ (gold data points), there is no skyrmion, the spin configuration is chiral (helical) as shown in Section 15.3. The single transition to the paramagnetic phase occurs at $T \simeq 2.3$ where the chiral ordering and the ferroelectric ordering are lost at the same time (see Fig. 15.13b).

Figure 15.13b shows the magnetic (purple) and ferroelectric (green) energies versus T for ($J^{mf} = -6$, $H^z = 0$). One sees the discontinuities of these curves at $T_c \simeq 2.3$, indicating the first-order transitions for both magnetic and ferroelectric at the same temperature. In fact, with such a strong J^{mf} the transitions in both

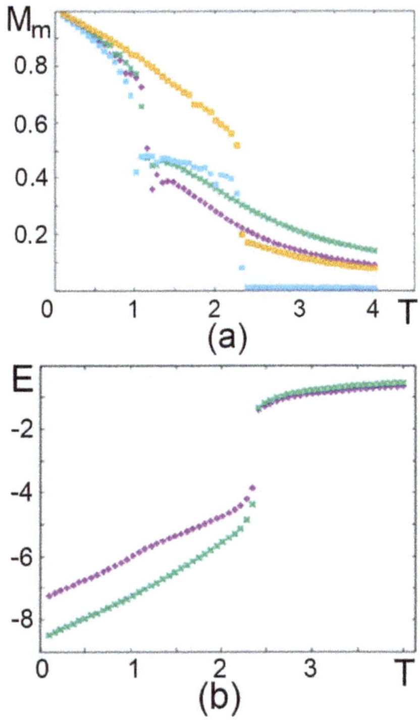

Figure 15.13 (a) Order parameter of magnetic film versus T. The purple and green dots correspond to M for ($J^{mf} = -2.5$, $H^z = 1$) and ($J^{mf} = -2.5$, $H^z = 1.5$), blue and gold dots correspond to M for ($J^{mf} = -6$, $H^z = 1$) and ($J^{mf} = -6$, $H^z = 0$). (b) Energies of magnetic (purple dots) and ferroelectric (green dots) subsystems versus T for ($J^{mf} = -6$, $H = 0$).

magnetic and ferroelectric films are driven by the interface, this explains the same T_c for both. The first-order transition observed here can be understood because the present system is a frustrated system due to the competing interactions. So far all frustrated non-collinear spin systems have been found possessing a first-order transition (see Ref. [85] and references in Ref. [251]).

Let us show the effect of an applied electric field. For the ferroelectric film, polarizations are along the z axis so that an applied electric field **E** along this direction will remove the phase transition: The order parameter never vanishes when $E \neq 0$, similar to the case of a ferromagnet in an applied magnetic field. This is seen in

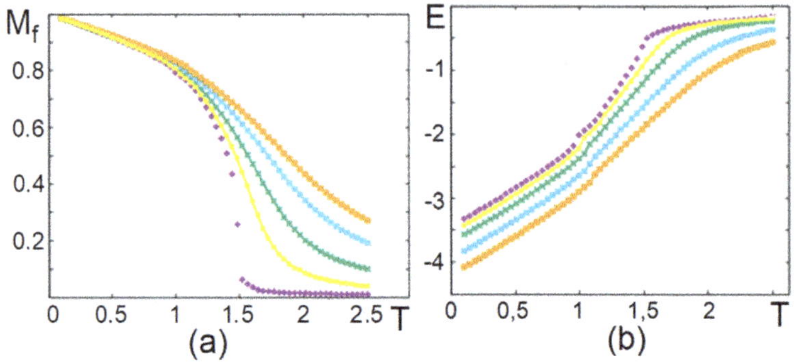

Figure 15.14 (a) Order parameter and (b) energy of ferroelectric film, versus temperature for $E = 0$ (purple dots), $E = 0.25$ (green line), $E = 0.5$ (blue line), $E = 0.75$ (gold line), $E = 1$ (yellow line). The interface magnetoelectric interaction is $J^{mf} = -1.2$

Fig. 15.14. Note that the energy has a sharp change of curvature for $E = 0$ indicating a transition, other energy curves with $E \neq 0$ do not show a transition. One notices some anomalies at $T \sim 1 - 1.1$ which are due to the effect of the magnetic transition in this temperature range.

15.4 Spin Waves in Zero Field

We have shown in the previous section Monte Carlo results for the phase transition in our superlattice model. Here let us show theoretically spin waves (SW) excited in the magnetic film in zero field, in some simple cases. The method we employ is the Green's function technique for non-collinear spin configurations which has been shown to be efficient for studying low-T properties of quantum spin systems such as helimagnets [89] and systems with a DM interaction [104].

In this section, we consider the same Hamiltonian supposed in Eqs. (15.4)–(15.10) but with quantum spins of amplitude 1/2.

As seen in the previous section, the spins lie in the xy planes, each on its quantization local axis lying in the xy plane (quantization axis being the ζ axis, see Fig. 15.15).

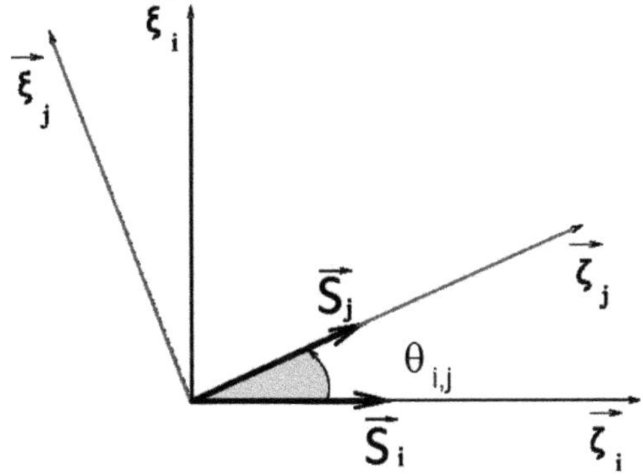

Figure 15.15 The spin quantization axes of $\mathbf{S}_i$ and $\mathbf{S}_j$ are $\hat{\zeta}_i$ and $\hat{\zeta}_j$, respectively, in the xy plane.

Expressing the spins in the local coordinates, one has

$$\mathbf{S}_i = S_i^{\xi_i}\hat{\xi}_i + S_i^{\eta_i}\hat{\eta}_i + S_i^{\zeta_i}\hat{\zeta}_i \tag{15.16}$$

$$\mathbf{S}_j = S_j^{\xi_j}\hat{\xi}_j + S_j^{\eta_j}\hat{\eta}_j + S_j^{\zeta_j}\hat{\zeta}_j \tag{15.17}$$

where the i and j coordinates are connected by the rotation

$$\hat{\xi}_j = \cos\theta_{ij}\hat{\zeta}_i + \sin\theta_{ij}\hat{\xi}_i$$

$$\hat{\zeta}_j = -\sin\theta_{ij}\hat{\zeta}_i + \cos\theta_{ij}\hat{\xi}_i$$

$$\hat{\eta}_j = \hat{\eta}_i$$

where $\theta_{ij} = \theta_i - \theta_j$ being the angle between $\mathbf{S}_i$ and $\mathbf{S}_j$.

As we have seen above, the GS spin configuration for one monolayer is periodically non-collinear. For two-layer magnetic film, the spin configurations in two layers are identical by symmetry. However, for thickness larger than 2, the interior layer have angles different from that on the interface layer. It is not our purpose to treat that case though it is possible to do so using the method described in Ref. [104]. We rather concentrate ourselves in the case of a monolayer in this section.

In the following, we consider the case of spin one-half $S = 1/2$. Expressing the total magnetic Hamiltonian $\mathcal{H}_M = \mathcal{H}_m + \mathcal{H}_{mf}$ in the local coordinates [104]. Writing $\mathbf{S}_j$ in the coordinates $(\hat{\xi}_i, \hat{\eta}_i, \hat{\zeta}_i)$, one gets the following exchange Hamiltonian from Eqs. (15.4)–(15.10):

$$\mathcal{H}_M = -\sum_{<i,j>} J^m \Bigg\{ \frac{1}{4}\left(\cos\theta_{i,j} - 1\right)\left(S_i^+ S_j^+ + S_i^- S_j^-\right)$$
$$+\frac{1}{4}\left(\cos\theta_{i,j} + 1\right)\left(S_i^+ S_j^- + S_i^- S_j^+\right)$$
$$+\frac{1}{2}\sin\theta_{i,j}\left(S_i^+ + S_i^-\right) S_j^z - \frac{1}{2}\sin\theta_{i,j} S_i^z \left(S_j^+ + S_j^-\right)$$
$$+\cos\theta_{i,j} S_i^z S_j^z \Bigg\}$$
$$+\frac{D}{4}\sum_{\langle i,j\rangle}[(S_i^+ + S_i^-)(S_j^+ + S_j^-)|\sin\theta_{i,j}| + 4S_i^z S_j^z |\sin\theta_{i,j}|] \tag{15.18}$$

where $D = J^{mf} P^z$. Note that $P^z = 1$ in the GS. At finite T we replace P^z by $< P^z >$. In the above equation, we have used standard notations of spin operators for easier recognition when using the commutation relations in the course of calculation, namely

$$(S_i^{\xi_i}, S_i^{\eta_i}, S_i^{\zeta_i}) \rightarrow (S_i^x, S_i^y, S_i^z)$$
$$(S_j^{\xi_j}, S_j^{\eta_j}, S_j^{\zeta_j}) \rightarrow (S_j^x, S_j^y, S_j^z) \tag{15.19}$$

where we understand that S_i^x is in fact $S_i^{x_i}$ and so on.

Note that the sinus terms of $\mathcal{H}_m$, the 3rd line of Eq. (15.18), are zero when summed up on opposite NN unlike the sinus term of the DM Hamiltonian H_{mf}, Eq. (15.10) which remains thanks to the choice of the DM vectors for opposite directions in Eq. [104].

15.4.1 Monolayer

In two dimensions (2D) there is no long-range order at finite temperature (T) for isotropic spin models with short-range interaction [231]. Therefore, to stabilize the ordering at finite T, it is useful to add an anisotropic interaction to stabilize the magnetic long-range ordering when the wave vector vanishes (see Fig. 15.16a, for instance). It is known that in 2D or in very thin films, the integrands to calculate $< M >$ at a finite T [see Eq. (15.36)] diverges as kdk/k^2 (since $E => k^2$ for the ferromagnetic mode when $k => 0$) while in 3D this is k^2dk/k^2 as $k => 0$ (no problem of divergence). In MC simulations shown in the previous section, the statistical average was done using stochastic random configurations generated by statistical probability (no possible mode of $k = 0$). So, we do not encounter such a mathematical divergence as in the SW calculation.

We use the following anisotropy between $\mathbf{S}_i$ and $\mathbf{S}_j$ which stabilizes the angle determined above between their local quantization axes S_i^z and S_j^z:

$$\mathcal{H}_a = -\sum_{<i,j>} K_{i,j} S_i^z S_j^z \cos\theta_{i,j} \tag{15.20}$$

where $K_{i,j}$ is supposed to be positive, small compared to J^m, and limited to NN. Hereafter we take $K_{i,j} = K$ for NN pair in the xy plane, for simplicity. The total magnetic Hamiltonian $\mathcal{H}_M$ is finally given by (using operator notations)

$$\mathcal{H}_M = \mathcal{H}_m + \mathcal{H}_{mf} + \mathcal{H}_a \tag{15.21}$$

We now define the following two double-time Green's functions in the real space:

$$\begin{aligned} G_{i,j}(t,t') &= \ll S_i^+(t); S_j^-(t') \gg \\ &= -i\theta(t-t') < \left[S_i^+(t), S_j^-(t')\right] > \qquad (15.22) \\ F_{i,j}(t,t') &= \ll S_i^-(t); S_j^-(t') \gg \\ &= -i\theta(t-t') < \left[S_i^-(t), S_j^-(t')\right] > \qquad (15.23) \end{aligned}$$

The equations of motion of these functions read

$$\begin{aligned} i\hbar\frac{dG_{i,j}(t,t')}{dt} &= < \left[S_i^+(t), S_j^-(t')\right] > \delta(t-t') \\ &\quad - \ll \left[\mathcal{H}_M, S_i^+\right]; S_j^- \gg \qquad (15.24) \\ i\hbar\frac{dF_{i,j}(t,t')}{dt} &= < \left[S_i^-(t), S_j^-(t')\right] > \delta(t-t') \\ &\quad - \ll \left[\mathcal{H}_M, S_i^-\right]; S_j^- \gg \qquad (15.25) \end{aligned}$$

For the $\mathcal{H}_m$ and $\mathcal{H}_a$ parts, the above equations of motion generate terms such as $\ll S_l^z S_i^{\pm}; S_j^- \gg$ and $\ll S_l^{\pm} S_i^{\pm}; S_j^- \gg$. These functions can be approximated by using the Tyablikov decoupling to reduce to the above-defined G and F functions:

$$\ll S_l^z S_i^{\pm}; S_j^- \gg \simeq < S_l^z > \ll S_i^{\pm}; S_j^- \gg \quad (15.26)$$

$$\ll S_l^{\pm} S_i^{\pm}; S_j^- \gg \simeq < S_l^{\pm} > \ll S_i^{\pm}; S_j^- \gg \simeq 0 \quad (15.27)$$

The last expression is due to the fact that transverse spin wave motions $< \; S_l^{\pm} \; >$ are zero with time. For the DM term, the commutation relations $[\mathcal{H}, S_i^{\pm}]$ give rise to the following term:

$$D \sum_l \sin\theta [\mp S_i^z (S_l^+ + S_l^-) \pm 2 S_i^{\pm} S_l^z] \quad (15.28)$$

This leads to the following type of Green's function:

$$\ll S_i^z S_l^{\pm}; S_j^- \gg \simeq < S_i^z > \ll S_l^{\pm}; S_j^- \gg \quad (15.29)$$

Note that we have used defined θ positively. The above equation is thus related to G and F functions [see Eq. (13.24)].

We use the following Fourier transforms in the xy plane of the G and F Green's functions:

$$G_{i,j}(t, t', \omega) = \frac{1}{\Delta} \int_{BZ} d\mathbf{k}_{xy} e^{-i\hbar\omega(t-t')} g(\omega, \mathbf{k}_{xz}) e^{i\mathbf{k}_{xy}\cdot(\mathbf{R}_i - \mathbf{R}_j)}$$

$$F_{i,j}(t, t', \omega) = \frac{1}{\Delta} \int_{BZ} d\mathbf{k}_{xy} e^{-i\hbar\omega(t-t')} f(\omega, \mathbf{k}_{xy}) e^{i\mathbf{k}_{xy}\cdot(\mathbf{R}_i - \mathbf{R}_j)}$$

where the integral is performed in the first xy Brillouin zone (BZ) of surface Δ and ω is the SW frequency. Let us define the SW energy as $E = \hbar\omega$ in the following.

For a monolayer, we have after the Fourier transforms

$$(E + A)g + Bf = 2 < S^z >$$
$$-Bg + (E - A)f = 0 \quad (15.30)$$

where A and B are

$$A = -J^m[8 < S^z > \cos\theta(1 + d) - 4 < S^z > \gamma(\cos\theta + 1)]$$
$$-4D\sin\theta < S^z > \gamma + 8D\sin\theta < S^z > \quad (15.31)$$

$$B = 4J^m < S^z > \gamma(\cos\theta - 1) - 4D\sin\theta < S^z > \gamma \quad (15.32)$$

where the reduced anisotropy is $d = K/J^m$ and $\gamma = (\cos k_x a + \cos k_y a)/2$, k_x and k_y being the wave vector components in the xy planes, a the lattice constant.

The SW energies are determined by the secular equation

$$(E + A)(E - A) + B^2 = 0$$
$$[E + A][E - A] + B^2 = 0$$
$$E^2 - A^2 + B^2 = 0$$
$$E = \pm\sqrt{(A + B)(A - B)} \qquad (15.33)$$

where $\pm$ indicate the left and right SW precessions. We see that

(1) if $\theta = 0$, we have B and the last two terms of A are zero. We recover then the ferromagnetic SW dispersion relation

$$E = 2ZJ^m < S^z > (1 - \gamma) \qquad (15.34)$$

where $Z = 4$ is the coordination number of the square lattice (taking $d = 0$),

(2) if $\theta = \pi$, we have $A = 8J^m < S^z >$ and $B = -8J^m < S^z > \gamma$. We recover then the antiferromagnetic SW energy

$$E = 2ZJ^m < S^z > \sqrt{1 - \gamma^2} \qquad (15.35)$$

(3) in the presence of a DM interaction, we have $0 < \cos\theta < 1$ $(0 < \theta < \pi/2)$. If $d = 0$, the quantity in the square root of Eq. (15.33) is always ≥ 0 for any θ. It is zero at $\gamma = 1$. We do not need an anisotropy d to stabilize the SW at $T = 0$. If $d \neq 0$ then it gives a gap at $\gamma = 1$.

We show in Fig. 15.16 the SW energy calculated from Eq. (15.33) for $\theta = 0.3$ radian ($\simeq 17.2°$) and 1 radian ($\simeq 57.30°$). The spectrum is symmetric for positive and negative wave vectors and for left and right precessions. Note that for small values of θ (i.e., small D) E is proportional to k^2 at low k (cf. Fig. 15.16a), as in ferromagnets. However, for strong θ, E is proportional to k as seen in Fig. 15.16b. This behavior is similar to that in antiferromagnets [87]. The change of behavior is progressive with increasing θ, no sudden transition from k^2 to k behavior is observed.

In the case of $S = 1/2$, the magnetization is given by (see technical details in chapter 8):

$$\langle S^z \rangle = \frac{1}{2} - \frac{1}{\Delta} \iint dk_x dk_y \left[\frac{1}{e^{E_i/k_B T} - 1} + \frac{1}{e^{-E_i/k_B T} - 1} \right] \qquad (15.36)$$

where for each $\mathbf{k}$ one has $\pm E_i$ values.

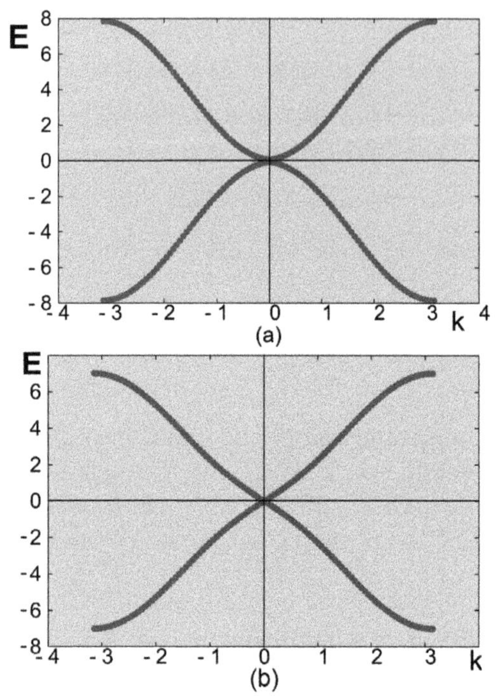

Figure 15.16 Spin wave energy $E(k)$ versus k ($k \equiv k_x = k_z$) for (a) $\theta = 0.3$ radian and (b) $\theta = 1$ in 2D at $T = 0$. See text for comments.

Since E_i depends on S^z, the magnetization can be calculated at finite temperatures self-consistently using the above formula.

It is noted that the anisotropy d avoids the logarithmic divergence at $k = 0$ so that we can observe a long-range ordering at finite T in 2D. We show in Fig. 15.17 the magnetization M ($\equiv < S^z >$) calculated by Eq. (15.36) for using $d = 0.001$. It is interesting to observe that M depends strongly on θ: At high T, larger θ yields stronger M. However, at $T = 0$ the spin length is smaller for larger θ due to the so-called spin contraction in antiferromagnets [see Eq. (8.61), Section 8.5]. As a consequence, there is a crossover of magnetizations with different θ at low T as shown in Fig. 15.17.

The spin length at $T = 0$ is shown in Fig. 15.18 for several θ.

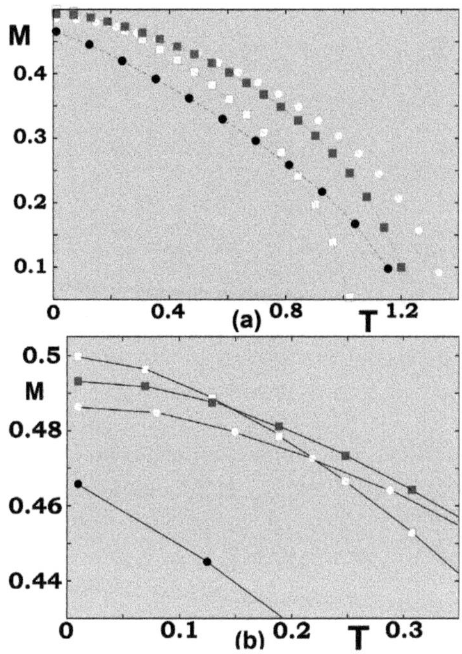

Figure 15.17 (a) Spin length $M =< S^z >$ versus temperature T for a 2D sheet with $\theta = 0.175$ (radian) (magenta void squares), $\theta = 0.524$ (blue filled squares), $\theta = 0.698$ (green void circles), $\theta = 1.047$ (black filled circles); (b) Zoom at low T to show magnetization crossovers.

15.4.2 Bilayer

We note that for magnetic bilayer between two ferroelectric films, the calculation similar to that of a monolayer can be done. By symmetry, spins between the two layers are parallel, the energy of a spin on a layer is

$$E_i = -4J^m S^2 \cos\theta - J^m S^2 + 4J^{mf} P^z S^2 \sin\theta \qquad (15.37)$$

where there are 4 in-plane NN and one parallel NN spin on the other layer. The interface coupling is with only one polarization instead of two (see Eq. (15.12)) for a monolayer for comparison.

The minimum energy corresponds to $\tan\theta = -J^{mf}/J^m$.

The calculation by Green's functions for a film with a thickness is straightforward: Writing Green's functions for each layer and making Fourier transforms in the xy planes, we obtain a system

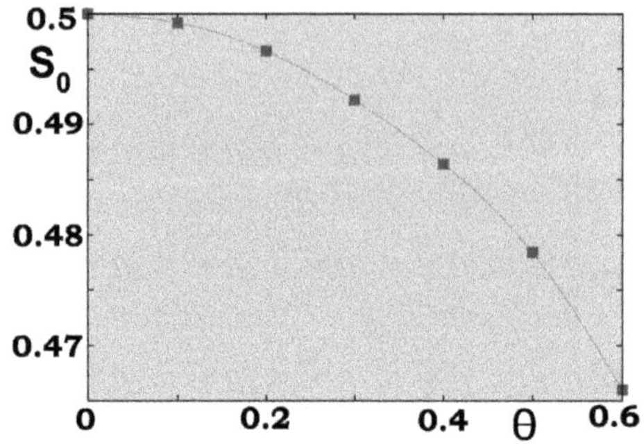

Figure 15.18 Spin length at temperature $T = 0$ for a monolayer versus θ (radian).

of coupled equations. For the details, the reader is referred to Ref. [89]. For a bilayer, the SW energy is the eigenvalues of the following matrix equation:

$$\mathbf{M}(E)\mathbf{h} = \mathbf{u}, \tag{15.38}$$

where

$$\mathbf{h} = \begin{pmatrix} g_{1,n'} \\ f_{1,n'} \\ g_{2,n'} \\ f_{2,n'} \end{pmatrix}, \qquad \mathbf{u} = \begin{pmatrix} 2\left\langle S_1^z \right\rangle \delta_{1,n'} \\ 0 \\ 2\left\langle S_2^z \right\rangle \delta_{2,n'} \\ 0 \end{pmatrix} \tag{15.39}$$

where $E = \hbar\omega$ and $\mathbf{M}(E)$ is given by

$$\begin{pmatrix} E + A_1 & B_1 & C_1 & 0 \\ -B_1 & E - A_1 & 0 & -C_1 \\ C_2 & 0 & E + A_2 & B_2 \\ 0 & -C_2 & -B_2 & E - A_2 \end{pmatrix} \tag{15.40}$$

with

$$A_1 = -J^m[8 < S_1^z > \cos\theta(1+d) - 4 < S_1^z > \gamma(\cos\theta + 1)] \\ -2J^m < S_2^z > -4D\sin\theta < S_1^z > \gamma + 8D\sin\theta < S_1^z > \tag{15.41}$$

$$A_2 = -J^m[8 < S_2^z > \cos\theta(1+d) - 4 < S_2^z > \gamma(\cos\theta + 1)] \\ -2J^m < S_1^z > -4D\sin\theta < S_2^z > +8D\sin\theta < S_2^z > \tag{15.42}$$

$$B_n = 4J^m < S_n^z > \gamma(\cos\theta - 1) - 4D\sin\theta < S_n^z > \gamma, \quad n = 1, 2 \tag{15.43}$$

$$C_n = 2J^m < S_n^z >, \quad n = 1, 2 \tag{15.44}$$

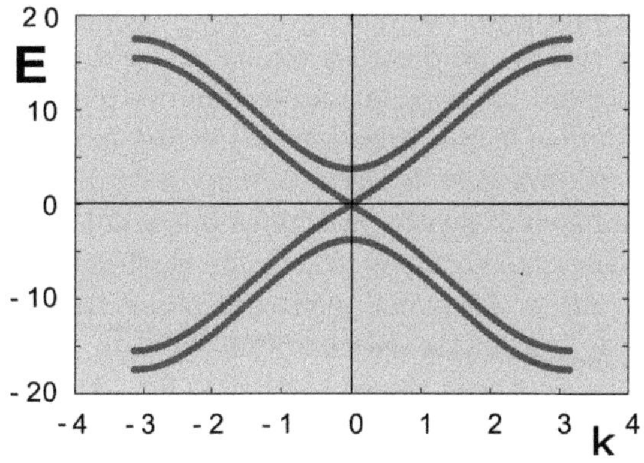

Figure 15.19 Spin wave energy E versus $k = k_x = k_y$ at $T = 0$ for a bilayer with $\theta = 0.6$ radian.

Note that by symmetry, one has $< S_1^z > = < S_2^z >$.

We show in Fig. 15.19 the SW spectrum of the bilayer case for a strong value $\theta = 0.6$ radian. There are two important points:

(i) The first mode has the $E \propto k$ antiferromagnetic behavior at the long wavelength limit for this strong θ.
(ii) The higher mode has $E \propto k^2$ which is the ferromagnetic wave due to the parallel NN spins in the z direction.

In conclusion of this section, we emphasize that the DM interaction affects strongly the SW mode at $k \to 0$. Quantum fluctuations in competition with thermal effects cause the crossover of magnetizations of different θ: In general, stronger θ yields stronger spin contraction at and near $T = 0$ so that the corresponding spin length is shorter. However, at higher T, stronger θ means stronger J^{mf} which yields stronger magnetization. It explains the crossover at moderate T.

15.5 Frustration Effect: $J_1 - J_2$ model

We consider in this section a superlattice composed of alternate "frustrated" magnetic films and "frustrated" ferroelectric films. The frustration due to competing interactions has been extensively investigated during the last four decades. The reader is referred to Ref. [85] for reviews on theories, simulations and experiments on various frustrated systems. In this section, we present the effect of the frustration in the presence of the DM interaction at the magnetoelectric interface. It turns out that the frustration gives rise to an enhancement of skyrmions created by the DM interaction in a field **H** applied perpendicularly to the films. Monte Carlo simulations are carried out to study the skyrmion phase transition in the superlattice as functions of the frustration strength.

The results have been shown in detail in Ref. [318]. We recall in the following some principal points.

15.5.1 Model

We use the same DM Hamiltonian for the interface coupling in Eqs. (15.10) and (15.11). For the magnetic and ferromagnetic Hamiltonians, H_m and H_f, we introduce the interactions between next-nearest neighbors (NNN) as follows:

$$H_m = -\sum_{i,j} J_{ij}^m \mathbf{S}_i \cdot \mathbf{S}_j - \sum_{i,k} J_{ik}^{2m} \mathbf{S}_i \cdot \mathbf{S}_k - \sum_i \mathbf{H} \cdot \mathbf{S}_i \qquad (15.45)$$

We shall take into account both the nearest neighbors (NN) interaction, denoted by J^m, and the NNN interaction denoted by J^{2m}. We consider $J^m > 0$ to be the same everywhere in the magnetic film. To introduce the frustration we shall consider $J^{2m} < 0$, namely antiferromagnetic interaction, the same everywhere. The external magnetic field **H** is applied along the z-axis which is perpendicular to the plane of the layers.

For the ferroelectric film, the Hamiltonian is given by

$$H_f = -\sum_{i,j} J_{ij}^f \mathbf{P}_i \cdot \mathbf{P}_j - \sum_{i,k} J_{ik}^{2f} \mathbf{P}_i \cdot \mathbf{P}_k \qquad (15.46)$$

where J_{ij}^f the interaction parameter between polarizations at sites i and j. Similar to the magnetic subsystem we will take the same $J_{ij}^f = J^f > 0$ for all NN pairs, and $J_{ij} = J^{2f} < 0$ for NNN pairs.

Let us emphasize that the bulk $J_1 - J_2$ magnetic model on the simple cubic lattice has been studied with Heisenberg spins [274] and the Ising model [151] where J_1 and J_2 are both antiferromagnetic (<0). The critical value $|J_2^c| = 0.25|J_1|$ above

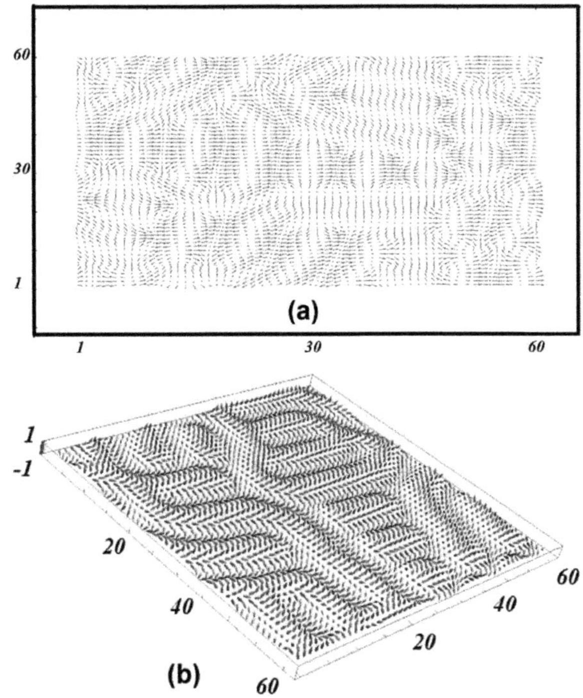

Figure 15.20 (a) 2D view of the GS configuration of the interface for $H = 0$ with $J^m = J^f = 1, J^{2m} = J^{2f} = -0.3, J^{mf} = -1.25$, (b) 3D view.

which the bipartite antiferromagnetic ordering changes into a frustrated ordering where a line is with spins up, and its neighboring lines are with spins down. In the case of $J^m > 0$ (ferro), and $J^{2m} < 0$ (antiferro), it is easy to show that the critical value where the ferromagnetic becomes antiferromagnetic is $J_c^{2m} = -0.5J^m$. Below this value, the antiferromagnetic ordering replaces the ferromagnetic ordering.

For the interface coupling we use the same notation as in the preceding sections, namely J^{mf}, independent of (i, j).

As before, since $\mathbf{P_k}$ is in the z direction, the DM vector is in the z direction. Thus, in the absence of an applied field the spins in the magnetic layers will lie in the xy plane to minimize the interface interaction energy.

15.5.2 Ground State

We note that in the case when the magnetic film has a frustration and a thickness, the angle between NN spins in each magnetic layer is different from that of the neighboring layers. The determination of the angles is analytically difficult. It is more convenient to use the numerical minimization method called "steepest descent method" to obtain the ground state (GS) spin configuration [248]. We use a sample size $N \times N \times L$. For most calculations, we select the lateral size $N \times N$ with $N = 60$ and we take $L_m = L_f = 4$ where L_m [L_f] is the thickness of magnetic (ferroelectric) layer. We use the periodic boundary conditions in the xy plane. As before, we take exchange parameters between NN spins and NN polarizations equal to 1, namely $J^m = J^f = 1$, for the simulations. We investigate the effects of the interaction parameters (J^{2m}, J^{2f}) and J^{mf}. We note that the steepest descent method calculates the real GS down to the value $J^{mf} = -1.25$. For values lower than this, the DM interaction is so strong that the angle θ tends to $\pi/2$. All magnetic exchange terms (scalar products) will be close to zero, the minimum total energy thus corresponds just to the DM energy.

Now we consider a case with the frustrated regime with $(J^{2f}, J^{2m}) \in (-0.4, 0)$, namely above the critical value -0.5 as mentioned above. The spin configuration in the case where $H = 0$ is shown in Fig. 15.20 for the interface magnetic layer. We observe here a stripe phase with long islands and domain walls. The inside magnetic layers have the same texture.

When H is increased, we observe the skyrmion crystal as seen in Fig. 15.21: the GS configuration of the interface and beneath magnetic layers obtained for $J^{mf} = -1.25$, with $J^{2m} = J^{2f} = -0.2$ and external magnetic field $H = 0.25$. A zoom of a skyrmion shown in Fig. 15.21c and the z-components across a skyrmion shown in Fig. 15.21d indicate that the skyrmion is of Bloch type.

At this field strength $H = 0.25$, if we increase the frustration, for example, $J^{2m} = J^{2f} = -0.3$, then the skyrmion structure is enhanced: We can observe a clear 3D skyrmion crystal structure not only in the interface layer but also in the interior layers. This is shown in Fig. 15.22, where the interface (top) and the second layer (bottom) are displayed.

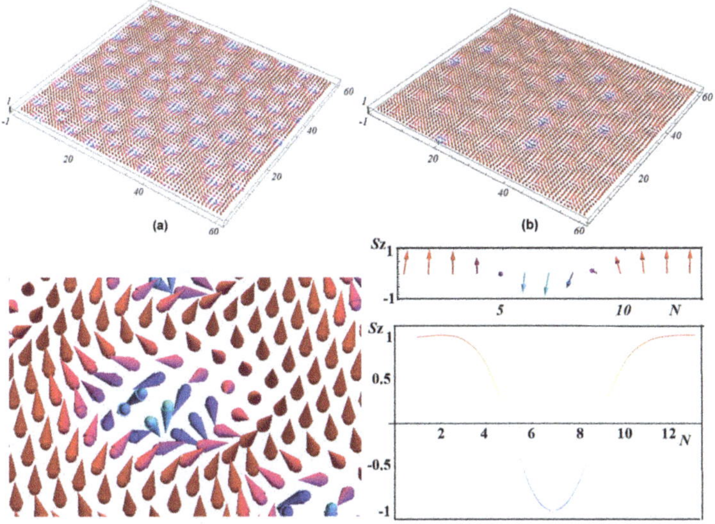

Figure 15.21 (a) 3D view of the GS configuration of the interface for moderate frustration $J^{2m} = J^{2f} = -0.2$, (b) 3D view of the GS configuration of the second magnetic layers, (c) zoom of a skyrmion on the interface layer: Red denotes up spin, four spins with clear blue color are down spin, other colors correspond to spin orientations between the two. The skyrmion is of the Bloch type, (d) z-components of spins across the skyrmion shown in (c). Other parameters: $J^{m} = J^{f} = 1$, $J^{mf} = -1.25$ and $H = 0.25$.

The highest value of frustration where the skyrmion structure can be observed is when $J^{2m} = J^{2f} = -0.4$ close to the critical value -0.5. We show this case in Fig. 15.23: The GS configuration of the interface (a) and second (interior) (b) magnetic layers are presented. Other parameters are the same as in the previous figures, namely $J^{mf} = -1.25$ and $H = 0.25$. We can observe a clear 3D skyrmion crystal structure in the whole magnetic layers, not only near the interface layer. Unlike the case where we do not take into account the interaction between NNN (see the previous sections and [317]), in the present case where the frustration is very strong we see that a large number of skyrmions are distributed over the whole magnetic layers with a certain periodicity close to a perfect crystal.

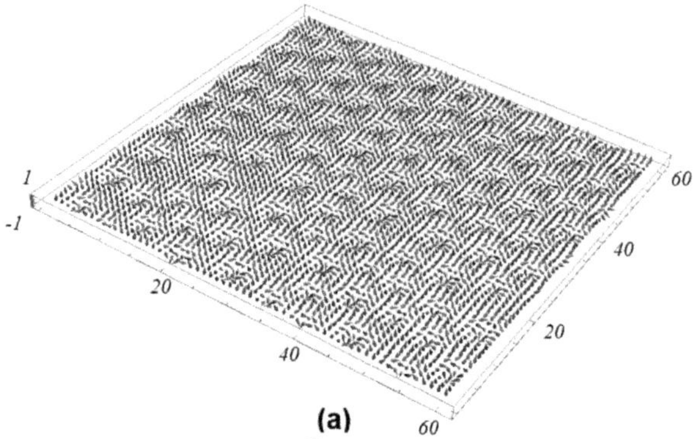

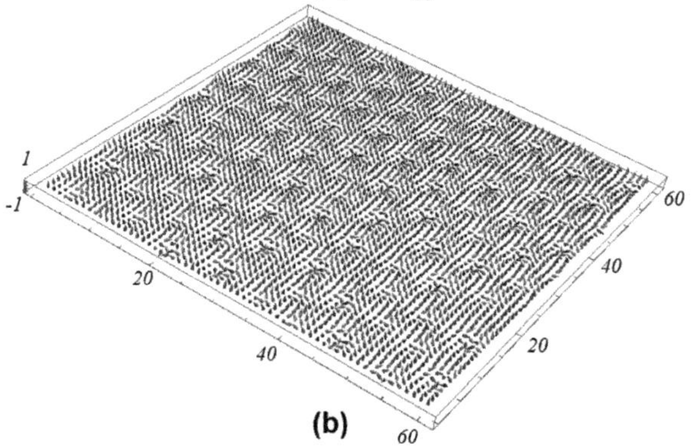

Figure 15.22 3D view of the GS configuration of (a) the interface, (b) the second layer, for stronger frustration $J^{2m} = J^{2f} = -0.3$. $J^{m} = J^{f} = 1$, $J^{2m} = J^{2f} = -0.3$, $J^{mf} = -1.25$ and $H = 0.25$.

Though we take the same value for J^{2m} and J^{2f} in the figures shown above, it is obvious that only the magnetic frustration J^{2m} is important for the skyrmion structure. The ferroelectric frustration affects only the stability of the polarizations at the interface. As long as J^{2f} does not exceed -0.5, the skyrmions are not affected by J^{2f} [318].

Now, if the magnetic frustration is not strong enough, the ferroelectric frustration plays an important role. Skyrmions disappear when $J^{2f} = -0.4$ [318]. We conclude that while magnetic frustration J^{2m} enhances the formation of skyrmions, the ferroelectric frustration J^{2f} in the weak magnetic frustration tends to

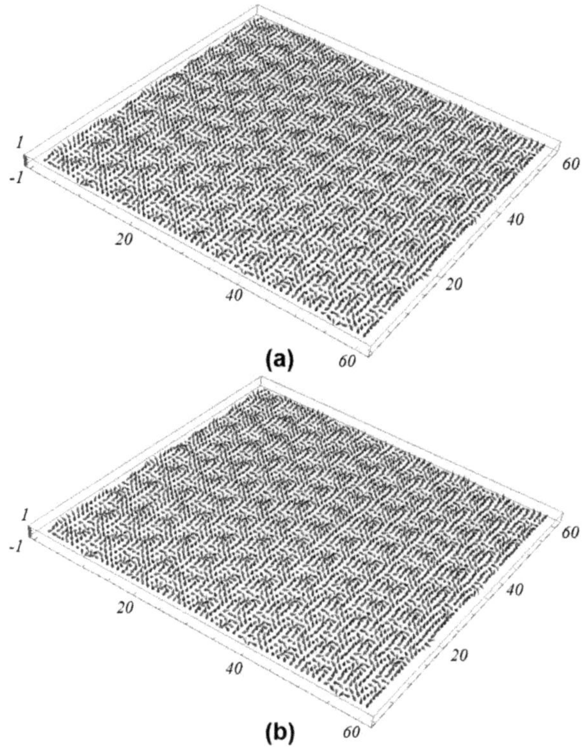

Figure 15.23 Strongest frustration $J^{2m} = J^{2f} = -0.4$ (a) 3D view of the GS configuration of the interface, (b) 3D view of the GS configuration of the second magnetic layers. Other parameters $J^{m} = J^{f} = 1$, $J^{mf} = -1.25$ and $H = 0.25$.

suppress skyrmions. The mechanism of these parameters when acting together seems to be very complicated.

15.5.3 Skyrmion Phase Transition

We use the same Metropolis algorithm [48, 199] as before with the size $N \times N \times L$ with $N = 60$ and thickness $L = L_m + L_f = 8$ (4 magnetic layers, 4 ferroelectric layers). Simulation times are 10^6 Monte Carlo steps (MCS) per spin for equilibrating the system and 10^6 MCS/spin for averaging. We calculate the internal energy and

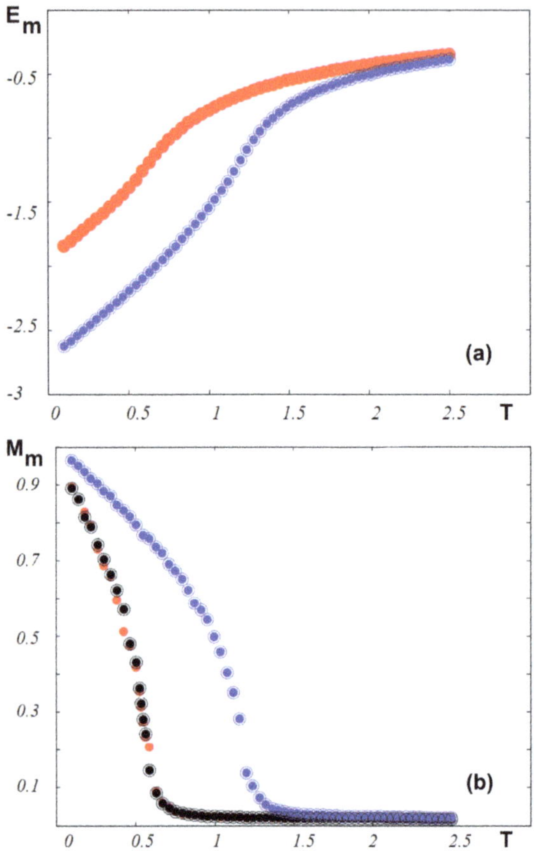

Figure 15.24 (a) Energy of the magnetic films versus temperature T for $(J^{2m} = J^{2f} = -0.4)$ (red), coinciding with the curve for $(J^{2m} = -0.4, J^{2f} = 0)$ (black, hidden behind the red curve). Blue curve is for $(J^{2m} = 0, J^{2f} = -0.4)$; (b) Order parameter of the magnetic films versus temperature T for $(J^{2m} = J^{2f} = -0.4)$ (red), $(J^{2m} = -0.4, J^{2f} = 0)$ (black), $(J^{2m} = 0, J^{2f} = -0.4)$ (blue). Other used parameters: $J^{mf} = -1.25$, $H = 0.25$.

the layer order parameters of the magnetic (M_m) and ferroelectric (M_f) films, defined in Eqs. (15.15) and (15.14).

In Fig.15.24, we show the magnetic energy and magnetic order parameter versus temperature in an external magnetic field, for various sets of NNN interaction. Note that the phase transition occurs at the energy curvature changes, namely at the maximum of the specific heat. The red curve in Fig.15.24a is for both sets ($J^{2m} = J^{2f} = -0.4$), ($J^{2m} = -0.4, J^{2f} = 0$). The change of curvature takes place at $T_c^m \simeq 0.60$. It means that the ferroelectric frustration does not affect the magnetic skyrmion transition at such a strong magnetic frustration ($J^{2m} = -0.4$). For ($J^{2m} = 0, J^{2f} = -0.4$), namely no magnetic frustration, the transition takes place at a much higher temperature $T_c^m \simeq 1.25$.

The magnetic order parameters shown in Fig. 15.24b confirm the skyrmion transition temperatures seen by the curvature change of the energy in Fig.15.24a.

15.6 Conclusion

We have studied in this chapter a new model for the interface coupling between a magnetic film and a ferroelectric film in a superlattice. This coupling has the form of a Dzyaloshinskii–Moriya (DM) interaction between a polarization and the spins at the interface.

The ground state shows uniform non-collinear spin configurations in zero field and skyrmions in an applied magnetic field. We have studied spin wave (SW) excitations in a monolayer and in a bilayer in zero field by the Green's function method. We have shown the strong effect of the DM coupling on the SW spectrum as well as on the magnetization at low temperatures.

Monte Carlo simulation has been used to study the phase transition occurring in the superlattice with and without applied field. Skyrmions have been shown to be stable at finite temperatures. We have also shown that the nature of the phase transition can be of second or first order, depending on the DM interface coupling.

We have also taken into account the frustration due to the NNN interactions in both magnetic and ferroelectric layers. As expected, the magnetic frustration enhances the creation of skyrmions while the ferroelectric frustration when strong enough destabilizes skyrmions if there is no strong magnetic frustration to resist. Note that the magnetic frustration reduces the transition temperature considerably.

The existence of skyrmions confined at the magneto-ferroelectric interface is very interesting. We believe that it can be used in transport applications in spintronic devices. A number of applications using skyrmions has been mentioned in the Introduction.

Chapter 16

Thin Films and Criticality

In this chapter, we study the critical behavior of magnetic thin films as a function of the film thickness. We use the ferromagnetic Ising model with the high-resolution multiple histogram Monte Carlo (MC) simulation. We show that though the 2D behavior remains dominant at small thicknesses, there is a systematic continuous deviation of the critical exponents from their 2D values. We explain these deviations using the concept of "effective" exponents suggested by Capehart and Fisher [53] in a finite-size analysis. The shift of the critical temperature with the film thickness obtained here by MC simulation is in excellent agreement with their prediction.

In the second part of the chapter, we show the crossover of the phase transition from first to second order in the frustrated Ising FCC antiferromagnetic film. This crossover occurs when the film thickness N_z is smaller than a value between 2 and 4 FCC lattice cells. The results are obtained with the high-performance Wang–Landau flat histogram technique, which allows to determine the first-order transition with efficiency.

The results shown in this chapter have been published in Refs. [270, 271].

Physics of Magnetic Thin Films: Theory and Simulation
Hung T. Diep

ISBN 978-981-4877-42-8 (Hardcover), 978-1-003-12110-7 (eBook)
www.jennystanford.com

16.1 Introduction

During the past 30 years, the physics of surfaces and objects of nanometric size has attracted immense interest. This is due to important applications in industry [36, 374]. An example is the so-called giant magneto-resistance (GMR) used in data storage devices, magnetic sensors, ... [18, 22, 134, 345]. In parallel to these experimental developments, much theoretical effort [35, 72, 73] has been devoted to the search of physical mechanisms lying behind the new properties found in nanoscale objects such as ultrathin films, ultrafine particles, quantum dots, spintronic devices, etc. This effort aimed not only at providing explanations for experimental observations but also at predicting new effects for future experiments.

The physics of two-dimensional (2D) systems is very exciting. Some of those 2D systems can be exactly solved: One famous example is the Ising model on the square lattice which has been solved by Onsager [265]. This model shows a phase transition at a finite temperature T_c given by $\sinh^2(2J/k_B T_c) = 1$ where J is the nearest-neighbor (NN) interaction. Another interesting result is the absence of long-range ordering at finite temperatures for the continuous spin models (XY and Heisenberg models) in 2D [231]. In general, three-dimensional (3D) systems for any spin models cannot be unfortunately solved. However, several methods in the theory of phase transitions and critical phenomena can be used to calculate the critical behaviors of these systems [380].

This chapter deals with the question of criticality in thin films, namely systems between 2D and 3D. Many theoretical studies have been devoted to thermodynamic properties of thin films, magnetic multilayers, ... [35, 72, 79, 80, 87, 252]. In spite of this, several points are still not yet understood. It has been known since a long time ago that the presence of a surface in magnetic materials can give rise to surface spin waves which are localized in the vicinity of the surface [78]. These localized modes may be acoustic with a low-lying energy or optical with a high energy, in the spin wave spectrum. Low-lying energy modes contribute to reduce in general surface magnetization at finite temperatures. One of the consequences is the surface disordering which may occur at a temperature lower than that for interior magnetization [75]. The existence of low-lying surface modes depends on the lattice structure, the surface orientation, the surface parameters, surface conditions (impurities, roughness, . . .), etc. There are two interesting cases: In the first case, a surface transition occurs at a temperature distinct from that of the interior spins, and in the second case, the surface transition coincides with the interior one, i.e., existence of a single transition. Theory of critical phenomena at surfaces [35, 72] and Monte Carlo (MC) simulations [200, 201] of critical behavior of the surface-layer magnetization at the extraordinary transition in the 3D Ising model have been carried out. These works suggested several scenarios in which the nature of the surface transition and the transition in thin films depends on many factors in particular on the symmetry of the Hamiltonian and on surface parameters.

In Chapters 9 and 10 while studying the frustration effects in thin films, we have calculated some critical exponents in those frustrated thin films. We found that the critical exponents have values somewhere between 2D and 3D universality classes. However, due to the presence of the frustration in those cases, we did not make a firm conclusion that the deviation of the critical exponents from the 2D values results uniquely from the film thickness. The work presented in this chapter is, therefore, performed on a ferromagnetic Ising thin film without frustration.

We confine ourselves here to the case of a simple cubic film with the Ising model. For our purpose, we suppose all interactions are the same everywhere even at the surface. This case is the simplest case where there is no surface-localized spin wave modes and there is only a single phase transition at a temperature for the whole system (no separate surface phase transition) [75, 78]. Other complicated cases will be left for future investigations. However, some discussions on this point for complicated surfaces have been reported in some of our previous papers [248, 249] and presented in Chapters 9 and 10.

In the case of a simple cubic film with the Ising model, Capehart and Fisher have studied the critical behavior of the susceptibility using a finite-size scaling analysis [53]. They showed that there is a crossover from 2D to 3D behavior as the film thickness increases. The so-called "effective" exponent γ has been shown to vary according to a scaling function depending both on the film thickness and the distance to the transition temperature. As will be seen below, the scaling suggested by Capehart and Fisher is in agreement with what we find here using extensive MC simulations.

We investigate in this chapter how the film thickness affects the critical exponents of the film as seen in simulations. Whatever the interpretation will be, the apparent deviations of the critical exponents from their 2D values are probably also seen in experiments. To carry out these purposes, we shall use MC simulations with highly accurate multiple-histogram technique presented in Chapter 6 (see original papers in Refs. [50, 110, 111]).

Section 16.2 is devoted to a description of the model and method. Results are shown and discussed in Section 16.3. Section 16.4 shows the crossover of the transition from the first-order the second-order order when the film thickness decreases. Concluding remarks are given in Section 16.5.

16.2 Model and Method

16.2.1 Model

Let us consider the Ising spin model on a film made from a ferromagnetic simple cubic lattice. The size of the film is $L \times L \times N_z$. We apply the periodic boundary conditions (PBC) in the xy planes to simulate an infinite xy dimension. The z direction is limited by the film thickness N_z. If $N_z = 1$ then one has a 2D square lattice.

The Hamiltonian is given by

$$\mathcal{H} = -\sum_{\langle i,j\rangle} J_{i,j}\sigma_i \cdot \sigma_j \tag{16.1}$$

where σ_i is the Ising spin of magnitude 1 occupying the lattice site i, $\sum_{\langle i,j\rangle}$ indicates the sum over the NN spin pairs σ_i and σ_j.

In the following, the interaction between two NN surface spins is denoted by J_s, while all other interactions are supposed to be ferromagnetic and all equal to $J = 1$ for simplicity. Let us note in passing that in the semi-infinite crystal the surface phase transition occurs at the bulk transition temperature when

$J_s \simeq 1.52J$. This point is called "extraordinary phase transition" which is characterized by some particular critical exponents [200, 201]. In the case of thin films, i. e. the thickness N_z is finite, not equal to 1, it has been theoretically shown that when $J_s = 1$ the bulk behavior is observed when the thickness becomes larger than a few dozens of atomic layers [78]: Surface effects are insignificant on thermodynamic properties such as the value of the critical temperature and the mean value of magnetization at a given T. When J_s is smaller than J, surface magnetization is destroyed at a temperature lower than that for bulk spins [75] (see Chapter 8). The criticality of a film with uniform interaction, i.e., $J_s = J$, has been studied by Capehart and Fisher as a function of the film thickness using a scaling analysis [53] and by MC simulations [55, 309]. The results by Capehart and Fisher indicate that as long as the film thickness is finite, the phase transition is strictly that of the 2D Ising universality class. However, they showed that at a temperature away from the transition temperature $T_c(N_z)$, the system can behave as a 3D one when the spin–spin correlation length $\xi(T)$ is much smaller than the film thickness, i.e., $\xi(T)/N_z \ll 1$. As T gets very close to $T_c(N_z)$, $\xi(T)/N_z \to 1$, the system undergoes a crossover to 2D criticality. We will return to this work for comparison with our results shown below.

16.2.2 Multiple-Histogram Technique

The multiple-histogram technique is known to reproduce with very high accuracy the critical exponents of second order phase transitions [50, 110, 111]. We recall it briefly here.

The overall probability distribution [111] at temperature T obtained from n independent simulations, each with N_j configurations, is given by

$$P(E, T) = \frac{\sum_{i=1}^{n} H_i(E) \exp[E/k_B T]}{\sum_{j=1}^{n} N_j \exp[E/k_B T_j - f_j]} \qquad (16.2)$$

where

$$\exp[f_i] = \sum_E P(E, T_i) \qquad (16.3)$$

The thermal average of a physical quantity A is then calculated by

$$\langle A(T)\rangle = \sum_E A\,P(E,T)/z(T) \tag{16.4}$$

in which

$$z(T) = \sum_E P(E,T) \tag{16.5}$$

Thermal averages of physical quantities are thus calculated as continuous functions of T, now the results should be valid over a much wider range of temperature than for any single histogram.

In MC simulations, one calculates the averaged order parameter $\langle M\rangle$ (M: magnetization of the system), averaged total energy $\langle E\rangle$, specific heat C_v, susceptibility χ, first-order cumulant of the energy C_U, and nth-order cumulant of the order parameter V_n for $n = 1$ and 2. These quantities are defined as

$$\langle E\rangle = \langle \mathcal{H}\rangle, \tag{16.6}$$

$$C_v = \frac{1}{k_B T^2}\left(\langle E^2\rangle - \langle E\rangle^2\right) \tag{16.7}$$

$$\chi = \frac{1}{k_B T}\left(\langle M^2\rangle - \langle M\rangle^2\right) \tag{16.8}$$

$$C_U = 1 - \frac{\langle E^4\rangle}{3\langle E^2\rangle^2} \tag{16.9}$$

$$V_n = \frac{\partial \ln M^n}{\partial(1/k_B T)} = \langle E\rangle - \frac{\langle M^n E\rangle}{\langle M^n\rangle} \tag{16.10}$$

Let us discuss the case where all dimensions can go to infinity. For example, consider a system of size L^d where d is the space dimension. For a finite L, the pseudo "transition" temperatures can be identified by the maxima of C_v and χ,. . . . These maxima do not in general take place at the same temperature. Only at infinite L that the pseudo "transition" temperatures of these respective quantities coincide at the real transition temperature $T_c(\infty)$. So when we work at the maxima of V_n, C_v and χ, we are in fact working at temperatures away from $T_c(\infty)$. This is an important point to bear in mind for the discussion given below.

For large values of L, the following scaling relations are expected (see details in Ref. [50]):

$$V_1^{\max} \propto L^{1/\nu}, \qquad V_2^{\max} \propto L^{1/\nu} \tag{16.11}$$

$$C_v^{\max} = C_0 + C_1 L^{\alpha/\nu} \tag{16.12}$$

and

$$\chi^{\max} \propto L^{\gamma/\nu} \tag{16.13}$$

at their respective "transition" temperatures $T_c(L)$, and

$$C_U = C_U[T_c(\infty)] + AL^{-\alpha/\nu} \tag{16.14}$$

$$M_{T_c(\infty)} \propto L^{-\beta/\nu} \tag{16.15}$$

and

$$T_c(L) = T_c(\infty) + C_A L^{-1/\nu} \tag{16.16}$$

where A, C_0, C_1 and C_A are constants. We estimate ν independently from $V_1^{\max}$ and $V_2^{\max}$. With this value we calculate γ from $\chi^{\max}$ and α from $C_v^{\max}$. Note that we can estimate $T_c(\infty)$ using the last expression. Then, using $T_c(\infty)$, we can calculate β from $M_{T_c(\infty)}$. The Rushbrooke scaling law $\alpha + 2\beta + \gamma = 2$ is then in principle verified.

Let us emphasize that the expressions Eqs. (16.11)–(16.16) are valid for large L. To be sure that L are large enough, one has to allow for corrections to scaling of the form, for example,

$$\chi^{\max} = B_1 L^{\gamma/\nu}(1 + B_2 L^{-\omega}) \tag{16.17}$$

$$V_n^{\max} = D_1 L^{1/\nu}(1 + D_2 L^{-\omega}) \tag{16.18}$$

where B_1, B_2, D_1 and D_2 are constants and ω is a correction exponent [112]. Similar forms exist also for the other exponents. Usually, these corrections are extremely small if L is large enough as is the case with today's large-memory computers. So, in general they do not alter the results using Eqs. (16.11)–(16.16).

16.2.3 The Case of Films with Finite Thickness

In the case of a thin film of size $L \times L \times N_z$, Capehart and Fisher [53] have showed that as long as the film thickness N_z is not allowed to go to infinity, there is a 2D-3D crossover if one does not work at the real transition temperature $T_c(L = \infty, N_z)$. Following Capehart and Fisher, let us define

$$\dot{t} = \frac{T - T_c(L = \infty, N_z)}{T_c(3D)} \tag{16.19}$$

$$x = N_z^{1/\nu_{3D}} \dot{t} \tag{16.20}$$

where ν_{3D} is the 3D ν exponent and $T_c(3D)$ the 3D critical temperature. When x is larger than a value x_0, i.e., at a temperature away from $T_c(L = \infty, N_z)$, the system behaves as a 3D one. While, when $x < x_0$ it should behave as a 2D one. This crossover was argued from a comparison of the correlation length in the z direction to the film thickness. As a consequence, if we work exactly at $T_c(L = \infty, N_z)$ we should observe the 2D critical exponents for finite N_z. Otherwise, we should observe the so-called "effective critical exponents" whose values are found between those of 2D and 3D cases. This point is fundamentally very important. There have been some attempts to verify it by MC simulations [309] but these results were not convincing due to their poor MC quality. In the following, we show with high-precision MC multiple-histogram technique that the prediction of Capehart and Fisher is really verified.

16.3 Results: Critical Exponents

The xy linear sizes $L = 20, 24, 30, \ldots, 80$ have been used in our simulations. Films of thickness from $N_z = 3$ up to 160 have been used to evaluate corrections to scaling.

In practice, we use first the standard MC simulations to localize for each size the transition temperatures $T_0^E(L)$ for specific heat and $T_0^m(L)$ for susceptibility. The equilibrating time is from 2×10^5 to 4×10^5 MC steps/spin and the averaging time is from 5×10^5 to 10^6 MC steps/spin. Next, we make histograms at eight different temperatures $T_j(L)$ around the transition temperatures $T_0^{E,m}(L)$ with 2×10^6 MC steps/spin, after discarding 10^6 MC steps/spin for equilibrating. Finally, we make again histograms at 8 different temperatures around the new transition temperatures $T_0^{E,m}(L)$ with 2×10^6 and 4×10^6 MC steps/spin for equilibrating and averaging time, respectively. Such an iteration procedure gives extremely good results for systems studied so far. The errors shown in the following have been estimated using statistical errors, which are very small thanks to our multiple-histogram procedure, and fitting errors given by fitting software.

We note that only ν is directly calculated from MC data. Exponent γ obtained from $\chi^{\max}$ and ν suffers little errors (systematic errors and errors from ν). Other exponents are obtained by MC data and several-step fitting. For example, to obtain α we have to fit $C_V^{\max}$ of Eq. 16.12 by choosing C_0, C_1 and by using the value of ν. So, in practice, in most cases, one calculates α or β from MC data and uses the Rushbrooke scaling law to calculate the remaining exponent.

Now, similar to the discussion given in Section 16.2.2, if we work at a distance away from $T_c(L = \infty, N_z)$ we should observe "effective critical exponents." This is the case because in the finite-size analysis using the multiple-histogram technique, we measure the maxima of V_n, C_V and χ which occur at different temperatures for a finite L. These temperatures, though close to, are not $T_c(L = \infty, N_z)$. To give a precision on this point, we show the values of these maxima and the corresponding temperatures for $N_z = 7$ in Table 16.1. For the value of $T_c(L = \infty, N_z = 7)$, see Table 16.2.

Given this fact, we emphasize that calculations using Eqs. (16.11)–(16.16) will give effective critical exponents except of course for the case $N_z = 1$ where the results correspond to real 2D critical exponents.

We show now the results obtained by MC simulations with the Hamiltonian (16.1). We have tested that all exponents do not change in the finite size scaling with $L \geq 30$. So most of results are shown for $L \geq 30$ except for ν where the lowest sizes $L = 20, 24$ can be used without modifying its value.

Let us show in Fig. 16.1 the layer magnetizations and their corresponding susceptibilities of the first three layers, in the case where $J_s = 1$. It is interesting to note that the magnetization of the surface layer is smaller than the magnetizations of the interior layers, as it has been shown theoretically by the Green's function method a long time ago [75, 78]. The surface spins have smaller local field due to the lack of neighbors, so thermal fluctuations will reduce more easily the surface magnetization with respect to the interior ones. The susceptibilities have their peaks at the same temperature, indicating a single transition.

Table 16.1 Maxima and temperatures at the maxima of V_n $(n = 1, 2)$, C_v and χ for various L with $N_z = 7$

L	C_v^{max}	χ^{max}	V_1^{max}	V_2^{max}	$T_c(C_v^{max})$	$T_c(\chi^{max})$	$T_c(V_1^{max})$	$T_c(V_2^{max})$
30	2.21115658	25.29532589	164.05948154	275.71581036	4.19027500	4.22755000	4.24277500	4.24900000
40	2.36517434	41.20958927	219.30094769	368.72462473	4.19305000	4.21895000	4.23025000	4.23500000
50	2.50496719	60.82008190	275.66203381	463.17327477	4.19275000	4.21340000	4.22210000	4.22600000
60	2.59177903	82.96529587	329.65536262	554.47606570	4.19270000	4.20940000	4.21710000	4.22045000
70	2.70129995	109.00528127	387.47245040	651.24905512	4.19250000	4.20640000	4.21260000	4.21530000
80	2.76931676	138.78113065	443.00488386	743.61068938	4.19220000	4.20410000	4.20965000	4.21205000

Table 16.2 Critical exponents, effective dimension and critical temperature at the infinite xy limit

N_z	ν	γ	α	β	d_{eff}	$T_c(L = \infty, N_z)$
1	0.9990 ± 0.0028	1.7520 ± 0.0062	0.00199 ± 0.00279	0.1266 ± 0.0049	2.0000 ± 0.0028	2.2699 ± 0.0005
3	0.9922 ± 0.0019	1.7377 ± 0.0035	0.00222 ± 0.00192	0.1452 ± 0.0040	2.0135 ± 0.0019	3.6365 ± 0.0024
5	0.9876 ± 0.0023	1.7230 ± 0.0069	0.00222 ± 0.00234	0.1639 ± 0.0051	2.0230 ± 0.0023	4.0234 ± 0.0028
7	0.9828 ± 0.0024	1.7042 ± 0.0087	0.00223 ± 0.00238	0.1798 ± 0.0069	2.0328 ± 0.0024	4.1939 ± 0.0032
9	0.9780 ± 0.0016	1.6736 ± 0.0084	0.00224 ± 0.00161	0.1904 ± 0.0071	2.0426 ± 0.0016	4.2859 ± 0.0022
11	0.9733 ± 0.0025	1.6354 ± 0.0083	0.00224 ± 0.00256	0.1995 ± 0.0088	2.0526 ± 0.0026	4.3418 ± 0.0032
13	0.9692 ± 0.0026	1.6122 ± 0.0102	0.00226 ± 0.00268	0.2059 ± 0.0092	2.0613 ± 0.0027	4.3792 ± 0.0034

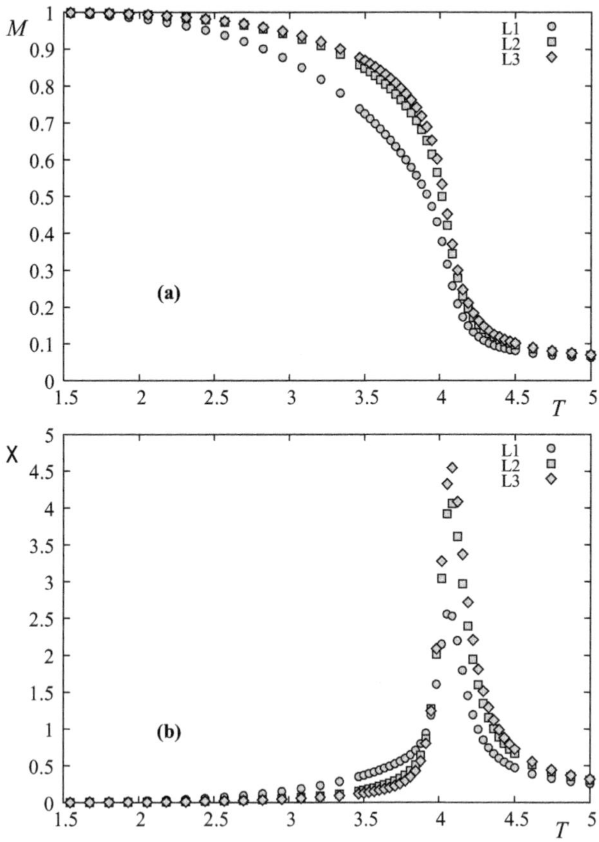

Figure 16.1 Layer magnetizations (a) and layer susceptibilities (b) versus T with $N_z = 5$ and $L = 24$.

Figure 16.2 shows total magnetization of the film and the total susceptibility. This indicates clearly that there is only one peak as said above.

16.3.1 Finite-Size Scaling

Let us show some results obtained from multiple histograms described above. Figure 16.3 shows the susceptibility and the first derivative V_1 versus T around their maxima for several sizes.

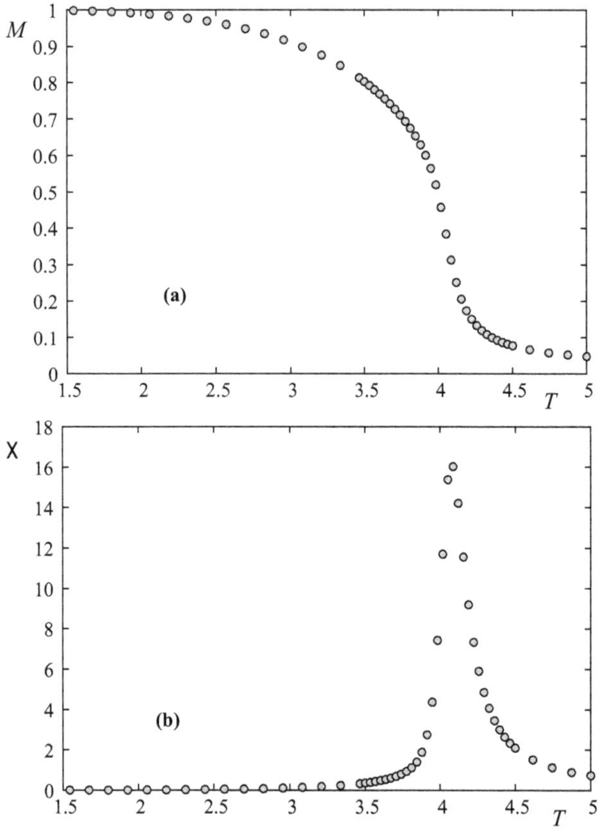

Figure 16.2 Total magnetization (a) and total susceptibility (b) versus T with $N_z = 5$ and $L = 24$.

We show in Fig. 16.4 the maximum of the first derivative of ln M with respect to $\beta = (k_B T)^{-1}$ versus L in the ln – ln scale for several film thicknesses up to $N_z = 13$. If we use Eq. (16.11) to fit these lines, i.e., without correction to scaling, we obtain $1/\nu$ from the slopes of the remarkably straight lines. These values are indicated on the figure. In order to see the deviation from the 2D exponent, we plot in Fig. 16.5 ν as a function of thickness N_z. We observe here a small but systematic deviation of ν from its 2D value ($\nu_{2D} = 1$) with increasing thickness. To show the precision of our method, we give here the results of $N_z = 1$. For $N_z = 1$, we have $1/\nu = 1.0010 \pm 0.0028$

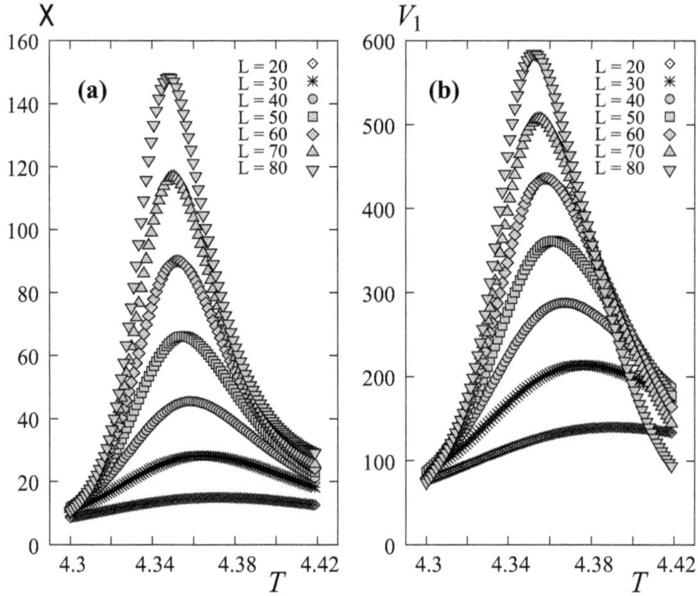

Figure 16.3 (a) Susceptibility and (b) V_1, as functions of T for several L with $N_z = 11$, obtained by multiple histogram technique.

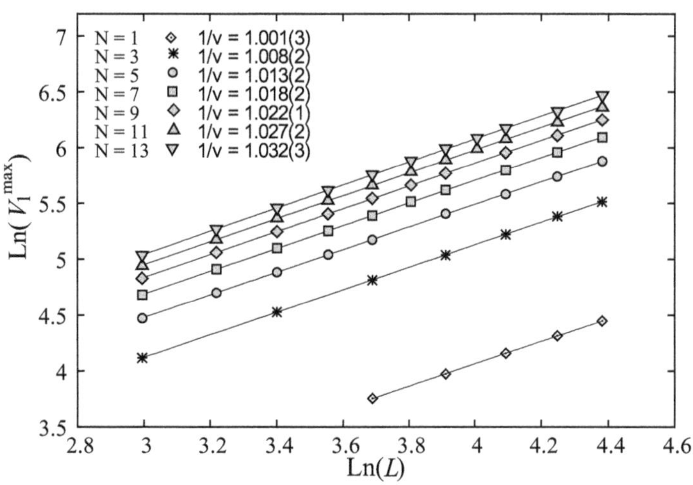

Figure 16.4 Maximum of the first derivative of $\ln M$ versus L in the $\ln - \ln$ scale. The slopes are indicated on the figure.

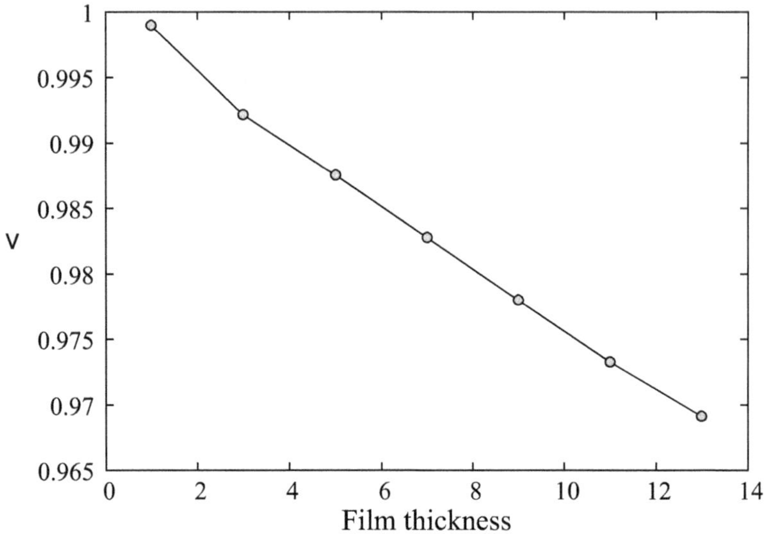

Figure 16.5 Effective exponent ν versus N_z.

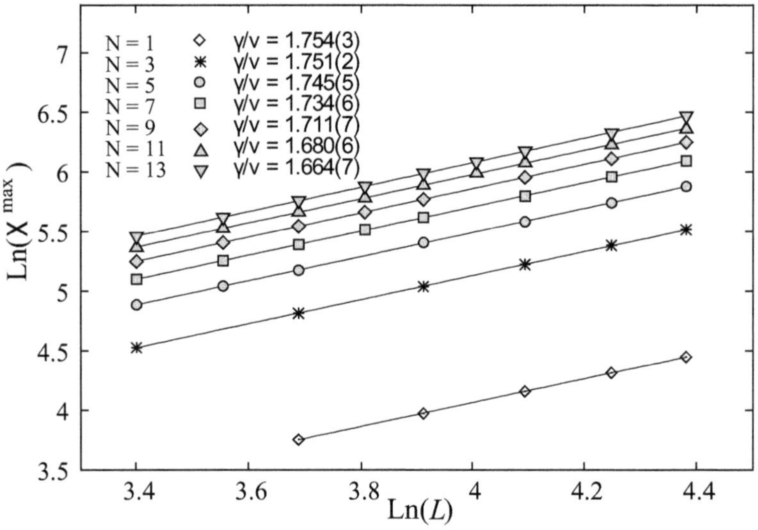

Figure 16.6 Maximum of susceptibility versus L in the ln – ln scale. The slopes are indicated on the figure.

which yields $\nu = 0.9990 \pm 0.0031$ and $\gamma/\nu = 1.7537 \pm 0.0034$ (see Figs. 16.6 and 16.7 below) yielding $\gamma = 1.7520 \pm 0.0062$. These results are in excellent agreement with the exact results $\nu_{2D} = 1$ and $\gamma_{2D} = 1.75$. The very high precision of our method is thus verified in the rather modest range of the system sizes $L = 20 - 80$ used in the present work. Note that the result of Ref. [309] gave $\nu = 0.96 \pm 0.05$ for $N_z = 1$ which is very far from the exact value.

The deviation of ν from the 2D value when N_z increases is due, as discussed earlier, to the crossover to 3D ($\dot{t}$ is not zero). Other exponents will suffer the same deviations as seen below.

We show in Fig. 16.6 the maximum of the susceptibility versus L in the $\ln - \ln$ scale for film thicknesses up to $N_z = 13$. We have used only results of $L \geq 30$. Including $L = 20$ and 24 will result, unlike the case of ν, in a decrease of γ of about one percent for $N_z \geq 7$. From the slopes of these straight lines, we obtain the values of effective γ/ν. Using the values of ν obtained above, we deduce the values of γ which are plotted in Fig. 16.7 as a function of thickness N_z. Unlike the case of ν, we observe here a stronger deviation of γ from its 2D value (1.75) with increasing thickness. This finding is somewhat interesting: The magnitude of the deviation from the 2D

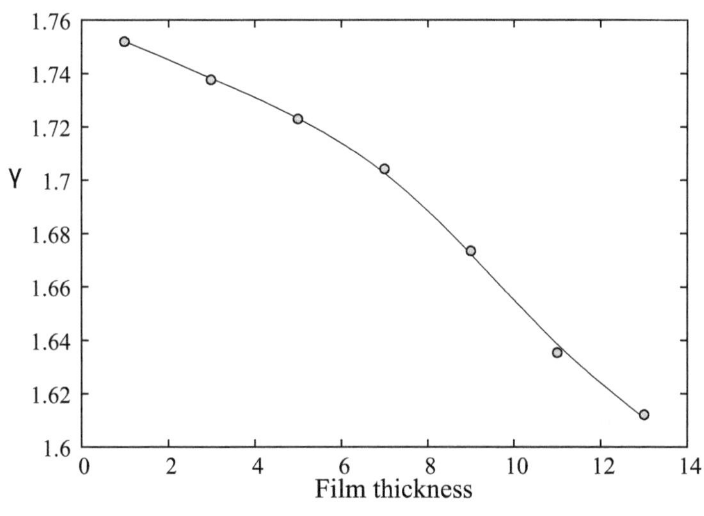

Figure 16.7 Effective exponent γ versus N_z.

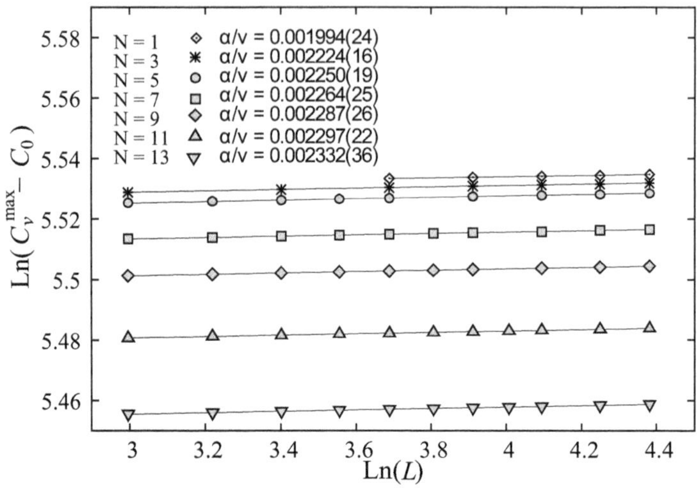

Figure 16.8 $\ln(C_v^{\max} - C_0)$ versus $\ln L$ for $N_z = 1, 3, 5, \ldots, 13$. The slope gives α/ν (see Eq. 16.12) indicated on the figure.

value may be different from one critical exponent to another: $\simeq 3\%$ for ν and $\simeq 8\%$ for γ when N_z goes from 1 to 13. We will see below that β varies even more strongly.

We show now in Fig. 16.8 the maximum of $C_v^{\max}$ versus L for $N_z = 1, 3, 5, \ldots, 13$. Note that for each N_z we had to look for C_0, C_1 and α/ν which give the best fit with data of $C_v^{\max}$. Due to the fact that there are several parameters which can induce a wrong combination of them, we impose that α should satisfy the condition $0 \leq \alpha \leq 0.11$ where the lower limit of α corresponds to the value of 2D case and the upper limit to the 3D case. In doing so, we get very good results shown in Fig. 16.8. From these ratios of α/ν we deduce α for each N_z. The values of α are shown in Table 16.2 for several N_z.

It is interesting to note that the effective exponents obtained above give rise to the effective dimension of thin film. This is conceptually not rigorous but this is what observed in experiments. Replacing the effective values of α obtained above in $d_{\text{eff}} = (2-\alpha)/\nu$ we obtain d_{eff} shown in Fig. 16.9.

We note that d_{eff} is very close to 2. It varies from 2 to $\simeq 2.061$ for N_z going from 1 to 13. The 2D character is thus dominant even with larger N_z. This supports the idea that the finite correlation in the z

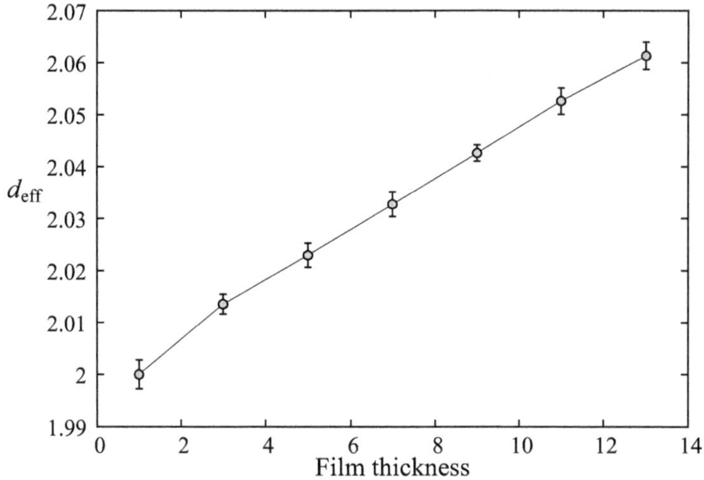

Figure 16.9 Effective dimension of thin film obtained by using effective exponents, as a function of thickness.

direction, though qualitatively causing a deviation, cannot strongly alter the 2D critical behavior. This point is interesting because, as said earlier, some thermodynamic properties may show already their 3D values at a thickness of about a few dozens of layers, but not the critical behavior. To show an example of this, let us plot in Fig. 16.10 the transition temperature at $L = \infty$ for several N_z, using Eq. 16.16 for each given N_z. As seen, $T_c(\infty)$ reaches already $\simeq 4.379$ at $N_z = 13$ while its value at 3D is 4.51 [70, 112]. A rough extrapolation shows that the 3D values is attained for $N_z \simeq 25$ while the critical exponents at this thickness are far away from the 3D ones.

Let us show the prediction of Capehart and Fisher [53] on the critical temperature as a function of N_z. Defining the critical-point shift as

$$\varepsilon(N_z) = [T_c(L = \infty, N_z) - T_c(3D)]\,/\,T_c(3D) \tag{16.21}$$

they showed that

$$\varepsilon(N_z) \approx \frac{b}{N_z^{1/\nu}}[1 + a/N_z] \tag{16.22}$$

where $\nu = 0.6289$ (3D value). Using $T_c(3D) = 4.51$, we fit the above formula with $T_c(L = \infty, N_z)$ taken from Table 16.2, we obtain

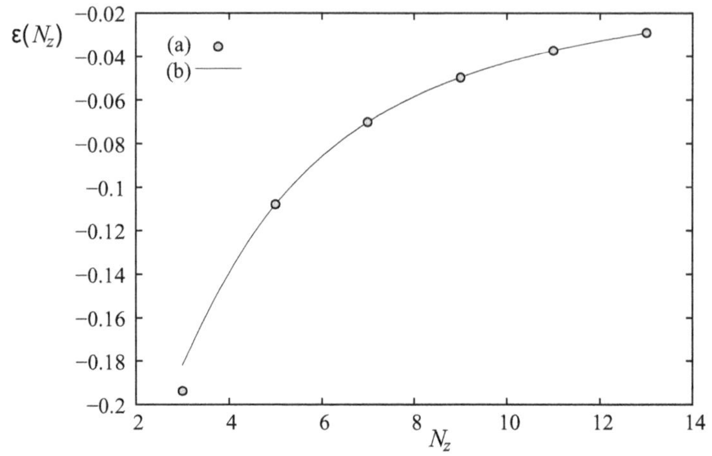

Figure 16.10 Critical temperature at infinite L as a function of the film thickness. Points are MC results, continuous line is the prediction of Capehart and Fisher, Eq. (16.22). The agreement is excellent.

$a = -1.37572$ and $b = -1.92629$. The MC results and the fitted curve are shown in Fig. 16.10. Note that if we do not use the correction factor $[1 + a/N_z]$, the fit is not good for small N_z. The prediction of Capehart and Fisher is thus very well verified.

We give here the precise values of $T_c(L = \infty, N_z)$ for each thickness. For $N_z = 1$, we have $T_c(L = \infty, N_z = 1) = 2.2699 \pm 0.0005$. Note that the exact value of $T_c(\infty)$ is 2.26919 by solving the equation $\sinh^2(2J/T_c) = 1$. Again here, the excellent agreement of our result shows the efficiency of the multiple histogram technique as applied in the present paper. The values of $T_c(L = \infty)$ for other N_z are summarized in Table 16.2.

Calculating now $M(L)$ at these values of $T_c(L = \infty, N_z)$ and using Eq. 16.15, we obtain β/ν for each N_z. For $N_z = 1$, we have $\beta/\nu = 0.1268 \pm 0.0022$ which yields $\beta = 0.1266 \pm 0.0049$ which is in excellent agreement with the exact result 0.125. Note that if we calculate β from $\alpha + 2\beta + \gamma = 2$, then $\beta = (2 - 1.75198 - 0.00199)/2 = 0.12302 \pm 0.0035$ which is in good agreement with the direct calculation within errors. We show in Fig. 16.11 the values of β obtained by direct calculation using Eq. 16.15. Note that the deviation of β from the 2D value when N_z varies from 1 to 13 is

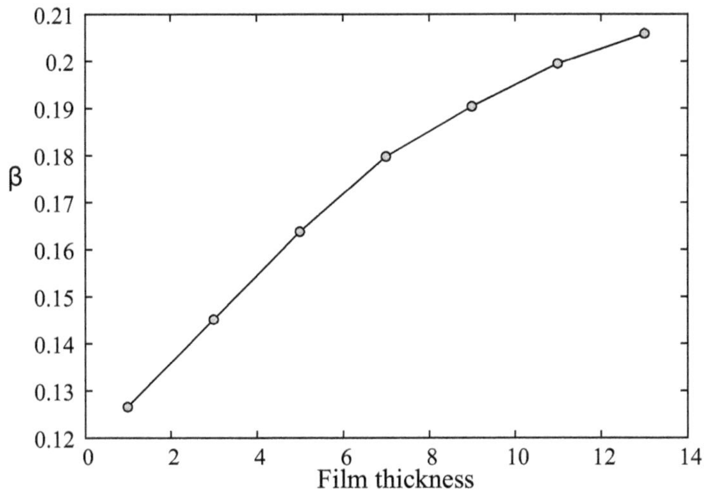

Figure 16.11 Effective exponent β, obtained by using Eq. 16.15, versus the film thickness.

due to the crossover effect discussed in Section 16.2.3. It represents about 60%. Remember that the 3D value of β is 0.3258 ± 0.0044 [112].

Finally, for convenience, let us summarize our results in Table 16.2 for $N_z = 1, 3, \ldots, 13$. Except for $N_z = 1$, all other cases are effective exponents discussed above. Due to the smallness of α, its value is shown with 5 decimals without rounding.

16.3.2 Larger Sizes and Correction to Scaling

We consider here the effects of larger L and of the correction to scaling. For the effect of larger L, we will extend our size up to $L = 160$, for just the case $N_z = 3$.

The results indicate that larger L does not change the results shown above. Figure 16.12a displays the maximum of V_1 as a function of L up to 160. Using Eq. (16.11), i.e., without correction to scaling, we obtain $1/\nu = 1.009 \pm 0.001$ which is to be compared to $1/\nu = 1.008 \pm 0.002$ using L up to 80. The change is, therefore, insignificant because it is at the third decimal i.e., at the error level. The same is observed for γ/ν as shown in Fig. 16.12b: $\gamma/\nu =$

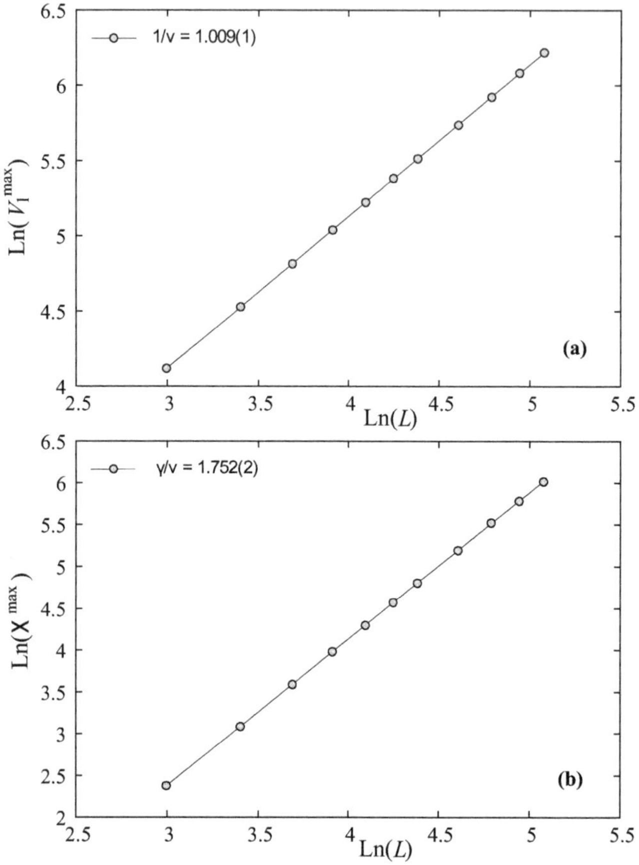

Figure 16.12 (a) V_1^{max} and (b) χ^{max} vs L up to 160 with $N_z = 3$.

1.752 ± 0.002 using L up to 160 instead of $\gamma/\nu = 1.751 \pm 0.002$ using L up to 80.

Now, let us allow for correction to scaling, i.e., we use Eq.(16.17) instead of Eq. (16.13) for fitting. We obtain the following values: $\gamma/\nu = 1.751 \pm 0.002$, $B_1 = 0.05676$, $B_2 = 1.57554$, $\omega = 3.26618$ if we use L = 70 to 160 (figure not shown). The value of γ/ν in the case of no scaling correction is 1.752 ± 0.002. Therefore, we can conclude that this correction is insignificant. The large value of ω explains the smallness of the correction.

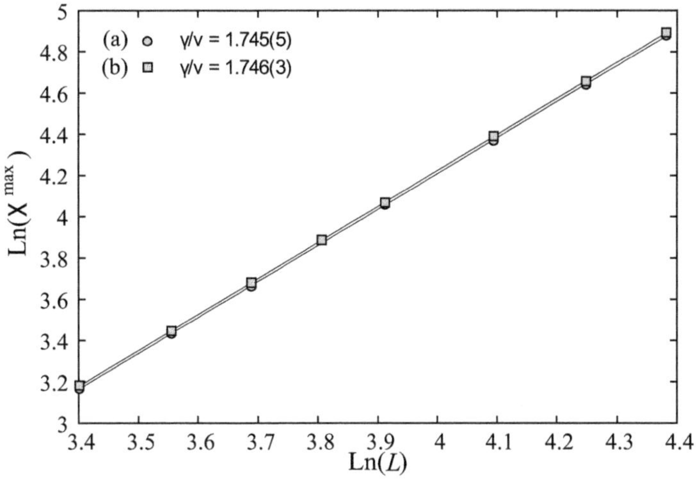

Figure 16.13 Maximum of susceptibility versus L in the ln − ln scale for $N_z = 5$ (a) without PBC in z direction (b) with PBC in z direction. The points of these cases cannot be distinguished in the figure scale. The slopes are indicated on the figure. See text for comments.

16.3.3 Role of Boundary Conditions

To close this section, let us touch upon the question: Does the absence of PBC in the z direction cause the deviation of the critical exponents? The answer is no: We have calculated ν and γ for $N_z = 5$ in both cases, with and without PBC in the z direction. The results show no significant difference between the two cases as seen in Figs. 16.14 and 16.13. We have found the same thing with $N_z = 11$ (not shown). So, we conclude that the fixed thickness will result in the deviation of the critical exponents, not from the absence of the PBC. This is somewhat surprising since we may think, incorrectly, that the PBC should mimic the infinite dimension so that we should obtain the 3D behavior when applying the PBC. As will be seen below, the 3D behavior is recovered only when the finite size scaling is applied in the z direction at the same time in the xy plane. To show this, we plot in Figs. 16.15 and 16.16 the results for the 3D case. Even with our modest sizes (up to $L = N_z = 21$, since it is not our purpose to treat the 3D case here), we obtain $\nu = 0.613 \pm 0.005$ and $\gamma = 1.250 \pm 0.005$ very close to their 3D best known values

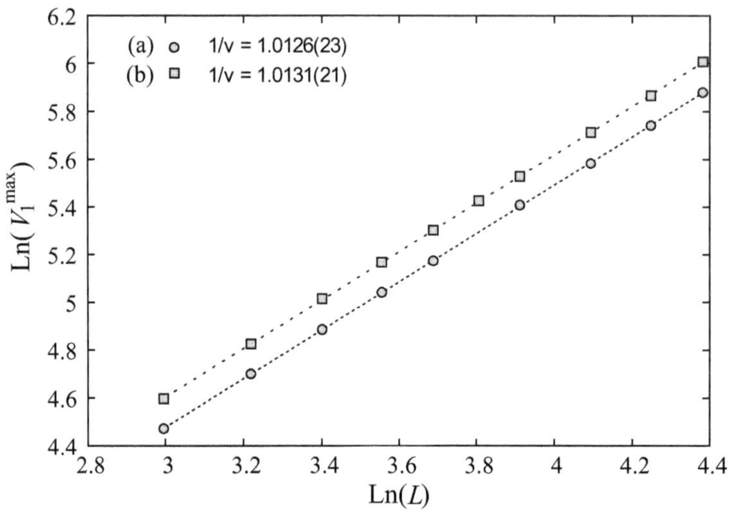

Figure 16.14 Maximum of the first derivative of ln M versus L in the ln − ln scale for $N_z = 5$ (a) without PBC in z direction (b) with PBC in z direction. The slopes are indicated on the figure. See text for comments.

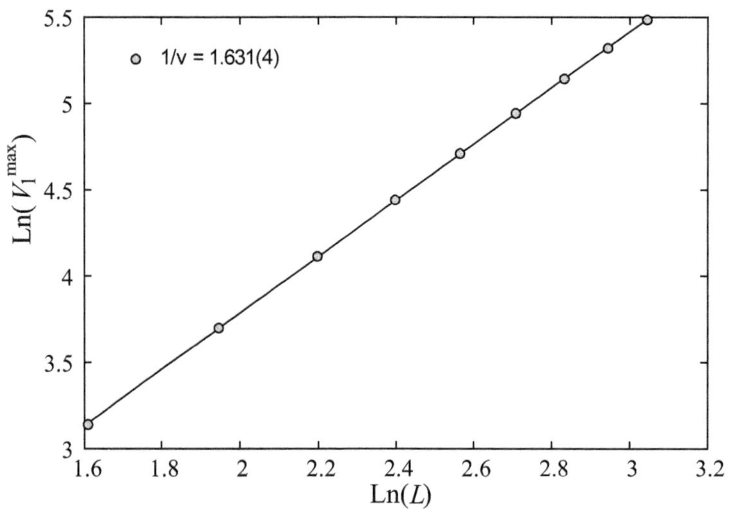

Figure 16.15 Maximum of the first derivative of ln M versus L in the ln − ln scale for 3D case.

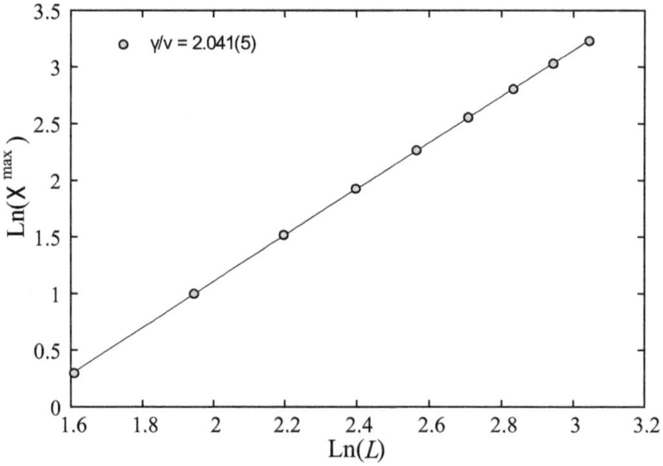

Figure 16.16 Maximum of susceptibility versus L in the ln − ln scale for 3D case.

$\nu_{3D} = 0.6302 \pm 0.0001$ from Ref. [70] and $\nu_{3D} = 0.6289 \pm 0.0008$ and $\gamma_{3D} = 1.2390 \pm 0.0025$ obtained by using $24 \leq L \leq 96$ given in Ref. [112].

16.4 Crossover from First- to Second-Order Transition in a Frustrated Thin Film

This section deals with the question whether or not the phase transition known in the bulk state changes its nature when the system is made as a thin film. In a recent work presented above, we have considered the case of a bulk second-order transition. We have shown that under a thin film shape, i.e., with a finite thickness, the transition shows effective critical exponents whose values are between 2D and 3D universality classes [270]. If we scale these values with a function of thickness as suggested by Capehart and Fisher [53] we should find, as long as the thickness is finite, the 2D universality class.

In this section, we study the effect of the film thickness in the case of a bulk first-order transition. The question to which we would like to answer is whether or not the first order becomes a second order when reducing the thickness. For that purpose we consider the face-centered cubic (FCC) Ising antiferromagnet. This system is fully frustrated with a very strong first-order transition.

On the one hand, effects of the frustration in spin systems have been extensively investigated during the last 30 years. In particular, by exact solutions, we have shown that frustrated spin systems have rich and interesting properties such as successive phase transitions with complicated nature, partial disorder, reentrance and disorder lines [67, 76]. Frustrated systems still challenge theoretical and experimental methods. For recent reviews, the reader is referred to Ref. [85].

We shall use the recent high-precision technique called "Wang–Landau" flat histogram Monte Carlo (MC) simulations to identify the order of the transition.

16.4.1 Model and Ground-State Analysis

It is known that the antiferromagnetic interaction between nearest-neighbor (NN) spins on the FCC lattice causes a very strong frustration. This is due to the fact that the FCC lattice is composed of tetrahedra each of which has four equilateral triangles. It is well-known [85] that it is impossible to fully satisfy simultaneously the three antiferromagnetic bond interactions on each triangle. In the case of the Ising model, the GS is infinitely degenerate for an infinite system size: On each tetrahedron, two spins are up and the other two are down. The FCC system is composed of edge-sharing tetrahedra. Therefore, there is an infinite number of ways to construct the infinite crystal. The minimum number of ways of such a construction is a stacking, in one direction, of uncorrelated antiferromagnetic planes. The minimum GS degeneracy of a L^3 FCC-cell system (L being the number of cells in each directions), is, therefore, equal to 3×2^{2L} where the factor 3 is the number of choices of the stacking direction, 2 the degeneracy of the antiferromagnetic spin configuration of each plane and $2L$ the number of atomic planes in one direction of the FCC crystal (the total number of spins is $N = 4L^3$). The GS degeneracy is, therefore, of the order of $2^{N^{1/3}}$. Note that at finite temperature, due to the so-called "order by disorder"

[146, 349], the spins will choose a long-range ordering. In the case of antiferromagnetic FCC Ising crystal, this ordering is an alternate stacking of up-spin planes and down-spin planes in one of the three direction. This has been observed also in the Heisenberg case [82] as well as in other frustrated systems [274].

The phase transition of the bulk frustrated FCC Ising antiferromagnet has been found to be of the first order [26, 272, 278, 280, 334]. Note that for the Heisenberg model, the transition is also found to be of the first order as in the Ising case [82, 137]. Other similar frustrated antiferromagnets such as the HCP XY and Heisenberg antiferromagnets [83] and stacked triangular XY and Heisenberg antiferromagnets [250, 251] show the same behavior.

Let us consider a film of FCC lattice structure with (001) surfaces. The Hamiltonian is given by

$$\mathcal{H} = -\sum_{\langle i,j\rangle} J_{i,j}\sigma_i \cdot \sigma_j \tag{16.23}$$

where σ_i is the Ising spin at the lattice site i, $\sum_{\langle i,j\rangle}$ indicates the sum over the NN spin pairs σ_i and σ_j.

In the following, the interaction between two NN on the surface is supposed to be antiferromagnetic and equal to J_s. All other interactions are equal to $J = -1$ for simplicity. Note that in a previous paper [249] we have studied the case of the Heisenberg model on the same FCC antiferromagnetic film as a function of J_s (see Chapter 9).

For Ising spins, the GS configuration can be determined in a simple way as follows: We calculate the energy of the surface spin in the two configurations shown in Fig. 16.17 where the film surface contains spins 1 and 2 while the beneath layer spins 3 and 4. In the ordering of type I (Fig. 16.17a), the spins on the surface (xy plane) are antiparallel and in the ordering of type II (Fig. 16.17b), they are parallel. Of course, apart from the overall inversion, for type I there is a degenerate configuration by exchanging the spins 3 and 4. To see which configuration is stable, we write the energy of a surface spin for these two configurations

$$E_I = -4|J_s|$$

$$E_{II} = 4|J_s| - 4|J| \tag{16.24}$$

One sees that $E_I \leq E_{II}$ when $J_s/J \geq 0.5$. In the following, we study the case $J_s = J = -1$ so that the GS configuration is of type I.

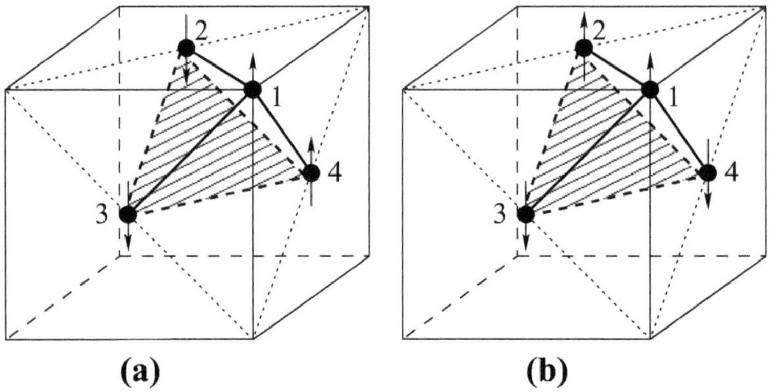

Figure 16.17 The ground state spin configuration of the FCC cell at the film surface: (a) ordering of type I for $J_s < -0.5|J|$, (b) ordering of type II for $J_s > -0.5|J|$.

16.4.2 Monte Carlo Results

In this paragraph, we show the results obtained by MC simulations with the Hamiltonian (16.23) using the high-precision Wang–Landau flat histogram technique [351] (see Chapter 6).

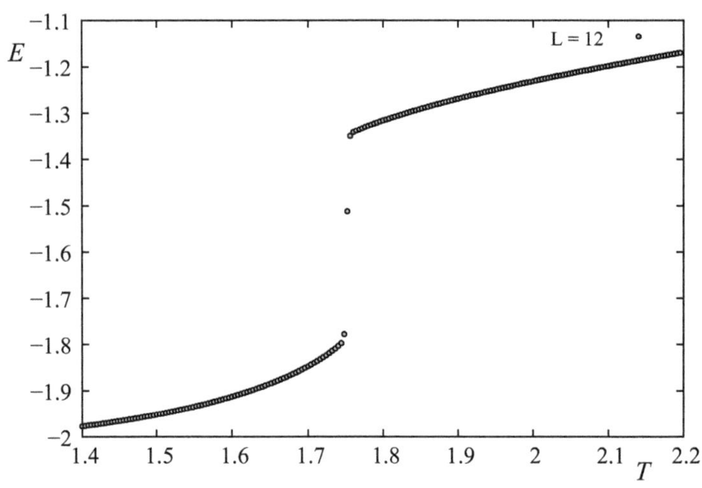

Figure 16.18 Bulk energy vs T for $L = N_z = 12$.

The film size used in our present work is $L \times L \times N_z$ where L is the number of cells in x and y directions, while N_z is that along the z direction (film thickness). We use here $L = 30, 40, \ldots, 150$ and $N_z = 2, 4, 8, 12$. Periodic boundary conditions are used in the xy planes. Our computer program was parallelized and run on a rack of several dozens of 64-bit CPU. $|J| = 1$ is taken as unit of energy in the following.

Before showing the results let us adopt the following notations. Sublattices 1 and 2 of the first FCC cell belongs to the surface layer, while sublattices 3 and 4 of the first cell belongs to the second layer (see Fig. 16.17a). In our simulations, we used N_z FCC cells, i.e., $2N_z$ atomic layers, and two symmetric film surfaces.

16.4.2.1 Crossover of the phase transition

As said earlier, the bulk FCC antiferromagnet with Ising spins shows a very strong first-order transition. This is seen in MC simulation even with a small lattice size as shown in Fig. 16.19.

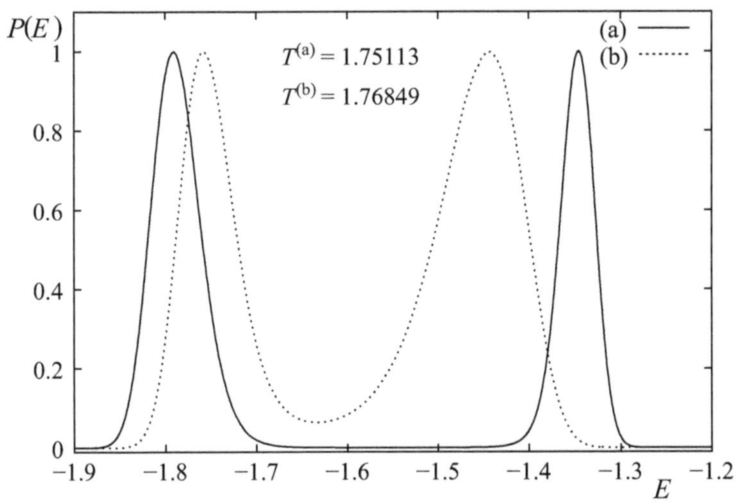

Figure 16.19 Bulk energy histogram for $L = N_z = 12$ with periodic boundary conditions in all three directions (a) and without PBC in z direction (b). The histogram was taken at the transition temperature T_c indicated on the figure.

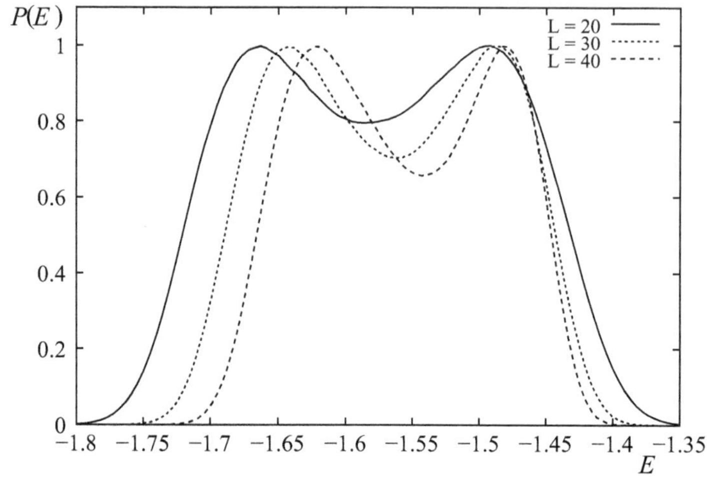

Figure 16.20 Energy histogram for $L = 20,\ 30,\ 40$ with film thickness $N_z = 4$ (8 atomic layers) at $T = 1.8218,\ 1.8223,\ 1.8227$, respectively.

Our purpose here is to see whether the transition becomes second order when we decrease the film thickness. As it turns out, the transition remains of first order down to $N_z = 4$ as seen by the double-peak energy histogram displayed in Fig. 16.20. Note that we do not need to go to larger L, the transition is clearly of first order already at $L = 40$.

In Fig. 16.21, we plot the latent heat ΔE as a function of thickness N_z. Data points are well fitted with the following formula:

$$\Delta E = A - \frac{B}{N_z^{d-1}}\left[1 + \frac{C}{N_z}\right] \qquad (16.25)$$

where $d = 3$ is the dimension, $A = 0.3370$, $B = 3.7068$, $C = -0.8817$. Note that the term N_z^{d-1} corresponds to the surface separating two domains of ordered and disordered phases at the transition. The second term in the brackets corresponds to a size correction. As seen in Fig. 16.21, the latent heat vanishes at a thickness between 2 and 3. This is verified by our simulations for $N_z = 2$. For $N_z = 2$, we find a transition with all second-order features: no discontinuity in energy (no double-peak structure) even when we go up to $L = 150$.

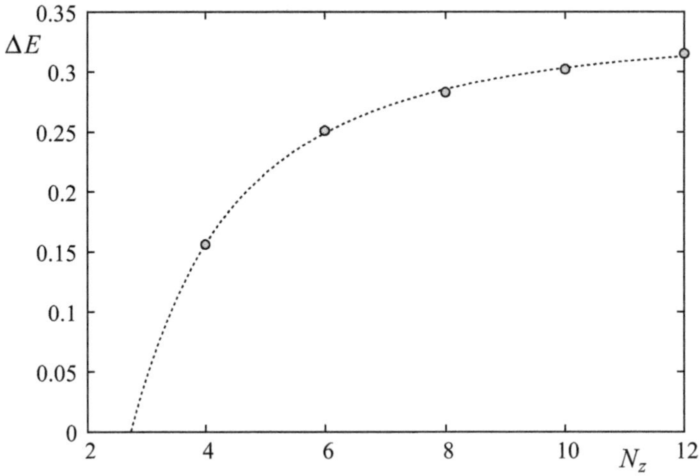

Figure 16.21 The latent heat ΔE as a function of thickness N_z.

Before showing in the following the results of $N_z = 2$, let us discuss the crossover. In the case of a film with finite thickness studied here, it appears that the first-order character is lost for very small N_z. A possible cause for the loss of the first-order transition is from the role of the correlation in the film. If a transition is of first order in 3D, i.e., the correlation length is finite at the transition temperature, then in thin films the thickness effect may be important: If the thickness is larger than the correlation length at the transition, than the first-order transition should remain. On the other hand, if the thickness is smaller than that correlation length, the spins then feel an "infinite" correlation length across the film thickness resulting in a second-order transition.

16.4.2.2 Film with 4 atomic layers ($N_z = 2$)

Let us show in Fig. 16.22 and Fig. 16.23 the energy and the magnetizations of sublattices 1 and 3 of the first two cells with $L = 120$ and $N_z = 2$.

It is interesting to note that the surface layer has larger magnetization than that of the second layer. This is not the case for non-frustrated films where the surface magnetization is always

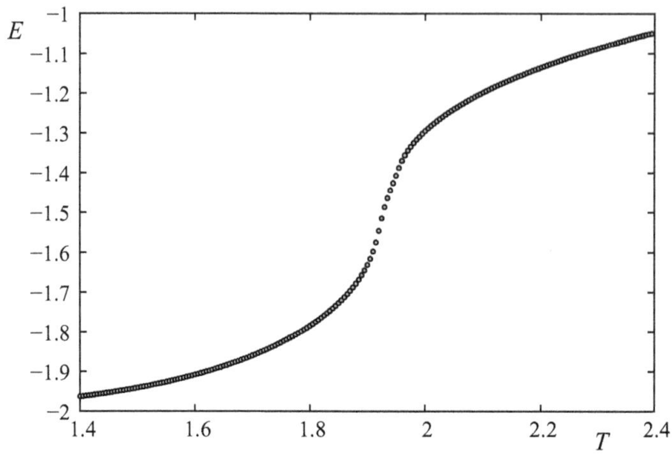

Figure 16.22 Energy versus temperature T for $L = 120$ with film thickness $N_z = 2$.

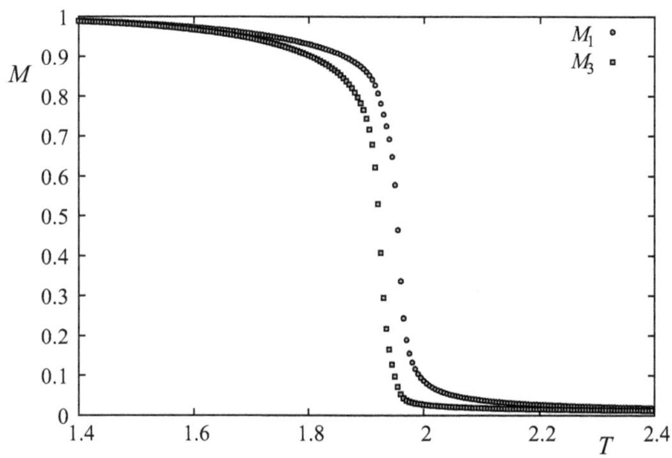

Figure 16.23 Sublattice magnetization for $L = 120$ with film thickness $N_z = 2$.

smaller than the interior ones because of the effects of low-lying energy surface-localized magnon modes [75, 78]. One explanation can be advanced: Due to the lack of neighbors surface spins are less frustrated than the interior spins. As a consequence, the surface spins maintain their ordering up to a higher temperature.

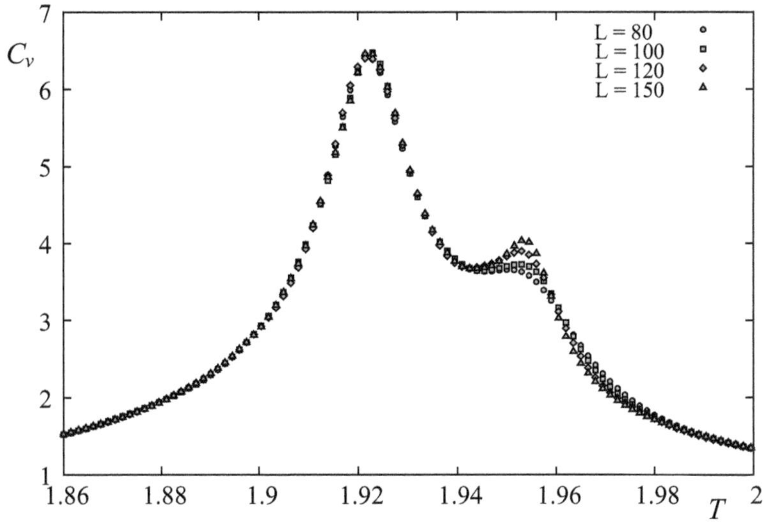

Figure 16.24 Specific heat are shown for various sizes L as a function of temperature.

Let us discuss finite-size effects in the transitions observed in Figs. 16.24 and 16.25. This is an important question because it is known that some apparent transitions are artifacts of small system sizes. To confirm further the observed transitions, we have made a study of finite-size effects on the layer susceptibilities by using the Wang–Landau technique [351].

We observe that there are two peaks in the specific heat: The first peak at $T_1 \simeq 1.927$, corresponding to the vanishing of the sublattice magnetization 3, does not depend on the lattice size while the second peak at $T_2 \simeq 1.959$, corresponding to the vanishing of the sublattice magnetization 1, does depend on L. Both histograms taken at these temperatures and the near-by ones show a Gaussian form indicating a non-first-order transition (see Fig. 16.26).

The fact that the peak at T_1 does not depend on L suggests two scenarios:

(i) The peak does not correspond to a real transition, since there exist systems where C_v shows a peak but we know that there is no transition just as in the case of 1D Ising chain.

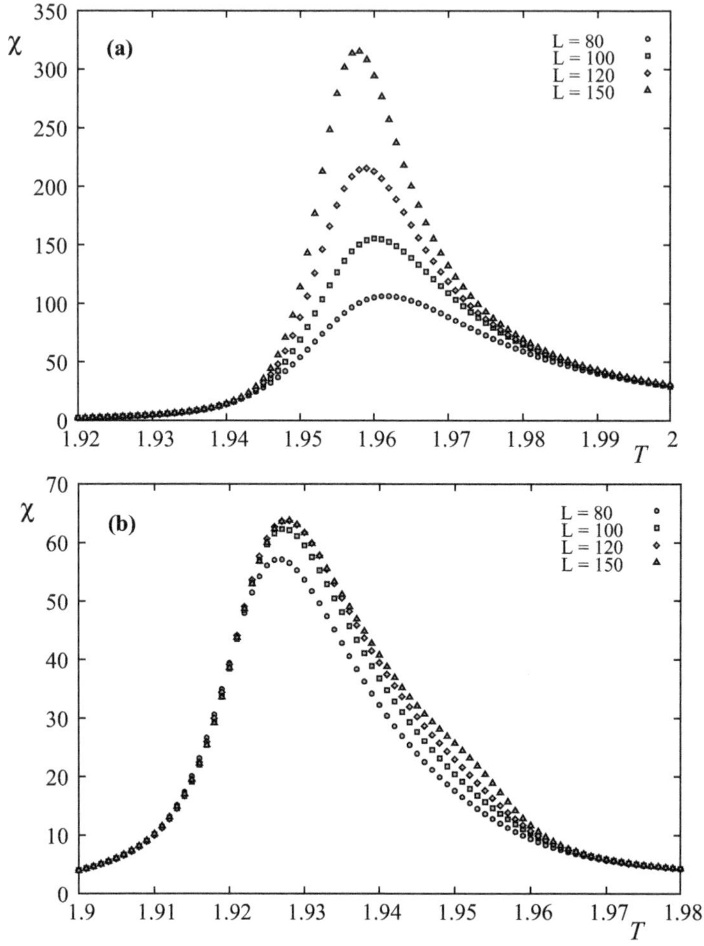

Figure 16.25 Susceptibilities of sublattice 1 (a) and 3 (b) are shown for various sizes L as a function of temperature.

(ii) The peak corresponds to a Kosterlitz–Thouless transition. To confirm this we need to check carefully several points such as the behavior of the correlation length, etc. This is a formidable task, which is not the scope of this chapter.

Whatever the scenario for the origin of the peak at T_1, we know that the interior layers are disordered between T_1 and T_2, while

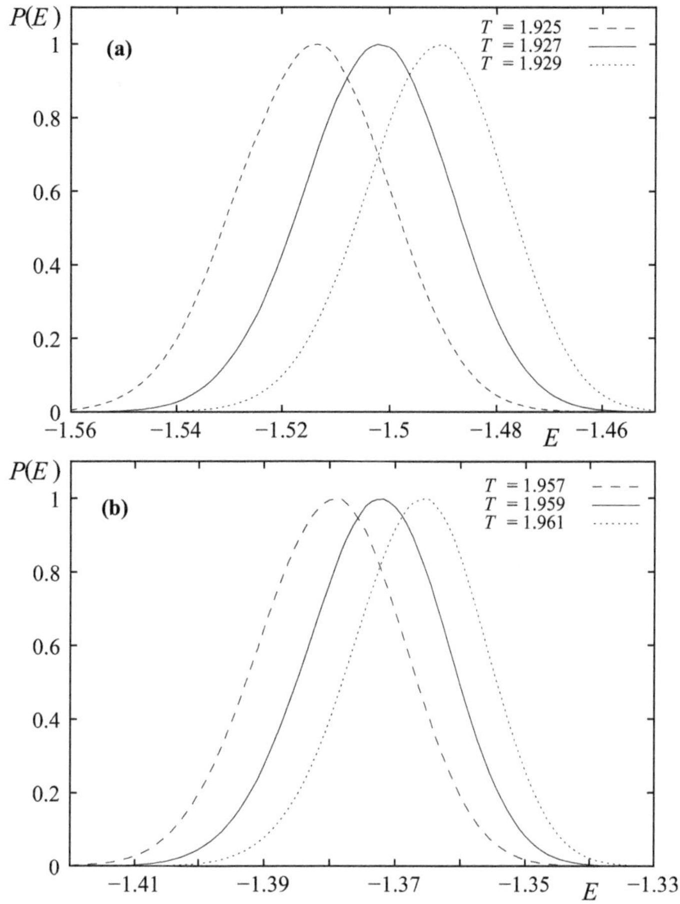

Figure 16.26 Energy histograms for $L = 120$ with film thickness $N_z = 2$ at the two temperatures (indicated on the figure) corresponding to the peaks observed in the specific heat. See text for comment.

the two surface layers are still ordered. Thus, the transition of the surface layers occurs while the disordered interior spins act on them as dynamical random fields. Unlike the true 2D random-field Ising model which does not allow a transition at finite temperature [160], the random fields acting on the surface layer are correlated. This explains why one has a finite-T transition here. Note that this situation is known in some exactly solved models where partial

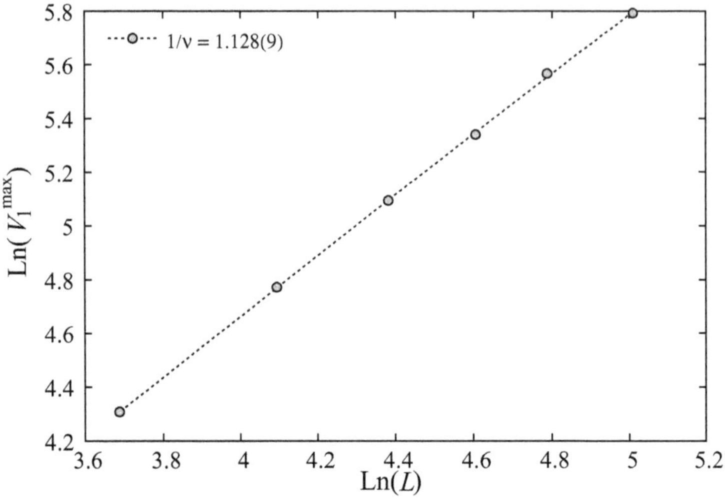

Figure 16.27 The maximum value of $V_1 = \langle E \rangle - \langle ME \rangle / \langle M \rangle$ versus L in the ln − ln scale. The slope of this straight line gives $1/\nu$.

disorder coexists with order at finite T [67, 76, 86]. However, it is not obvious to foresee what the universality class of the transition is at T_2. The theoretical argument of Capehart and Fisher [53] does not apply in the present situation because one does not have a single transition here, unlike the case of simple cubic ferromagnetic films studied before [270]. So, we wish to calculate the critical exponents associated with the transition at T_2.

The exponent ν can be obtained as follows. We calculate as a function of T the magnetization derivative with respect to $\beta = (k_B T)^{-1}$: $V_1 = \left\langle (\ln M)' \right\rangle = \langle E \rangle - \langle ME \rangle / \langle M \rangle$ where E is the system energy and M the sublattice order parameter. We identify the maximum of V_1 for each size L. From the finite-size scaling we know that $V_1^{\max}$ is proportional to $L^{1/\nu}$ [112]. We show in Fig. 16.27 the maximum of V_1 versus $\ln L$ for the first layer. We find $\nu = 0.887 \pm 0.009$. Now, using the scaling law $\chi^{\max} \propto L^{\gamma/\nu}$, we plot $\ln \chi^{\max}$ versus $\ln L$ in Fig. 16.28. The ratio of the critical exponents γ/ν is obtained by the slope of the straight line connecting the data points of each layer. From the value of ν we obtain $\gamma = 1.542 \pm 0.005$. These values do not correspond neither to 2D nor 3D Ising models ($\gamma_{2D} = 1.75$,

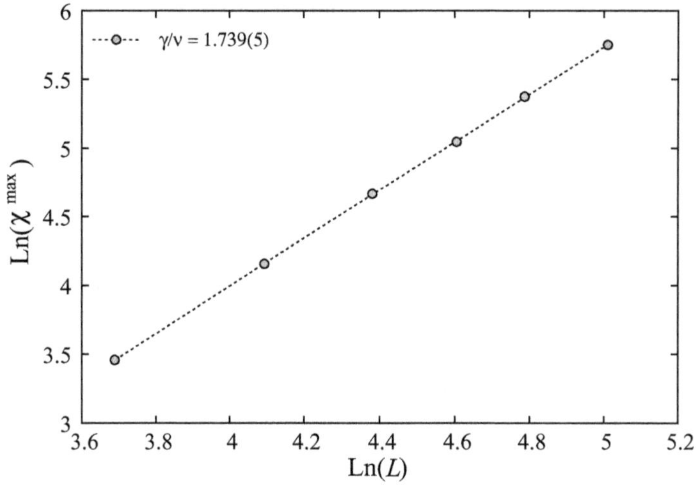

Figure 16.28 Maximum sublattice susceptibility $\chi^{\max}$ versus L in the ln − ln scale. The slope of this straight line gives γ/ν.

$\nu_{2D} = 1$, $\gamma_{3D} = 1.241$, $\nu_{3D} = 0.63$). We note, however, that if we think of the weak universality where only ratios of critical exponents are concerned [335], then the ratios of these exponents $1/\nu = 1.128$ and $\gamma/\nu = 1.739$ are not far from the 2D ones which are 1 and 1.75, respectively.

16.5 Concluding Remarks

We have considered a simple system, namely the Ising model on a simple cubic thin film, in order to clarify the point whether or not there is a continuous deviation of the 2D exponents with varying film thickness. From results obtained by the highly accurate multiple-histogram technique shown above, we conclude that the critical exponents in thin films show a continuous deviation from their 2D values as soon as the thickness departs from 1. This deviation stems from a deep physical mechanism: Capehart and Fisher [53] have argued that if one works exactly at the critical temperature $T_c(L = \infty, N_z)$ then the critical exponents should be those of 2D universality class as long as the film thickness is finite. At $T_c(L = \infty, N_z)$, the correlation in the z direction ξ remains finite while those in the xy planes become infinite. Hence, ξ is irrelevant to the criticality. This yields, therefore, the 2D behavior. However, when the system is away from $T_c(L = \infty, N_z)$, as is the case in numerical simulations using finite sizes, the system may have a 3D behavior as long as $\xi \ll N_z$. This should yield a deviation of 2D critical exponents. The results we obtained in this paper verify this picture. In addition, the prediction of Capehart and Fisher [53] for the shift of the critical temperature with the film thickness is in a perfect agreement with our simulations. Note, furthermore, that (i) the deviations of the exponents from their 2D values are very different in magnitude: while ν and α vary very little over the studied range of thickness, γ and specially β suffer stronger deviations, (ii) with a fixed thickness $N_z \neq 1$, the same "effective" exponents are observed, within errors, in simulations with and without periodic boundary condition in the z direction, (iii) to obtain the 3D behavior, the finite size scaling should be applied simultaneously in the three directions, i.e., all dimensions should be allowed to go to infinity. If we do not apply the scaling in the z direction, we will not obtain 3D behavior even with a very large, but fixed, thickness and even with periodic boundary condition in the z direction, (iv) with regard to the critical behavior, thin films behave as 2D systems but with effective critical exponents whose values deviate from those of 2D universality.

For a film with a first-order transition when the thickness is thick enough, we have shown by the high-performance Wang–Landau technique that there is a crossover to a second-order phase transition when the thickness goes down to a few atomic layers.

Chapter 17

Spin Resistivity in Thin Films

In this chapter, we show the recent results on the spin resistivity in magnetically ordered materials obtained by Monte Carlo simulations. We discuss its behavior as a function of temperature in various types of crystal: ferromagnetic, antiferromagnetic and frustrated spin systems. In the model used for simulations, we take into account the interaction between itinerant spins and that between lattice spins and itinerant spins. We also include a chemical potential term, as well as an electric field. We study in particular the behavior of the spin resistivity at and near the magnetic phase transition where the effect of the magnetic ordering is strongest. In ferromagnetic crystals, the spin resistivity shows a sharp peak very similar to the magnetic susceptibility. This can be understood if one relates the spin resistivity to the spin–spin correlation as suggested in a number of theories. The dependence of the shape of the peak on physical parameters such as carrier concentration, magnetic field strength, relaxation time, etc., is discussed. In antiferromagnets, the peak is not so pronounced and in some cases it is absent. Its direct relationship to the spin–spin correlation is not obvious. As for frustrated spin systems with strong first-order transition, the spin resistivity shows a discontinuity at the phase transition. To show the efficiency of the simulation method, we compare our results with

Physics of Magnetic Thin Films: Theory and Simulation
Hung T. Diep

ISBN 978-981-4877-42-8 (Hardcover), 978-1-003-12110-7 (eBook)
www.jennystanford.com

recent experimental data performed on semiconducting MnTe of NiAs structure. We observe a very good agreement with experiments on the spin resistivity in the whole range of temperature.

The results of this chapter are part of the results which have been published in Refs. [3–6, 151, 218–220].

17.1 Introduction

The study of the behavior of the resistivity is one of the fundamental tasks in materials science. This is because the transport properties occupy the first place in electronic devices and applications. The resistivity has been studied since the discovery of the electron a century ago by the simple Drude theory using the classical free particle model with collisions due to atoms in the crystal. The following relation is established between the conductivity σ and the electronic parameters e (charge) and m (mass):

$$\sigma = \frac{ne^2\tau}{m} \tag{17.1}$$

where τ is the electron relaxation time, namely the average time between two successive collisions. In more sophisticated treatments of the resistivity where various interactions are taken into account, this relation is still valid provided two modifications (i) the electron mass is replaced by its effective mass which includes various effects due to interactions with its environment (ii) the relaxation time τ is not a constant but dependent on collision mechanisms. The first modification is very important, the electron can have a "heavy" or "light" effective mass which modifies its mobility in crystals. The second modification has a strong impact on the temperature dependence of the resistivity: τ depends on some power of the electron energy, this power depends on the diffusion mechanisms such as collisions with charged impurities, neutral impurities, magnetic impurities, phonons, magnons, etc. As a consequence, the relaxation time averaged over energy, $< \tau >$, depends differently on T according to the nature of the collision source. The properties of the total resistivity stem thus from different kinds of diffusion processes. Each contribution has in general a different temperature dependence. Let us summarize the most important contributions to the total resistivity $\rho_t(T)$ at low temperature (T) in the following expression:

$$\rho_t(T) = \rho_0 + A_1T^2 + A_2T^5 + A_3 \ln \frac{\mu}{T} \quad (17.2)$$

where A_1, A_2 and A_3 are constants. The first term is T-independent, the second term proportional to T^2 represents the scattering of itinerant spins at low T by lattice spin waves. Note that the resistivity caused by a Fermi liquid is also proportional to T^2. The T^5 term corresponds to low-T resistivity in metals. This is due to the scattering of itinerant electrons by phonons. Note that at high T, metals show a linear-T dependence. The ln term is the resistivity due to the quantum Kondo effect caused by a magnetic impurity at very low T.

We are interested here in the spin resistivity ρ of magnetic materials. This subject has been investigated intensively both experimentally and theoretically for more than five decades. The rapid development of the field is due mainly to many applications in particular in spintronics.

Experiments have been performed in many magnetic materials including metals, semiconductors and superconductors. One interesting aspect of magnetic materials is the existence of a magnetic phase transition from a magnetically ordered phase to the paramagnetic (disordered) state. Very recent experiments such as those performed on the following compounds show different forms of anomaly of the magnetic resistivity at the magnetic phase transition temperature: ferromagnetic $SrRuO_3$ thin films [368], Ru-doped induced ferromagnetic $La_{0.4}Ca_{0.6}MnO_3$ [215], antiferromagnetic ϵ-$(Mn_{1-x}Fe_x)_{3.25}Ge$ [98], semiconducting $Pr_{0.7}Ca_{0.3}MnO_3$ thin films [377], superconducting $BaFe_2As_2$ single crystals [352], and $La_{1-x}Sr_xMnO_3$ [308]. Depending on the material, ρ can show a sharp peak at the magnetic transition temperature T_C [227] or just only a change of its slope, or an inflexion point. The latter case gives rise to a peak of the differential resistivity $d\rho/dT$ [269, 324].

As for theories, the T^2 magnetic contribution in Eq. (17.2) has been obtained from the magnon scattering by Kasuya [178]. However, at high T in particular in the region of the phase transition, much less has been known. de Gennes and Friedel [69] proposed the idea that the magnetic resistivity results from the spin–spin correlation so it should behave as the magnetic susceptibility, thus it should diverge at T_C. Fisher and Langer [117], and Kataoka [179] have suggested that the range of spin–spin correlation changes the shape of ρ near the phase transition. The resistivity due to magnetic impurities has been calculated by Zarand et al. [375] as a function of the Anderson's localization length. This parameter expresses in fact a kind the correlation sphere induced around each impurity. Their result shows that the resistivity peak depends on this parameter, in agreement with the spin–spin correlation idea.

The absence of Monte Carlo (MC) simulation in the literature on the spin transport has motivated our recent works: We have studied the spin current in ferromagnetic [3–5] and antiferromagnetic [6, 218–220] materials by MC simulations. The behavior of ρ as a function of T has been shown to be in agreement with main experimental features and theoretical investigations mentioned above.

In this chapter, we give a review of these works and outline the most important aspects and results. We consider in some details the case of MnTe where our simulation is in excellent agreement with experiments.

In Section 17.2, we show our basic model. We describe our MC method in Section 17.3. Results on ferromagnets and antiferromagnets are shown and discussed in Section 17.4. The spin resistivity of frustrated spin systems is presented in Section 17.5. Surface effects on the spin resistivity in a multilayer are shown in Section 17.6. The case of MnTe is considered in Section 17.7. Concluding remarks are given in Section 17.8.

17.2 Model

The model used in our MC simulation is very general. The itinerant spins move in a crystal whose lattice sites are occupied by localized spins. The itinerant spins and the localized spins may be of Ising, XY or Heisenberg models. Their interaction is usually limited to nearest neighbors (NN) but this assumption is not necessary. It can be ferromagnetic or antiferromagnetic.

Our purpose here is to study the effect of the magnetic transition on ρ. This transition occurs at a high temperature where it is known that the quantum nature of itinerant electron spins does not make significant additional effects with respect to the classical spin model. Therefore, to simplify the task, we consider here the classical spin model.

17.2.1 Interactions

We consider a crystal of a given lattice structure where each lattice site is occupied by a spin. The interaction between the lattice spins is given by the following Hamiltonian:

$$\mathcal{H}_l = -\sum_{(i,j)} J_{i,j}\mathbf{S}_i \cdot \mathbf{S}_j \tag{17.3}$$

where $\mathbf{S}_i$ is the spin localized at lattice site i of Ising, XY or Heisenberg model, $J_{i,j}$ the exchange integral between the spin pair $\mathbf{S}_i$ and $\mathbf{S}_j$ which is not limited to the interaction between nearest-neighbors (NN). Hereafter, except otherwise stated, we take $J_{i,j} = J$ for NN spin pairs, for simplicity. $J > 0$ (< 0) denotes ferromagnetic (antiferromagnetic) interaction. The system size is $L_x \times L_y \times L_z$ where $L_i (i = x, y, z)$ is the number of lattice cells in the i direction. Periodic boundary conditions (PBC) are used in all directions.

We define the interaction between itinerant spins and localized lattice spins as follows:

$$\mathcal{H}_r = -\sum_{i,j} I_{i,j}\sigma_i \cdot \mathbf{S}_j \tag{17.4}$$

where σ_i is the spin of the i-th itinerant electron and $I_{i,j}$ denotes the interaction which depends on the distance between electron i and spin $\mathbf{S}_j$ at lattice site j. For simplicity, we suppose the following interaction expression:

$$I_{i,j} = I_0 e^{-\alpha r_{ij}} \tag{17.5}$$

where $r_{ij} = |\mathbf{r}_i - \mathbf{r}_j|$, I_0 and α are constants. We use a cut-off distance D_1 for the above interaction. In the same way, the interaction between itinerant electrons is defined by

$$\mathcal{H}_m = -\sum_{i,j} K_{i,j}\sigma_i \cdot \sigma_j \tag{17.6}$$

$$K_{i,j} = K_0 e^{-\beta r_{ij}} \tag{17.7}$$

with $K_{i,j}$ being the interaction between electrons i and j, limited in a sphere of radius D_2. The choice of the constants K_0 and β will be discussed below.

Note that the choice of an exponential law does not affect the general feature of our results presented in this chapter because the short cut-off distance used here limits the interaction to a small number of neighbors, typically to next nearest neighbors (NNN), so the choice of another law such as a power law, or even discrete interaction values, for such a small cut-off will not make a qualitative difference in the results.

Itinerant electrons move under an electric field applied along the x axis. The PBC ensure that the electrons that leave the system at one end are to be reinserted at the other end. These boundary conditions are used in order to conserve the average density of itinerant electrons. One has

$$\mathcal{H}_E = -e\epsilon.\ell \tag{17.8}$$

where e is the electronic charge, ϵ an applied electric field and ℓ a displacement vector of an electron.

Since the interaction between itinerant electron spins is attractive, we need to add a kind of "chemical potential" in order to avoid a possible collapse of electrons into some points in the crystal and to ensure a homogeneous spatial distribution of electrons during the simulation. The chemical potential term is given by

$$\mathcal{H}_c = D[n(\mathbf{r}) - n_0] \tag{17.9}$$

where $n(\mathbf{r})$ is the concentration of itinerant spins in the sphere of D_2 radius, centered at $\mathbf{r}$, n_0 the average concentration, and D a constant parameter.

17.2.2 Choice of Parameters and Units

As said earlier, our model is very general. Several kinds of materials such as metals, semiconductors, insulating magnetic materials, etc., can be studied with this model, provided an appropriate choice of the parameters. For example, non-magnetic metals correspond to $I_{i,j} = K_{i,j} = 0$ (free conduction electrons). Magnetic semiconductors correspond to the choice of parameters K_0 and I_0 so as the energy of an itinerant electron due to the interaction $\mathcal{H}_r$ should be much lower than that due to $\mathcal{H}_m$, namely itinerant electrons are more or less bound to localized atoms. Note that $\mathcal{H}_m$ depends on the concentration of itinerant spins: For example, the dilute case yields a small $\mathcal{H}_m$. We make simulations for typical values of parameters which correspond to semiconductors. The choice of the parameters has been made after numerous test runs. We describe the principal requirements which guide the choice:

(i) We choose the interaction between lattice spins as unity, i.e., $|J| = 1$.
(ii) We choose the interaction between an itinerant and its surrounding lattice spins so as its energy E_i in the low T region is the same order of magnitude with that between lattice spins. To simplify, we take $\alpha = 1$. This case corresponds to a semiconductor, as said earlier.
(iii) The interaction between itinerant spins is chosen so that this contribution to the itinerant spin energy is smaller than E_i in order to highlight the effect of the lattice ordering on the spin current. To simplify, we take $\beta = 1$.
(iv) The choice of D is made in such a way to avoid the formation of clusters of itinerant spins (agglomeration) due to their attractive interaction [Eq. (17.7)].
(v) The electric field is chosen not so strong in order to avoid its dominant effect that would mask the effects of thermal fluctuations and of the magnetic ordering.
(vi) The density of the itinerant spins is chosen in a way that the contribution of the interactions between themselves is much weaker than E_i, as said above in the case of semiconductors.

Within the above requirements, a variation of each parameter does not change qualitatively the results shown below. Only the variation of D_1 in some antiferromagnets does change the results (see Ref. [219]).

The energy is measured in the unit of $|J|$. The temperature is expressed in the unit of $|J|/k_B$. The distance (D_1 and D_2) is in the unit of the lattice constant a. Real units will be used in Section 17.7 for comparison with experiments.

17.3 Simulation Method

Using the Metropolis algorithm, we first equilibrate the lattice at a given temperature T without itinerant electrons. When equilibrium is reached, we randomly add N_0 polarized itinerant spins into the lattice. Each itinerant electron interacts with lattice spins in a sphere of radius D_1 centered at its position, and with other itinerant electrons in a sphere of radius D_2. We next equilibrate the itinerant spins using the following updating. We calculate the energy E_{old} of an itinerant electron taking into account all interactions described above. Then we perform a trial move of length ℓ taken in an arbitrary direction with random modulus in the interval $[R_1, R_2]$ where $R_1 = 0$ and $R_2 = a$ (NN distance), a being the lattice constant. Note that the move is rejected if the electron falls in a sphere of radius r_0 centered at a lattice spin or at another itinerant electron. This excluded space emulates the Pauli exclusion. We calculate the new energy E_{new} and use the Metropolis algorithm to accept or reject the electron displacement. We choose another itinerant electron and begin again this procedure. When all itinerant electrons are considered, we say that we have made a MC sweeping, or one MC step/spin. We have to repeat a large number of MC steps/spin to reach a stationary transport regime. We then perform the averaging to determine physical properties such as magnetic resistivity, electron velocity, energy, etc., as functions of temperature. We define the dimensionless spin resistivity ρ as

$$\rho = \frac{1}{n_e} \tag{17.10}$$

where n_e is the number of itinerant electron spins crossing a unit slice perpendicular to the x direction per unit of time. An example with real units is shown in Section 17.7.

In order to have sufficient statistical averages on microscopic states of both the lattice spins and the itinerant spins, we use what we call "multi-step averaging procedure": after averaging the resistivity over N_1 steps for "each" lattice spin configuration, we thermalize again the lattice with N_2 steps in order to take another disconnected lattice configuration. Then we take back the averaging of the resistivity for N_1 steps for the new lattice configuration. We repeat this cycle for N_3 times, usually several hundreds of thousands times. The total MC steps for averaging is about 4×10^5 steps per spin in our simulations. This procedure reduces strongly thermal fluctuations observed in our previous work [4].

Of course, the larger N_2 and N_3 are, the better the statistics become. The question is what is the correct value of N_1 for averaging with one lattice spin configuration at a given T? This question is important because this is related to the relaxation time τ_L of the lattice spins compared to that of the itinerant spins, τ_I. The two extreme cases are (i) $\tau_L \simeq \tau_I$, one should take $N_1 = 1$, namely the lattice spin configuration should change with each move of itinerant spins (ii) $\tau_L \gg \tau_I$, in this case, itinerant spins can travel in the same lattice configuration for many times during the averaging.

In order to choose a right value of N_1, we consider the following temperature dependence of τ_L in non-frustrated spin systems. The relaxation time is expressed in this case as [154, 267]

$$\tau_L = \frac{A}{|1 - T/T_C|^{z\nu}} \tag{17.11}$$

where A is a constant, ν the correlation critical exponent, and z the dynamic exponent which depend on the spin model and space dimension. For 3D Ising model, $\nu = 0.638$ and $z = 2.02$. From this expression, we see that as T tends to T_C, τ_L diverges. In the critical region around T_C the system encounters thus the so-called "critical slowing down": the spin relaxation is extremely long due to the divergence of the spin–spin correlation. When we take into account the temperature dependence of τ_L, the shape of the resistivity is modified strongly at T_C where τ_L is very long, and in the paramagnetic phase where the relaxation time is very short due to rapid thermal fluctuations. On the other hand, at low T, τ_L does not modify ρ because in the ordered phase the spin landscape from one microscopic state to another does not change significantly to affect the motion of the itinerant spin (see the discussion in Ref. [220]).

In simulations, we consider a film with a thickness of N_z cubic cells in the z direction. Each of the xy planes contains $N_x \times N_y$ cells. The periodic boundary conditions are used on the xy planes to ensure that the itinerant electrons that leave the system at the second end are to be reinserted at the first end. For the z direction, we use the mirror reflection at the two surfaces. These boundary conditions conserve thus the average density of itinerant electrons. Dynamics of itinerant electrons is created by an electric field applied along the x axis.

17.4 Spin Resistivity in Ferromagnets and Antiferromagnets

In ferromagnets, experimental data mentioned above show a peak at T_C. The peak is related to the critical slowing down where the relaxation time diverges. Direct MC simulations in the case of Ising spin give a pronounced peak at T_C as shown in Fig. 17.1 in agreement with experiments. Note that ρ increases at low T. The reasons for this are multiple: It can stem from the freezing or crystallization of itinerant spins at low T or just from the smallness of the number of conduction electrons in such a low-T region. The shape of ρ depends on many factors: lattice structure, various interactions encountered by itinerant spins, electron concentration, relaxation time, spin model, magnetic-field amplitude, etc. For example, a decrease in the interaction between itinerant spins K_0 will reduce the increase of ρ as $T \to 0$, an applied magnetic field will decrease the peak height, the larger carrier concentration will reduce ρ in particular at T_C. All of these have been discussed in Ref. [218]. We note a strong effect of the temperature dependence of τ_L on ρ for $T \geq T_C$. This is very important because τ_L depends intrinsically on the material via ν and z.

For a quantitative comparison with experiments for a given material, it is necessary to take into account the specific parameters of that material. This is what we do in Section 17.7.

In antiferromagnets much less is known because there have been very few theoretical investigations which have been carried out. Haas [138] has shown that while in ferromagnets the resistivity ρ shows a sharp peak at the magnetic transition of the lattice spins, in antiferromagnets there is no such a peak. We found that the peak exists in antiferromagnets but it is less pronounced as seen in Fig. 17.1. The alternate change of sign of the spin–spin correlation with distance may have something to do with the absence of a

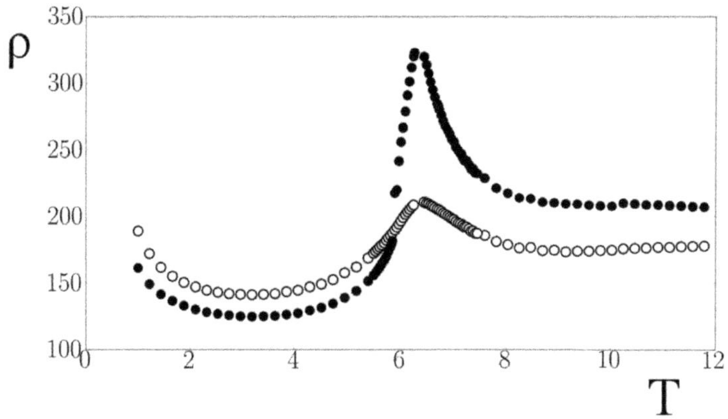

Figure 17.1 BCC ferromagnetic and antiferromagnetic films: Resistivity R with temperature-dependent relaxation for ferro- (black circles) and antiferromagnet (white circles) in arbitrary unit versus temperature T, in zero magnetic field, with electric field $\epsilon = 1$, $I_0 = 2$, $K_0 = 0.5$, $A = 1$.

sharp peak. We have tested, for example, the effect of the cut-off distance D_1 [219]: When D_1 increases, it will include successively up-spin shells and down-spin shells in the sphere of radius D_1. As a consequence, the difference between the numbers of up and down spins in the sphere oscillates with varying D_1, making an oscillatory behavior of ρ at small D_1, unlike in ferromagnets. It is interesting to note that in the presence of an itinerant spin, the ferromagnet and its antiferromagnet counterpart are no more invariant by the local Mattis transformation ($J_{ij} \to -J_{ij}$, $\mathbf{S}_j \to -\mathbf{S}_j$).

Note that we can calculate the spin resistivity using Boltzmann's equation combined with Monte Carlo simulations. The method consists of two steps:

(i) to write Boltzmann's equation in terms of the cluster sizes and cluster numbers at a given T,
(ii) to calculate the cluster numbers and the cluster sizes by Monte Carlo simulations using Hoshen–Kopelmann's algorithm [156]. Inserting the results into Boltzmann's equation, we obtain the relaxation time which allows us to calculate the spin resistivity. This has been done in Refs. [5, 6].

17.5 Spin Resistivity in Frustrated Systems

17.5.1 Simple Cubic $J_1 - J_2$ Model

We consider the simple cubic lattice shown in Fig. 17.2. The Hamiltonian is given by

$$\mathcal{H} = -J_1 \sum_{(i,j)} \mathbf{S}_i \cdot \mathbf{S}_j - J_2 \sum_{(i,m)} \mathbf{S}_i \cdot \mathbf{S}_m \tag{17.12}$$

where $\mathbf{S}_i$ is the Ising spin at the lattice site i, $\sum_{(i,j)}$ is made over the NN spin pairs with interaction J_1, while $\sum_{(i,m)}$ is performed over the NNN pairs with interaction J_2. We are interested in the frustrated regime. Therefore, hereafter we suppose that $J_1 = -J$ ($J > 0$, antiferromagnetic interaction, and $J_2 = -\eta J$ where η is a positive parameter. The ground state (GS) of this system is easy to obtain either by minimizing the energy, or by comparing the energies of different spin configurations, or just a numerical minimizing by a steepest descent method [248]. We obtain the antiferromagnetic configuration shown by the upper figure of Fig. 17.3 for $|J_2| < 0.25|J_1|$, or the configuration shown in the lower figure for $|J_2| > 0.25|J_1|$. Note that this latter configuration is 3-fold degenerate by choosing the parallel NN spins on x, y or z axis. With the permutation of black and white spins, the total degeneracy is thus 6.

The phase transition in the case of the Heisenberg model in the frustrated region ($|J_2| > 0.25|J_1|$) has been found to be of first order [274]. The system is very unstable due to its large degeneracy. We find that the case of the Ising spin shows an even stronger first-order transition [151]. It is interesting to note that the resistivity

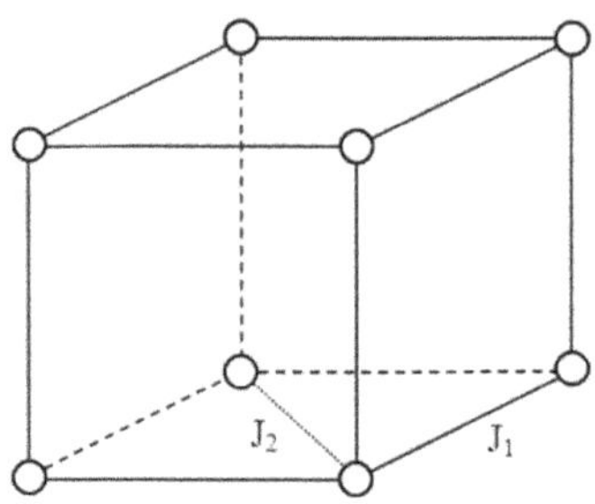

Figure 17.2 Simple cubic lattice with nearest and next-nearest neighbor interactions, J_1 and J_2, indicated.

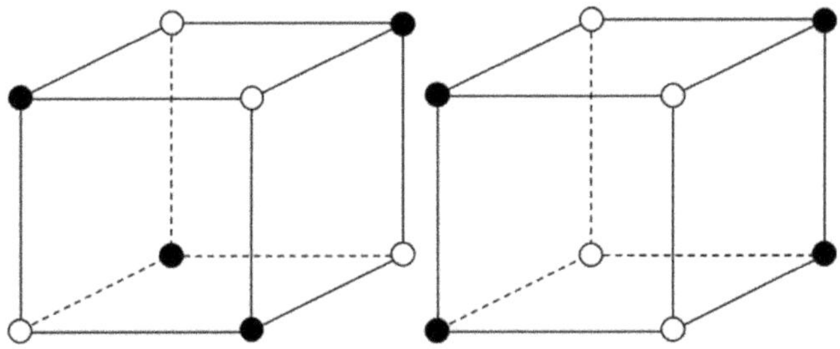

Figure 17.3 Simple cubic lattice. Up-spins: white circles, down-spins: black circles. Top: Ground state when $|J_2| < 0.25|J_1|$. Bottom: Ground state when $|J_2| > 0.25|J_1|$.

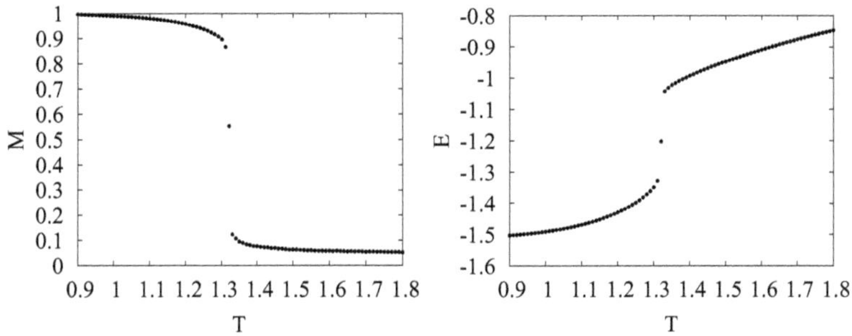

Figure 17.4 Left: Sublattice magnetization M versus T, Right: Energy versus T, for $|J_2| = 0.26|J_1|$, $N_x = N_y = 20$, $N_z = 6$.

of itinerant spins in systems with a first-order transition undergoes a discontinuity at T_C just as the system energy and the order parameter. We show ρ in Fig. 17.7 for several cut-off distance D_1. One observes here that ρ can jump or fall at the transition depending on the interaction range D_1. The resistivity discontinuity has been confirmed in another system with first-order transition, the frustrated FCC antiferromagnet [219]. This seems to be a general rule.

We show first the result of the lattice alone, namely without itinerant spins. The lattice in the frustrated region, i.e., $|J_2/J_1| > 0.25$, shows a strong first-order transition as seen in Fig. 17.4: The sublattice magnetization and the energy per spin as functions of T for $J_2 = -0.26|J_1|$ for the lattice size $N_x = N_y = 20$, $N_z = 6$

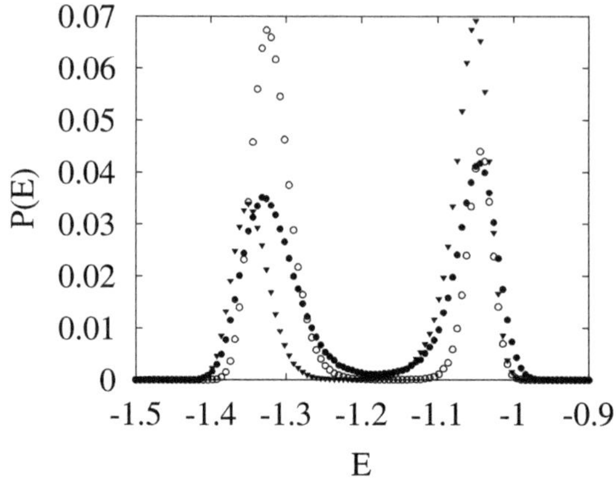

Figure 17.5 Energy histogram taken at the transition temperature T_C for $J_2 = -0.26|J_1|$: black circles are for $N_x = N_y = 20$, $N_z = 6$, $T_C = 1.320$, void circles for $N_x = N_y = 30$, $N_z = 6$, $T_C = 1.320$ and black triangles for $N_x = N_y = 20$, $N_z = 10$, $T_C = 1.305$. Other parameters are $I_0 = K_0 = 0.5$, $D_1 = 0.8a$, $D_2 = a$, $D = 1$, $\epsilon = 1$.

show a discontinuity at the transition temperature. To check further the first-order nature of the transition, we have calculated the energy histogram at the transition temperature T_C. This is shown in Fig. 17.5. The double-peak structure indicates the coexistence of the ordered and disordered phases at T_C. The distance between two peaks represents the latent heat.

Now we consider the lattice with the presence of itinerant spins. As far as the interaction between itinerant spins is attractive, we need a chemical potential to avoid the collapse of the system. The strength of the chemical potential D depends on K_0. We show in Fig. 17.6 the collapse phase diagram which allows to choose for a given K_0, an appropriate value of D.

We show now the main result on the spin resistivity versus T for $|J_2| = 0.26|J_1|$ for several values of D_1. Other parameters are the same as in Fig. 17.4. As said in Section 17.2.2, within the physical constraints, the variation of most of the parameters does not change qualitatively the physical effects observed in simulations, except for the parameter D_1. Due to the AF ordering, increasing D_1 means that we include successively neighboring down and up spins

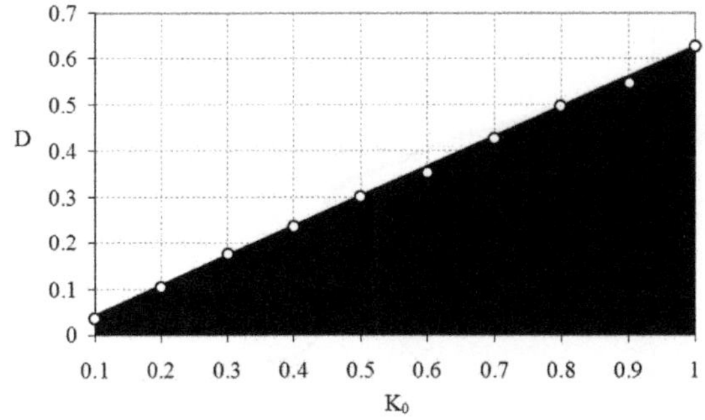

Figure 17.6 Phase diagram in the plane (K_0, D). The collapse region is in black, for $|J_2| = 0.26|J_1|$. Other parameters are $D_1 = D_2 = a$, $I_0 = 0.5$, $\epsilon = 1$.

surrounding a given itinerant spin. As a consequence, the energy of the itinerant spin oscillates with varying D_1, giving rise to the change of behavior of ρ: ρ can make a down fall or an upward jump at T_C depending on the value of D_1 as shown in Fig. 17.7. Note the discontinuity of ρ at T_C. This behavior has been observed and analyzed in terms of the averaged magnetization in the sphere of radius D_1 in the frustrated FCC antiferromagnet [219].

17.5.2 Fully Frustrated Face-Centered Cubic Antiferromagnet

We present here some results of the spin resistivity in a film of fully frustrated antiferromagnetic face-centered cubic lattice with Ising spins [219]. We show that the spin resistivity versus temperature exhibits a discontinuity at the phase transition temperature: an upward jump or a downward fall, depending on how many parallel and antiparallel localized spins interacting with a given itinerant spin. The surface effects as well as the difference of two degenerate states on the resistivity are analyzed. Comparison with non-frustrated antiferromagnets is shown to highlight the frustration effect. We also show and discuss the results of the Heisenberg spin model on the same lattice.

We consider a thin film of FCC lattice structure where each lattice site is occupied by an Ising spin whose values are ± 1.

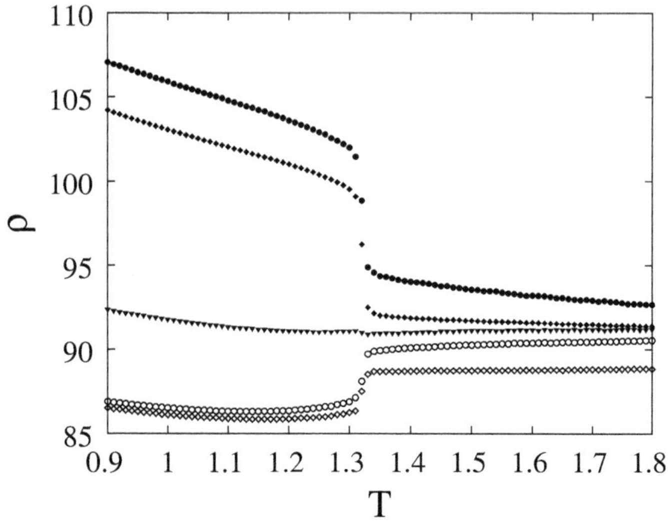

Figure 17.7 Spin resistivity versus T for $|J_2| = 0.26|J_1|$ for several values of D_1: from up to down $D_1 = 0.7a, 0.8a, 0.94a, a, 1.2a$. Other parameters are $N_x = N_y = 20$, $N_z = 6$, $I_0 = K_0 = 0.5$, $D_2 = a$, $D = 1$, $\epsilon = 1$.

The Heisenberg spin model is considered in Section 17.5.4. The interaction between the lattice spins is limited to nearest-neighbor (NN) pairs with the following Hamiltonian:

$$\mathcal{H}_l = -\sum_{(i,j)} J_{i,j} \mathbf{S}_i \cdot \mathbf{S}_j \quad (17.13)$$

where $\mathbf{S}_i$ is an Ising spin, $J_{i,j}$ the exchange integral between the NN spin pair $\mathbf{S}_i$ and $\mathbf{S}_j$. Hereafter we take $J_{i,j} = J_s$ for surface spins and $J_{i,j} = J$ for other NN spin pairs. We consider here the antiferromagnetic interaction $J_s, J < 0$ for the rest of this chapter. The system size is $N_x \times N_y \times N_z$ where N_x is the number of FCC cells in the x direction, etc. Periodic boundary conditions (PBC) are used in the x and y directions while the surfaces perpendicular to the z axis are free. The film thickness is N_z.

The FCC antiferromagnet is a fully frustrated system which is composed of tetrahedra each of which has four equilateral triangles. We know that it is impossible to fully satisfy simultaneously the three antiferromagnetic bond interactions on each triangle. As a consequence, the bulk lattice has an infinite ground-state

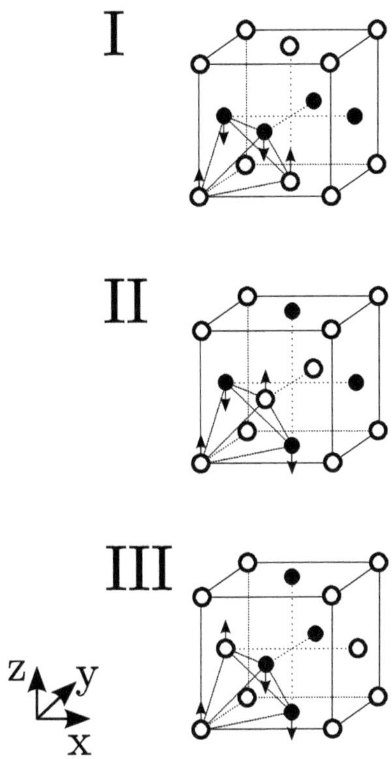

Figure 17.8 Ground state spin configuration of the FCC cell at the film surface (basal *xy* plane). The horizontal (vertical) axis is the *x* (*z*) axis. Top: ground state when $|J_s| < 0.5|J|$. Middle and bottom: first and second degenerate ground states when $|J_s| > 0.5|J|$.

degeneracy [85]. In the case of a thin film, the surface spin configuration depends on J_s as shown in Fig. 17.8 [271].

For $|J_s| < 0.5|J|$, the ground state is composed of ferromagnetic *xy* planes antiferromagnetically stacked in the *z* direction a shown in the upper figure of Fig. 17.8. For $|J_s| > 0.5|J|$, the ground state is two-fold degeneracy as shown in the middle and lower figures of Fig. 17.8. The difference of these two configurations is that the middle figure is an alternate stacking of up- and down-spin planes in the *y* direction while the lower figure is an alternate stacking of up- and down-spin planes in the *x* direction. These degenerate states are not equivalent in the spin transport in the *x* direction as seen below: In the first degenerate state, the itinerant spins move in the x direction between an up-spin plane and a down-spin plane, while in the second degenerate state the itinerant spins meet successively an up-spin plane and a down-spin plane perpendicular to their trajectories. We will present our results for these two cases separately.

In our simulations, we use the lattice size $N_x = N_y = 20$ and $N_z = 8$.

For studying the spin transport, we consider $N_0 = (N_x \times N_y \times N_z)/2$ itinerant spins (one electron per two FCC unit cells). Except otherwise stated, we choose interactions $I_0 = K_0 = 0.5$, $D_1 \in [0.6a; 2a]$, $D_2 = a$, $D = 0.35$, $\epsilon = 1$, $N_0 = 1600$, and $r_0 = 0.05a$. A discussion on the effect of a variation of each of these parameters was given in 17.2.2.

Note, however, that due to the form of the interaction given by Eq. (17.7), the itinerant spins have a tendency to form compact clusters to gain energy. This tendency is neutralized more or less by the concentration gradient term, or chemical potential, given by Eq. (17.9). The value of D has to be chosen so as to avoid a collapse of itinerant spins. We show in Fig. 17.9 the phase diagram in the space (K_0, D). The limit depends, of course, on the values of D_1 and D_2.

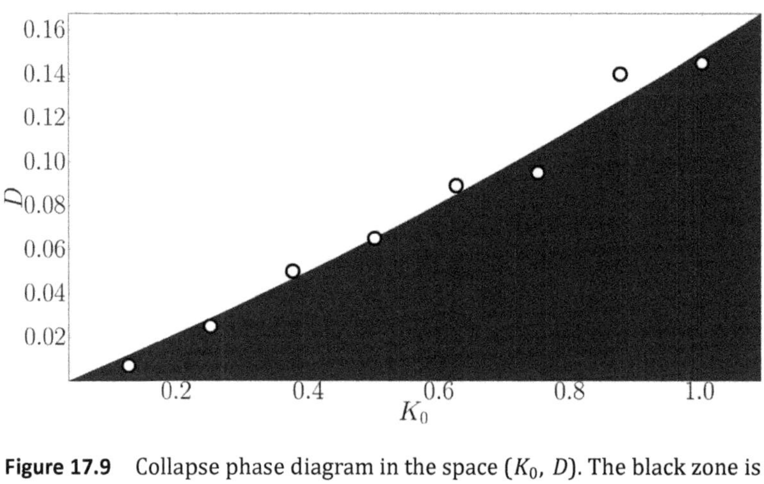

Figure 17.9 Collapse phase diagram in the space (K_0, D). The black zone is the collapse region. $D_1 = D_2 = a$. See text for comments.

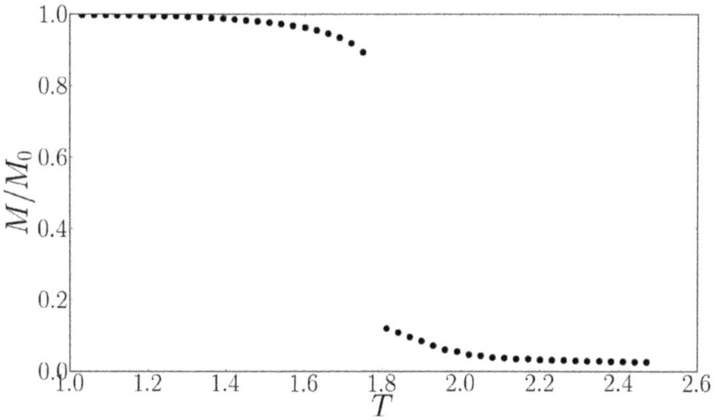

Figure 17.10 Staggered magnetization of antiferromagnetic FCC thin film of thickness $N_z = 8$ versus T. The transition temperature $T_C \simeq 1.79$.

17.5.2.1 Results for the Ising case

We show in Fig. 17.10 the staggered magnetization of the lattice as a function of T. As seen here the transition is of first order with a discontinuity at $T_C \simeq 1.79$. Note that the Ising antiferromagnetic FCC thin film shows a first-order transition down to a thickness of about four atomic layers [271].

We consider the first degenerate configuration shown in the middle figure of Fig. 17.8 with $J_s = J = -1$. In Fig. 17.11, we show the spin resistivity ρ versus T for two typical values of D_1. In all cases resistivity ρ is small for low T then increases with increasing T. At T_c, it undergoes a discontinuity upward jump. After transition, the resistivity decreases slowly to the same value for all D_1 in paramagnetic phase.

In Fig. 17.12 we show the spin resistivity ρ versus T for two typical values of D_1 using the second degenerate ground state (bottom figure of Fig. 17.8). Here we see that depending on D_1, the jump at the transition can be upward or downward.

In Fig. 17.13 we show the resistivity of the first and second degenerate states at a given $D_1 = a$, for comparison. The upward and downward jumps are seen. This difference is due to the positions of down spins on the trajectory of the itinerant spins [see the discussion below Eq. (17.13)].

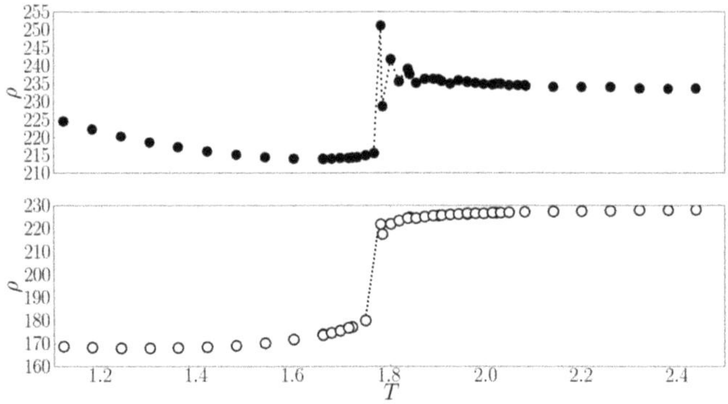

Figure 17.11 Resistivity of thin film of size $N_x = N_y = 20$ and $N_z = 8$ for $N_0 = 1600$ itinerant spins versus T for $D_1 = a$ (black circles) and $D_1 = 1.25a$ (white circles), a being the lattice constant. Case of the first degenerate state (middle figure of Fig. 17.8). $J_s = J = -1.0$, $I_0 = K_0 = 0.5$, $D = 0.35$.

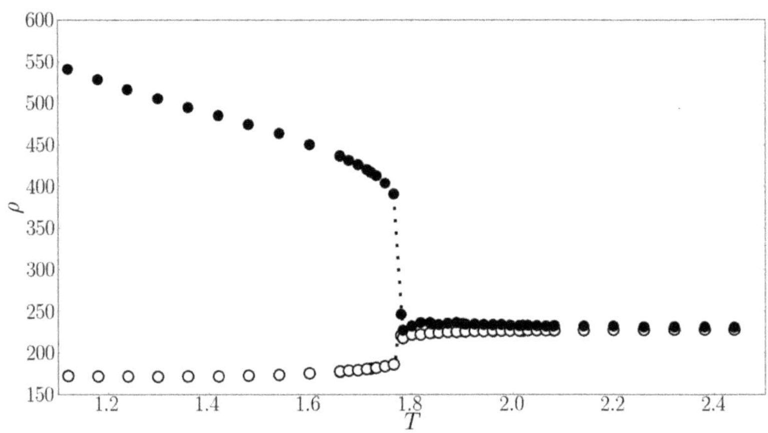

Figure 17.12 Resistivity versus temperature in the case of second degenerate state (bottom figure of Fig. 17.8) for $D_1 = a$ (black circles) and $D_1 = 1.25a$ (white circles) with $N_z = 8$, $N_0 = 1600$, $J_s = J = -1.0$, $I_0 = K_0 = 0.5$, $D = 0.35$.

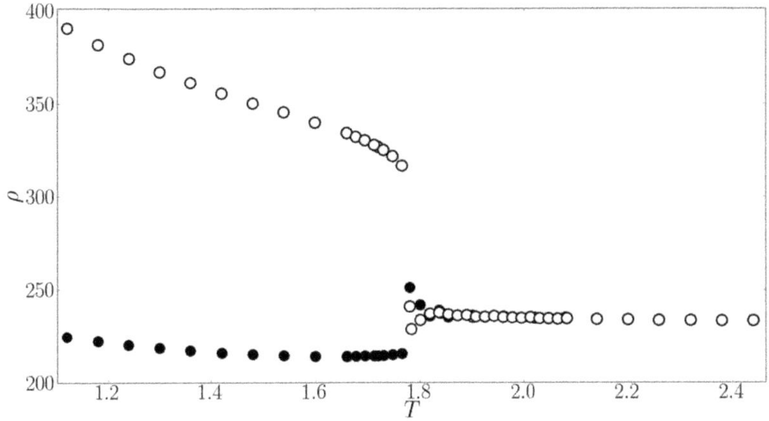

Figure 17.13 Resistivity versus T for first (black points) and second degenerate (white points) configurations for $D_1 = a$ with $N_z = 8$, $N_0 = 1600$, $J_s = J = -1.0$, $I_0 = K_0 = 0.5$, $D = 0.35$.

17.5.3 Surface Effects

In order to enhance the surface effect, in addition to a small value of J_s we allow the exchange interaction between a surface spin and its neighbors in the beneath layer to be J_p which will be taken to be small in magnitude. We show in Fig. 17.14 the surface magnetization and the magnetizations of the interior layers as functions of T for $J_s = J_p = -0.5$ and $J = -1$. As seen here, the surface transition takes place at a lower temperature $T_1 \simeq 1.2$ while interior layers become disordered at $T_2 \simeq 1.8$. As a consequence, one expects that the surface fluctuations at T_1 will induce an anomaly in ρ in addition to that at T_2. This is shown in Fig. 17.15. Note that the increase of ρ at low T is an effect of a freezing of itinerant spins at low T as discussed above.

17.5.4 Results for the Heisenberg Case

In this section, we presently briefly the results on the same lattice with the Heisenberg spin model. Itinerant spins are the same as used above, namely polarized Ising spins. This assumption allows to outline only the effect of the continuous nature of the Heisenberg lattice spin on the resistivity. The full Hamiltonian with different

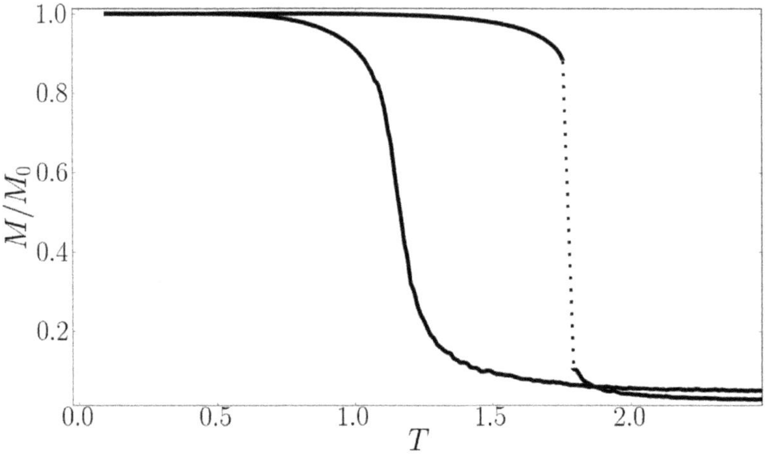

Figure 17.14 Layer magnetizations versus T for $J_s = J_p = -0.5$ and $J = -1$. Other parameters: $D_1 = a$, $N_z = 8$, $N_0 = 1600$, $I_0 = K_0 = 0.5$, $D = 0.35$. The surface transition is at $T_1 \simeq 1.2$. The vertical dotted line is a guide to the eye indicating the discontinuous fall of interior layer magnetization.

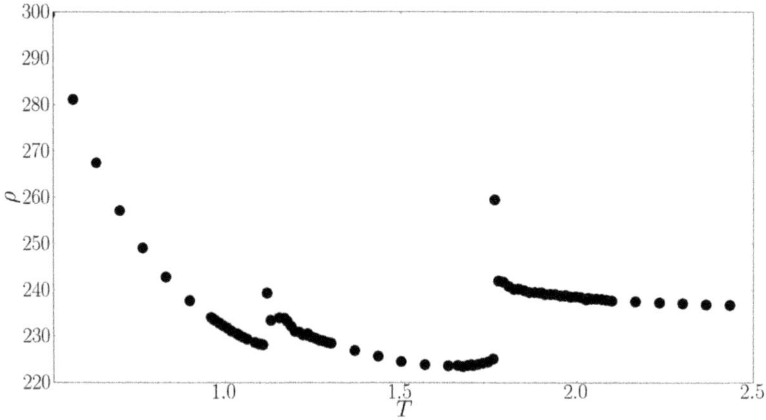

Figure 17.15 Resistivity versus temperature T in the case shown in Fig. 17.14. There are two anomalies occurring, respectively, at the surface and bulk transition temperatures.

kinds of interaction is assumed as above except the exchange interaction between lattice spins. This is given by

$$\mathcal{H} = -\sum_{\langle i,j\rangle} J_{i,j}\mathbf{S}_i \cdot \mathbf{S}_j - A\sum_{\langle i,j\rangle} S_i^z S_j^z \tag{17.14}$$

where $\mathbf{S}_i$ is the Heisenberg spin at the site i and A an Ising-like anisotropy which is assumed to be negative to favor an antiparallel spin ordering on the z axis. When A is zero, one has the isotropic Heisenberg model. In order to have at phase transition at a non-zero T, we should take a non-zero value for A because it is known, by the theorem of Mermin–Wagner [231], that for vector spin models there is no long-range ordering at finite temperatures in two dimensions. The small thickness considered here is, in a phase-transition point of view, equivalent to a two-dimensional system. Except A, note that we use the same assumptions as in Eq. (17.13).

Let us show now in Fig. 17.16 the resistivity as a function of T for two typical values of D_1. As seen, depending on the value of D_1, ρ undergoes a sharp increase or decrease at T_C. At some values such as that corresponding to the upper curve of the upper figure, the resistivity can go across a large region of fluctuations without a sharp jump. So in experiments, care should be taken to interpret similar behavior if any. Note that the second degenerate configuration yields always a larger resistivity than in the first one, as observed in the Ising case in the previous section.

The effect of A on the resistivity is not very important in the reasonable range [0.1, 1.5]: Except the fact that T_C varies with A, for instance $T_C \simeq 0.65$ for $A = 0.5$ and $T_C \simeq 0.55$ for $A = 0.1$, the discontinuity of ρ at T_C diminishes only slightly with decreasing A.

17.5.5 Remarks

We have shown in this section that the spin resistivity ρ of the fully frustrated FCC antiferromagnet is quite different from that of ferromagnets [4] and non-frustrated antiferromagnets [218] shown in previous section. ρ does not show a peak at the magnetic phase transition temperature. It shows instead a discontinuous jump at the transition temperature T_C. The jump depends on the numbers of parallel and antiparallel localized spins which interact with an

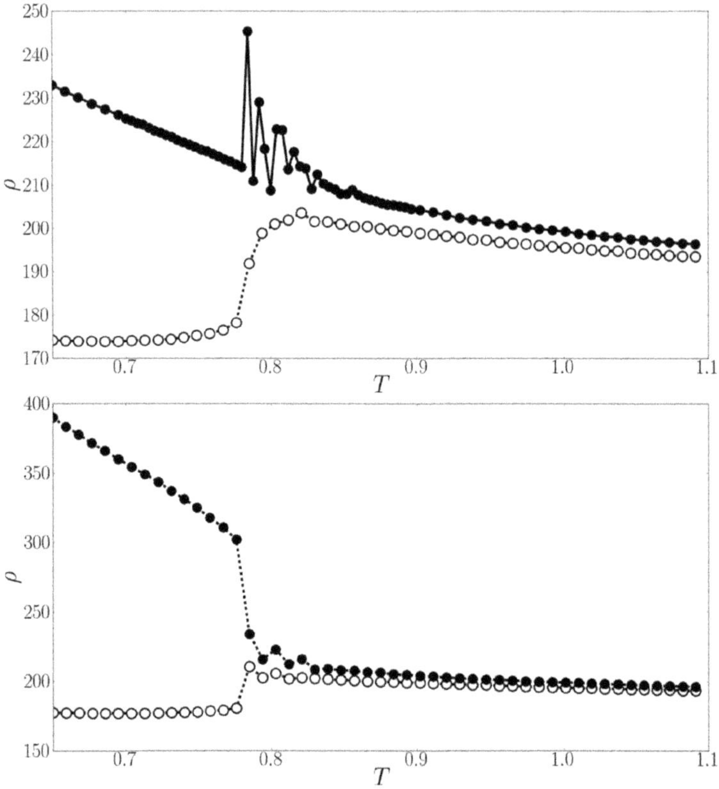

Figure 17.16 Heisenberg case. Resistivity of thin film of size $N_x = N_y = 20$ and $N_z = 8$ for $N_0 = 1600$ itinerant spins versus T for $D_1 = a$ (black circles) and $D_1 = 1.25a$ (white circles) in unit of the lattice constant a for first (upper) and second (lower) degenerate states. $A = -1, J_s = J = -1.0$, $I_0 = K_0 = 0.5$, $D = 0.35$.

itinerant spin. After transition, the resistivity tends to a saturation value independent of D_1. The abrupt behavior of ρ at T_C in the antiferromagnetic FCC Ising lattice is an effect of the frustration which causes a first-order transition of the lattice magnetic ordering leading to a discontinuity of ρ at T_C.

We are not aware of experiments performed on spin transport in materials with first-order magnetic transition. Our result is thus a prediction which would be useful for future experiments. Note, however, that for electrical transport, the electrical resistivity shows a discontinuity at a metal-insulator "first-order" transition in $PrNiO_3$ [128] and $NdNiO_3$ [129]. The magnetic resistivity found in this chapter has also a discontinuity behavior at a magnetic "first-order" transition. This similarity shows that the resistivity is closely related to the nature of the phase transition, whatever its origin (magnetic, insulator-metal, ...) may be. The mapping between the two cases, however, is not the scope of this chapter.

We have also shown that the surface disordering causes a peak of the resistivity at the surface transition temperature.

In the Heisenberg model, the spin continuous degrees of freedom weaken the first-order transition, yielding in general a reduction of the critical temperature and a less abrupt change of the resistivity at the transition.

As a last remark, let us emphasize that the behavior of the spin resistivity at T_C is quite different from one antiferromagnet to another. It depends on many factors such as the lattice structure, the interaction range, the spin model and the instability (in particular due to frustration) of the spin ordering. We have studied here the effects of some of them, but a throughout understanding needs much more investigations and analysis.

17.6 Surface Effects in a Multilayer

We see so far that when there is a magnetic phase transition, the spin resistivity undergoes an anomaly. In magnetic thin films, when there is a surface phase transition at a temperature T_S different from that of the bulk one (T_c), we expect two peaks of ρ one at T_S and the other at T_c. We show here an example of a thin film composed of three sub-films: The middle film of four atomic layers between two surface films of five layers. The lattice sites are occupied by Ising spins interacting with each other via nearest-neighbor ferromagnetic interaction. Let us suppose the interaction between spins in the outside films be J_S and that in the middle film be J. The inter-film interaction is J. The lattice structure is face-centered cubic. In order to enhance surface effects, we suppose in addition $J_S << J$. We show in Fig. 17.17 the layer magnetization as well as the layer susceptibility. We observe that the outside films

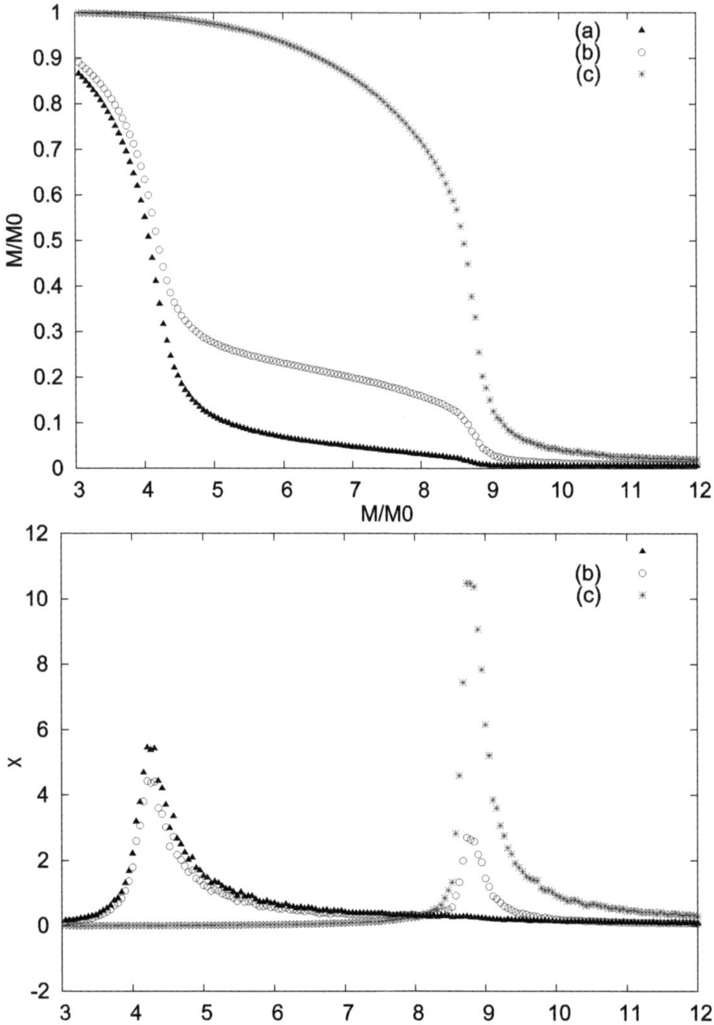

Figure 17.17 Top: Magnetization versus T in the case where the system is made of three films: The first and the third have five layers with a weaker interaction J_s, while the middle has four layers with interaction $J = 1$. We take $J_s = 0.2J$. Black triangles: magnetization of the surface films, stars: magnetization of the middle film, void circles: total magnetization. Bottom: Susceptibility versus T. Black triangles: susceptibility of the surface films, stars: susceptibility of the middle films, void circles: total susceptibility. See text for comments.

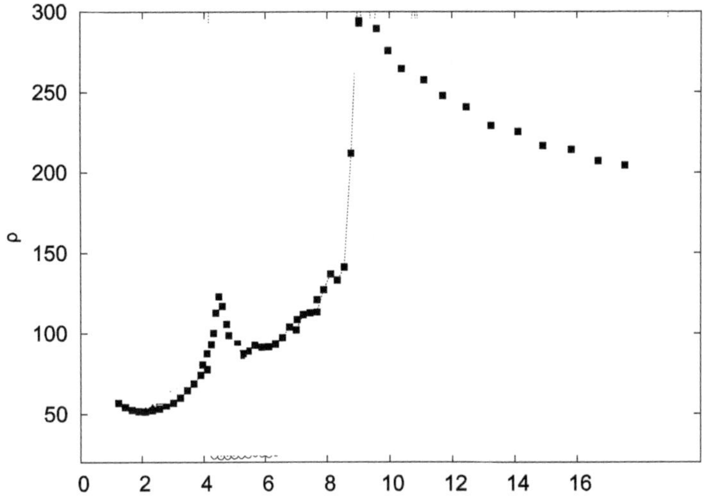

Figure 17.18 Resistivity ρ in arbitrary unit versus T of the system described in the previous figure's caption.

undergo a phase transition at a temperature far below the transition temperature of the middle film. As stated above, a phase transition induces an anomaly in the spin resistivity: The two phase transitions observed in Fig. 17.17 give rise to two peaks of ρ shown in Fig. 17.18. The surface peak of ρ has been also seen in a frustrated film [219].

17.7 The Case of MnTe

The pure MnTe has either the zinc blende structure [148] or the hexagonal NiAs one shown in Fig. 17.19. We confine ourselves in the latter case where the Néel temperature is $T_N = 310$ K [339]. Hexagonal MnTe is a crossroad semiconductor with a big gap (1.27 eV) and a room-temperature carrier concentration of $n = 4.3 \times 10^{17}\text{cm}^{-3}$ [8, 236]. Without doping, MnTe is non-degenerate. The behavior of ρ in MnTe as a function of T has been experimentally shown [13, 59, 101, 142, 209]. The hexagonal is composed of ferromagnetic xy hexagonal planes antiferromagnetically stacked in the c direction. The NN distance in the c direction is $c/2 \simeq 3.36$ Å shorter than the in-plane NN distance which is $a = 4.158$ Å. Neutron

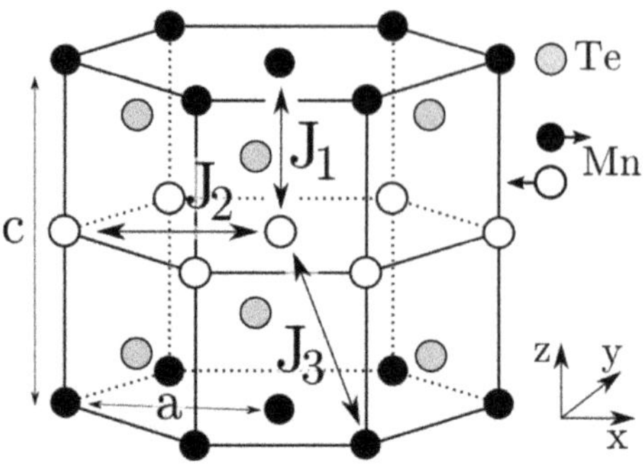

Figure 17.19 Structure of MnTe of NiAs type is shown. Antiparallel spins are shown by black and white circles. NN interaction is marked by J_1, next NN interaction by J_2, and third NN one by J_3.

scattering experiments show that the main exchange interactions between Mn spins in MnTe are (i) interaction between NN along the c axis with the value $J_1/k_B = -21.5 \pm 0.3$ K, (ii) ferromagnetic exchange $J_2/k_B \approx 0.67 \pm 0.05$ K between in-plane neighboring Mn (they are next NN by distance), (iii) third NN antiferromagnetic interaction $J_3/k_B \simeq -2.87 \pm 0.04$ K. The spins are lying in the xy planes perpendicular to the c direction with a small in-plane easy-axis anisotropy D [339]. We note that the values of the exchange integrals given above have been deduced from experimental data by fitting with a formula obtained from a free spin wave theory [339]. Other fittings with mean-field theories give slightly different values: $J_1/k_B = -16.7$ K, $J_2/k_B = 2.55$ K and $J_3/k_B = -0.28$ K [236].

The Hamiltonian is given by

$$\mathcal{H} = -J_1 \sum_{(i,j)} \mathbf{S}_i \cdot \mathbf{S}_j - J_2 \sum_{(i,m)} \mathbf{S}_i \cdot \mathbf{S}_m - J_3 \sum_{(i,k)} \mathbf{S}_i \cdot \mathbf{S}_k - D_a \sum_i (S_i^x)^2 \tag{17.15}$$

where $\mathbf{S}_i$ is the Heisenberg spin at the lattice site i, $\sum_{(i,j)}$ is made over the NN spin pairs $\mathbf{S}_i$ and $\mathbf{S}_j$ with interaction J_1, while $\sum_{(i,m)}$ and $\sum_{(i,k)}$ are made over the NNN and third NN neighbor pairs with interactions J_2 and J_3, respectively. $D_a > 0$ is an anisotropy constant which favors the in-plane x easy-axis spin configuration. The Mn spin is experimentally known to be of the Heisenberg model with magnitude $S = 5/2$ [339].

The interaction between an itinerant spin and surrounding Mn spins in semiconducting MnTe is written as

$$\mathcal{H}_i = -\sum_n I(\mathbf{r}-\mathbf{R}_n)\sigma \cdot \mathbf{S}_n \tag{17.16}$$

where $I(\mathbf{r}-\mathbf{R}_n) > 0$ is a ferromagnetic exchange interaction between itinerant spin σ at $\mathbf{r}$ and Mn spin $\mathbf{S}_n$ at lattice site $\mathbf{R}_n$. The sum on lattice spins $\mathbf{S}_n$ is limited at cut-off distance $D_1 = a$. We use here the Ising model for the electron spin. In doing so, we neglect the quantum effects which are, of course, important at very low temperature but not in the transition region at room temperature where we focus our attention. We suppose the following distance dependence of $I(\mathbf{r}-\mathbf{R}_n)$:

$$I(\mathbf{r}-\mathbf{R}_n) = I_0 \exp[-\alpha(\mathbf{r}-\mathbf{R}_n)] \tag{17.17}$$

where I_0 and α are constants. We choose $\alpha = 1$ for convenience. The choice of I_0 should be made so that the interaction $\mathcal{H}_i$ yields an energy much smaller than the lattice energy due to $\mathcal{H}$ (see the discussion on the choice of variables given above). Note that the cut-off distance is rather short so that the obtained results shown below still keep a general character which does not depend on the choice of exponential form. Since in MnTe the carrier concentration is $n = 4.3 \times 10^{17}\text{cm}^{-3}$, very low with respect to the concentration of its surrounding lattice spins $\simeq 10^{22}\text{cm}^{-3}$, we do not take into account the interaction between itinerant spins.

As said before, the values of the exchange interactions deduced from experimental data depend on the model Hamiltonian, in particular the spin model, as well as the approximations. Furthermore, in semiconductors, the carrier concentration is a function of T. In our model, there is, however, no interaction between itinerant spins. Therefore, the number of itinerant spins used in the simulation is important only for statistical average: The larger the number of itinerant spins, the better the statistical average. The current obtained is proportional to the number of itinerant spins but there are no extra physical effects. Using the exchange integrals slightly modified with respect to the ones given above, we have calculated ρ of the hexagonal MnTe. The result of ρ is shown in Fig. 17.20. Note that with J_3 slightly larger in magnitude than the value deduced from experiments, we find $T_N = 310$ K in excellent agreement with

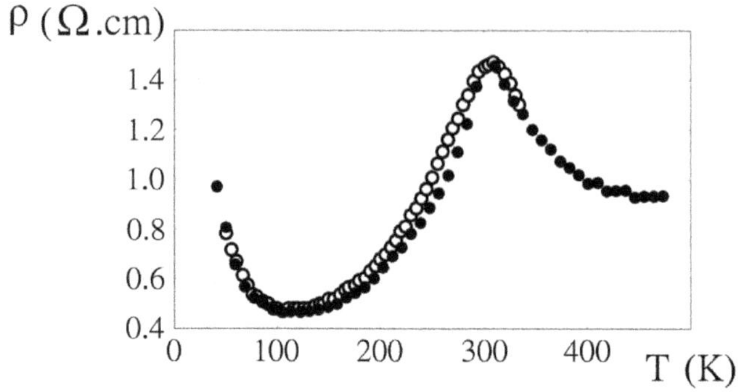

Figure 17.20 Spin resistivity ρ versus temperature T. Black circles are from Monte Carlo simulation, white circles are experimental data taken from He et al. The parameters used in the simulation are $J1 = -21.5$K, $J_2 = 2.55$ K, $J_3 = -9$ K, $I_0 = 2$ K, $D_a = 0.12$ K, $D_1 = a = 4.148$ Å, $\epsilon = 2 \times 10^5$ V/m, $L = 30a$ (lattice size L^3).

experiments. Furthermore, we observe that ρ shows a pronounced peak and coincides with the experimental data. The values we used to obtain that agreement are $A = 1$ and Heisenberg critical exponents $\nu = 0.707$, $z = 1.97$ [267]. In the temperature regions below $T < 140$ K and above T_N the MC result is also in excellent agreement with experiment, unlike in our previous work [6] using Boltzmann's equation.

Using the value of ρ, we obtain the relaxation time of itinerant spin equal to $\tau_I \simeq 0.1$ ps, and the mean free path equal to $\bar{l} \simeq 20$ Å, at the critical temperature.

17.8 Conclusion

We have shown in this chapter how MC simulations can be used to produce properties of spin transport in magnetic materials. The method is very general, it can be easily applied to a wide range of materials from ferromagnets to antiferromagnets of different lattices and spin models. The results of the spin resistivity ρ as a function of temperature under different situations can be obtained and compared to experiments. We were concentrated in the magnetic phase transition region where theories failed to predict correct behaviors of ρ. This is due to the fact that the magnetic resistivity is intimately related to the spin–spin correlation which is very different from one material to another. This correlation, as we know in the domain of phase transition and critical phenomena, governs the nature of the transition: phase transitions of second order of different universality classes and phase transitions of first order. Needless to say, the nature of the phase transition affects the behavior of ρ as seen above: different shapes of ρ and discontinuity at T_C, etc. We have, for a good demonstration of the efficiency of our method, studied the case of MnTe where experimental data are recently available for the whole temperature range. Our result is in excellent agreement with experiments: It reproduces the correct Néel temperature as well as the shape of the peak at the phase transition.

Part III

Solutions to Exercises and Problems

Chapter 18

Solutions to Exercises and Problems

18.1 Solutions to Problems of Chapter 1

Problem 1. Orbital and spin moments of an electron:

Guide: The state of an electron in an atomic orbital is defined by four quantum numbers n, l, m_l, m_s where $n = 1, 2, 3, \cdots$; $l = 0, 1, 2, \cdots, n-1$; $m_l = -l, -l+1, \cdots, l-1, l$; $m_s = -s, s$ with $s = 1/2$. The orbital angular momentum is $\mathbf{L}$ with eigenvalues $\hbar m_l$, the spin angular momentum is $\mathbf{S}$ with eigenvalue $\hbar m_s$. The orbital magnetic moment is $\mathbf{M}_l = -\mu_B \mathbf{L}$, and the spin magnetic moment is $\mathbf{M}_s = -g\mu_B \mathbf{S}$ where $g = 2.0023 \simeq 2$ (Landé factor or electron spectroscopic factor). The total magnetic moment of an electron is thus $\mathbf{M}_t = -\mu_B(\mathbf{L} + g\mathbf{S})$.

Problem 2. Zeeman effect:

Guide:

(a) The number of Fe atoms in 1 m^3: $N = \frac{7970\times 6.025\times 10^{26}}{56} = 8.58 \times 10^{28}$ (the Avogadro number per kilogram is 6.025×10^{26})
The magnetic moment of a Fe atom is thus
$M = \frac{1.7\times 10^6}{8.58\times 10^{28}}$ Am^2 $= 1.98 \times 10^{-23} = 2.49 \times 10^{-29}$ JA/m $= 2.14\mu_B$ (Bohr magneton $\mu_B = 9.27 \times 10^{-24}$ Joule/Tesla $= 1.16 \times 10^{-29}$ JA/m).

Physics of Magnetic Thin Films: Theory and Simulation
Hung T. Diep

ISBN 978-981-4877-42-8 (Hardcover), 978-1-003-12110-7 (eBook)
www.jennystanford.com

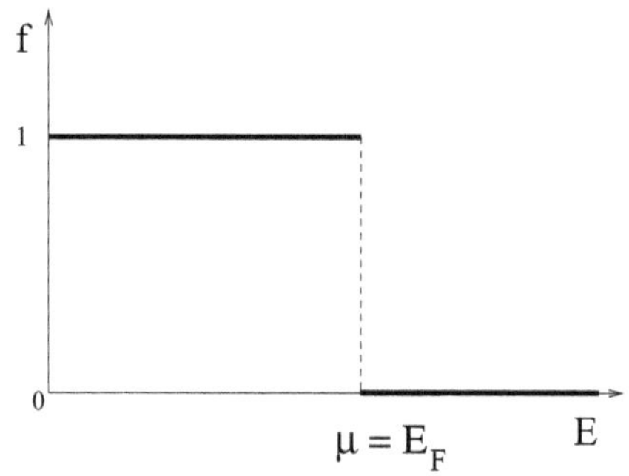

Figure 18.1 Fermi–Dirac distribution function at $T = 0$ versus energy E. The scale of E is arbitrary, μ is taken equal to 1.

(b) The energy due to the Zeeman effect is $\Delta E = \mu_B B = \mu_B \mu_0 H$ ($B = \mu_0 H$: magnetic field). We have $\Delta E = 0.9273 \times 10^{-23} \mu_0 H$ Joules (μ_0, vacuum permeability, $= 1.257 \times 10^{-6}$ H/m).
For $\mu_0 H = 0.5$ Tesla, $\Delta E = 0.464 \times 10^{-23}$ J
For $\mu_0 H = 1$ Tesla, $\Delta E = 0.927 \times 10^{-23}$ J
For $\mu_0 H = 2$ Tesla, $\Delta E = 1.85 \times 10^{-23}$ J.
The variation of the frequency: $\Delta\nu = \nu - \nu_0 = \Delta E/h$.

Problem 3. Fermi–Dirac distribution for free-electron gas:

Solution: At $T = 0$, namely $\beta = \infty$, we see that $f = 1$ for $E \leq \mu$, and $f = 0$ for $E > \mu$: The electrons occupy all energy levels up to μ. Each energy level has two electrons, one with up spin and the other with down spin. One defines the Fermi level E_F by $E_F = \mu$, namely the highest energy level which is occupied at $T = 0$. The function f at $T = 0$ is shown in Fig. 18.1.
At $T \neq 0$, the Fermi–Dirac distribution function is shown in Fig. 18.2. Electrons occupy all levels with decreasing f for increasing E.

Problem 4. Sommerfeld's expansion:

Solution: We show that

$$I = \int_0^\mu h(E)dE + \frac{\pi^2}{6}(k_B T)^2 h^{(1)}(E)|_{E=\mu} + \frac{7\pi^4}{360}(k_B T)^4 h^{(3)}(E)|_{E=\mu} + \cdots \qquad (18.1)$$

where $h^{(n)}(E)|_{E=\mu}$ is the n-th derivative of $h(E)$ at $E = \mu$.

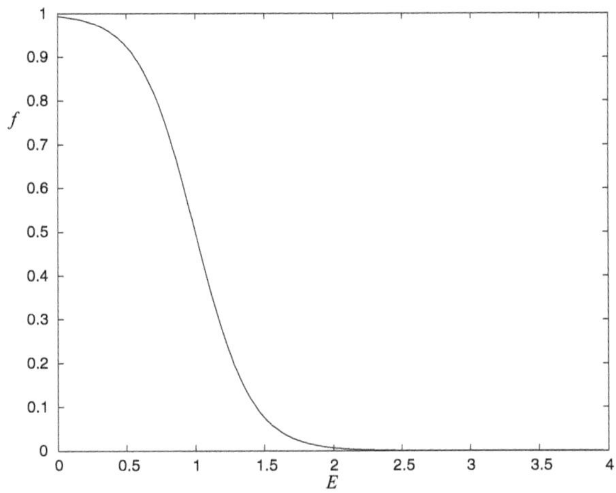

Figure 18.2 Fermi–Dirac distribution function at $T \neq 0$ versus energy E. The scale of E is arbitrary, μ is taken equal to 1.

Demonstration: We have

$$I = \int_{-\infty}^{\infty} h(E) f(E) dE \tag{18.2}$$

We define $g(E) = \int_{-\infty}^{E} dE h(E)$. Integration of I by parts gives

$$\begin{aligned} I = \int_{-\infty}^{\infty} h(E) f(E) dE &= [g(E) f(E)]_{-\infty}^{\infty} \\ &- \int_{-\infty}^{\infty} dE g(E) \frac{\partial f(E)}{\partial E} \\ &= 0 + \int_{-\infty}^{\infty} dE g(E) \left(-\frac{\partial f(E)}{\partial E} \right) \end{aligned} \tag{18.3}$$

where we have used $f \to 0$ when $E \to \infty$, and $g(E) \to 0$ when $E \to -\infty$.

At low T, the function $-\frac{\partial f(E)}{\partial E}$ is significant only near μ (slope of $f(E)$). This justifies an expansion of $g(E)$ around μ:

$$g(E) = g(\mu) + \sum_{n=1}^{\infty} \left[\frac{(E-\mu)^n}{n!} \right] \left[\frac{d^n g(E)}{dE^n} \right]_{E=\mu} \tag{18.4}$$

Replacing this series in (18.3), we get

$$I = g(\mu) + \sum_{n=1}^{\infty} \int_{-\infty}^{\infty} dE \frac{(E-\mu)^n}{n!} \left[\frac{d^n g(E)}{dE^n}\right]_{E=\mu} \left(-\frac{\partial f(E)}{\partial E}\right)$$

$$= g(\mu) + \sum_{n=1}^{\infty} \int_{-\infty}^{\infty} dE \frac{(E-\mu)^{2n}}{(2n)!} \left[\frac{d^{2n-1} h(E)}{dE^{2n-1}}\right]_{E=\mu}$$

$$\times \left(-\frac{\partial f(E)}{\partial E}\right)$$

where only terms of even power are non-zero, odd terms being zero because the integrands are odd functions with symmetric limits [note that $\frac{\partial f(E)}{\partial E}$ is an even function with respect to $(E-\mu)$]. Putting $x = \beta(E-\mu)$, we obtain

$$I = g(\mu) + \sum_{n=1}^{\infty} c_{2n} (k_B T)^{2n} \left[\frac{d^{2n-1} h(E)}{dE^{2n-1}}\right]_{E=\mu} \tag{18.5}$$

where

$$c_{2n} = \frac{1}{(2n)!} \int_{-\infty}^{\infty} x^{2n} \left[-\frac{d}{dx} \frac{1}{e^x + 1}\right] dx \tag{18.6}$$

Integration by parts gives

$$c_{2n} = \frac{1}{(2n)!} \left[-[x^{2n} \frac{1}{e^x+1}]_{-\infty}^{\infty} + 2n \int_{-\infty}^{\infty} x^{2n-1} \frac{1}{e^x+1} dx\right]$$

$$= \frac{1}{(2n)!} \left[0 + 4n(1 - 2^{1-2n}) \Gamma(2n) \zeta(2n)\right] \tag{18.7}$$

where we have used the formula

$$\int_0^{\infty} x^{n-1} \frac{1}{e^x+1} dx = (1 - 2^{1-n}) \Gamma(n) \zeta(n) \tag{18.8}$$

We have

$$\Gamma(n+1) = \int_0^{\infty} t^n \exp(-t) dt = n!$$

$$\Gamma(n+1) = n\Gamma(n)$$

$$\zeta(x) = \sum_{n=1}^{\infty} \frac{1}{n^x}$$

with $\zeta(2) = \frac{\pi^2}{6}$ and $\zeta(4) = \frac{\pi^4}{90}$. Rarely, we need to go beyond $2n = 4$ in the Sommerfeld's expansion.

Problem 5. Pauli paramagnetism:

Solution: Energies of spin in a magnetic field **B** applied in the z direction: $E_{\uparrow} = E - \mu_B B$ and $E_{\downarrow} = E + \mu_B B$. The resulting magnetic moment is $M = \mu_B(N_{\uparrow} - N_{\downarrow})$ which is written as

$$M = \mu_B \left[\int_{\mu_B B}^{\infty} dE \frac{\rho(E)}{e^{\beta(E-\mu_B B-\mu)}+1} - \int_{-\mu_B B}^{\infty} dE \frac{\rho(E)}{e^{\beta(E+\mu_B B-\mu)}+1} \right]$$
$$= \mu_B \int_0^{\infty} dE \left[\rho(E+\mu_B B) - \rho(E-\mu_B B)\right] f(E,T,\mu) \tag{18.9}$$

where we changed $E \to E \pm \mu_B B$. For weak fields, we can expand the density of states around E (energy in zero field): $\rho(E \pm \mu_B B) \simeq \rho(E) \pm \mu_B B\,[\rho'(E)]_E$. We get $M \simeq 2\mu_B^2 B \int_0^{\infty} dE \rho'(E) f(E,T,\mu)$. We deduce

$$\chi = \frac{dM}{dB} \simeq 2\mu_B^2 \int_0^{\infty} dE \rho'(E) f(E,T,\mu) \tag{18.10}$$

At low T, we can make a Sommerfeld's expansion [see (A.58)] for this integral

$$\chi \simeq 2\mu_B^2 \left[\int_0^{\mu} dE \rho'(E) + \frac{\pi^2}{6}\rho''(\mu)(k_B T)^2 \right]$$
$$= 2\mu_B^2 \left[\rho(\mu) - A\frac{\pi^2}{48\mu^{3/2}}(k_B T)^2 \right]$$

where we have used (A.44): $\rho(E) = AE^{1/2}$. Using (A.60), we obtain

$$\chi = 2\mu_B^2 \rho(E_F) \left[1 - \frac{\pi^2}{12}\left(\frac{k_B T}{E_F}\right)^2 \right] \tag{18.11}$$

The first term is independent of T as we have found in (1.19). The second term depends on T^2.

At high T, $f \simeq e^{-\beta(E-\mu)}$, (18.10) becomes

$$\chi \simeq 2\mu_B^2 \int_0^{\infty} dE \rho'(E) e^{-\beta(E-\mu)}$$
$$= 2\mu_B^2 \left[\rho'(E) e^{-\beta(E-\mu)}|_0^{\infty} + \beta \int_0^{\infty} dE \rho(E) e^{-\beta(E-\mu)} \right]$$
$$= 2\mu_B^2 \beta \left[0 + N/2\right] = \frac{N\mu_B^2}{k_B T} \tag{18.12}$$

where N is the total number of electrons of spins $\uparrow$ and $\downarrow$. This results is called "Curie's law."

Remark: We have used (A.44) without factor 2 of the spin degeneracy because we distinguish in the calculation each kind of spin.

Problem 6. Paramagnetism of free atoms for arbitrary **J**:

Solution: The average total magnetic moment in the magnetic field **B** applied in the z direction is

$$M^z = g\mu_B \sum_{i=1}^{N} J_i^z \tag{18.13}$$

where N is the total number of atoms. J_i^z is the z component of the moment $\mathbf{J}_i$ of the i-th atom. The Zeeman energy of the magnetic moment of the i-th atom in the field is $H_i = -\mathbf{M}_i \cdot \mathbf{B} = -M_i^z B$.

The average value $< J_i^z >$ is calculated by the canonical description (see Appendix A) as follows:

$$< J_i^z >= \frac{\sum_{J_i^z=-J}^{J} J_i^z \mathrm{e}^{-\beta H_i}}{Z_i} \tag{18.14}$$

where $\beta = \frac{1}{k_B T}$ and Z_i the partition function defined by

$$\begin{aligned} Z_i &= \sum_{J_i^z=-J}^{J} \exp(\beta B M_i^z) = \sum_{J_i^z=-J}^{J} \exp(\beta g\mu_B B J_i^z) \\ &= \frac{\sinh[\beta g\mu_B B J\,(J+\frac{1}{2})]}{\sinh[\frac{1}{2}\beta g\mu_B B J\,]} \end{aligned} \tag{18.15}$$

where we have used the formula of geometric series. We obtain

$$\sum_{J_i^z=-J}^{J} J_i^z e^{-\beta H_i} = \frac{\partial}{\partial\alpha} \sum_{J_i^z=-J}^{J} \mathrm{e}^{\beta g\mu_B J_i^z} \quad (\alpha \equiv \beta g\mu_B)$$

$$= \frac{\partial}{\partial\alpha} Z_i$$

$$= \frac{(J+\frac{1}{2})\cosh(J+\frac{1}{2})\alpha \sinh\frac{\alpha}{2} - \frac{1}{2}\sinh(J+\frac{1}{2})\alpha\cosh\frac{\alpha}{2}}{\sinh^2\frac{\alpha}{2}}$$

Thus

$$< J^z >= J\, B_J(x) \tag{18.16}$$

$B_J(\cdots)$ is the Brillouin function defined by

$$B_J(x) = \frac{2J+1}{2J} \coth \frac{(2J+1)x}{2J} - \frac{1}{2J} \coth \frac{x}{2J} \quad (18.17)$$

where

$$x = \beta g \mu_B J B \quad (18.18)$$

We get

$$\overline{m} = \frac{< M^z >}{V} = g\mu_B \sum_{i=1}^{N} < J_i^z >= \frac{N g \mu_B J}{V} B_J \left(\frac{g\mu_B J B}{k_B T} \right) \quad (18.19)$$

At high temperatures, one has $B_J(x) \simeq \frac{J+1}{3J} x - \cdots$. Thus,

$$\chi = \frac{m}{B} \simeq \frac{N}{V} (g\mu_B)^2 \frac{J(J+1)}{3} \frac{1}{k_B T} \quad (18.20)$$

This is the Curie's $1/T$ law. At low T, one has $B_J(x) \simeq 1 - \frac{1}{J} \exp(-x/J)$. One gets

$$\overline{m} \simeq \frac{N}{V} g\mu_B J \left[1 - \frac{1}{J} \exp(-x/J) \right] \rightarrow_{T \to 0} \frac{N}{V} g\mu_B J \quad (18.21)$$

The value at $T = 0$ corresponds to the saturated value of $\overline{m}$, namely $\frac{N}{V} g\mu_B J$.

Problem 7. Langevin's theory of diamagnetism:

Solution: The first explanation of the diamagnetism has been given by the theory of Langevin using the classical mechanics:

(a) We have the relation $m = iA$

(b) We have for an electron, $m = eA/\tau$ where τ is the period (time necessary to make a full circular motion), e electron charge. With $\tau = 2\pi r/v$ (r: radius, v: velocity) and $A = \pi r^2$, we have the orbital magnetic moment written as $m = evr/2$.

(c) The variation of the magnetic flux ϕ induced by $\mathbf{B}$ gives rise to an electric field

$$E = -\frac{1}{L} \frac{d\phi}{dt} = -\frac{1}{L} \frac{d(AB)}{dt} = -\frac{A}{L} \frac{dB}{dt} \quad (L = 2\pi r)$$

Acceleration:

$$a = \frac{dv}{dt} = \frac{eE}{m_e} = -\frac{er}{2m_e} \frac{dB}{dt} = -\frac{\mu_0 er}{2m_e} \frac{dH}{dt}$$

Integrating this relation, we have

$$\int_{v_1}^{v_2} dv = v_2 - v_1 = -\frac{\mu_0 er}{2m_e} H$$

The variation of the magnetic moment of the electron is thus

$$\Delta m = er(v_2 - v_1)/2 = -\frac{\mu_0 e^2 r^2 H}{4m_e}$$

The negative sign indicates the diamagnetic character.

(d) We project the orbit of radius r on a plane perpendicular to the field: the radius of the projected orbit $R = r\cos\theta$. We replace, in the above result of Δm, r by $r\cos\theta$. The average on all directions is obtained by integrating on θ:

$$\Delta m = -\frac{\mu_0 e^2 H}{4m_e}\int r^2 \cos^2\theta \sin\theta d\theta$$
$$= -\frac{\mu_0 e^2 r^2 H}{4m_e}\left[-\frac{\cos^3\theta}{3}\right]_0^{\pi}$$
$$= -\frac{\mu_0 e^2 r^2 H}{6m_e} \quad (18.22)$$

(e) If there are Z electrons in an atom:

$$\Delta\mathcal{M} = -N\frac{\mu_0 Z e^2 r^2 H}{4m_e}$$

where N is the number of atoms in a volume unit: $N = N_A\rho/M$ (N_A: Avogadro number). The resulting susceptibility is

$$\chi = \Delta\mathcal{M}/H = -N\frac{\mu_0 Z e^2 r^2 H}{4m_e} < 0.$$

Note: See Section 1.4 for a quantum treatment.

Problem 8. Langevin's theory of paramagnetism:

Solution: The case of discrete spins of magnitude 1/2 has been studied in Section 1.2. Here we study the case of continuous spins (Langevin's theory).

Langevin's theory: The Maxwell–Boltzmann's probability for a state of energy $E = -\mathbf{m}\cdot\mathbf{B}$ (Zeeman energy) is

$$p(E) = C\exp(-\beta E) = \exp(\beta\mathbf{m}\cdot\mathbf{B}) \quad (18.23)$$

where C is the normalization constant.
In an isotropic material, magnetic moments $\mathbf{m}$ are distributed in random directions. The number of moments in an elementary volume is

$$dn = C2\pi \sin\theta d\theta \exp(\beta m B \cos\theta) \tag{18.24}$$

The total number of moments in a volume unit is

$$N = C2\pi \int_0^\pi \sin\theta d\theta \exp(\beta m B \cos\theta),$$

thus

$$C = N/2\pi \int_0^\pi \sin\theta d\theta \exp(\beta m B \cos\theta).$$

The component along the z axis of the total resulting magnetic moment, namely magnetization, is

$$\begin{aligned} M &= \int_0^\pi m \cos\theta dn \\ &= \frac{Nm \int_0^\pi \cos\theta \sin\theta d\theta \exp(\beta m B \cos\theta)}{\int_0^\pi \sin\theta d\theta \exp(\beta m B \cos\theta)} \\ &= Nm\mathcal{L}\left(\frac{mB}{k_B T}\right) \\ &= Nm\mathcal{L}\left(\frac{\mu_0 m H}{k_B T}\right) \end{aligned} \tag{18.25}$$

where we used $x = \cos\theta$ ($dx = -\sin\theta d\theta$) for integration, and $\mathcal{L}(y) \equiv \coth(y) - \frac{1}{y}$. μ_0, vacuum permeability, is equal to $1,257 \times 10^{-6}$ H/m. For weak fields, an expansion of the Langevin function $\mathcal{L}(y)$ gives $M = N\mu_0 m^2 H/(3k_B T)$, leading to the Curie's law $\chi = M/H = N\mu_0 m^2/(3k_B T) > 0$ (paramagnetism).

18.2 Solutions to Problems of Chapter 2

Problem 1. Ising antiferromagnet: order parameter

Answer: We take the case of a chain of N Ising spins. The order parameter is defined by $M_s = \frac{1}{N}\sum_i (-1)^i S_i$ (staggered magnetization) where S_i is the spin at the lattice site i.

Problem 2. Potts model:

Answer:

(a) Order parameter of q-state Potts model:

$$p = \left[q \frac{\max(M_1, M_2, \cdots, M_q)}{N} - 1 \right] \frac{1}{q-1} \quad (18.26)$$

where M_i is the number of spins in the state i ($i = 1, \cdots, q$) and N the number of spins in the system. In an ordered state, only one of the states is present, M_i is equal to N, so $p = 1$, and in the disordered state, all M_i are equal ($= N/q$), so $p = 0$.

(b) The ground state and its degeneracy when $J > 0$: When $J > 0$, in the ground state all lattice sites have the same Potts value.

(c) If $J < 0$: The interaction of two different values has lower energy (zero). If $q = 2$ we see that the ground state is a configuration of alternating spins: This is the "antiferromagnetic ordering." The degeneracy is 2 (permutation of the two values of q).

For $q = 3$ ($i = 1, 2, 3$), the ground state is constructed by choosing sequences of diagonal lines 1-2-3-1-2-3 ... or 1-3-2-1-3-2 ..., namely any sequence of diagonal lines with no adjacent similar numbers. An example is shown in Fig. 18.3. There are 3 ways for choosing a number for the first diagonal line, 2 ways for each of the following lines. Hence, the number of configurations in the ground state (degeneracy) is $3 \times (2)^{L-1} \times 2 \propto 2^L$, L being the number of diagonal lines (equal to the linear lattice size). The last factor 2 is to take into account the fact that there are two diagonal lines in the square lattice. This construction generates a semi-ordering: There is an ordering on each diagonal line but no ordering on the second diagonal line perpendicular to the first one.

There is another way to construct the ground state which is completely disordered: Let us consider the square lattice defined on the xy plane. Each lattice site is defined by two indices (i, j). We fill the lattice sites

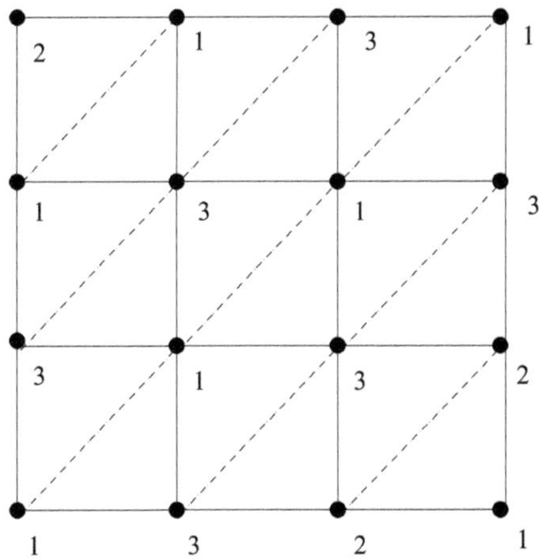

Figure 18.3 An example of the construction of a ground-state configuration which has an order on each diagonal line (discontinued lines) but no order on the perpendicular diagonal lines.

line by line starting from $j = 0$, from $i = 0$ to $i = L$ (from left to right, bottom to top). The lattice sites on the first x line ($j = 0$) can be randomly filled with three values 1, 2, 3, never similar values at two adjacent sites. On the next line ($j = 1$), the lattice site at (i, j) has its neighbor at $(i, j - 1)$ (below, on the previous line): there are two ways to choose its value which should be different from that at $(i, j-1)$. Once its value is chosen, we go next to its neighbor at $(i + 1, j)$: Its value should be different from that at (i, j) and that at $(i + 1, j - 1)$ (see Fig. 18.4, left panel). If the value at (i, j) is equal to the value at $(i + 1, j - 1)$, then we have two possible values for $(i + 1, j)$. If they are different, then we have only one possible value left (no choice) for $(i+1, j)$. We go to the next lattice site on the same line, namely the site at $(i + 2, j)$, and we proceed in the same manner: We do not have any problem which can stop the pursuit of the construction of the line. We go to the next line at $j + 1$ and continue the construction from left to right on the line. We see that the ground states obtained

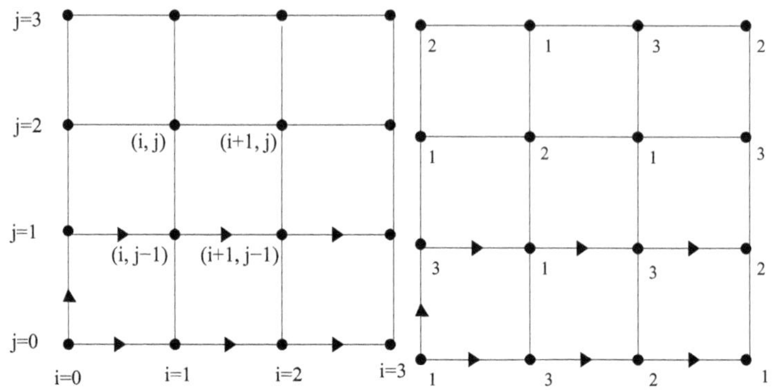

Figure 18.4 Left: indexation of lattice sites for description in the text. Right: example of a random ground-state configuration constructed in the way described in the text.

by such a way do not have a long-range order as seen in the example shown in Fig. 18.4 (right panel). Note that there are 2^N to place the numbers on the first line in the way described above. On each of the following lines, a number of the sites have only one choice as said above: Only $\alpha_i N$ sites ($\alpha_i < 1$) have two choices. So, the random ground states have a degeneracy is of the order of $2^{N+\alpha_1 N+\alpha_2 N+\cdots} = 2^{aN^2}$ ($a < 1$). It is much larger than the degeneracy 2^N of the ordered ground states shown in Fig. 18.3.

(d) The Potts model is equivalent to the Ising model when $q = 2$: There are two states for each site of energy $-J$ and 0 for Potts model, and $\pm J$ for Ising model. There is only a shift of energy in the calculation. Physical results are identical.

Problem 3. Domain walls:

Solution: The interaction between two neighboring spins is $E = -2J\,\mathbf{S}_i \cdot \mathbf{S}_j = -2J\,S^2\cos(\theta_i - \theta_j)$ where θ_i is the angle between the x axis and $\mathbf{S}_i$. In the ferromagnetic state, the angle $\phi = (\theta_i - \theta_j) = 0$. In a domain wall which has N-spin thickness, ϕ is π/N. If N is large as it is the frequent case, ϕ is small so that $E \simeq -2J\,S^2(1 - \frac{1}{2}\phi^2) = -2J\,S^2 + J\,S^2\phi^2$. The first term is the ferromagnetic-state energy and the second term is the energy of the spin deviation. The total energy of the wall is $\Delta = NJS^2\phi^2 = NJS^2(\pi/N)^2 = JS^2\pi^2/N$.

Problem 4. Bragg–Williams approximation:

Solution:

(a) Entropy: One has $N_+ = N(1+X)/2$, $N_- = N(1-X)/2$, with $0 \leq X \leq 1$. The number of configurations (microscopic states) is $W = N!/N_+!N_-!$. Using the Stirling formula (see Appendix A) $\ln n! \simeq n\ln n - n$ for large n, one has

$$\begin{aligned}
S &= k_B \ln W = k_B \ln\left[\frac{N!}{N_+!N_-!}\right] \\
&= k_B[N\ln N - N - N_+\ln N_+ + N_+ - N_-\ln N_- + N_-] \\
&= k_B\left\{N\ln N - \frac{N(1+X)}{2}\ln\left[\frac{N(1+X)}{2}\right]\right. \\
&\quad \left. - \frac{N(1-X)}{2}\ln\left[\frac{N(1-X)}{2}\right]\right\} \\
&= -k_B N\left\{\frac{1+X}{2}\ln\left[\frac{1+X}{2}\right]\right. \\
&\quad \left. + \frac{1-X}{2}\ln\left[\frac{(1-X)}{2}\right]\right\}
\end{aligned} \tag{18.27}$$

(b) The probability to have an up spin at a lattice site: $p_+ = N_+/N$. The number of up-up spin pairs is thus

$$N_{up-up} = \frac{1}{2}zN_+p_+ = \frac{1}{8}zN(1+X)^2$$

because there are z bonds around each site. The factor 1/2 is to remove the double counting of each bond. In the same manner, the number of down-down spin pairs and the number of antiparallel spin pairs are

$$N_{down-down} = \frac{1}{2}zN_-p_- = \frac{1}{8}zN(1-X)^2$$
$$N_{antip} = \frac{1}{2}z(N_+p_+ + N_-p_-) = \frac{1}{4}zN(1-X^2)$$

(c) The energy of the crystal is

$$E = -J\left(N_{up-up} + N_{down-down} - N_{antip}\right) = -\frac{1}{2}zJNX^2$$

(d) The free energy

$$F = E - TS$$
$$= -\frac{1}{2}zJ\,NX^2 + k_BT\,N\left\{\frac{1+X}{2}\ln\left[\frac{1+X}{2}\right]\right.$$
$$\left.+\frac{1-X}{2}\ln\left[\frac{1-X}{2}\right]\right\} \quad (18.28)$$

The minimum of F corresponds to $\partial F/\partial X = 0$. This gives

$$\partial F/\partial X = zNJ\,X - \frac{1}{2}Nk_BT\,\ln\frac{1+X}{1-X} = 0$$

$$\frac{2zJ}{k_BT}X = \ln\frac{1+X}{1-X}$$

$$\exp\left[\frac{2zJ}{k_BT}X\right] = \frac{1+X}{1-X}$$

$$X = \tanh\left[\frac{zJ}{k_BT}X\right] \quad (18.29)$$

This equation is equivalent to the mean-field equation (2.12) in the case where $S = \pm 1$.

(e) Using an expansion of $\tanh[\frac{zJ}{k_BT}X]$ when $X \to 0$ up to third order in X and proceeding in the same manner as for Eq. (2.23), one obtains $T_c = zJ/k_B$. When $T > T_c$, one has $X = 0$. The entropy (18.27) is then $S = k_BN\ln 2$. This result can be directly obtained by the simple following argument: In the disordered phase, each spin is independent with two states ± 1. There are N sites so the total number of system spin configurations is just $W = 2^N$. The entropy is then $S = k_B\ln W = k_BN\ln 2$.

Remark: The Bragg–Williams approximation was initially used for binary alloys where an A atom is represented by an up spin, and a B atom by a down spin. The mixing of A and B atoms is favored if J is negative: This is equivalent to an antiferromagnetic ordering where an A atom (up spin) is surrounded by B atoms (down spins) at low temperatures.

Problem 5. Binary alloys by spin language:

Solution:

(a) Since $\epsilon > \phi$, the energy of a spin is lower when it is surrounded by neighbors of the other kind. This yields a perfect antiferromagnetic ordering at $T = 0$.

(b)
- One has $0 \leq N_{\uparrow,I} = N(1+x)/4 \leq N/2$; hence, $-1 \leq x \leq 1$. When $x = 0$, $N_{\uparrow,I} = N/4$: The system is in the disordered state. The number of $\uparrow$-spins occupying sites II is $N_{\uparrow,II} = N/2 - N_{\uparrow,I} = N(1-x)/4$.
  For down spins (B atoms), one has $N_{\downarrow,II} + N_{\uparrow,II} = N/2$; hence, $N_{\downarrow,II} = N(1+x)/4$. One deduces $N_{\downarrow,I} = N/2 - N_{\downarrow,II} = N(1-x)/4$.
  One considers the case $x > 0$ in the following:
- The probability for a $\uparrow$-spin to be at a site of the type I is
  $P(\uparrow, I) = N_{\uparrow,I}/(N/2) = (1+x)/2$, and that at a site of the type II is
  $P(\uparrow, II) = N_{\uparrow,II}/(N/2) = (1-x)/2$.
  In the same way, for a $\downarrow$-spin one has
  $P(\downarrow, I) = N_{\downarrow,I}/(N/2) = (1-x)/2$ and $P(\downarrow, II) = N_{\downarrow,II}/(N/2) = (1+x)/2$.
- Let $N_{\uparrow,\uparrow}$, $N_{\downarrow,\downarrow}$, and $N_{\uparrow,\downarrow}$ be the numbers of $\uparrow\uparrow$, $\downarrow\downarrow$ and $\uparrow\downarrow$ pairs. The probability to have a $\uparrow\uparrow$ pair is
  $P(\uparrow, \uparrow) = P(\uparrow, I)P(\uparrow, II) + P(\uparrow, II)P(\uparrow, I) = (1-x^2)/2$; hence, $N_{\uparrow,\uparrow} = NP(\uparrow, \uparrow) = N(1-x^2)/2$.
  In the same way, one has: $N_{\downarrow,\downarrow} = N(1-x^2)/2$.
  For a $\uparrow\downarrow$ pair, the probability is

$$\begin{aligned}P(\uparrow, \downarrow) &= P(\uparrow, I)P(\downarrow, II) + P(\uparrow, II)P(\downarrow, I)\\ &\quad + P(\downarrow, I)P(\uparrow, II) + P(\downarrow, II)P(\uparrow, I)\\ &= (1+x)^2/4 + (1-x)^2/4 + (1-x)^2/4\\ &\quad + (1+x)^2/4 = 1 + x^2 \qquad (18.30)\end{aligned}$$

  Therefore, $N_{\uparrow,\downarrow}=NP(\uparrow, \downarrow)=N(1+x^2)$.
- One has $E = [N_{\uparrow,\uparrow} + N_{\downarrow,\downarrow}]\epsilon + N_{\uparrow,\downarrow}\phi$. Thus,

$$E = N(\epsilon + \phi) - N(\epsilon - \phi)x^2 \qquad (18.31)$$

- When E is given, x^2 is determined. One takes $x > 0$. Let $\Omega(E)$ be the number of microscopic states of energy equal to E. To calculate $\Omega(E)$, one calculates the number of ways to choose $N(1+x)/4$ $\uparrow$-spins among $N/2$ sites of the type I and at the same time to choose $N(1+x)/4$ sites among $N/2$ sites of the type II to place them. One has

$$\Omega(E) = \left[C_{N/2}^{N(1+x)/4}\right]^2 = \left[\frac{(N/2)!}{[N(1+x)/4]![N(1-x)/4]!}\right]^2 \quad (18.32)$$

Using the Stirling formula, one writes

$$\begin{aligned}\ln\Omega(E) &= 2\{\ln(N/2)! - \ln[N(1+x)/4]! \\ &\quad - \ln[N(1-x)/4]!\} \\ &\simeq 2\{(N/2)\ln(N/2) - N/2 - [N(1+x)/4] \\ &\quad \times \ln[N(1+x)/4] \\ &\quad + N(1+x)/4 - [N(1-x)/4] \\ &\quad \times \ln[N(1-x)/4] + N(1-x)/4\} \\ &= 2\{(N/2)\ln(N/2) - [N(1+x)/4] \\ &\quad \times \ln[N(1+x)/4] \\ &\quad - [N(1-x)/4]\ln[N(1-x)/4]\}\end{aligned} \quad (18.33)$$

The entropy is given by $S = k_B \ln\Omega$.

- The temperature T is calculated by (see Appendix A):

$$\begin{aligned}T^{-1} &= \frac{\partial S}{\partial E} = k_B \frac{\partial \ln\Omega(E)}{\partial x}\frac{\partial x}{\partial E} \\ &= \frac{k_B}{4(\epsilon-\phi)x}\{\ln[N(1+x)/4] - \ln[N(1-x)/4]\} \\ &= \frac{k_B}{4(\epsilon-\phi)x}\ln\frac{1+x}{1-x}\end{aligned}$$

One gets

$$\begin{aligned}\ln\frac{1+x}{1-x} &= \frac{4(\epsilon-\phi)x}{k_B T} \\ \frac{1+x}{1-x} &= \exp\left[\frac{4(\epsilon-\phi)x}{k_B T}\right] \\ x &= \tanh\left[\frac{2(\epsilon-\phi)x}{k_B T}\right]\end{aligned} \quad (18.34)$$

This equation is of the mean-field equation type [see (2.12)]. We have $x = 1$ at $T = 0$ and $x = 0$ at $T = \infty$. Between these limits, there exists a temperature below which x is not zero. An expansion at small x gives

$$x \simeq 2(\epsilon - \phi)x/(k_B T) - [2(\epsilon - \phi)x/(k_B T)]^3/3$$
$$x[1 - 2(\epsilon - \phi)/(k_B T)] = -[2(\epsilon - \phi)/(k_B T)]^3 x^3 \quad (18.35)$$

If $x \neq 0$, one can simplify the two sides to get

$$2(\epsilon - \phi)/(k_B T) - 1 = [2(\epsilon - \phi)/(k_B T)]^3 x^2 \quad (18.36)$$

Since $2(\epsilon - \phi)/(k_B T) > 0$, the right-hand side is positive. This relation is satisfied if on the left-hand side one has $2(\epsilon - \phi)/(k_B T) - 1 > 0$ namely $T < 2(\epsilon - \phi)/k_B \equiv T_c$. T_c is the critical temperature.
If we return to the binary alloy language, we say we have an ordered binary alloy structure when $T < T_c$ and a disordered structure for $T > T_c$.

Problem 6. Critical temperature of ferrimagnet:

Solution: We make an expansion of (2.73) and (2.74) when $< S_A^z >$ and $< S_B^z >$ are small in the same manner as for (2.20). We have

$$< S_A^z > = a < S_B^z > - b < S_B^z >^3 + \cdots \quad (18.37)$$
$$< S_B^z > = c < S_A^z > - d < S_A^z >^3 + \cdots \quad (18.38)$$

where

$$a = \frac{S_A(S_A+1)}{3}\frac{CJ_1}{k_B T} \quad (18.39)$$
$$b = \frac{[S_A^2 + (S_A+1)^2]S_A(S_A+1)}{90}\left(\frac{CJ_1}{k_B T}\right)^3 \quad (18.40)$$
$$c = \frac{S_B(S_B+1)}{3}\frac{CJ_1}{k_B T} \quad (18.41)$$
$$d = \frac{[S_B^2 + (S_B+1)^2]S_B(S_B+1)}{90}\left(\frac{CJ_1}{k_B T}\right)^3 \quad (18.42)$$

Replacing (18.38) in (18.37), we write

$$< S_A^z > (ac - 1) = (ad + bc^3) < S_A^z >^3 \qquad (18.43)$$

For $< S_A^z > \neq 0$, we can simplify it on both sides. The remaining equation is

$$(ac - 1) = (ad + bc^3) < S_A^z >^2 \qquad (18.44)$$

The right-hand side of this equation is positive; therefore, $ac - 1 > 0$. Replacing the coefficients a and c we obtain

$$k_B T < \frac{CJ_1}{3}\sqrt{S_A(S_A+1)S_B(S_B+1)} \equiv k_B T_N \qquad (18.45)$$

This means that non-zero solutions of $< S_A^z >$ are found only below T_N. Note that replacing (18.37) in (18.38) gives the same solution.

Problem 7. Improvement of mean-field theory:
Solution:

(a) We write

$$\mathbf{S}_1 \cdot \mathbf{S}_2 = S_1^z S_2^z + (S_1^+ S_2^- + S_1^- S_2^+)/2$$

The states of two spins 1/2 are

$$\phi_1 = |1/2, 1/2 >, \quad \phi_2 = |1/2, -1/2 >,$$

$$\phi_3 = |-1/2, 1/2 >, \quad \phi_4 = |-1/2, -1/2 > .$$

To calculate

$$[-2J\,[S_1^z S_2^z + (S_1^+ S_2^- + S_1^- S_2^+)/2] - D[(S_1^z)^2 + (S_2^z)^2] - B(S_1^z + S_2^z)]|\phi_i >$$

we use

$$S^{\pm}|jm> = [j(j+1) - m(m \pm 1)]^{1/2}\hbar|j, m \pm 1 > \quad (j = 1/2, m = \pm 1/2)$$

$$S^z|m> = \hbar m|m> \qquad (18.46)$$

We obtain a matrix 4×4. A simple diagonalization gives the following eigenvalues for the two-spin cluster:

$\epsilon_1 = -J/2 - D/2 - B\ (\uparrow\uparrow)$, $\epsilon_2 = 3J/2 - D/2\ (\uparrow\downarrow - \downarrow\uparrow)$, $\epsilon_3 = -J/2 - D/2\ (\uparrow\downarrow + \downarrow\uparrow)$, $\epsilon_4 = -J/2 - D/2 + B(\downarrow\downarrow)$ (we have taken $\hbar = 1$).

(b) We put the cluster of two spins $\mathbf{S}_i$ and $\mathbf{S}_j$ in a lattice: it has $(Z - 1)$ neighbors. We treat the interaction of the 4 cluster configurations found above in zero field $(B = 0)$ with these neighbors by the mean-field theory. The energy of the cluster in the crystal depends on the embedded cluster spin configurations, they are in increasing energies:
$\phi_1 = (\uparrow\uparrow) \rightarrow E_1 = -J/2 - 2J(Z-1) < S^z >$
$(\phi_2 + \phi_3)/2 = (\uparrow\downarrow + \downarrow\uparrow) \rightarrow E_2 = -J/2$
$\phi_4 = (\downarrow\downarrow) \rightarrow E_3 = -J/2 + 2J(Z-1) < S^z >$
$(\phi_2 - \phi_3)/2 = (\uparrow\downarrow - \downarrow\uparrow) \rightarrow E_4 = 3J/2$
We consider the cluster of two spins as a superspin with the z component $S^z = (S_i^z + S_j^z)/2$. We have

$$< S^z >= Tr\frac{1}{2}(S_i^z + S_j^z)\exp(-\beta E)/Tr\exp(-\beta E)$$

where

$$\begin{aligned}\text{Tr}\exp(-\beta E) &= \exp(\beta J/2)\exp(\beta X) + \exp(\beta J/2)\\ &\quad + \exp(\beta J/2)\exp(-\beta X) + \exp(-\beta 3J/2)\\ &\quad (X \equiv 2J(Z-1) < S^z >)\end{aligned}$$

and

$$\begin{aligned}\text{Tr}\frac{1}{2}(S_i^z + S_j^z)\exp(-\beta E)] &= \frac{1}{2}\exp(\beta J/2)\exp(\beta X) + 0\\ &\quad -\frac{1}{2}\exp(\beta J/2)\exp(-\beta X) + 0\\ &= \exp(\beta J/2)\sinh\beta X\end{aligned}$$

Hence,

$$< S^z >= \frac{\sinh\beta X}{2\left[\cosh\beta X + \exp(-\beta J)\cosh\beta J\right]} \tag{18.47}$$

We see that $< S^z >= 0$ is a solution of this equation. An expansion around $< S^z >= 0$ gives

$$\begin{aligned}&2 < S^z > \left[\frac{-3 + 2\beta(Z-1)J - e^{-2\beta J}}{2}\right]\\ &= \beta^2 4(Z-1)^2 J^2 < S^z >^3\end{aligned} \tag{18.48}$$

The solution $< S^z > \neq 0$ is possible if

$$-3 + 2\beta(Z-1)J - e^{-2\beta J} > 0.$$

T_c is obtained by solving $-3+2\beta_c(Z-1)J - e^{-2\beta_c J} = 0$ where $\beta_c = (k_B T_c)^{-1}$. We obtain

$$e^{-2J/k_B T_c} + 3 - 2(Z-1)J/k_B T_c = 0 \qquad (18.49)$$

Problem 8. Interaction between next-nearest neighbors in mean-field treatment:

Solution:

(a) All spins are parallel at $T = 0$. All interactions are fully satisfied.

(b) The hypothesis of the mean-field theory: All neighboring spins of a spin are replaced by an average value which is used to calculate the value of the spin under consideration.

(c) The energy of a spin at $T = 0$ is $E = -Z_1 J_1 - Z_2 J_2$ where Z_1 and Z_2 are the numbers of nearest neighbors and of next-nearest neighbors, respectively. For a body-centered cubic lattice, $Z_1 = 8$, $Z_2 = 6$. E is the energy which maintains the spin ordering: The lower it is, the higher the temperature is needed to destroy the ordering. Thus, the stronger J_2 is, the higher the transition temperature becomes.

(d) The same calculation as that in the chapter by replacing CJ with $Z_1 J_1 + Z_2 J_2$ in Eqs. (2.5)–(2.7) and in the following equations to obtain the final mean-field equation.

(e) The critical temperature is obtained by replacing CJ in Eq. (2.23) by $Z_1 J_1 + Z_2 J_2$.

(f) Now we suppose $J_2 < 0$. When $|J_2| \gg J_1$, it is obvious that the J_2 interaction imposes the antiferromagnetic ordering to make the overall energy negative. The spins on the cube corners form an antiferromagnetic sublattice, the centered spins form another antiferromagnetic sublattice, independent of the first one. Since each spin has 4 up neighbors and 4 down neighbors (make a figure to convince yourself), its interaction energy with nearest neighbors is zero, independent of J_1. The energy of such a spin configuration is thus $E_{Antif} = Z_2 J_2 = -Z_2 |J_2|$.

If $|J_2| \ll J_1$ then the ferromagnetic state is more favorable. Its energy is $E_{Ferro} = -Z_1 J_1 - Z_2 J_2 = -Z_1 J_1 + Z_2 |J_2|$.

The critical value of $|J_2|$ below which the ferromagnetic state is stable is determined by solving $E_{Ferro} < E_{Antif}$. We have $|J_2^c| = \frac{Z_1 J_1}{2Z_2}$ or $J_2^c = -\frac{Z_1 J_1}{2Z_2}$. When $|J_2| < |J_2^c|$ (or $J_2 > J_2^c$) we have the ferromagnetic ordering. Otherwise, we have the antiferromagnetic one.

Problem 9. We repeat Problem 7 in the case of an antiferromagnet:

Solution: With $J < 0$: the changes with respect to the ferromagnetic case are

(i) in question (a): no change

(ii) in question (b): we have the inverse order of energies $E_4 < E_3 < E_2 < E_1$, because $J < 0$. The remaining calculation is exactly the same. The four configurations embedded in the crystal give the following energies:

ϕ_1: $E_1 = -J/2$

$(\phi_2 + \phi_3)/2$: $E_2 = -J/2 - 2J(Z-1) < S^z >$

ϕ_4: $E_3 = -J/2$

$(\phi_2 - \phi_3)/2$: $E_4 = 3J/2 + 2J(Z-1) < S^z >$

When putting these energies in the calculation of the average $< (S_i^z + S_j^z)/2 >$, be careful to use the energy of the corresponding crystal-field spin configuration in the argument of the exponential and to use the correct sign of each neighboring spin. Make a draw to help. We will have

$$e^{+2J/k_B T_c} + 3 + 2(Z-1)J/k_B T_c = 0 \qquad (18.50)$$

It is the same as the ferromagnetic result, bearing in mind that $J < 0$ here.

18.3 Solutions to Problems of Chapter 3

Problem 1. Demonstration of (3.51)–(3.52):

Demonstration: In (3.50), by replacing $\epsilon_{\vec{k}} \simeq 2J\, S(ka)^2$ and using (3.46), we have

$$\sum_{\mathbf{k}} \epsilon_{\mathbf{k}} < n_{\mathbf{k}} > \simeq \frac{N}{(2\pi)^2} \int_0^{\infty} 2J\, S(ka)^2 \sum_{l=1}^{l=\infty} e^{-l\beta 2J\, S(ka)^2} k^2 dk \tag{18.51}$$

Putting $x = l\beta 2J\, Sk^2$ $(a = 1)$ and integrating, we obtain (3.51) and then (3.52).

Problem 2. Chain of Heisenberg spins:

Solution:

(a) One has

$$\omega = 2J_1SZ(1 - \cos(ka)) + 2J_2SZ(1 - \cos(2ka))$$

$(Z = 2$, number of neighbors)

(b) If $J_2 < 0$,

$$\omega = 2J_1SZ(1 - \cos(ka)) - 2|J_2|SZ(1 - \cos(2ka))$$

We plot ω versus k. We see that ω is strongly affected by J_2 when $k \to 0$. Analytically, we take the derivative of ω with respect to k, we have

$$\begin{aligned} d\omega/dk &= 2J_1SZa\sin(ka) - 4a|J_2|SZ\sin(2ka) \\ &= 2SZa[J_1\sin(ka) - 4|J_2|\sin(ka)\cos(ka)] \\ &= 2SZa\sin(ka)[J_1 - 4|J_2|\cos(ka)] \end{aligned}$$

This derivative is zero at $k = 0$ (uniform mode) and at $\cos(ka) = \frac{J_1}{4|J_2|} = -\frac{J_1}{4J_2}$ $(J_2 < 0)$. This second case is called "soft mode" because the slope (stiffness) of ω is zero at this value of k. We have a helimagnetic ordering for $J_2 < -J_1/4$ (see Sections 3.4 and 5.7.3).

Problem 3. Heisenberg spin systems in two dimensions:

Solution:

(a) $\omega = 2J\,SZ(1 - \gamma_k)$ where $Z = 4$ (number of nearest neighbors in a square lattice), $\gamma_k = (\cos(k_xa) + \cos(k_ya))/2$. $\omega \to 2J\, S(ka)^2$ when $\mathbf{k} \to 0$.

(b) $<S^z> = 1/2 - A\int_{ZB} \frac{2\pi kdk}{\exp(\beta\omega)-1}$ [A: constant, see (3.43)–(3.43)]. The most important contribution to the integral comes from the small k region where $\omega \to 2JS(ka)^2$. We have $<S^z> \simeq 1/2 - A\int_{ZB} \frac{2\pi kdk}{1+\beta JS(ka)^2-1} \simeq 1/2 - A\int_{ZB} \frac{2\pi kdk}{\beta JS(ka)^2}$. This integral diverges at $k = 0$; hence, $< S^z >$ is not defined if $T \neq 0$. There is no long-range order for $T \neq 0$ in 2D (see the rigorous theorem of Mermin–Wagner in Ref. [231]).

Note: In 3D, we replace in the integral $2\pi kdk$ by $4\pi k^2dk$. The integral does not diverge at $k = 0$. The long-range ordering exists at $T \neq 0$ in 3D.

Problem 4. Demonstration of Eqs. (3.131)–(3.133):

Demonstration: We have

$$a_{\mathbf{k}} = \alpha_{\mathbf{k}} \cosh\theta_k - \alpha^+_{-\mathbf{k}} \sinh\theta_k \tag{18.52}$$

$$a^+_{\mathbf{k}} = \alpha^+_{\mathbf{k}} \cosh\theta_k - \alpha_{-\mathbf{k}} \sinh\theta_k \tag{18.53}$$

where we can show that $\alpha_{\mathbf{k}}$ and $\alpha^+_{\mathbf{k}}$ obey the boson commutation relations (see similar demonstration in Problem 10). Replacing these expressions in the Hamiltonian (3.125), we have

$$\begin{aligned}\mathcal{H} &= -NSJ(\mathbf{Q}) + \frac{S}{2}\sum_{\mathbf{k}}[A(\mathbf{k},\mathbf{Q})(a_{\mathbf{k}}a^+_{\mathbf{k}} + a^+_{\mathbf{k}}a_{\mathbf{k}})\\ &\quad + B(\mathbf{k},\mathbf{Q})(a_{\mathbf{k}}a_{-\mathbf{k}} + a^+_{\mathbf{k}}a^+_{-\mathbf{k}})]\\ &= -NSJ(\mathbf{Q}) + \frac{S}{2}\sum_{\mathbf{k}}\{\sqrt{A(\mathbf{k},\mathbf{Q})^2 - B(\mathbf{k},\mathbf{Q})^2}[\alpha_{\mathbf{k}}\alpha^+_{\mathbf{k}}\alpha^+_{\mathbf{k}}\alpha_{\mathbf{k}}]\\ &\quad + [B(\mathbf{k},\mathbf{Q})\cosh 2\theta_k - A(\mathbf{k},\mathbf{Q})\sinh 2\theta_k][\alpha_{\mathbf{k}}\alpha_{-\mathbf{k}} + \alpha^+_{\mathbf{k}}\alpha^+_{-\mathbf{k}}]\}\end{aligned}$$

The Hamiltonian is diagonal if the second term in the curly brackets $\{\cdots\}$ is zero, namely

$$\tanh(2\theta_k) = \frac{B(\mathbf{k},\mathbf{Q})}{A(\mathbf{k},\mathbf{Q})} \tag{18.54}$$

Omitting the constant term, we have

$$\mathcal{H} = \frac{S}{2}\sum_{\mathbf{k}} \hbar\omega_{\mathbf{k}}[\alpha^+_{\mathbf{k}}\alpha_{\mathbf{k}} + \alpha_{\mathbf{k}}\alpha^+_{\mathbf{k}}] \tag{18.55}$$

where the energy of the magnon of mode $\mathbf{k}$ is

$$\hbar\omega_{\mathbf{k}} = \sqrt{A(\mathbf{k},\mathbf{Q})^2 - B(\mathbf{k},\mathbf{Q})^2} \tag{18.56}$$

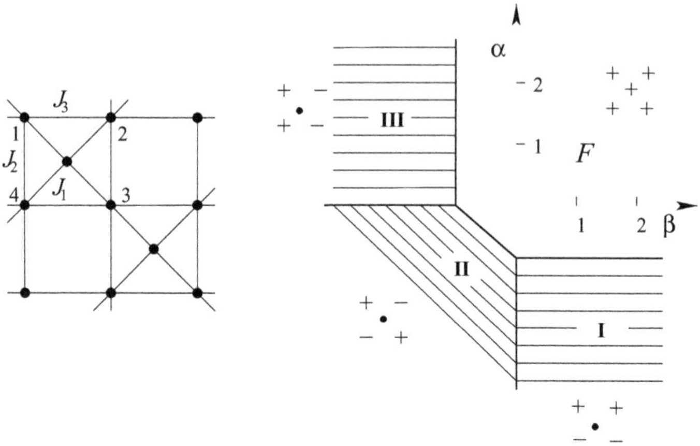

Figure 18.5 Left: Union-Jack lattice: diagonal, vertical and horizontal bonds denote the interactions J_1, J_2 and J_3, respectively. Right: Phase diagram of the ground state shown in the plane ($\alpha = J_2/J_1$, $\beta = J_3/J_1$). Heavy lines separate different phases and spin configuration of each phase is indicated (up, down and free spins are denoted by +, − and o, respectively). The three kinds of partially disordered phases and the ferromagnetic phase are denoted by I, II, III and F, respectively.

Problem 5. "Union-Jack" lattice:

Solution: We write the energy expression for each kind of configuration. Then, we compare two by two to determine the frontier between them. The result is shown in Fig. 18.5. See details in Ref. [67].

Problem 6. Ground state of the triangular antiferromagnet with XY spins:

Solution: In the case of the triangular plaquette, suppose that spin $\mathbf{S}_i$ ($i = 1, 2, 3$) of magnitude S makes an angle θ_i with the **Ox** axis. Writing E and minimizing it with respect to the angles θ_i, one has

$$E = J\left(\mathbf{S}_1 \cdot \mathbf{S}_2 + \mathbf{S}_2 \cdot \mathbf{S}_3 + \mathbf{S}_3 \cdot \mathbf{S}_1\right)$$
$$= J\,S^2\left[\cos(\theta_1 - \theta_2) + \cos(\theta_2 - \theta_3) + \cos(\theta_3 - \theta_1)\right],$$
$$\frac{\partial E}{\partial \theta_1} = -J\,S^2\left[\sin(\theta_1 - \theta_2) - \sin(\theta_3 - \theta_1)\right] = 0,$$

$$\frac{\partial E}{\partial \theta_2} = -J\,S^2\,[\sin(\theta_2 - \theta_3) - \sin(\theta_1 - \theta_2)] = 0,$$

$$\frac{\partial E}{\partial \theta_3} = -J\,S^2\,[\sin(\theta_3 - \theta_1) - \sin(\theta_2 - \theta_3)] = 0.$$

A solution of the last three equations is $\theta_1 - \theta_2 = \theta_2 - \theta_3 = \theta_3 - \theta_1 = 2\pi/3$. One can also write

$$E = J\,(\mathbf{S}_1 \cdot \mathbf{S}_2 + \mathbf{S}_2 \cdot \mathbf{S}_3 + \mathbf{S}_3 \cdot \mathbf{S}_1) = -\frac{3}{2}J\,S^2 + \frac{J}{2}(\mathbf{S}_1 + \mathbf{S}_2 + \mathbf{S}_3)^2.$$

The minimum of E corresponds to $\mathbf{S}_1 + \mathbf{S}_2 + \mathbf{S}_3 = 0$ which yields the 120° structure. This is also true for Heisenberg spins.

Problem 7. Ground state of Villain's model:

Solution: The energy of a plaquette of the 2D Villain's model with XY spins defined in Fig. 18.6 with $\mathbf{S}_1$ and $\mathbf{S}_2$ linked by the antiferromagnetic interaction η, is written as

$$H_p = \eta \mathbf{S}_1 \cdot \mathbf{S}_2 - \mathbf{S}_2 \cdot \mathbf{S}_3 - \mathbf{S}_3 \cdot \mathbf{S}_4 - \mathbf{S}_4 \cdot \mathbf{S}_1 \qquad (18.57)$$

where $(\mathbf{S}_i)^2 = 1$. The variational method gives

$$\delta[H_p - \frac{1}{2}\sum_{i=1}^{4} \lambda_i (\mathbf{S}_i)^2] = 0 \qquad (18.58)$$

By symmetry, $\lambda_1 = \lambda_2 \equiv \lambda$, $\lambda_3 = \lambda_4 \equiv \mu$. We have

$$\lambda \mathbf{S}_1 - \eta \mathbf{S}_2 + \mathbf{S}_4 = 0 \qquad (18.59)$$

$$-\eta \mathbf{S}_1 + \lambda \mathbf{S}_2 + \mathbf{S}_3 = 0 \qquad (18.60)$$

$$\mathbf{S}_2 + \mu \mathbf{S}_3 + \mathbf{S}_4 = 0 \qquad (18.61)$$

$$\mathbf{S}_1 + \mathbf{S}_3 + \mu \mathbf{S}_4 = 0 \qquad (18.62)$$

Hence,

$$(\lambda - \mu)(\mathbf{S}_1 + \mathbf{S}_2) + (\mathbf{S}_3 + \mathbf{S}_4) = 0 \qquad (18.63)$$

$$(\mathbf{S}_1 + \mathbf{S}_2) + (\mu + 1)(\mathbf{S}_3 + \mathbf{S}_4) = 0 \qquad (18.64)$$

We deduce

$$\mu = -\left[\frac{1+\eta}{\eta}\right]^{1/2} \qquad (18.65)$$

$$\lambda = \eta\mu = -[\eta(1+\eta)]^{1/2} \qquad (18.66)$$

To calculate the angle between two spins, for instance $\mathbf{S}_1$ and $\mathbf{S}_4$, we write

$$(\lambda \mathbf{S}_1 + \mathbf{S}_4)^2 = (-\eta \mathbf{S}_2)^2$$

Hence,

$$\mathbf{S}_1 \cdot \mathbf{S}_4 = \cos\theta_{14} = \frac{1}{2\lambda}(\eta^2 - \lambda^2 - 1) = \frac{1}{2}\left[\frac{\eta+1}{\eta}\right]^{1/2}$$

We find in the same manner,

$$\cos\theta_{23} = \cos\theta_{34} = \cos\theta_{41} = \frac{1}{2}\left[\frac{\eta+1}{\eta}\right]^{1/2}$$

We have

$$\theta_{14} = \theta_{23} = \theta_{34} = \theta_{41} \equiv \theta$$

Note that $|\theta_{12}| = 3|\theta|$. These solutions exist if $|\cos\theta| \leq 1$, namely $\eta > \eta_c = 1/3$. When $\eta = 1$, we have $\theta = \pi/4$, $\theta_{12} = 3\pi/4$.

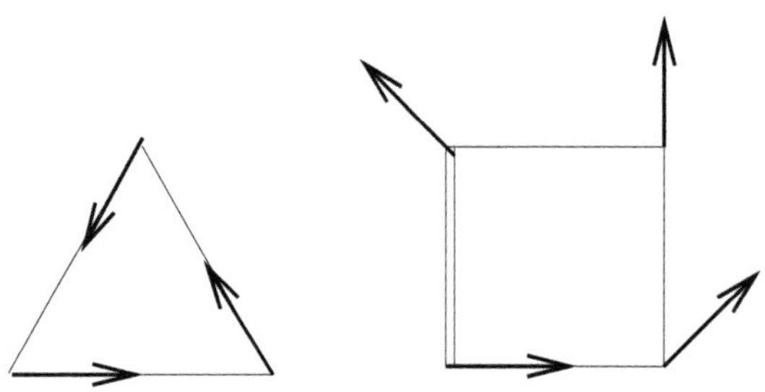

Figure 18.6 Examples of frustrated spin systems. Left: antiferromagnetic triangular lattice with vector spins (XY or Heisenberg spins), Right: Villain's model with XY spins.

Problem 8. Uniaxial anisotropy:

Answer:

(a) To follow the method of the chapter.

(b) Yes, because the integral does not diverge any more at $k = 0$ in the presence of d.

Problem 9. Commutation relations of Holstein–Primakoff operators:

Solution: The operators a^+ and a defined in the Holstein–Primakoff approximation respect rigorously the commutation relations between the spin operators:

$$[S_l^+, S_m^-] = 2S_l^z\delta_{lm} \quad \text{and} \quad [S_l^z, S_m^\pm] = \pm S_l^\pm\delta_{lm}$$

Demonstration: Replacing the spin operators by the Holdstein-Primakoff operators, one has

$$\begin{aligned}[S_l^+, S_m^-] &= 2S[f_l(S)a_l, a_m^+ f_m(S)] \\ &= 2S[f_l(S)a_l a_m^+ f_m(S) - a_m^+ f_m(S) f_l(S) a_l]\end{aligned}$$

If $l = m$, one has

$$\begin{aligned}[S_l^+, S_l^-] &= 2S[f(S)aa^+ f(S) - a^+ f(S) f(S) a] \\ &= 2S\left[\left(1 - \frac{a^+a}{2S}\right)^{1/2}(a^+a+1)\left(1 - \frac{a^+a}{2S}\right)^{1/2}\right. \\ &\quad \left. -a^+\left(1 - \frac{a^+a}{2S}\right)a\right] \\ &= 2S\left[\left(1 - \frac{a^+a}{2S}\right)(a^+a+1) - a^+a + \frac{a^+a^+aa}{2S}\right] \\ &= 2S\left[1 - \frac{a^+a}{S}\right] \\ &= 2[S - a^+a] = 2S_l^z \end{aligned} \tag{18.67}$$

If $l \neq m$, one obtains in the same manner $[S_l^+, S_m^-] = 0$. For the second relation, when $l = m$ one has

$$\begin{aligned}[S_l^z, S_l^+] &= \sqrt{2S}[(S - a^+a)fa - fa(S - a^+a)] \\ &= -\sqrt{2S}[faS - faa^+a - Sfa + a^+afa] \\ &= -\sqrt{2S}[-faa^+a + fa^+aa] = -\sqrt{2S}[-faa^+a \\ &\quad + f(aa^+ - 1)a] \\ &= \sqrt{2S}fa = S_l^+ \end{aligned} \tag{18.68}$$

If $l \neq m$, one obtains $[S_l^z, S_m^\pm] = 0$. Similarly, one has $[S_l^z, S_m^-] = -S_l^-\delta_{lm}$.

Remark: One has used $a^+af = fa^+a$ in the above demonstration of (18.67) because

$$a^+af = a^+a\left(1 - \tfrac{a^+a}{2S}\right)^{1/2} = \left(a^+aa^+a - \tfrac{a^+aa^+aa^+a}{2S}\right)^{1/2}$$
$$= \left(1 - \tfrac{a^+a}{2S}\right)^{1/2} a^+a = fa^+a.$$

Problem 10. Operators defined in (3.74)–(3.77) obey the commutation relations:

Demonstration: We have

$$\begin{aligned}[\alpha_{\mathbf{k}}, \alpha^+_{\mathbf{k}'}] &= [a_{\mathbf{k}} \cosh\theta_k + b^+_{\mathbf{k}} \sinh\theta_k, a^+_{\mathbf{k}'} \cosh\theta'_k + b_{\mathbf{k}'} \sinh\theta'_k] \\ &= \cosh\theta_k \cosh\theta'_k[a_{\mathbf{k}}, a^+_{\mathbf{k}'}] + \cosh\theta_k \sinh\theta'_k[a_{\mathbf{k}}, b^+_{\mathbf{k}'}] \\ &\quad + \sinh\theta_k \cosh\theta'_k[b^+_{\mathbf{k}}, a^+_{\mathbf{k}'}] + \sinh\theta_k \sinh\theta'_k[b^+_{\mathbf{k}}, b_{\mathbf{k}'}] \\ &= \cosh\theta_k \cosh\theta'_k\delta(k, k') + 0 + 0 \\ &\quad - \sinh\theta_k \sinh\theta'_k\delta(k, k') \\ &= [\cosh^2\theta_k - \sinh^2\theta_k]\delta(k, k') = \delta(k, k') \end{aligned} \tag{18.69}$$

The same demonstration is done for the other relations.

Problem 11. Magnon soft mode:

Demonstration: The magnon spectrum (3.113) becomes unstable when the interaction between next-nearest neighbors defined in ϵ, Eq. (3.107), is larger than a critical constant.

The spectrum becomes unstable when one of its frequencies tends to zero: This mode is termed as "soft mode." Numerically, we plot (3.113) versus **k** for various values of ϵ and determine its critical value. Analytically, we see that interaction J_2 affects modes near $k_x = k_y = k_z = \pi/a$. To increase J_2 makes the frequencies of these modes decrease. The first mode to become zero occurs at

$\epsilon = \epsilon_c = \frac{2}{3}\frac{1-|\alpha|}{1+|\alpha|}$

18.4 Solutions to Problems of Chapter 4

Problem 1. Proofs of (4.13):

Solution:

We consider the following integral in the complex plane:

$$\int_C \frac{e^{-iz(t-t')}}{z + i\epsilon} dz \tag{18.70}$$

where C is an integral contour chosen in the plane. This integral has a pole at $z_0 = -i\epsilon$ on the imaginary axis in the lower half plane. If $t - t' > 0$, we choose the contour in the lower half plane including the pole z_0: $C = -R \rightarrow +R + C'$ where $\pm R$ are on the real axis and C' is the half circle going from R to $-R$ in the lower half plane. Using the theorem of residues we write

$$\int_C \frac{e^{-iz(t-t')}}{z+i\epsilon}dz = \int_{-R}^{R} \frac{e^{-ix(t-t')}}{x+i\epsilon}dx + \int_{C'} \frac{e^{-iz(t-t')}}{z+i\epsilon}dz$$
$$= -2\pi i \text{Residue} \quad (18.71)$$

where the minus sign come from the sense of the contour and the only pole lying inside the contour is z_0:

$$\text{Residue} = \lim_{z\rightarrow z_0} \frac{e^{-iz(t-t')}}{z+i\epsilon} \times (z - z_0)$$
$$= e^{-iz_0(t-t')} = e^{-\epsilon(t-t')}$$
$$= 1 \ (\lim \ \epsilon \rightarrow 0)$$

Taking the limit $R \rightarrow \infty$, we can show that the integral $\int_{C'\rightarrow\infty} \frac{e^{-iz(t-t')}}{z+i\epsilon}dz$ goes to zero. Therefore,

$$\lim_{\epsilon\rightarrow 0^+} \int_{-\infty}^{\infty} \frac{e^{-ix(t-t')}}{x+i\epsilon}dx = -2\pi i \ \text{for} \ t - t' > 0 \quad (18.72)$$

Now, if $t - t' < 0$, we choose the contour C in the upper half plane. The right-hand side of (18.71) is zero because there is no pole inside C. We have thus

$$\lim_{\epsilon\rightarrow 0^+} \int_{-\infty}^{\infty} \frac{e^{-ix(t-t')}}{x+i\epsilon}dx = 0 \ \text{for} \ t - t' < 0 \quad (18.73)$$

Combining (18.72) and (18.73) and using the Heavyside function $\Theta(t - t') = 1$ if $t - t' > 0, = 0$ if $t - t' < 0$, we obtain the formula (4.13):

$$\Theta(t - t') = \lim_{\epsilon\rightarrow 0^+} \frac{i}{2\pi} \int_{-\infty}^{\infty} \frac{e^{-ix(t-t')}}{x+i\epsilon}dx \quad (18.74)$$

Problem 2. Demonstration of Eq. (4.22): We have (4.21)

$$G_{lm}(t - t') = \ll S_l^+(t); S_m^-(t') \gg$$
$$= -i\theta(t - t') < \left[S_l^+(t), S_m^-(t')\right] > \quad (18.75)$$

Noting that $\theta(t) = \int_{-\infty}^{t} \delta(t)dt$ so that $d\theta(t)/dt = \delta(t)$, we write

$$
\begin{aligned}
i\frac{dG_{lm}(t-t')}{dt} &= \frac{d\theta(t-t')}{dt} < \left[S_l^+(t), S_m^-(t')\right] > \\
&\quad + \theta(t-t')\frac{d < \left[S_l^+(t), S_m^-(t')\right] >}{dt} \\
&= \delta_{tt'} 2 < S_l^z > \delta_{lm} \\
&\quad - \ll \left[\mathcal{H}, S_l^+\right](t); S_m^-(t') \gg \\
&= 2 < S_l^z > \delta_{lm}\delta_{tt'} \\
&\quad - \ll \left[\mathcal{H}, S_l^+\right](t); S_m^-(t') \gg \qquad (18.76)
\end{aligned}
$$

where we have used in the second equality a commutation relation for the first term and the equation of motion $id\hat{O}/dt = -[\mathcal{H}, \hat{O}]$ for the second term. We calculate now the commutator $A = [\mathcal{H}, S_l^+]$. Using $\mathcal{H}$ of (3.31) without the factor 2 of J and without the applied field term, we have

$$
\begin{aligned}
A &= -\frac{1}{2}J \sum_{<l',m'>} \{[S_{l'}^z S_{m'}^z + \frac{1}{2}(S_{l'}^+ S_{m'}^- + S_{l'}^- S_{m'}^+), S_l^+]\} \\
&= -\frac{1}{2}J \sum_{<l',m'>} \{[S_{l'}^z S_{m'}^z, S_l^+] + \frac{1}{2}[S_{l'}^+ S_{m'}^-, S_l^+] \\
&\quad + \frac{1}{2}[S_{l'}^- S_{m'}^+, S_l^+]\} \\
&= -\frac{1}{2}J \sum_{<l',m'>} \{[S_{l'}^z, S_l^+]S_{m'}^z + S_{l'}^z[S_{m'}^z, S_l^+] + \frac{1}{2}[S_{l'}^+, S_l^+]S_{m'}^- \\
&\quad + \frac{1}{2}S_{l'}^+[S_{m'}^-, S_l^+] + \frac{1}{2}[S_{l'}^-, S_l^+]S_{m'}^+ + \frac{1}{2}S_{l'}^-[S_{m'}^+, S_l^+]\} \\
&= -\frac{1}{2}J \sum_{<l',m'>} \{S_l^+ S_{m'}^z \delta_{ll'} + S_{l'}^z S_l^+ \delta_{m'l} + 0 - S_{l'}^+ S_l^z \delta_{lm'} \\
&\quad - S_l^z \delta_{ll'} S_{m'}^+ + 0\} \\
&= -\frac{1}{2}J \left[\sum_{m'} S_l^+ S_{m'}^z + \sum_{l'} S_{l'}^z S_l^+ - \sum_{l'} S_{l'}^+ S_l^z - \sum_{m'} S_l^z S_{m'}^+\right] \\
&= -J \sum_{l'} [S_{l'}^z S_l^+ - S_l^z S_{l'}^+] \qquad (18.77)
\end{aligned}
$$

where the pre-factor $\frac{1}{2}$ was added to remove the double counting due to the double sum and where we have used

$[AB, C] = [A, C]\,B + A\,[B, C]$ and

$$\left[S_l^+, S_m^-\right] = 2S_l^z\delta_{lm}$$
$$\left[S_m^z, S_l^\pm\right] = \pm S_l^\pm\delta_{lm}$$

Note that in the last line of Eq. (18.77) we gathered the sums by changing the dummy variables and permuted operators such as $S_l^+S_{l'}^z = S_{l'}^zS_l^+$ because the indices l and l' indicate two different neighboring sites [in the Hamiltonian, l' and m' are neighboring sites, one of these sites is equal to l because of the delta functions in the 4th equality of Eq. (18.77)]. Inserting (18.77) into (18.76) and putting $l' = l + \rho$ where ρ are the vectors connecting l to its neighbors, we obtain Eq. (4.22):

$$\begin{aligned} i\hbar\frac{dG_{lm}(t)}{dt} &= 2 < S_l^z > \delta_{lm}\delta(t) - \ll \left[\mathcal{H}, S_l^+\right](t); S_m^- \gg \\ &= 2 < S_l^z > \delta_{lm}\delta(t) \\ &\quad -J\sum_\rho \ll S_l^z(t)S_{l+\rho}^+(t) - S_l^+(t)S_{l+\rho}^z(t); S_m^- \gg \end{aligned}$$

Problem 3. Helimagnet by Green's function method:

Solution: We consider a crystal of simple cubic lattice with Heisenberg spins of amplitude 1/2. The interaction J_1 between nearest neighbors is ferromagnetic. Suppose that along the y axis there exists an antiferromagnetic interaction J_2 between next nearest neighbors, in addition to J_1.

(a) In the xz plane the only interaction is J_1, so the spins in the plane are parallel. Along the y axis, the competition between in the ferromagnetic J_1 and the antiferromagnetic J_2 can give rise to a helical structure (see Section 3.4). Let θ be the helical angle between two nearest neighboring spins in the y direction. The energy of a spin is written as

$$E = -4J_1 - 2J_1\cos\theta + 2|J_2|\cos(2\theta) \qquad (18.78)$$

where the first term is the energy in the xz plane, the second and third terms are the energy in the y

direction. Minimizing E with respect to θ we have

$$\frac{dE}{d\theta} = 0 = 2J_1 \sin\theta - 4|J_2|\sin(2\theta)$$

$$0 = 2J_1 \sin\theta - 8|J_2|\sin\theta\cos\theta$$

$$= 2\sin\theta(J_1 - 4|J_2|\cos\theta) \tag{18.79}$$

The solutions are

(i) $\sin\theta = 0$, namely $\theta = 0$ (solution 1), $\theta = \pi$ (solution 2)

(ii) $\cos\theta = \frac{J_1}{4|J_2|}$ (solution 3) if $-1 \leq \frac{J_1}{4|J_2|} \leq 1$, namely $\frac{|J_2|}{J_1} \geq \frac{1}{4} \equiv \alpha_c$. This solution corresponds to the helimagnetic configuration.

We can compare the energies of the three solutions

$$E_1 = -4J_1 - 2J_1 + 2|J_2|$$

$$E_2 = -4J_1 + 2J_1 + 2|J_2|$$

$$E_3 = -4J_1 - 2\frac{J_1^2}{4|J_2|} + 2|J_2|\left(2\frac{J_1^2}{16|J_2|^2} - 1\right)$$

We see that E_2 is a maximum, and

$$E_1 < E_3 \text{ when } \frac{|J_2|}{J_1} < \alpha_c$$

$$E_3 < E_1 \text{ when } \frac{|J_2|}{J_1} > \alpha_c$$

(b) Let θ be the helical angle between two nearest neighboring spins in the y direction. The Hamiltonian in terms of θ is given by (3.124) (without the anisotropy term):

$$\mathcal{H} = -\frac{1}{4}\sum_{(i,j)} J\left(\mathbf{R}_{ij}\right)\{(S_i^+S_j^- + S_i^-S_j^+)[1 + \cos(\theta \cdot \mathbf{R}_{ij})]$$
$$- (S_i^+S_j^+ + S_i^-S_j^-)[1 - \cos(\theta \cdot \mathbf{R}_{ij})]$$
$$+ 4S_i^zS_j^z\cos(\theta \cdot \mathbf{R}_{ij})$$
$$+ 2[(S_i^+ + S_i^-)S_j^z - S_i^z(S_j^+ + S_j^-)]\sin(\theta \cdot \mathbf{R}_{ij})\} \tag{18.80}$$

where $\mathbf{R}_{ij}$ is the distance vector between the two spins i and j, θ is the vector of magnitude θ perpendicular to the angle plane (xz). In the present problem, we have

(i) $\cos(\theta \cdot \mathbf{R}_{ij}) = 0$ for nearest neighbors i and j belonging to the xz plane,

(ii) $\cos(\theta \cdot \mathbf{R}_{ij}) = \cos\theta$ for nearest neighbors i and j on the y axis (lattice constant=1),

(iii) $\cos(\theta \cdot \mathbf{R}_{ij}) = \cos(2\theta)$ for next nearest neighbors i and j on the y axis (distance $R_{ij} = 2$).

(c) We define two Green's functions by (4.69)–(4.70) which lead to, using the RPA decoupling scheme,

$$i\hbar\frac{dG_{m\ell}(t-t')}{dt} = 2 < S_k^z > \delta_{m\ell}\delta(t)$$
$$-\frac{1}{2}\sum_{m'} J_{m,m'}[< S_k^z > (\cos\theta_{m,m'} - 1)$$
$$\times F_{m'\ell}(t-t')$$
$$+ < S_k^z > (\cos\theta_{m,m'} + 1)G_{m'\ell}(t-t')$$
$$-2 < S_{m'}^z > \cos\theta_{m,m'}G_{m\ell}(t-t')] \tag{18.81}$$

$$i\hbar\frac{dF_{m\ell}(t-t')}{dt} = \frac{1}{2}\sum_{m'} J_{m,m'}[< S_m^z > (\cos\theta_{m,m'} - 1)$$
$$\times G_{m'\ell}(t-t')$$
$$+ < S_m^z > (\cos\theta_{m,m'} + 1)F_{m'\ell}(t-t')$$
$$-2 < S_{m'}^z > \cos\theta_{m,m'}F_{m\ell}(t-t')] \tag{18.82}$$

Using the Fourier transforms and summing on neighbors with their corresponding angles mentioned above, we obtain

$$\mathbf{M}(\omega)\,\mathbf{g} = \mathbf{u}, \tag{18.83}$$

where

$$\mathbf{g} = \begin{pmatrix} g_{n,n'} \\ f_{n,n'} \end{pmatrix}, \qquad \mathbf{u} = \begin{pmatrix} 2\langle S^z\rangle \delta_{n,n'} \\ 0 \end{pmatrix}, \tag{18.84}$$

and

$$\mathbf{M}(\omega) = \begin{pmatrix} \omega + A & B \\ -B & \omega - A \end{pmatrix}, \tag{18.85}$$

where

$$A = J_1 < S^z > \Big[Z(\gamma_{\mathbf{k}} - 1) + (\cos\theta + 1)\cos k_y$$
$$-2\cos\theta + \frac{|J_2|}{J_1}[\cos(2\theta) + 1]\cos(2k_y)$$
$$-2\frac{|J_2|}{J_1}\cos(2\theta)\Big]$$
$$B = J_1 < S^z > (\cos\theta - 1)\cos k_y$$
$$-|J_2| < S^z > [\cos(2\theta) - 1]\cos(2k_y) \quad (18.86)$$

in which, $Z = 4$ is the number of nearest neighbors in the xz plane, $\theta = \arccos(J_1/4|J_2|)$, and $\gamma_{\mathbf{k}} = [2\cos(k_x) + 2\cos(k_z)]/Z$.
Non-trivial solutions of (18.83) impose that

$$\det \begin{vmatrix} \omega + A & B \\ -B & \omega - A \end{vmatrix} = 0$$

We have

$$0 = (\omega + A)(\omega - A) + B^2 \quad (18.87)$$
$$\omega = \pm\sqrt{A^2 - B^2} \quad (18.88)$$

We show in Fig. 18.7 the spin wave spectrum (18.88) as a function of the wave vector in the helical direction k_y, for $\theta = \pi/3$ ($|J_2|/J_1 = 0.5$) at $k_x = k_y = 0$. We see that $\omega = 0$ at θ.

(d) Note that

(i) when $\cos\theta = 1$ (ferromagnet), taking $J_2 = 0$, we have

$$A = J_1 < S^z > \Big[4(2\cos k_x + 2\cos k_z)/4 - 4$$
$$+ 2\cos k_y - 2\Big]$$
$$= J_1 < S^z > \Big[2\cos k_x + 2\cos k_z + 2\cos k_y - 6\Big]$$
$$= J_1 < S^z > Z'(\gamma'_{\mathbf{k}} - 1)$$

The expression (18.88) is reduced to the dispersion relation (4.32) for ferromagnets $\omega = Z'J_1\langle S^z\rangle(1 - \gamma'_{\mathbf{k}})$ where $Z' = 6$ and $\gamma'_{\mathbf{k}} = \big[2\cos(k_x) + 2\cos k_y + 2\cos(k_z)\big]/Z'$ [we have in this case $B = 0$ in (18.88)].

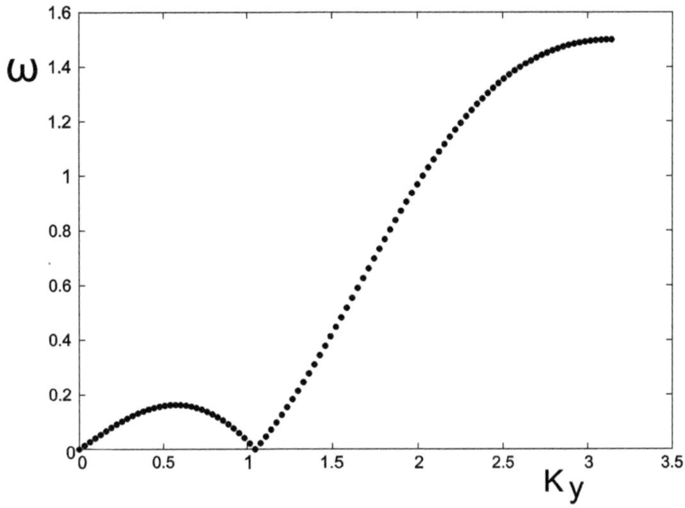

Figure 18.7 Spin wave spectrum versus the wave vector k_y in the simple cubic lattice with a helical structure in the y axis, in the case $\theta = \pi/3$ (namely $|J_2|/J_1 = 0.5$), and $k_x = k_z = 0$.

(ii) when $\cos\theta = -1$ (collinear antiferromagnets), for $J_2 = 0$, we have, by returning to (18.81)–(18.82) to calculate A and B ($J_1 < 0$),

$$A = |J_1| < S^z > Z'$$
$$B = |J_1| < S^z > Z'\gamma'_{\mathbf{k}} \qquad (18.89)$$

Eq. (18.88) then gives the dispersion relation (4.59) of antiferromagnets: $\omega = \pm Z'|J_1| \langle S^z \rangle \sqrt{1 - (\gamma'_{\mathbf{k}})^2}$.

Problem 4. Green's function method for a system of Ising spins $\pm 1/2$ in an applied magnetic field in one dimension:

Solution: We consider the following Hamiltonian:

$$\mathcal{H} = -2J \sum_n S_n^z S_{n+1}^z - g\mu_B H \sum_n S_n^z \qquad (18.90)$$

where $S_n^z = \pm 1/2$ is the z component of the spin given by the Pauli matrices. $2S_n^z = \sigma_n^z = \pm 1$ [Eqs. (1.3)–(1.5)]. The Ising model corresponds to the assumption that only the z component is taken into account. For simplicity, we often take $S_z = \pm 1$, the results are, however, identical to a constant factor.

Consider the site i. We define the following Green's function:

$$G_i(t) \equiv \ll S_i^+(t); S_i^- \gg \quad (18.91)$$

The equation of motion of $G_i(t)$ is ($\hbar = 1$)

$$i\frac{dG_i(t)}{dt} = 2 < S_i^z > \delta(t)$$
$$+2J \ll S_i^+(t)[S_{i+1}^z(t) + S_{i-1}^z(t)]; S_i^- \gg$$
$$+g\mu_B H \ll S_n^+(t); S_n^- \gg \quad (18.92)$$

This equation generates the following Green's function:

$$G_{i+1}(t) = \ll S_i^+(t)[S_{i+1}^z(t) + S_{i-1}^z(t)]; S_i^- \gg \quad (18.93)$$

Equation (18.92) becomes

$$i\frac{dG_i(t)}{dt} = 2 < S_i^z > \delta(t) + 2G_{i+1}(t) + g\mu_B H G_i(t) \quad (18.94)$$

The equation of motion of $G_{i+1}(t)$ generates the following Green's function:

$$G_{i+2}(t) \equiv \ll S_i^+(t)S_{i-1}^z(t)S_{i+1}^z(t); S_i^- \gg \quad (18.95)$$

We write the equation of motion of $G_{i+2}(t)$ but we shall neglect Green's functions of higher orders. We have

$$i\frac{dG_i(t)}{dt} = 2 < S_i^z > \delta(t) + 2G_{i+1}(t) + g\mu_B H G_i(t)$$
$$i\frac{dG_{i+1}(t)}{dt} = 2 < S_i^z(S_{i-1}^z + S_{i+1}^z > \delta(t) + J\, G_i(t)$$
$$+4J\, G_{i+2}(t) + g\mu_B H G_{i+1}(t)$$
$$i\frac{dG_{i+2}(t)}{dt} = 2 < S_{i-1}^z S_i^z S_{i+1}^z > \delta(t) + \frac{J}{2} G_i(t)$$
$$+g\mu_B H G_{i+2}(t)$$

The Fourier transforms $G_k(t) = \int_{-\infty}^{\infty} G_k(E)e^{-iEt}dE$ ($k = i$, $i+1, i+2$) yield, putting $E' = E - g\mu_B H$,

$$E'G_i(E') - 2G_{i+1}(E') = \frac{1}{\pi}x_1$$
$$-J\, G_i(E') + E'G_{i+1}(E') - 4J\, G_{i+2}(E') = \frac{2}{\pi}x_2$$
$$-\frac{J}{2}G_{i+1}(E') + E'G_{i+2}(E') = \frac{1}{\pi}x_3$$

where

$$x_1 = < S_i^z >,$$
$$2x_2 = < S_i^z(S_{i-1}^z + S_{i+1}^z) >,$$
$$x_3 = < S_i^z S_{i-1}^z S_{i+1}^z > .$$

The solutions of these equations are

$$G_i(E') = \frac{1}{\pi}\left[\frac{x_1/2 - 2x_3}{E'} + \frac{x_1/4 + x_2 + x_3}{E' - 2J} + \frac{x_1/4 - x_2 + x_3}{E' + 2J}\right]$$

$$G_{i+1}(E') = \frac{1}{\pi}\left[\frac{x_1/4 + x_2 + x_3}{E' - 2J} + \frac{-x_1/4 + x_2 - x_3}{E' + 2J}\right]$$

$$G_{i+2}(E') = \frac{1}{\pi}\left[\frac{-x_1/8 + x_3/2}{E'} + \frac{x_1/16 + x_2/4 + x_3/4}{E' - 2J} + \frac{x_1/16 - x_2/4 + x_3/4}{E' + 2J}\right]$$

The spectral theorem [see Eq. (4.39)] gives

$$< S_i^- S_i^+ > = \frac{x_1 - 4x_3}{e^{\beta g\mu_B H} - 1} + \frac{x_1/2 + 2x_2 + 2x_3}{e^{\beta(g\mu_B H + 2J)} - 1} + \frac{x_1/2 - 2x_2 + 2x_3}{e^{\beta(g\mu_B H - 2J)} - 1}$$

$$< S_i^- S_i^+ (S_{i-1}^z + S_{i+1}^z > = \frac{x_1/2 + 2x_2 + 2x_3}{e^{\beta(g\mu_B H + 2J)} - 1} + \frac{-x_1/2 + 2x_2 - 2x_3}{e^{\beta(g\mu_B H - 2J)} - 1}$$

$$< S_i^- S_i^+ S_{i-1}^z S_{i+1}^z > = \frac{-x_1/4 + x_3}{e^{\beta g\mu_B H} - 1} + \frac{x_1/8 + x_2/2 + x_3/2}{e^{\beta(g\mu_B H + 2J)} - 1} + \frac{x_1/8 - x_2/2 + x_3/2}{e^{\beta(g\mu_B H - 2J)} - 1}$$

We expand these expressions at small values of H. At the first order in H, we have

$$x_1 - 4x_3 = \beta g\mu_B H\left[\frac{1}{2} - 2x_2 \coth\beta J\right] \quad (18.96)$$

$$\left[-1 + \frac{1}{2}\coth\beta J\right]x_1 + 2x_3\coth\beta J = \frac{\beta g\mu_B H}{\sinh^2\beta J}x_2 \quad (18.97)$$

$$x_1 - 4x_3 = 2\beta g\mu_B H[-x_2' + x_2\coth\beta J] \quad (18.98)$$

where $x_2' =< S_{i-1}^z S_{i+1}^z >$.
We have three equations for four unknowns x_1, x_2, x_3 and x_2'. It is not possible to solve them. However, when $H = 0$, we have $x_1 = x_3 = 0$, namely there is no ordering at $T \neq 0$ in one dimension (remember $x_1 =< S_i^z >$), as we will show by the renormalization group in Section 5.4.2.
To go further, we calculate the correlation function between the spin at the site 0 and the spin at the site n: We consider the following Green's functions:

$$G_{0,n} = \ll S_0^+(t)S_n^z(t); S_0^- \gg$$
$$G_{1,n} = \ll S_0^+(t)[S_{-1}^z(t) + S_1^z(t)]S_n^z(t); S_0^- \gg$$
$$G_{2,n} = \ll S_0^+(t)S_{-1}^z(t)S_1^z(t)S_n^z(t); S_0^- \gg$$

Following the same method as above, we obtain

$$2x_{1,n} - 8x_{3,n} = \beta g\mu_B H[x_1 - 4x_{2,n}\coth\beta J]$$
$$< S_1^z S_n^z > + < S_{-1}^z S_n^z > = (x_{1,n} + 4x_{3,n})\coth\beta J - 2\frac{\beta g\mu_B H}{\sinh^2\beta J}x_{2,n}$$
$$4x_{3,n} - x_{1,n} = 2\beta g\mu_B H[< S_{-1}^z S_1^z S_n^z > - x_{2,n}\coth\beta J]$$

where $x_{1,n} =< S_0^z S_n^z >$, $2x_{2,n} =< S_0^z(S_{-1}^z + S_1^z) >$, $x_{3,n} =< S_0^z S_{-1}^z S_1^z S_n^z >$
When $H = 0$, we have $x_{1,n} = 4x_{3,n}$. Hence,

$$< S_{-1}^z S_n^z > + < S_1^z S_n^z >= 2 < S_0^z S_n^z > \coth\beta J \quad (18.99)$$

We take a solution of the form $< S_m^z S_n^z >= AX^{n-m}$ where A is a constant: when $m = n$, we have $< S_n^z S_n^z >= A = 1/4$.

Setting $n = 0, m = -1, m = 1$ and $m = 0$ in (18.99) we have

$$X + \frac{1}{X} = 2\coth\beta J \qquad (18.100)$$

The solution of this equation is $X = A\tanh\frac{\beta J}{2}$, namely

$$< S_0^z S_n^z >= \frac{1}{4}\left[\tanh\frac{\beta J}{2}\right]^n \qquad (18.101)$$

We see that, for $n = 0$, we have $< S_0^z S_0^z >= 1/4$ as expected for a spin $1/2$. For $n = 1$, we have

$$x_2 =< S_0^z S_1^z >= \frac{1}{4}\tanh\frac{\beta J}{2} \qquad (18.102)$$

Substituting this in (18.96) and (18.97) we obtain

$$x_1 =< S_0^z >= \beta g\mu_B H e^{\beta J} \qquad (18.103)$$

The susceptibility is thus

$$\chi = Ng\mu_B < S_0^z > /H = \beta(g\mu_B)^2 e^{\beta J} = \frac{(g\mu_B)^2 e^{\beta J}}{k_B T} \qquad (18.104)$$

This result is also obtained by the transfer matrix method [see (5.90)]. Note that the argument of the exponential is βJ because of the factor 2 in the Hamiltonian and spins $1/2$: To find the argument of the exponential of χ in Problem 5.6 [Eq. (5.90)] of Section 5.9, we divide J by 2 and multiply by 4 (inverse of the square of spin magnitude).

Problem 5. Effect of next-nearest neighbor interaction:

Guide: We follow the same calculation in Section 4.2 in adding the interaction J_2 in the equations (4.22), (4.26)–(4.33). We obtain, instead of (4.29),

$$\hbar\omega G_{\mathbf{k}}(\omega) = \frac{< S^z >}{\pi} + J < S^z > ZG_{\vec{k}}(\omega)\left[1 - \frac{1}{Z}\sum_{\rho} e^{i\mathbf{k}\cdot\rho}\right]$$
$$+J_2 < S^z > Z_2 G_{\mathbf{k}}(\omega)\left[1 - \frac{1}{Z_2}\sum_{\vec{\rho}_2} e^{i\mathbf{k}\cdot\rho_2}\right]$$

where Z is the number of nearest neighbors and Z_2 that of next-nearest neighbors. We have

$$G_{\mathbf{k}}(\omega) = \frac{< S^z >}{\pi} \times$$

$$\frac{1}{\hbar\omega - ZJ \ < S^z > (1 - \gamma(\mathbf{k})) - Z_2 J_2 < S^z > (1 - \gamma_2(\mathbf{k}))}$$

$$= \frac{< S^z >}{\pi} \frac{1}{\hbar\omega - \epsilon_{\mathbf{k}}} \tag{18.105}$$

where

$$\gamma(\mathbf{k}) = \frac{1}{Z} \sum_{\rho} \mathrm{e}^{i\vec{k}\cdot\rho} \tag{18.106}$$

$$\gamma_2(\mathbf{k}) = \frac{1}{Z_2} \sum_{\vec{\rho}_2} \mathrm{e}^{i\mathbf{k}\cdot\rho_2} \tag{18.107}$$

and

$$\epsilon_{\mathbf{k}} = ZJ \ < S^z > (1 - \gamma(\vec{k})) + Z_2 J_2 < S^z > (1 - \gamma_2(\mathbf{k})) \tag{18.108}$$

Problem 6. Magnon spectrum in Heisenberg triangular antiferromagnet: Green's function method

Solution: Using the ground state of a triangular lattice determined in Problem 6 of Section 3.6 for the angle $\theta_{k,k'}$ between two neighboring spins, the Fourier transformations of Eqs. (4.71)–(4.72) can be written as

$$\mathbf{M}(\omega)\,\mathbf{g} = \mathbf{u}, \tag{18.109}$$

where

$$\mathbf{g} = \begin{pmatrix} g_{n,n'} \\ f_{n,n'} \end{pmatrix}, \qquad \mathbf{u} = \begin{pmatrix} 2\left\langle S^z \right\rangle \delta_{n,n'} \\ 0 \end{pmatrix}, \tag{18.110}$$

and

$$\mathbf{M}(\omega) = \begin{pmatrix} A^+ & B \\ -B & A^- \end{pmatrix}, \tag{18.111}$$

where

$$A^{\pm} = \omega \pm \left[\frac{1}{2} J \left\langle S^z \right\rangle (Z\gamma_{\mathbf{k}}) (\cos\theta + 1) - J \left\langle S^z \right\rangle Z \cos\theta\right]$$

$$B = \frac{1}{2} J \left\langle S^z \right\rangle (\cos\theta - 1)(Z\gamma_{\mathbf{k}}), \tag{18.112}$$

in which, $Z = 6$ is the number of nearest neighbors, $\theta = 2\pi/3$ the angle between two neighboring spins, and $\gamma_{\mathbf{k}} = \left[2\cos(k_x a) + 4\cos(k_x a/2)\cos\left(k_y a\sqrt{3}/2\right)\right]/Z$.
Non-trivial solutions of (18.109) impose

$$\det\begin{vmatrix} A^+ & B \\ -B & A^- \end{vmatrix} = 0$$

We have

$$0 = A^+A^- + B^2$$

$$\omega = \pm\sqrt{\left[\frac{1}{2}J\langle S^z\rangle(Z\gamma_{\mathbf{k}})(\cos\theta+1) - J\langle S^z\rangle Z\cos\theta\right]^2 - B^2}$$

$$= \pm ZJ\langle S^z\rangle$$

$$\times\sqrt{\left[\frac{1}{2}\gamma_{\mathbf{k}}(\cos\theta+1) - \cos\theta\right]^2 - \left[\frac{1}{2}(\cos\theta-1)\gamma_{\mathbf{k}}\right]^2}$$

$$= \pm ZJ\langle S^z\rangle\sqrt{\cos\theta(1-\gamma_{\mathbf{k}})(\cos\theta-\gamma_{\mathbf{k}})} \qquad (18.113)$$

We show in Fig. 18.8 the spin wave frequency ω versus k_y in the first Brillouin zone for $k_x = 0$.

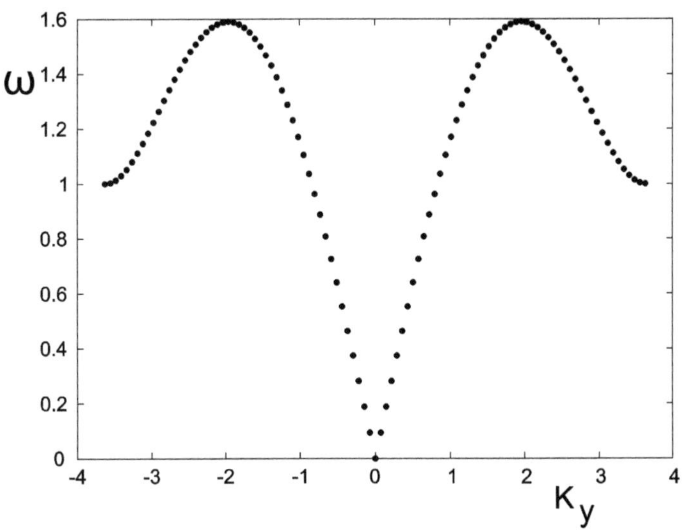

Figure 18.8 Spin wave spectrum versus the wave vector k_y in the case of a triangular antiferromagnet ($S = 1/2, k_x = 0, J = 1$).

Note that when $\cos\theta = 1$, the expression (18.113) is reduced to $\omega = ZJ\langle S^z\rangle(1-\gamma_{\mathbf{k}})$ which is the dispersion relation (4.32) for ferromagnets. When $\cos\theta = -1$, Eq. (18.113) becomes $\omega = ZJ\langle S^z\rangle\sqrt{1-\gamma_{\mathbf{k}}^2}$ which is precisely the dispersion relation (4.59) of collinear antiferromagnets. Using the formula (3.97) for $T = 0$, we have

$$M \simeq \frac{N}{2}(S - \Delta S) \tag{18.114}$$

where $\Delta S = \sum_{\mathbf{k}} \sinh^2\Theta_k$ is independent of T. Θ_k is given by (3.83)

$$\tanh(2\Theta_k) = \gamma_{\mathbf{k}} \tag{18.115}$$

Writing $\sinh\Theta_k$ as a function of $\tanh(2\Theta_k)$, one can calculate numerically the sum in ΔS using the wave vectors in the first Brillouin zone. We have the result $\Delta S \simeq 0.21$, namely the spin length at $T = 0$ is reduced from $S = 1/2$ to 0.29 (compared to 0.303 for a square antiferromagnet). This strong zero-point contraction is due to the frustration of the triangular antiferromagnet. Other more precise methods taking into account higher-order fluctuations should yield a smaller value for the spin length at $T = 0$.

Remark: The integral should be performed inside the first Brillouin zone which is a hexagon.

18.5 Solutions to Problems of Chapter 5

Problem 1. Chain of Ising spins by exact method:

We write the Hamiltonian as

$$\mathcal{H} = -J\sum_{n=1}^{N}\sigma_n\sigma_{n+1} \tag{18.116}$$

with $\sigma_{N+1} = \sigma_1$. The partition function is written as

$$\begin{aligned} Z &= \sum_{\sigma_1=\pm1}\sum_{\sigma_2=\pm1}\cdots\sum_{\sigma_N=\pm1} e^{\beta\sum_{n=1}^{N}\sigma_n\sigma_{n+1}} \\ &= \sum_{\sigma_1=\pm1}\sum_{\sigma_2=\pm1}\cdots\sum_{\sigma_N=\pm1}\prod_{n=1}^{N} e^{\beta\sigma_n\sigma_{n+1}} \end{aligned} \tag{18.117}$$

where $\beta = J/k_BT$. Since $\sigma_n\sigma_{n+1} = \pm 1$, we have the following identity (by verification):

$$e^{\beta\sigma_n\sigma_{n+1}} = \cosh\beta + \sigma_n\sigma_{n+1}\sinh\beta \tag{18.118}$$

Equation (18.117) becomes

$$\begin{aligned} Z &= \sum_{\sigma_1=\pm1}\sum_{\sigma_2=\pm1}\cdots\sum_{\sigma_N=\pm1}[\cosh\beta + \sigma_1\sigma_2\sinh\beta] \\ &\quad\times[\cosh\beta + \sigma_2\sigma_3\sinh\beta]\cdots \\ &= \sum_{\sigma_1=\pm1}\sum_{\sigma_2=\pm1}\cdots\sum_{\sigma_N=\pm1}[(\cosh\beta)^N \\ &\quad+(\cosh\beta)^{N-1}\sinh\beta(\sigma_2\sigma_3) \\ &\quad+\cdots+\cosh\beta(\sinh\beta)^{N-1}(\sigma_3\sigma_4)(\sigma_4\sigma_5)\cdots \\ &\quad\cdots(\sigma_{N+1}\sigma_1)+(\sinh\beta)^N(\sigma_1\sigma_2)(\sigma_2\sigma_3)\cdots(\sigma_{N+1}\sigma_1)] \end{aligned} \tag{18.119}$$

Except the first and the last terms, all other terms of the sum in the square brackets $[\cdots]$ are zero because in each term there is one σ which appears once in the factor giving rise, when summed up, two terms of opposite signs. The first term in (18.119) does not depend on σ, it gives $2^N(\cosh\beta)^N$. The last term yields $2^N(\sinh\beta)^N$ because each σ appears twice in its factor. We have

$$Z = 2^N[(\cosh\beta)^N + (\sinh\beta)^N] \tag{18.120}$$

For $T \neq 0$ ($\beta \neq \infty$), we have $\cosh\beta > \sinh\beta$. With $N \gg 1$, we can neglect $(\sinh\beta)^N$ compared to $(\cosh\beta)^N$. Thus,

$$Z = 2^N(\cosh\beta)^N\left[1+\left(\frac{\sinh\beta}{\cosh\beta}\right)^N\right] = 2^N(\cosh\beta)^N \tag{18.121}$$

The free energy is

$$F = -k_BT\ln Z = -Nk_BT\ln[2\cosh(J/k_BT)]$$

The average energy is thus [see (A.10)]

$$\overline{E} = -\frac{\partial\ln Z}{\partial\beta} = -NJ\tanh(\beta J) \tag{18.122}$$

The heat capacity is

$$C = dE/dT = Nk_B\left[\frac{k_BT}{J}\cosh\frac{J}{k_BT}\right]^{-2} \tag{18.123}$$

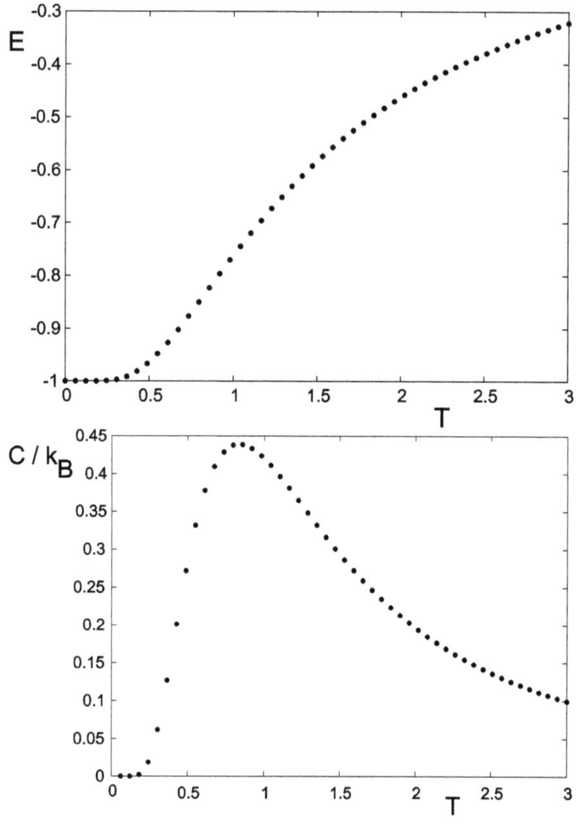

Figure 18.9 Energy E (top) and heat capacity C (bottom) per spin versus temperature T. $J\,/\,k_B = 1$ has been used.

The energy and the heat capacity per spin are shown in Fig. 18.9. We see that C has a maximum but no divergence. Thus, there is no phase transition at finite T.

Problem 2. Chain of Ising spins by micro-canonical method:

(a) In the ground state all spins are parallel, the system energy is thus $E = -J\,N$. The ground-state is two-fold degenerate: All spins are $\uparrow$ or $\downarrow$.

(b) The first excited state: One spin is reversed $\cdots\ \uparrow\uparrow\uparrow\downarrow\uparrow\uparrow\uparrow\ \cdots$ or a block of spins is reversed $\cdots\ \uparrow\uparrow\uparrow\downarrow\downarrow\downarrow\uparrow\uparrow\uparrow\ \cdots$. The energies of the two states are

identical with two "unsatisfied bonds" of energy $+J$. The energy of the system in the first excited state is thus $E(2) = -J(N-2) + 2J = -J(N - 2 \times 2)$ where 2 is the number of unsatisfied bonds. The degeneracy of the first excited state is equal to the number of ways to choose two unsatisfied bonds among N bonds of the system. This is given by $g(2) = 2C_N^2$ where the factor 2 results from the global reversal of the whole system.
One deduces that the energy $E(2n)$ of a state in which there are $2n$ unsatisfied bonds is $E(2n) = -J(N - 2n) + 2nJ = -J(N - 2 \times 2n)$. The degeneracy is $g(2n) = 2C_N^{2n}$.
The energy of the system is maximum when all bonds are unsatisfied, namely $2n = N$ so that $E_{\max} = 0$. The degeneracy of this state is $g = 2C_N^N = 2$.

(c) For a given energy $E(2n)$, the entropy is

$$\begin{aligned} S &= k_B \ln g(2n) \\ &= k_B[\ln 2 + \ln N! - \ln(2n)! - ln(N-2n)!] \\ &\simeq k_B[\ln 2 + N \ln N - 2n \ln 2n - (N-2n)\ln(N-2n)] \end{aligned}$$

where the Stirling formula has been used. The microcanonical temperature is

$$\begin{aligned} T^{-1} &= \frac{\partial S}{\partial E} = \frac{\partial S}{\partial n}\frac{\partial n}{\partial E} \\ &= \frac{k_B}{2J} \ln \frac{N-2n}{2n} \end{aligned} \quad (18.124)$$

from which,

$$\frac{N-2n}{2n} = \exp(2J/k_B T) = \frac{N}{2n} - 1 = 1/x - 1 \quad (18.125)$$

where x is the percentage of unsatisfied bonds $x = \frac{2n}{N}$. One obtains

$$x = \frac{1}{1 + \exp(2J/k_B T)} \quad (18.126)$$

At low temperatures, $x \to 0$: All spins are parallel (the system is ferromagnetic). At high temperatures, $x \to 1/2$: Half of the bonds are unsatisfied, the system is disordered (paramagnetic).

Problem 3. Chain of Ising spins by canonical method:
One considers again the system defined in Problem 2 but in the canonical situation: It is maintained at the temperature T.

(a) The partition function is given by

$$\begin{aligned} Z &= \sum_{n=0}^{N} g(2n) \exp(-\beta E(2n)) \\ &= 2 \sum_{n=0}^{N} C_N^{2n} \exp[\beta J (N - 4n)] \\ &= 2 \exp(\beta J N) \sum_{n=0}^{N} C_N^{2n} \exp(-\beta 4 J n) \end{aligned} \quad (18.127)$$

In combining the two following Newton relations, one has

$$\begin{aligned} (1+u)^N + (1-u)^N &= \sum_{n=0}^{N} C_N^n u^n [1 + (-1)^n] \\ &= 2 \sum_{n'=0}^{N/2} C_N^{2n'} u^{2n'} \end{aligned} \quad (18.128)$$

since all odd terms in n are canceled out. Putting $u = \exp(-\beta 2J)$, and using (18.128), one rewrites (18.127) as

$$\begin{aligned} Z &= 2 \exp(\beta J N) \sum_{n=0}^{N} C_N^{2n} \exp(-\beta 4 J n) \\ &= \exp(\beta J N) \left[[1 + \exp(-\beta 2J)]^N \right. \\ &\quad \left. + [1 - \exp(-\beta 2J)]^N \right] \\ &= \left[[\exp(\beta J) + \exp(-\beta J)]^N \right. \\ &\quad \left. + [\exp(\beta J) - \exp(-\beta J)]^N \right] \\ &= 2^N \left[\cosh^N(\beta J) + \sinh^N(\beta J) \right] \end{aligned} \quad (18.129)$$

This relation of Z is exact (see Problem 1).

(b) The system energy is thus

$$\overline{E} = -\frac{\partial \ln E}{\partial \beta} = -\frac{\partial}{\partial \beta} \ln 2^N [\cosh^N(\beta J) + \sinh^N(\beta J)]$$

$$= -J\,N \left[\frac{\cosh^{N-1}(\beta J)\sinh(\beta J) + \sinh^{N-1}(\beta J)\cosh(\beta J)}{\cosh^N(\beta J) + \sinh^N(\beta J)} \right]$$

(c) The average percentage $\overline{x}$ of unsatisfied bonds is

$$\overline{x} = \frac{2\overline{n}}{N} = \frac{1}{2} + \frac{\overline{E}}{J\,N}$$

$$= \frac{1}{2}\left[1 - \frac{\cosh^{N-1}(\beta J)\sinh(\beta J) + \sinh^{N-1}(\beta J)\cosh(\beta J)}{\cosh^N(\beta J) + \sinh^N(\beta J)} \right]$$

$$= \frac{e^{-\beta J}}{2}\left[\frac{\cosh^{N-1}(\beta J) - \sinh^{N-1}(\beta J)}{\cosh^N(\beta J) + \sinh^N(\beta J)} \right]$$

At low temperatures, $\overline{x} \to 0$. At high temperatures, $\overline{x} \to 1/2$. One finds again here the results using the micro-canonical method found in Problem 2.

The curves $\overline{E}$ and the calorific capacity C are plotted in Problem 1 (Fig. 18.9).

Problem 4. Low- and high-temperature expansions of the Ising model on the square lattice:

Solutions: We have

$$\mathcal{H} = -J \sum_{<i,j>} \sigma_i\,\sigma_j \qquad (18.130)$$

where the sum is performed over nearest neighbors and $\sigma_{i(j)} = \pm 1$.

(a) The partition function

$$Z = \sum_{\sigma_1=\pm 1,\cdots} \exp(K \sum_{<i,j>} \sigma_i\,\sigma_j)$$

$$= \sum_{\sigma_1=\pm 1,\cdots} \prod_{<ij>} \exp(K\sigma_i\,\sigma_j) \qquad (18.131)$$

where $K = J/(k_B T)$. In the ground state (GS), we have $E_0 = -2J\,N = -N_b J$ and $Z = 2\exp(N_b K)$ where the factor 2 comes from the GS degeneracy (reversing all spins), $N_b = 2N$ is the total number of links.

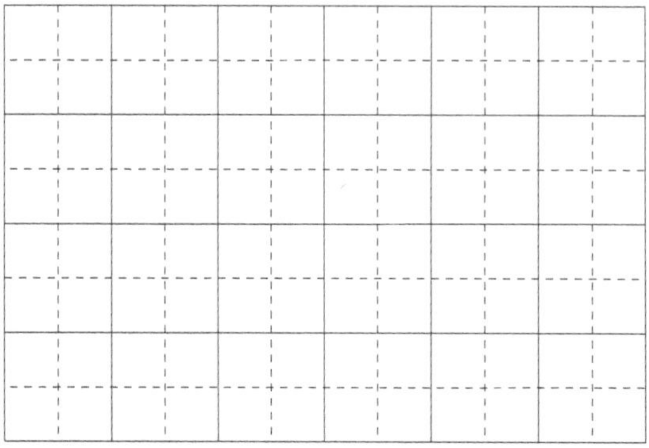

Figure 18.10 Real square lattice (solid lines) and its dual lattice (broken lines).

Let us construct the dual lattice by drawing the links (broken lines) perpendicular to the real links (solid lines) as shown in Fig. 18.10. The case of the square lattice is special: Its dual lattice formed by the broken lines is also a square lattice. This is not the case in general: For example, the dual lattice of the triangular lattice is the honeycomb lattice.

(b) Low-temperature expansion:

- One reversed spin: There are four broken links around it, degeneracy $= N$ (the number of choices of a spin among N spins), the energy is increased from $-4J$ to $+4J$ so that $E = E_0 + 8J$ (see Fig. 18.11).
- Two reversed neighbors: Six broken links, degeneracy $= N_b$ (the number of choices of a link among N_b links), the energy is increased from $-6J$ to $+6J$: $E = E_0 + 12J$.
- Three reversed neighboring spins: Eight broken links for both configurations (trimer with two links on a line, or trimer with two perpendicular links), degeneracy $= 6N$ (there are N choices of the central site and six choices to form a trimer with the central site: one straight trimer along x axis and the second

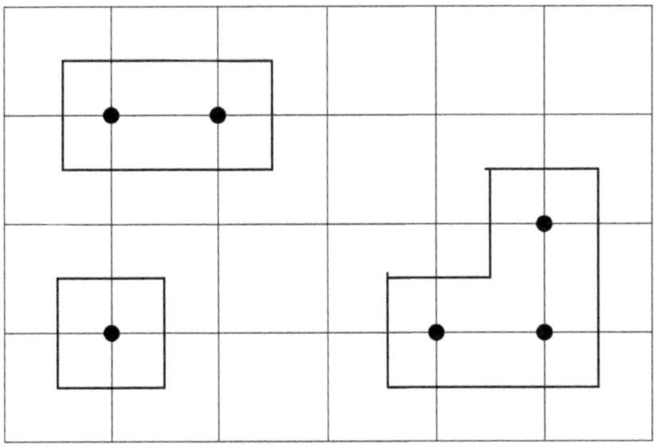

Figure 18.11 Graphs (heavy lines) crossing broken links for one, two three reversed spin clusters. The reversed spins are shown by black circles, other spins are not shown.

straight trimer along y axis, and four perpendicular trimers), $E = E_0 + 16J$.

Calculation of Z with the first excited states of energies $E_0 + 8J$, $E_0 + 12J$, $E_0 + 16J$: To do that, we have to find all excited states for each level. The first two levels concern one and two reversed neighboring spins as found above. However, the level $E_0 + 16J$ corresponds not only to the excited trimer shown above but also to two other cases: a cluster of four sites forming a square, two disconnected reversed spins. Both cases have 8 broken links. The first case has a degeneracy of N (the number of choices of the first site of the square), the second case has a degeneracy of $N(N-5)$ (the number of choices of the first reversed spin is N, the number of choices of the second disconnected reversed spin is $N-5$ where 5 is the number of spins concerned by the first reversed spins which are to be avoided). We finally have

$$Z = 2e^{2NK}[1 + Ne^{-8K} + 2Ne^{-12K} \\ +[6N + N + N(N-5)]e^{-16K} + \cdots] \tag{18.132}$$

(c) We draw a path P which encircles each cluster of reversed spins: This path crosses the broken links around the cluster. For such a closed path, we can verify that it crosses an even number of broken links as follows. Imagine a rectangular path, the number of broken links is even. Now, including an additional site anywhere around the path will add two additional broken links. Excluding a site inside the path will reduce the number of broken links by 2. Such a construction shows clearly that any closed path crosses an even number of broken links.
Let $\ell(P)$ be the number of broken links crossed by the path. Since the variation of energy when breaking a link is $\Delta E = J - (-J) = 2J$. A path crossing $\ell(P)$ broken links corresponds to $\Delta E = +2\ell(P)\,J$, so that the system energy is $E(P) = E_0 + \Delta E = E_0 + 2\ell(P)\,J$. The partition functions is thus

$$Z = \sum_P e^{-E(P)/k_BT} = 2e^{N_bK} \sum_P e^{-2K\ell(P)} \qquad (18.133)$$

(d) High-temperature expansion: Using Eq. (18.118) we write the partition function as

$$\begin{aligned} Z &= \sum_{\sigma_1=\pm1,\cdots} \prod_{<ij>} \exp(K\sigma_i\,\sigma_j) \\ &= \sum_{\sigma_1=\pm1,\cdots} \prod_{<ij>} (\cosh K + \sigma_i\,\sigma_j \sinh K) \\ &= (\cosh K)^{N_b} \sum_{\sigma_1=\pm1,\cdots} \prod_{<ij>} (1 + \sigma_i\,\sigma_j \tanh K) \end{aligned}$$

Since there are N_b links, there are N_b factors in the product. We draw a link between two nearest sites in each factor (this link is in the real space). We expand the product, we see that if a given spin appears an odd number of times in a term of the resulting polynomial, then when summing on its values, this term gives two opposite values yielding a zero contribution. Each non-zero term contains an even number of times of each spin, so that the spin factor in front of $\tanh K$ is equal to 1. The power of $\tanh K$ for this term is nothing but the number of links of spins in front of it. Consider, for example, the term $x = \sigma_1\sigma_2\ \sigma_2\sigma_3\ \sigma_3\sigma_{10}\ \sigma_{10}\sigma_9\ \sigma_9\sigma_8\ \sigma_8\sigma_1 \tanh^6 K$ shown in Fig. 18.12. This term has six links each of which connects two spins. Each spin appears twice because there are two links emanating from it. Hence, $x = \tanh^6 K$. The corresponding graph is shown by the lower left graph in Fig. 18.12. Therefore, we can replace the polynomial by the sum

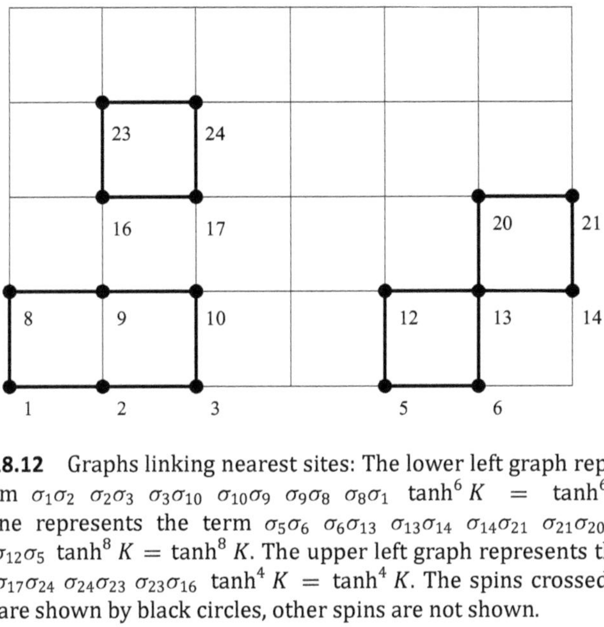

Figure 18.12 Graphs linking nearest sites: The lower left graph represents the term $\sigma_1\sigma_2\ \sigma_2\sigma_3\ \sigma_3\sigma_{10}\ \sigma_{10}\sigma_9\ \sigma_9\sigma_8\ \sigma_8\sigma_1\ \tanh^6 K = \tanh^6 K$, the right one represents the term $\sigma_5\sigma_6\ \sigma_6\sigma_{13}\ \sigma_{13}\sigma_{14}\ \sigma_{14}\sigma_{21}\ \sigma_{21}\sigma_{20}\ \sigma_{20}\sigma_{13}\ \sigma_{13}\sigma_{12}\ \sigma_{12}\sigma_5\ \tanh^8 K = \tanh^8 K$. The upper left graph represents the term $\sigma_{16}\sigma_{17}\ \sigma_{17}\sigma_{24}\ \sigma_{24}\sigma_{23}\ \sigma_{23}\sigma_{16}\ \tanh^4 K = \tanh^4 K$. The spins crossed by the graphs are shown by black circles, other spins are not shown.

$$Z = 2^N(\cosh K)^{N_b} \sum_P (\tanh K)^{\ell(P)} \tag{18.134}$$

where 2^N comes from the sum $\sum_{\sigma_1=\pm1,\cdots}$ and $\ell(P)$ is the number of links in the closed graph P.

(e) Duality:

The partition functions Z in (18.133) and (18.134) have the same structure: Since the prefactors are non-singular, the summations over the closed paths determine the singularity of Z. Note that $\ell(P)$ in (18.133) corresponds to the path drawn in the dual lattice (see Fig. 18.11). While, $\ell(P)$ in (18.134) corresponds to the path drawn in the real lattice (see Fig. 18.12). Nevertheless, these two kinds of path have the same structure with an even number of links in each path. Therefore, we can connect the two Z by fixing

$$e^{-2K^*} = \tanh K \qquad (18.135)$$

$$\text{Hence,}\quad K^* = -\frac{1}{2}\ln\tanh K \qquad (18.136)$$

where K^* corresponds to the low-T phase and K to the high-T phase. The above relation (18.136) is called the "duality" condition which connects the low- and the high-T phases. We deduce from (18.134) and (18.133) the following relation between the high-temperature $Z(K)$ and the low-temperature $Z(K^*)$:

$$Z(K) = 2^N(\cosh K)^{N_b}\sum_P(\tanh K)^{\ell(P)}$$

$$Z(K) = 2^N(\cosh K)^{N_b}e^{-N_bK^*}Z(K^*) \qquad (18.137)$$

where $\sum_P(\tanh K)^{\ell(P)}$ has been replaced by $\sum_P e^{-2K^*\ell(P)}$ and then by $Z(K^*)/(2e^{N_bK^*}) = Z(K^*)/(e^{N_bK^*})$ from (18.133) (the factor 2 in the denominator is neglected because N is large).

(f) The critical temperature of the Ising model on the square lattice is obtained by using the duality. We have

$$\begin{aligned}\sinh 2K &= 2\sinh K\cosh K = 2\tanh K\cosh^2 K\\ &= \frac{2\tanh K}{1-\tanh^2 K}\\ &= \frac{2e^{-2K^*}}{1-e^{-4K^*}} = \frac{2}{e^{2K^*}-e^{-2K^*}}\\ &= \frac{1}{\sinh 2K^*} \qquad (18.138)\end{aligned}$$

from which

$$\sinh 2K\ \sinh 2K^* = 1 \qquad (18.139)$$

This relation is symmetric with respect to K and K^*: when K^* increases, K decreases, and vice versa. If the system undergoes a single phase transition, it should undergo at the same point of K and K^*, namely $K_c^* = K_c$. To satisfy (18.139), we should have

$$\sinh 2K_c = \sinh 2K_c^* = 1 \tag{18.140}$$

Thus,

$$\begin{aligned}
\frac{e^{2K_c} - e^{-2K_c}}{2} &= 1 \\
e^{2K_c} - 2e^{-2K_c} - 2 &= 0 \\
e^{4K_c} - 1 - 2e^{2K_c} &= 0 \\
X^2 - 2X - 1 &= 0 \ (X = e^{2K_c}) \\
X &= 1 + \sqrt{2} \ \text{(positive solution)} \\
K_c &= \frac{\ln(1+\sqrt{2})}{2}
\end{aligned} \tag{18.141}$$

Therefore, $k_B T_c/J = 1/K_c \simeq 2.27$ which is the exact Onsager's solution.

Problem 5. Critical temperatures of the triangular lattice and the honeycomb lattice by duality:
Solutions: We have

$$\mathcal{H} = -J \sum_{<i,j>} \sigma_i \sigma_j \tag{18.142}$$

where the sum is performed over nearest neighbors on the triangular lattice and $\sigma_{i(j)} = \pm 1$.
The dual lattice by construction is a honeycomb lattice shown by the broken lines in Fig. 18.13.
First, we follow the same method as for the square lattice in Problem 4 above: Writing the partition functions of the triangular lattice using low-temperature expansion with graphs on links of the dual lattice and high-temperature expansion with graphs on links of the real lattice, we obtain a relation similar to Eq. (18.137)

$$Z_t(K) = 2^N (\cosh K)^{N_b} e^{-N_b K} Z_h(K^*) \tag{18.143}$$

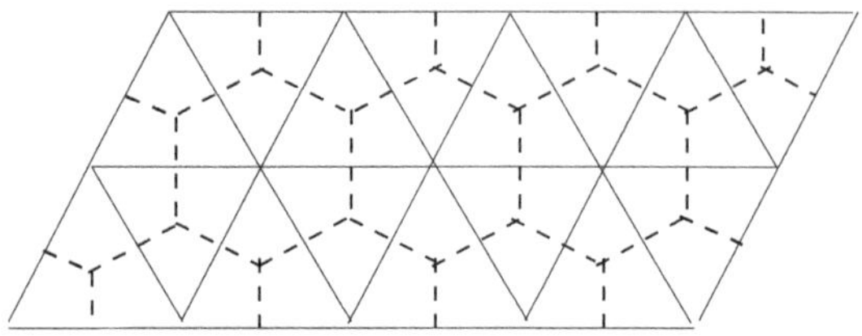

Figure 18.13 Triangular lattice (solid lines) and its dual honeycomb lattice (broken lines).

where $Z_t(K)$ denotes the low-temperature partition function of the triangular lattice, $Z_h(K^*)$ the high-temperature dual (honeycomb) lattice and $N_b = 3N$ is the total number of links.

Now, we can relate the two partition functions in another relation as follows. For convenience, the dual lattice is shifted as shown in Fig. 18.14. We consider the plaquette defined by three sites 1,2 and 3 with a site at the center. We calculate the following quantity:

$$\begin{aligned} A_p(K^*) &= \sum_{\sigma_0=\pm 1} \exp[K^*\sigma_0(\sigma_1+\sigma_2+\sigma_3)] \\ &= \sum_{\sigma_0=\pm 1} \cosh^3 K^*(1+\sigma_0\,\sigma_1 \tanh K^*) \\ &\quad \times (1+\sigma_0\,\sigma_2 \tanh K^*) \times (1+\sigma_0\,\sigma_3 \tanh K^*) \\ &= 2\cosh^3 K^*[1+\tanh^2 K^*(\sigma_1\,\sigma_2+\sigma_2\,\sigma_3+\sigma_3\,\sigma_1)] \end{aligned} \tag{18.144}$$

where we used the remark before Eq. (18.134) to expand the product of the first equality.

Now, we consider the plaquette defined by three sites 1, 2 and 3 of the triangular lattice. We calculate the following quantity:

$$\begin{aligned} B_p(K^+) &= \exp[K^+(\sigma_1\,\sigma_2+\sigma_2\,\sigma_3+\sigma_3\,\sigma_1)] \\ &= \cosh^3 K^+(1+\sigma_1\,\sigma_2 \tanh K^+)(1+\sigma_2\,\sigma_3 \tanh K^+) \\ &\quad (1+\sigma_3\,\sigma_1 \tanh K^+) \\ &= \cosh^3 K^+[1+\tanh^3 K^+ + (\tanh K^+ + \tanh^2 K^+) \\ &\quad \times(\sigma_1\,\sigma_2+\sigma_2\,\sigma_3+\sigma_3\,\sigma_1)] \end{aligned}$$

$$= (\cosh^3 K^+ + \sinh^3 K^+)[1 + \frac{\sinh 2K^+}{2\cosh 2K^+ - \sinh 2K^+}$$
$$\times(\sigma_1\,\sigma_2 + \sigma_2\,\sigma_3 + \sigma_3\,\sigma_1)] \tag{18.145}$$

If we set the factors of $(\sigma_1\,\sigma_2 + \sigma_2\,\sigma_3 + \sigma_3\,\sigma_1)$ in (18.144) and (18.145) to be equal, then we have

$$\tanh^2 K^* = \frac{\sinh 2K^+}{2\cosh 2K^+ - \sinh 2K^+} \tag{18.146}$$

so that

$$\sum_{\sigma_0=\pm 1} \exp[K^*\sigma_0(\sigma_1 + \sigma_2 + \sigma_3)] = \frac{2\cosh^3 K^*}{\cosh^3 K^+ + \sinh^3 K^+}$$
$$\times \exp[K^+(\sigma_1\,\sigma_2 + \sigma_2\,\sigma_3 + \sigma_3\,\sigma_1)] \tag{18.147}$$

This relation connects the two dual lattices. To find the full partition functions, it suffices to sum over all spins of the plaquette and to take the product over all plaquettes on each side of the above equation, we then have

$$Z_h(K^*) = \left(\frac{2\cosh^3 K^*}{\cosh^3 K^+ + \sinh^3 K^+}\right)^N Z_t(K^+) \tag{18.148}$$

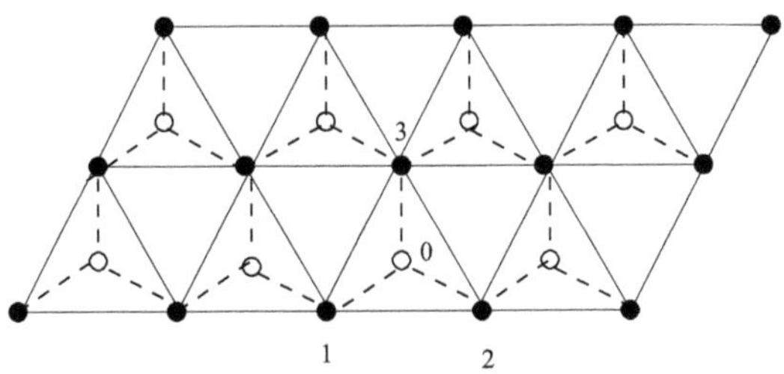

Figure 18.14 Triangular lattice (solid lines) and its dual honeycomb lattice (broken lines) shifted for convenience: The four sites are numbered from 0 to 3 for calculating the partition function.

Replacing $Z_h(K^*)$ given by (18.148) in (18.143), we have

$$Z_t(K) = 2^N(\cosh K)^{3N}e^{-3NK^*}$$
$$\left(\frac{2\cosh^3 K^*}{\cosh^3 K^+ + \sinh^3 K^+}\right)^N Z_t(K^+) \tag{18.149}$$

Using (18.146) to eliminate K^* in the above equation, and after some algebra, we get

$$\frac{Z_t(K)}{(\sinh 2K)^{N/2}} = \frac{Z_t(K^+)}{(\sinh 2K^+)^{N/2}} \tag{18.150}$$

Let us calculate the critical temperature of the triangular lattice. We note that by graph constructions for low- and high-temperatures, we obtained Eq. (18.135) which is very general, independent of the lattice structure: It was established using the square lattice, but all arguments leading to it are also valid for the triangular lattice. Using Eq. (18.135) to eliminate K^* in Eq. (18.146), and after some algebra, we obtain

$$(e^{4K} - 1)(e^{4K^+} - 1) = 4 \tag{18.151}$$

As before in the case of the square lattice, this relation shows that if K increases, K^+ decreases, and vice versa. The transition should occur at the same critical temperature $K_c = K_c^+$. The solution of Eq. (18.151) at K_c is thus

$$e^{4K_c} - 1 = e^{4K_c^+} - 1 = 2, \quad \text{hence } k_B T_c/J = 3.640$$

To find the critical temperature of the honeycomb lattice, we follow the same method as above: We obtain

$$\cosh 2K_c = 2, \quad \text{hence } k_B T_c/J = 1.518.$$

18.6 Solutions to Problems of Chapter 8

Problem 1. Surface magnon:

Solution: In the ferromagnetic case, we just write the equation of motion for S_m^+. We use next the Fourier transform in the xy plane. We obtain then, for $n > 2$,

$$(E - E_n)U_n = \left[4\gamma_1(\mathbf{k}_\parallel)(U_{n-1} + U_{n+1}) + \epsilon(U_{n-2} + U_{n+2})\right] \tag{18.152}$$

and, for the first two layers,

$$(E - E_1)U_1 = \left[4\gamma_1(\mathbf{k}_\parallel)U_2 + \epsilon U_3\right] \tag{18.153}$$

$$(E - E_2)U_2 = \left[4\gamma_1(\mathbf{k}_\parallel)(U_1 + U_3) + \epsilon U_4\right] \tag{18.154}$$

where $E = \frac{\hbar\omega}{J_1 S}$, $\epsilon = \frac{J_2}{J_1}$, and

$$E_n = 8 - 6\epsilon[1 - \gamma_2(\mathbf{k}_\parallel)](n = 2, 3, \cdots) \quad (18.155)$$

$$E_1 = 8 - 5\epsilon[1 - \gamma_2(\mathbf{k}_\parallel)] \quad (18.156)$$

$$\gamma_1(\mathbf{k}_\parallel) = \cos\left(\frac{k_x a}{2}\right)\cos\left(\frac{k_y a}{2}\right) \quad (18.157)$$

$$\gamma_2(\mathbf{k}_\parallel) = \frac{1}{2}[\cos(k_x a) + \cos(k_y a)] \quad (18.158)$$

To study bulk modes, we replace $U_{n\pm1} = U_n \exp(\pm i k_z a/2)$ in Eq. (18.152) to obtain the energy of the bulk mode of wave vector $(\mathbf{k}_\parallel, k_z)$:

$$E = E_n - \left[8\gamma_1(\mathbf{k}_\parallel)\cos(k_z a/2) + 2\epsilon\cos(k_z a)\right]$$

For surface modes, we replace $U_{n\pm1} = U_1\phi^n$ in Eqs. (18.152)–(18.154) to obtain surface-mode energy E and the damping factor ϕ.

Problem 2. Critical next-nearest-neighbor interaction:
Solution: If J_1 and J_2 are both ferromagnetic (>0), the ferromagnetic state is stable. However, if J_2 becomes negative, the ferromagnetic state becomes unstable beyond a critical value of $\epsilon = |J_2|/J_1$. For an infinite crystal, the critical value is obtained by setting the energy E of the lowest magnon mode equal to zero. This corresponds to the instability due to a soft mode. The stable state is no more ferromagnetic for $\epsilon > \epsilon_c$ with $\epsilon_c = \frac{4}{3}$, namely $J_2 < -\frac{4}{3}J_1$

Problem 3. Uniform magnetization approximation:
Solution: If we replace all $< S_n^z >$ by a unique value M, we see that all elements of the matrix **M** in Section 8.4 are proportional to M. The eigenvalues E_i obtained by solving $\det\mathbf{M} = 0$ is, therefore, proportional to M.

Problem 4. Multilayers: critical magnetic field
Solution: Without applied field, the spin configuration is A(up spin)$-B$(down spin)-C(up spin). In the very strong field applied along z, the B spins all turn up. We calculate the critical field beyond which the spin configuration is that state.

The in-plane exchange energy for a B spin does not change whether the B film is in up or down state. We consider a column of spins in the z direction. The energy of the spins of B when they are antiparallel to $\mathbf{H}$ is

$$E_{AF} = -J_1(N_1 - 1) + J_s - HN_1 - J_2(N_2 - 2) + J_s \\ + HN_2 - J_3(N_3 - 1) - HN_3.$$

The energy of B spins when they are parallel to $\mathbf{H}$ is

$$E_F = -J_1(N_1 - 1) - J_s - HN_1 - J_2(N_2 - 2) - J_s \\ - HN_2 - J_3(N_3 - 1) - HN_3$$

(J_s is negative). We see that $E_F < E_{AF}$ when $H > H_c = \frac{2J_s}{N_2}$ (we have taken $J_1 = J_2 = J_3$).

Problem 5. Mean-field theory for thin films:

Solution: We assume the Hamiltonian $\mathcal{H} = -J\sum_{i,j} S_i S_j$ where $S_i = \pm 1$ (Ising spin at the lattice site i). We suppose a simple cubic lattice with a (001) surface. Using the mean-field theory we have the average values of the spins in the three layers

$$< S_1 > = \tanh\left[\beta J\left(Z < S_1 > + < S_2 >\right)\right]$$
$$< S_2 > = \tanh\left[\beta J\left(Z < S_2 > + < S_1 > + < S_3 >\right)\right]$$
$$< S_3 > = \tanh\left[\beta J\left(Z < S_3 > + < S_2 >\right)\right]$$

where $\beta = 1/(k_B T)$ and $Z = 4$ the number of nearest neighbors in the xy plane. By symmetry $< S_1 >=< S_3 >$; therefore, we have only two equations to solve. Numerically, it can be easily done in a self-consistent manner.

Problem 6. Holstein–Primakoff method:

Guide: We can modify the equations (3.41)–(3.43) for a semi-infinite crystal: We write the equation of motion for a spin in each layer. We obtain a set of difference equations. We use next the Fourier transform in the xy plane. We can use (3.41)–(3.43) to make some comments: The lower the energy $\epsilon_{\mathbf{k}}$ is, the larger $< n_{\mathbf{k}} >$ becomes [see (3.41)]. As a consequence, M becomes smaller [see (3.43)]. Thus, the low-energy surface modes lower the layer magnetization near the surface with a stronger effect for surface magnetization because of the lack of neighbors.

Problem 7. Frustrated surface: surface spin rearrangement

Solution: If there is only the surface, then the surface spins form a planar 120° configuration (see Section 5.7.3). We suppose this structure lies in the xy plane. Now, when the beneath layer acts on the surface spins with a ferromagnetic interaction, the surface spins on a triangle turn into the z direction to satisfy partially the ferromagnetic interaction: Let α be the projection on the (xy) plane the angle between two neighboring surface spins and β be the angle between a surface spin and its neighbor in the second layer (see Fig. 18.15). We have

$$\alpha_{1,2} = 0, \; \alpha_{2,3} = \frac{2\pi}{3}, \; \alpha_{3,1} = \frac{4\pi}{3} \tag{18.159}$$

The energy of a cell formed by a surface triangle and the beneath triangle is, for spins 1/2,

$$H_p = -\frac{9J}{2} - \frac{3J}{2}\cos\beta - \frac{9J_s}{2}\cos^2\beta + \frac{9J_s}{4}\sin^2\beta. \tag{18.160}$$

The minimization of this energy gives

$$\frac{\partial H_p}{\partial \beta} = \frac{27J_s}{2}\cos\beta\sin\beta + \frac{3J}{2}\sin\beta = 0 \tag{18.161}$$

Hence,

$$\cos\beta = -\frac{J}{9J_s}. \tag{18.162}$$

This solution is possible for $J_s < -J/9$. For $J_s > -J/9$, the stable state is ferromagnetic.

Note that we can use the steepest-descent method described in Chapter 9 to determine numerically the ground state of classical spin systems.

Problem 8. Ferrimagnetic film:

Solution: Let us take into account the difference between the average layer magnetizations, namely $< S_m^z > = < S_1^z >$ if m belongs to the first layer, $< S_l^z > = < S_2^z >$ if l belongs to the second layer, etc. Modifying (8.5)–(8.6) for the first two layers, we have

$$(E - E_1)U_1 = - < S_1^z > \left[4\gamma_1(\mathbf{k}_{\parallel})U_2 + \epsilon U_3\right]$$
$$(E + E_2)U_2 = - < S_2^z > \left[4\gamma_1(\mathbf{k}_{\parallel})(U_1 + U_3) + \epsilon U_4\right]$$

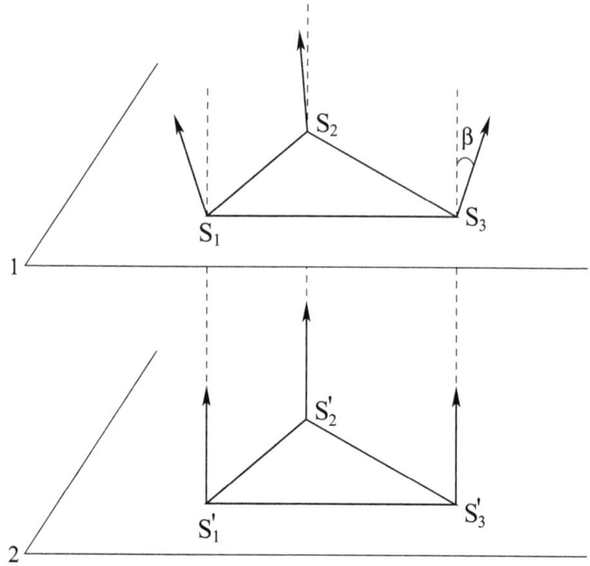

Figure 18.15 Spin configuration of the ground state. The projection on the (xy) plane of the angle between two surface spins is $\alpha = 120°$. The angle between a surface spin and the beneath spin is β.

We can write, in the same manner, the two equations for layers 3 and 4 as

$$(E - E_3)U_3 = - < S_3^z > \left[4\gamma_1(\mathbf{k}_\parallel)(U_2 + U_4) + \epsilon U_1\right]$$
$$(E + E_4)U_4 = - < S_4^z > \left[4\gamma_1(\mathbf{k}_\parallel)U_3 + \epsilon U_2\right]$$

where the notations are defined as

$$E_1 = 4 < S_2^z > -4\epsilon < S_1^z > \left[1 - \gamma_2(\mathbf{k}_\parallel)\right] - \epsilon < S_3^z >$$
$$E_2 = -4(< S_1^z > + < S_3^z >) - 4\epsilon < S_2^z > \left[1 - \gamma_2(\mathbf{k}_\parallel)\right] - \epsilon < S_4^z > \quad (18.163)$$
$$E_3 = 4(< S_2^z > + < S_4^z >) - 4\epsilon < S_3^z > \left[1 - \gamma_2(\mathbf{k}_\parallel)\right] - \epsilon < S_1^z > \quad (18.164)$$
$$E_4 = -4 < S_3^z > -4\epsilon < S_4^z > \left[1 - \gamma_2(\mathbf{k}_\parallel)\right] - \epsilon < S_2^z >$$

Note that $< S_n^z >$ (n: layer n) is positive for $n = 2, 4$ (A sublattice) and negative for $n = 1, 3$ (B sublattice). Other notations in the above equations were defined after (8.6). We can put the above equations under a matrix form

$$\begin{pmatrix} x_{11} & x_{12} & x_{13} & 0 \\ x_{21} & x_{22} & x_{23} & x_{24} \\ x_{31} & x_{32} & x_{33} & x_{34} \\ 0 & x_{42} & x_{43} & x_{44} \end{pmatrix} \begin{pmatrix} U_1 \\ U_2 \\ U_3 \\ U_4 \end{pmatrix} = 0$$

where x represents a non-zero elements. Non-trivial solutions impose that the determinant is zero. Solving numerically det $|\cdots| = 0$ we obtain the energy eigenvalues, and by replacing them into the above matrix equation we get the spin wave amplitudes. If $U_1 = U_2 = U_3 = U_4$, then the corresponding energy is a bulk mode. It is a surface mode otherwise. Note that the two surfaces of the 4-layer film are not symmetric. We should have four distinct modes. For $\epsilon = 0$ and $k_x = k_y = 0$, we have $\gamma_1(\mathbf{k}_\parallel) = \cos(\frac{k_x a}{2})$ $\cos(\frac{k_y a}{2}) = 1$ and $\gamma_2(\mathbf{k}_\parallel) = \frac{1}{2}[\cos(k_x a) + \cos(k_y a)] = 1$. We write for this case the matrix as follows:

$$\begin{pmatrix} x_{11} & x_{12} & 0 & 0 \\ x_{21} & x_{22} & x_{23} & 0 \\ 0 & x_{32} & x_{33} & x_{34} \\ 0 & 0 & x_{43} & x_{44} \end{pmatrix} \begin{pmatrix} U_1 \\ U_2 \\ U_3 \\ U_4 \end{pmatrix} = 0$$

where non-zero elements are

$$x_{11} = E - 4 < S_1^z >, \; x_{12} =< S_1^z >$$
$$x_{21} = 4 < S_2^z >, \; x_{22} = E - 4(< S_1^z > + < S_3^z >),$$
$$x_{23} = 4 < S_2^z >$$
$$x_{32} = 4 < S_3^z >, \; x_{33} = E - 4(< S_2^z > + < S_4^z >),$$
$$x_{34} = 4 < S_3^z >$$
$$x_{43} = < S_4^z >, \; x_{44} = E - 4 < S_3^z >$$

This matrix can be diagonalized without difficulty.
For the case $k_x = k_y = \pi/a$, we proceed in the same manner.

Appendix A

Introduction to Statistical Physics

A.1 Introduction

Statistical physics and quantum physics constitute the foundation of the modern physics. They provide methods to study properties of matter.

Methods of statistical physics allow us to study macroscopic properties of large systems using microscopic mechanisms and structures proposed by quantum mechanics. Thanks to a combination of quantum mechanics and statistical mechanics, we have seen since the second half of the 20-th century spectacular discoveries and progress in modern physics, in particular in the field of condensed matter, which have radically changed our way of life. For the fundamentals of statistical physics and its application to condensed matter, the reader is referred to the book *Statistical Physics: Fundamentals and Application to Condensed Matter* [88].

This Appendix aims at recalling elements of statistical physics which are used throughout this book.

Statistical physics for systems at equilibrium is based on one single postulate called "the fundamental postulate" introduced in the case of an isolated system at equilibrium. The complete properties of an isolated system are deduced from this postulate. Other systems,

Physics of Magnetic Thin Films: Theory and Simulation
Hung T. Diep

ISBN 978-981-4877-42-8 (Hardcover), 978-1-003-12110-7 (eBook)
www.jennystanford.com

not isolated but in some special conditions, can be studied from methods derived from the fundamental postulate. One class of such systems includes systems in contact with a very large heat reservoir: a system of this class has a constant temperature fixed by the heat reservoir. The method used to study this class of systems is called the "canonical description" A second class of systems includes systems in contact with a large reservoir of heat and particles. A system of this class has a constant temperature and a constant chemical potential given by the reservoir. The number of particles of the system is not constant. This number fluctuates around a mean value when the system is at equilibrium. It is an "open system." The method used to study this class of systems is called the "grand-canonical description."

We consider a system of particles. In statistical physics the most fundamental quantity is the statistical entropy defined by

$$S = -k_B \sum_l P_l \ln P_l \tag{A.1}$$

where P_l is the probability of the microscopic state l of the system and k_B the Boltzmann constant. We shall use the statistical entropy to express various physical quantities in the following.

A.2 Isolated Systems: Microcanonical Description

A.2.1 Fundamental Postulate

A system is said "isolated" when it has no interaction with the remaining universe. It is obvious that such a definition is not rigorous: we should understand that interactions are so small that they are not observable and the parameters imposed on the system from the outside world such as energy E, volume V and number of particles N, are constant for all time.

The accessible microscopic states of an isolated system are the states which obey the external constraints. Let Ω be the total number of accessible microscopic states. The fundamental postulate of the statistical mechanics of systems at equilibrium states that

"All accessible microscopic states of an isolated system at equilibrium have the same probability."

According the above postulate, we have

$$P_l = \frac{1}{\Omega} \tag{A.2}$$

for any accessible microscopic state l. The above probability is called "microcanonical probability." Microscopic states which verify this probability constitute a "microcanonical ensemble." The description of properties of a system using the above probability is called "microcanonical description."

Statistical entropy S of an isolated system is thus

$$\begin{aligned} S &= -k_B \sum_l P_l \ln P_l = k_B \sum_l \frac{1}{\Omega} \ln \Omega \\ &= k_B \frac{1}{\Omega} \ln \Omega \sum_l 1 = k_B \ln \Omega \end{aligned} \tag{A.3}$$

Equation (A.3) is called "microcanonical entropy."

The microcanonical temperature T is defined by

$$\frac{1}{T} = \frac{\partial S}{\partial E} \tag{A.4}$$

The microcanonical pressure p is defined by

$$\frac{p}{T} = -\frac{\partial S}{\partial V} \tag{A.5}$$

The microcanonical chemical potential μ is defined by

$$\frac{\mu}{T} = \frac{\partial S}{\partial N} \tag{A.6}$$

We can show that these definitions correspond to physical quantities of the same names in thermodynamics. We can also show that the spontaneous evolution toward equilibrium of an isolated system when an external constraint is removed is always accompanied by an increase of statistical entropy S. Equilibrium is reached when S is maximum (see demonstration in Ref. [88]).

A.2.2 Applications

A.2.2.1 Two-level systems

We consider an isolated system of N independent, discernible particles. We suppose that $N \gg 1$. Each particle has two energy levels ϵ_1 and ϵ_2. The total energy of the system is equal to E. Using the microcanonical description, we calculate the total number of accessible microscopic states $\Omega(E)$, the microcanonical entropy $S(E)$ and the microcanonical temperature T as follows:

- The total number of accessible microscopic states $\Omega(E)$:
  System energy: $E = N_1\epsilon_1 + N_2\epsilon_2$ = constant, N_1=number of particles on ϵ_1, N_2=number of particles on ϵ_2, we have $N = N_1 + N_2$ = constant; hence, $E = N_1\epsilon_1 + (N - N_1)\epsilon_2$. As E is constant, N_1 is thus determined. The total number of accessible microscopic states $\Omega(E)$ is equal to the number of ways to choose N_1 particles among N for the level ϵ_1. Thus, $\Omega = \frac{N!}{N_1!N_2!} = \frac{N!}{N_1!(N-N_1)!}$.
- The microcanonical entropy $S(E)$:
  $S = k_B \ln \Omega = k_B[\ln N! - \ln N_1! - \ln(N - N_1)!$. Using the Stirling formula for $N \gg 1$: $\ln N! \simeq N \ln N - N$, we have $S \simeq k_B[N \ln N - N - N_1 \ln N_1 + N_1 - (N - N_1)\ln(N - N_1) + (N - N_1)] = k_B[N \ln N - N_1 \ln \frac{N_1}{N-N_1} - N \ln(N - N_1)]$
- The microcanonical temperature T as a function of ϵ_1 and ϵ_2:
  $T^{-1} = \frac{\partial k_B \ln \Omega}{\partial E} = \frac{\partial k_B \ln \Omega}{\partial N_1}\frac{\partial N_1}{\partial E} = \frac{1}{\epsilon_1 - \epsilon_2} k_B[-\ln N_1 - 1 + \ln(N - N_1) + 1] = \frac{k_B}{\epsilon_1 - \epsilon_2} \ln \frac{N-N_1}{N_1}$; hence, $\frac{\epsilon_1-\epsilon_2}{k_B T} = \ln \frac{N-N_1}{N_1}$ or $\frac{N_1}{N-N_1} = \exp(-\beta(\epsilon_1 - \epsilon_2)$.
  We obtain the relation of N_1 and N_2 as functions of T, ϵ_1 and ϵ_2: $N_1 = N\frac{\exp(-\beta(\epsilon_1-\epsilon_2)}{1+\exp(-\beta(\epsilon_1-\epsilon_2)} = N\frac{\exp(2\beta\epsilon)}{1+\exp(2\beta\epsilon)}$
- With $\epsilon(>0) \equiv -\epsilon_1 = \epsilon_2$, we have
  At low T, $N_1 \to N$; hence, $N_2 \to 0$. This result is obvious since particles occupy low-energy level at $T = 0$. At high T, $N_1 \to N/2$; hence, $N_2 \to N/2$: particles are equally distributed on the two levels.

A.2.2.2 Classical ideal gas

We consider a classical ideal gas of N particles, of volume V. The gas is isolated with a total energy E. Using the microcanonical description, we calculate

- The microcanonical entropy:
  The classical phase space (see Section A.6 below) has $6N$ dimensions ($3N$ for particle positions and $3N$ for particle

momenta). The number of states of energy $\leq E$ is the number of states in the sphere of radius $p = \sqrt{2mE}$

$$\mathcal{N} = \int_V d\mathbf{r}_1 \int_V d\mathbf{r}_2 \cdots \int_{p_1^2+p_2^2+\cdots+p_N^2=2mE} d\mathbf{p}_1 d\mathbf{p}_2 \cdots$$

$$= V^N \int_{p_1^2+p_2^2+\cdots+p_N^2=2mE} d\mathbf{p}_1 d\mathbf{p}_2 \cdots \tag{A.7}$$

where $\mathbf{p}_i$ is the momentum of the i-th particle. The integration on $\mathbf{p}_i$ gives the volume of the sphere in $3N$ dimensions which is proportional to the radius to the power of $3N$, namely $V^N(2mE)^{3N/2}$ where V^N is the integration over N positions. The number of states of energy equal to E, namely $\Omega(E)$, is the number of states lying on the surface of the sphere of radius $V^N(2mE)^{3N/2}$, we have $\Omega(E) \propto V^N E^{3N/2-1} \simeq V^N E^{3N/2}$ (because $N \gg 1$). The microcanonical entropy is thus $S = A \ln V^N E^{3N/2}$ where A is a constant,

- The microcanonical temperature T:
  We have $T^{-1} = \frac{\partial k_B \ln \Omega}{\partial E} = k_B \frac{3N}{2E}$; hence, $E = \frac{3}{2} N k_B T$, namely the result of classical thermodynamics.
- The microcanonical pressure p:
  We have $\frac{p}{T} = \frac{\partial k_B \ln \Omega}{\partial V} = k_B N/V$; hence, $pV = N k_B T$. This is the equation of state found in thermodynamics using the kinetic theory of gas.

A.3 Systems at Constant Temperature: Canonical Description

When a system is in contact with a heat reservoir much larger than the system, the reservoir imposes on the system its temperature T. The system energy is no more constant, it fluctuates by heat exchange with the reservoir. The equilibrium is reached when the system temperature is equal to that of the reservoir. If we know T, we can calculate the main properties of the system as seen in the following.

The probability of the microscopic state l in the canonical situation is given by

$$P_l = \frac{e^{-\beta E_l}}{Z(T, V, N)} \tag{A.8}$$

where $\beta = \frac{1}{k_B T}$, and

$$Z = \sum_l e^{-\frac{E_l}{k_B T}} \tag{A.9}$$

We call Z the "partition function." This function depends on external variables imposed on the system such as T, V (system volume) and N (number of particles).

The probability (A.8) is called "canonical probability." The ensemble of microscopic states obeying this probability is called "canonical ensemble." The description of properties of the system using (A.8) is called "canonical description." We see below that we can express various physical quantities in terms of Z:

- *Average energy and heat capacity:*
  The average energy $\overline{E}$ of the system is

$$\begin{aligned}\overline{E} &= \sum_l E_l P_l \\ &= \sum_l \frac{E_l e^{-\frac{E_l}{k_B T}}}{Z} \\ &= -\frac{1}{Z}\frac{\partial}{\partial \beta}\sum_l e^{-\beta E_l} = -\frac{1}{Z}\frac{\partial Z}{\partial \beta} \\ &= -\frac{\partial \ln Z}{\partial \beta}\end{aligned} \tag{A.10}$$

  The heat capacity is

$$\begin{aligned}C_V &= \frac{d\overline{E}}{dT} = \frac{d}{dT}\left[-\frac{\partial \ln Z}{\partial \beta}\right] \\ &= -\frac{\partial^2 \ln Z}{\partial \beta^2}\left(\frac{d\beta}{dT}\right) \\ &= \frac{1}{k_B T^2}\frac{\partial^2 \ln Z}{\partial \beta^2}\end{aligned} \tag{A.11}$$

- *Canonical entropy:*
  Replacing P_l by (A.8) in (A.1), we have

$$S = -k_B \sum_l \frac{e^{-\beta E_l}}{Z}(-\beta E_l - \ln Z)$$

$$= k_B \left[\beta \overline{E} + \ln Z \sum_l \frac{e^{-\beta E_l}}{Z}\right]$$

$$= \frac{\overline{E}}{T} + k_B \ln Z \tag{A.12}$$

- *Free energy:*
  The free energy F is defined by

$$F = -k_B T \ln Z \tag{A.13}$$

  As Z, F is a function of T, V and N. This definition allows us to write

$$S = \frac{\overline{E}}{T} - \frac{F}{T} \tag{A.14}$$

  or more often,

$$F = \overline{E} - TS \tag{A.15}$$

- *Canonical pressure p:*
  p is defined by

$$p = -\frac{\partial F}{\partial V} \tag{A.16}$$

- *Canonical chemical potential μ:*
  μ is defined by

$$\mu = \frac{\partial F}{\partial N} \tag{A.17}$$

We can show that a system in a canonical situation tends to equilibrium, when an external constraint is removed, in the sense of decreasing F during the spontaneous evolution. The system reaches equilibrium when F is minimum [88].

A.3.1 Applications

A.3.1.1 Two-level systems

We consider the two-level system in A.2.2.1 using the canonical description at T. We calculate the following:

- The partition function Z:
  We have $Z = z^N$ where z is the partition function of one particle, $z = \sum_i \exp(-\beta\epsilon_i) = \exp(-\beta\epsilon_1) + \exp(-\beta\epsilon_2) = 2\cosh(\beta\epsilon)$.
- The average energy $\overline{E}$ and the heat capacity C_V:
  $\overline{E} = -\frac{\partial \ln Z}{\partial \beta} = -N\epsilon\frac{\sinh(\beta\epsilon)}{\cosh(\beta\epsilon)} = -N\epsilon\tanh(\beta\epsilon)$.
  $C_V = \frac{d\overline{E}}{dT} = \frac{N}{k_B T^2}\frac{\epsilon^2}{\cosh^2(\beta\epsilon)}$
- The number of particles in each level:
  We have $\overline{E} = N_1\epsilon_1 + N_2\epsilon_2 = \epsilon(-N_1 + N_2) = \epsilon(-N_1 + N - N_1) = \epsilon(-2N_1 + N)$. Using $\overline{E}$ given above, we have $-N\epsilon\tanh(\beta\epsilon) = \epsilon(N - 2N_1)$; hence, $N_1 = \frac{N}{2}[1 + \tanh(\beta\epsilon)]$.
- At low T, $N_1 \to N$ (because $\tanh(\beta\epsilon) \to 1$); hence, $N_2 \to 0$. At high T, $N_1 \to N/2$ (because $\tanh(\beta\epsilon) \to 0$); hence, $N_2 \to N/2$. We have the same results as by the microcanonical description.

A.3.1.2 Classical ideal gas

We calculate with the canonical description the partition function, the average energy and the pressure as follows:

- The partition function $Z = z^N/N!$ (undiscernible particles):

$$\begin{aligned} z &= \frac{1}{h^3}\int_V d\mathbf{r}\int d\mathbf{p}\exp(-\beta p^2/2m) \\ &= \frac{V}{h^3}\int_{-\infty}^{\infty} dp_x \exp(-\beta p_x^2/2m)\int_{-\infty}^{\infty} dp_y \exp(-\beta p_y^2/2m) \\ &\quad \times \int_{-\infty}^{\infty} dp_z \exp(-\beta p_z x^2/2m) \\ &= \frac{V}{h^3}\left(\sqrt{2\pi m/\beta}\right)^3 \end{aligned} \tag{A.18}$$

  using $\int_{-\infty}^{\infty} du \exp(-au^2) = \sqrt{\pi/a}$.
- $\overline{E} = -\frac{\partial \ln Z}{\partial \beta} = \frac{3}{2\beta}N$; hence, $\overline{E} = \frac{3Nk_BT}{2}$.
- $p = k_B T\frac{\partial \ln Z}{\partial V} = Nk_B T/V$.

We recover here the same result obtained by the microcanonical description given in the previous section.

A.4 Open Systems at Constant Temperature: Grand-Canonical Description

When a system is in contact with a reservoir of heat and particles much larger than the system, the reservoir imposes on the system its temperature T and its chemical potential μ. The system is in the "grand-canonical situation." The energy and the number of particles of the system fluctuate by exchange of heat and particles with the reservoir. The system reaches equilibrium when its temperature is equal to T and its chemical potential equal to μ of the reservoir.

The grand-canonical probability of the microscopic state l is given by

$$P_l = \frac{e^{-\beta(E_l-\mu N_l)}}{\mathcal{Z}} \tag{A.19}$$

where

$$\mathcal{Z} = \sum_l e^{-\beta(E_l-\mu N_l)} \tag{A.20}$$

$\mathcal{Z}$ is called "grand-partition function." The ensemble of microscopic states obeying the probability (A.19) is called "grand-canonical ensemble."

$\mathcal{Z}$ plays an important role in the calculation of principal properties of the system. As $\ln \mathcal{Z}$ appears often in the calculation, we define a new function J, called "grand potential," by

$$J = -k_B T \ln \mathcal{Z} \tag{A.21}$$

We can express the following quantities as functions of J :

- *Average number of particles:*

$$\begin{aligned} \overline{N} = \sum_l N_l P_l &= \frac{1}{\mathcal{Z}}\frac{1}{\beta}\frac{\partial}{\partial\mu}\sum_l e^{-\beta(E_l-\mu N_l)} \\ &= k_B T \frac{\partial \ln \mathcal{Z}}{\partial \mu} \\ &= -\frac{\partial J}{\partial \mu} \end{aligned} \tag{A.22}$$

- *Average energy:*

$$\begin{aligned}\overline{E} - \mu\overline{N} &= \sum_l (E_l - \mu N_l) P_l \\ &= -\frac{1}{\mathcal{Z}}\frac{\partial}{\partial\beta}\sum_l e^{-\beta(E_l - \mu N_l)} \\ &= -\frac{\partial \ln \mathcal{Z}}{\partial \beta} \\ &= \frac{\partial}{\partial\beta}(\beta J) \\ &= J + \beta\frac{\partial J}{\partial\beta} \end{aligned} \tag{A.23}$$

We deduce

$$\begin{aligned}\overline{E} &= J + \beta\frac{\partial J}{\partial\beta} + -\mu\frac{\partial J}{\partial\mu} \\ &= J + \left(\mu\frac{\partial}{\partial\mu} + \beta\frac{\partial}{\partial\beta}\right) J \end{aligned} \tag{A.24}$$

- *Grand-canonical pressure p:*
  p is defined by

$$p = -\frac{\partial J}{\partial V} \tag{A.25}$$

- *Grand-canonical entropy:*
  Using P_l of (A.19), we have

$$\begin{aligned} S &= -k_B \sum_l P_l \ln P_l \\ &= -k_B \frac{1}{\mathcal{Z}}\sum_l \mathrm{e}^{-\beta(E_l - \mu N_l)}[-\beta(E_l - \mu N_l - \ln \mathcal{Z}] \\ &= k_B\beta(\overline{E} - \mu\overline{N}) + k_B \ln \mathcal{Z}) \\ &= \frac{1}{T}(\overline{E} - \mu\overline{N}) + k_B \ln \mathcal{Z}) \\ &= \frac{1}{T}\left(J + \beta\frac{\partial J}{\partial\beta}\right) - \frac{J}{T} \\ &= \frac{1}{k_B T^2}\frac{\partial J}{\partial\beta} \\ &= -\frac{\partial J}{\partial T} \end{aligned} \tag{A.26}$$

We can show that in the spontaneous evolution when an external constraint is removed, the system tends to equilibrium in the sense of decreasing J. The new equilibrium is attained when J is minimum (see demonstration in Ref. [88]).

A.4.1 Applications

We consider again a classical ideal gas studied above by the microcanonical and canonical descriptions. We show below that the grand-canonical description gives the same results. We calculate the grand-partition function, the average number of particles of the system, the energy and the pressure as follows:

- The grand-partition function:
  $\mathcal{Z} = \sum_N e^{N\beta\mu} Z(T, N, V)$ where Z is the partition function. We have $Z = z^N/N! =$; therefore,
  $\mathcal{Z} = \sum_{N=0}^{\infty} e^{N\beta\mu} z^N/N! = \sum_{N=0}^{\infty} (\lambda z)^N/N! = \exp(\lambda z)$ where $\lambda = e^{\beta\mu}$ (fugacity).
- The average number of particles:
  $\bar{N} = \frac{1}{\beta}\frac{\partial \ln \mathcal{Z}}{\partial \mu} = \frac{1}{\beta}\frac{\partial \lambda z}{\partial \mu}$
  $= \frac{1}{\beta} z \frac{\partial e^{\beta\mu}}{\partial \mu} = z e^{\beta\mu} = \lambda z.$
  We have
  $\bar{E} - \mu\bar{N} = -\frac{\partial \ln \mathcal{Z}}{\partial \beta} = -\frac{\partial \lambda z}{\partial \beta}$
  $= -\frac{\partial (e^{\beta\mu} z)}{\partial \beta} = -\mu\lambda z - \lambda\frac{\partial z}{\partial \beta}$
  $= -\mu\bar{N} + \frac{3\bar{N}k_B T}{2}$; hence, $\bar{E} = \frac{3\bar{N}k_B T}{2}$.
- $p = -\frac{\partial J}{\partial V} = k_B T\frac{\partial J}{\partial V} = k_B T\frac{\partial \ln \mathcal{Z}}{\partial V}$
  $= k_B T\lambda\frac{\partial z}{\partial V} = k_B T\lambda z/V$ [see z in (A.18)], hence, $pV = \bar{N}k_B T$.

We obtain thus the same equation of state for the gas as before.

A.5 Fermi–Dirac and Bose–Einstein Statistics

We can use the grand-canonical description to demonstrate the Fermi–Dirac and Bose–Einstein distributions.

We consider a system of identical, independent and indiscernible particles. In such a hypothesis, each particle has the same "list" of individual states. We can define each individual state by the number of particles present in that state. Let k be an individual state of energy ϵ_k and n_k its number of particles. The total energy and the total number of particles in the microscopic state l of the system are

$$E_l = \sum_k n_k \epsilon_k \tag{A.27}$$

$$N_l = \sum_k n_k \tag{A.28}$$

The grand-partition function reads

$$\begin{aligned} \mathcal{Z} &= \sum_l \mathrm{e}^{-\beta(E_l - \mu N_l)} \\ &= \sum_{\{n_k\}} \mathrm{e}^{-\beta \sum_k n_k(\epsilon_k - \mu)} \end{aligned} \tag{A.29}$$

where, for a given distribution of particles $\{n_k\}$ on individual states, we make the sum in the argument of the exponential, then we repeat for another distribution $\{n'_k\}$ until all distributions have been considered. This procedure is equivalent to the sum on the states l. In doing so, we can express $\mathcal{Z}$ as

$$\mathcal{Z} = \prod_k z_k \tag{A.30}$$

where

$$z_k = \sum_{n_k} \mathrm{e}^{-\beta n_k(\epsilon_k - \mu)} \tag{A.31}$$

It is noted that the sum in z_k is performed on all possible values of n_k for the level ϵ_k. We distinguish two cases:

- Bosons (particles of spin 0 or integer):
  In this case, $n_k = 0, 1, 2, \cdots$ (no limit). We have

$$\begin{aligned} z_k^{BE} &= \sum_{n_k=0}^{\infty} \mathrm{e}^{-\beta n_k(\epsilon_k - \mu)} \\ &= \frac{1}{1 - \mathrm{e}^{-\beta(\epsilon_k - \mu)}} \end{aligned} \tag{A.32}$$

- Fermions (particles of spin half-integer):
  In this case, $n_k = 0, 1$. We have

$$\begin{aligned} z_k^{FD} &= \sum_{n_k=0}^{1} \mathrm{e}^{-\beta n_k(\epsilon_k - \mu)} \\ &= 1 + \mathrm{e}^{-\beta(\epsilon_k - \mu)} \end{aligned} \tag{A.33}$$

To calculate the average number of particles $\overline{n}_k$ of the state of energy ϵ_k, we use (A.22), (A.28) and (A.30):

$$\overline{N} = \sum_k \overline{n}_k = -\frac{\partial J}{\partial \mu}$$

$$= k_B T \sum_k \frac{\partial \ln z_k}{\partial \mu} \tag{A.34}$$

We obtain

$$\overline{n}_k = k_B T \frac{\partial \ln z_k}{\partial \mu} \tag{A.35}$$

Using (A.32) and (A.33), we have, for the boson case,

$$f(\epsilon)^{BE} \equiv \overline{n}_k^{BE} = \frac{1}{e^{\beta(\epsilon_k - \mu)} - 1} \tag{A.36}$$

and for the fermion case,

$$f(\epsilon)^{FD} \equiv \overline{n}_k^{FD} = \frac{1}{e^{\beta(\epsilon_k - \mu)} + 1} \tag{A.37}$$

The distributions $f(\epsilon)^{BE}$ and $f(\epsilon)^{FD}$ are called "Bose–Einstein and Fermi–Dirac distributions," respectively.

A.6 Phase Space: Density of States

A.6.1 Definition

The phase space is defined by the number of degrees of freedom which characterize the microscopic states of the system. In a quantum case, each state is defined by some quantum numbers. For example, each of the microscopic states of a free particle in a box is defined by a wave vector **k** which is quantified by the boundary conditions, the state of an electron in an atomic orbital is given by four quantum numbers (n, l, m_l, m_s). In the case of a system of classical particles each of the microscopic states of the system is defined by the momentum and the position of each particle, $\mathbf{p}_i$ and $\mathbf{r}_i$. These variables constitute the phase space. The sum on the microscopic states is taken over all of these variables.

We consider a system of N classical particles in three dimensions. The number of degrees of freedom is $6N$ because each particle is defined by 6 variables: three of $\mathbf{p}_i$ and three of $\mathbf{r}_i$. We attribute a "volume" of $6N$ dimensions for each microscopic state in the phase space. The elementary volume occupied by a particle is chosen as small as allowed by the uncertainty principle of quantum mechanics. The smallest elementary volume occupied by a particle is equal to $(\sqrt{h})^6 = h^3$ where h is the Planck constant. This choice is made to discretize the classical "continuous" phase space in order to count the number of states: it suffices to divide a chosen volume in the phase space by the elementary volume to find the number of states contained in that volume. To find results for classical particles, we let $h \to 0$ at the end of the calculation.

The sum on the microscopic states is written as an integral in the classical phase space as follows:

$$\sum_k \cdots = \frac{1}{h^{3N}} \int d\mathbf{p}_1 \int d\mathbf{p}_2 \cdots \int d\mathbf{p}_N \int d\mathbf{r}_1 \int d\mathbf{r}_2 \cdots \int d\mathbf{r}_N \cdots \tag{A.38}$$

We consider now a quantum system. When the size of the system is large (thermodynamic limit) we can consider the energy as a continuous variable. We can replace the sum on discrete microscopic states $\mathbf{k}$ by an integral on the energy ϵ but we have to take into account the degeneracy of each energy. For a continuous energy, the degeneracy is the density of states. We write

$$\overline{E} = \sum_k \overline{n}_k \epsilon_k = \int_{\epsilon_0}^{\infty} d\epsilon \rho(\epsilon) \overline{n}_k \epsilon \tag{A.39}$$

where ϵ_0 is the lowest energy and $\rho(\epsilon)$ the density of states. According to the studied case, we replace in this integral $\overline{n}_k$ by $\overline{n}_k^{BE}$ or $\overline{n}_k^{FD}$.

A.6.2 Density of States of a Free Particle in Three Dimensions

For a free electron in a box of linear dimension L in three dimensions, the Schrödinger equation with the periodic boundary conditions gives the following solution for the energy:

$$\epsilon(\mathbf{k}) = \frac{\hbar^2 k^2}{2m} \tag{A.40}$$

with

$$k_i = \frac{2\pi n_i}{L} \quad (i = x, y, z) \tag{A.41}$$

where $n_i = 0, \pm 1, \pm 2, \ldots$.

Let $\mathcal{N}(\epsilon)$ be the number of microscopic states of energy $\leq \epsilon$. The number of states between ϵ and $\epsilon + \Delta\epsilon$ is thus $\mathcal{N}(\epsilon + \Delta\epsilon) - \mathcal{N}(\epsilon)$. The density of states $\rho(\epsilon)$ is defined by

$$\rho(\epsilon) = lim_{\Delta\epsilon \to 0} \frac{\mathcal{N}(\epsilon + \Delta\epsilon) - \mathcal{N}(\epsilon)}{\Delta\epsilon} = \frac{d\mathcal{N}}{d\epsilon} \tag{A.42}$$

This is the number of states of energy ϵ, namely the degeneracy in the continuous energy case. We calculate $\mathcal{N}(\epsilon)$: in the phase space defined by (k_x, k_y, k_z) where each state is defined by a point (k_x, k_y, k_z), the volume of each state is $(2\pi/L)^3$ [see Eq. (A.41)]. The number of microscopic states of energy $\leq \epsilon$ is the volume of the sphere of radius $k = \sqrt{\frac{2m\epsilon}{\hbar^2}}$ divided by the volume of a state:

$$\mathcal{N}(\epsilon) = \frac{4\pi}{3} k^3 \frac{L^3}{8\pi^3} = \frac{L^3}{6\pi^2} \left(\frac{2m}{\hbar^2}\right)^{3/2} \epsilon^{3/2} \tag{A.43}$$

We deduce

$$\rho(\epsilon) = \frac{\Omega}{4\pi^2} \left(\frac{2m}{\hbar^2}\right)^{3/2} \epsilon^{1/2} \tag{A.44}$$

where $\Omega = L^3$ the system volume, and m the particle mass. If the particle has a spin s then the density of states is

$$\rho(\epsilon) = (2s + 1) \frac{\Omega}{4\pi^2} \left(\frac{2m}{\hbar^2}\right)^{3/2} \epsilon^{1/2} \tag{A.45}$$

$(2s + 1)$ is the spin degeneracy. For electrons, $s = 1/2$.

A.7 Properties of a Free Fermi Gas at $T = 0$

At $T = 0$, $f = 1$ for $E < \mu_0 = E_F$, and $f = 0$ for $E > \mu_0$. One replaces, therefore, the upper limit in the integrals (A.54) and (A.56) by E_F and one replaces $f(E) = 1/[\exp(\beta(E - \mu)) + 1]$ by 1.

A.7.1 Fermi Energy

Equation (A.54) becomes

$$N = (2S+1)\int_0^{E_F} \rho(E)dE \tag{A.46}$$

Using $\rho(E) = AE^{1/2}$ [see (A.45)], one obtains

$$N = \frac{2}{3}BE_F^{3/2} \tag{A.47}$$

where

$$B = (2S+1)A = (2S+1)\frac{V}{4\pi^2}\left(\frac{2m}{\hbar^2}\right)^{3/2} \tag{A.48}$$

from which,

$$E_F = \frac{\hbar^2}{2m}\left(\frac{6\pi^2}{2S+1}\frac{N}{V}\right)^{2/3} \tag{A.49}$$

using B given by (A.48).

This result shows that the Fermi energy depends on the density of fermions $n = \frac{N}{V}$ of the gas. Since the Fermi energy is the chemical potential at $T = 0$, the chemical potential in general is closely related to the particle density.

A.7.2 Total Average Kinetic Energy

The total average kinetic energy at $T = 0$ is

$$\overline{E}_0 = (2S+1)\int_0^{E_F} E\rho(E)dE \tag{A.50}$$

from which, one has

$$\overline{E}_0 = (2S+1)A\int_0^{E_F} E^{3/2}dE = \frac{2}{5}BE_F^{5/2} \tag{A.51}$$

Using (A.47) one gets

$$\overline{E}_0 = \frac{3}{5}NE_F \tag{A.52}$$

One sees that the energy of a free Fermi gas is not zero $T = 0$, in contrast to the case of a classical ideal gas.

A.8 Properties of a Free Fermi Gas at Low Temperatures

The exact formula for the total number of particles for a large system is:

$$N = (2S+1)\frac{L^3}{(2\pi)^3}\int d\mathbf{k}\frac{1}{e^{\beta(E_k-\mu)}+1} \tag{A.53}$$

or

$$N = (2S+1)\int_0^\infty dE\rho(E)\frac{1}{e^{\beta(E-\mu)}+1} \tag{A.54}$$

The average energy is

$$\overline{E} = (2S+1)\frac{L^3}{(2\pi)^3}\int d\mathbf{k}\frac{E_k}{e^{\beta(E_k-\mu)}+1} \tag{A.55}$$

or

$$\overline{E} = (2S+1)\int_0^\infty dE\rho(E)\frac{E}{e^{\beta(E-\mu)}+1} \tag{A.56}$$

A.8.1 Sommerfeld's Expansion

One considers the following integral:

$$I = \int_0^\infty h(E)f(E)dE \tag{A.57}$$

where $h(E)$ is a function with finite derivative with respect to E at any order. At low temperatures, one can show that this integral can be expanded in powers of T as follows (see Exercise 4 of Chapter 1):

$$\begin{aligned} I = &\int_0^\mu h(E)dE + \frac{\pi^2}{6}(k_BT)^2h^{(1)}(E)|_{E=\mu} \\ &+\frac{7\pi^4}{360}(k_BT)^4h^{(3)}(E)|_{E=\mu}+\cdots \end{aligned} \tag{A.58}$$

where $h^{(n)}(E)|_{E=\mu}$ is the n-th derivative of $h(E)$ with respect to E, taken at $E=\mu$.

A.8.2 Chemical Potential, Average Energy and Calorific Capacity

Using (A.58) for the integral in (A.54) one obtains to the order of T^2:

$$N = \frac{2}{3}B\mu^{3/2} + \frac{\pi^2}{6}(k_B T)^2 \frac{1}{2} B\mu^{-1/2} + O(T^4) \tag{A.59}$$

where B is given by (A.48). Since N does not vary with T, by equalizing (A.47) and (A.59) one obtains

$$\mu = E_F \left[1 - \frac{\pi^2}{12}\left(\frac{k_B T}{E_F}\right)^2 - O(T^4)\right] \tag{A.60}$$

This equation shows that the chemical potential μ is a function of temperature.

In the same manner, using (A.58) for (A.56) with $h(E) = E\rho(E)$, one gets

$$\overline{E} = \overline{E}_0 \left[1 + \frac{5\pi^2}{12}\left(\frac{k_B T}{E_F}\right)^2 + O(T^4)\right] \tag{A.61}$$

where μ has been replaced by (A.60). The average energy thus increases as T^2 at low temperatures.

The calorific capacity of a free Fermi gas at low T is, therefore,

$$C_V = \frac{d\overline{E}}{dT} = \frac{1}{3}\pi^2 \rho(E_F) k_B^2 T = \gamma T \tag{A.62}$$

One sees that C_V is linear in T. Note that for a classical ideal gas $C_V = 3Nk_B/2$, independent of T.

A.9 Free Fermi Gas at the High-Temperature Limit

When T is very high, one shows that the free Fermi gas becomes a classical ideal gas: the quantum nature disappears at high temperatures.

The energy (A.56) becomes at high T

$$\overline{E} \simeq (2S+1)\int_0^\infty dE \rho(E) E \mathrm{e}^{-\beta(E-\mu)} \tag{A.63}$$

where one has neglected 1 in the denominator because $e^{\beta(E-\mu)} \gg 1$ at high T. Replacing $\rho(E)$ by (A.45) and putting $u = \beta E$, one has

$$\begin{aligned}\overline{E} &= (2S+1)A\int_0^\infty dE\,E^{3/2}e^{-\beta(E-\mu)}\\ &= (2S+1)e^{\beta\mu}A\frac{1}{\beta^{5/2}}\int_0^\infty du u^{3/2}e^{-u}\\ &= (2S+1)e^{\beta\mu}A\frac{1}{\beta^{5/2}}\Gamma(5/2)\end{aligned} \tag{A.64}$$

where one has used the definition of the Γ function. Similarly, Eq. (A.54) becomes at high T

$$\begin{aligned}N &\simeq (2S+1)\int_0^\infty dE\rho(E)e^{-\beta(E-\mu)}\\ &= (2S+1)e^{\beta\mu}A\frac{1}{\beta^{3/2}}\int_0^\infty du u^{1/2}e^{-u}\\ &= (2S+1)e^{\beta\mu}A\frac{1}{\beta^{3/2}}\Gamma(3/2)\end{aligned} \tag{A.65}$$

From these two equations, one finds

$$\frac{\overline{E}}{N} = \frac{\Gamma(5/2)}{\beta\Gamma(3/2)} = \frac{3k_BT}{2} \tag{A.66}$$

This is the equation obtained for a classical ideal gas. The free Fermi gas loses, thus, the quantum nature at high temperatures.

Appendix B

Second Quantization

The method of second quantization is very useful in the study of systems of weakly interacting particles. In particular, the second quantization is an efficient tool to describe collective elementary excitations such as phonons and magnons.

B.1 First Quantization: Symmetric and Antisymmetric Wave Functions

One considers a system of N identical, indiscernible particles. The Hamiltonian is written as

$$\mathcal{H} = \sum_i H_i + \frac{1}{2}\sum_{i,j} V(\mathbf{r}_i, \mathbf{r}_j) \tag{B.1}$$

where H_i is a single-particle term such as the kinetic energy of the particle i and $V(\mathbf{r}_i, \mathbf{r}_j)$ the interaction between two particles at $\mathbf{r}_i$ and $\mathbf{r}_j$.

In the first quantization, the postulate on the symmetry of the wave function allows us to distinguish bosons and fermions: For bosons, the permutation of two particles in their states does not change the sign of the corresponding wave function, while

Physics of Magnetic Thin Films: Theory and Simulation
Hung T. Diep

ISBN 978-981-4877-42-8 (Hardcover), 978-1-003-12110-7 (eBook)
www.jennystanford.com

for fermions the permutation does change its sign. One of the consequences of the symmetry postulate is that in the case of bosons an individual state can contain any number of particles while in the case of fermions, each individual state can have zero or one particle (see Appendix A). This is known as the Pauli principle. In the method of second quantization, it is the symmetry of the operators which allows one to distinguish bosons and fermions as seen in the following.

B.2 Second Quantization: Representation of Microstates by Occupation Numbers

Since the particles are identical and indiscernible, one can imagine that they have the same "list" of individual states: Each of them takes one state of the list. A state i is characterized by some physical parameters such as the wave vector and the spin state, $\mathbf{k}_i$ and σ_i ($i = 1, \cdots, N$). This state of the list is occupied by n_i particles. One can define a microstate of the system by the numbers of particles $\{n_i\}$ in the individual states $(\mathbf{k}_i, \sigma_i)$, $i = 1, \cdots, N$. All possible different particle distributions $\{n_i\}$ constitute the ensemble of microstates of the system. One says that the system is defined by a "state vector" given by

$$|\Psi\rangle = |n_1, n_2, \cdots, n_k, \cdots, n_N\rangle \tag{B.2}$$

where n_k is the number of particles occupying the individual state k. This state vector replaces the wave function of the Schrödinger equation. As for wave functions, one imposes that the state vectors form a complete set of orthogonal states. One has

$$\langle\Psi'|\Psi\rangle = \langle n'_1, \cdots, n'_k, \cdots|n_1, \cdots, n_k, \cdots\rangle = \delta_{n'_1,n_1}\cdots\delta_{n'_k,n_k}\cdots \tag{B.3}$$

where $\delta_{n'_k,n_k}$ is the Kronecker symbol.

B.2.1 The Case of Bosons

One introduces the operators a_k and a_k^+ defined by the following relations:

$$a_k^+|\Psi\rangle = \sqrt{n_k+1}|\cdots, n_k+1, \cdots\rangle \tag{B.4}$$

$$a_k|\Psi\rangle = \sqrt{n_k}|\cdots, n_k - 1, \cdots\rangle \tag{B.5}$$

As seen in the kets, operator a_k^+ creates a particle while operator a_k destroys a particle in the state k when they operate on $|\Psi>$. For this reason, they are called creation and annihilation operators, respectively. By the above definitions, one sees that

$$a_k^+ a_k|\Psi\rangle = n_k|\cdots, n_k, \cdots\rangle \tag{B.6}$$

This relation shows that operator $a_k^+ a_k$ has the eigenvalue n_k which is the number of particles in the state k. One calls, therefore, $a_k^+ a_k$ operator "occupation number."

In addition, using (B.4) and (B.5), one gets

$$a_k a_k^+|\Psi\rangle = (n_k + 1)|\cdots, n_k, \cdots\rangle \tag{B.7}$$

Comparing this relation to (B.6), one obtains

$$a_k a_k^+ - a_k^+ a_k = 1 \tag{B.8}$$

Now, if $k \neq k'$, by using (B.3) one has

$$\begin{aligned}\langle\Psi|a_{k'} a_k^+|\Psi\rangle &= \sqrt{(n_k + 1)n_{k'}}\langle\ldots, n_k, n_{k'}, \ldots|\ldots, n_k{+}1, n_{k'}{-}1, \ldots\rangle \\ &= 0 \qquad\qquad (B.9)\\ \langle\Psi|a_k^+ a_{k'}|\Psi\rangle &= \sqrt{(n_k + 1)n_{k'}}\langle\ldots, n_k, n_{k'}, \ldots|\ldots, n_k{+}1, n_{k'}{-}1, \ldots\rangle \\ &= 0 \qquad\qquad (B.10)\end{aligned}$$

Combining with (B.8) one can write

$$[a_{k'}, a_k^+] \equiv a_{k'} a_k^+ - a_k^+ a_{k'} = \delta_{k,k'} \tag{B.11}$$

One can show in the same manner that for arbitrary k and k', one has

$$[a_{k'}, a_k] = 0\ , \quad [a_{k'}^+, a_k^+] = 0 \tag{B.12}$$

Relations (B.11) and (B.12) are called "commutation relations."

Hamiltonian (B.1) in the case of bosons can be written in the second quantization as (see demonstration in Ref. [88])

$$\hat{\mathcal{H}} = \sum_{p,r}\langle r|H|p\rangle a_r^+ a_p + \frac{1}{2}\sum_{pqrs}\langle rs|V|pq\rangle a_r^+ a_s^+ a_p a_q \tag{B.13}$$

where

$$\langle r|H|p\rangle = \int d\mathbf{r}\phi_r^*(\mathbf{r})H(\mathbf{r})\phi_p(\mathbf{r}) \tag{B.14}$$

$$\langle rs|V|pq\rangle = \iint d\mathbf{r}d\mathbf{r}'\phi_r^*(\mathbf{r})\phi_s^*(\mathbf{r}')V(\mathbf{r}, \mathbf{r}')\phi_p(\mathbf{r})\phi_q(\mathbf{r}') \tag{B.15}$$

The wave function $\phi_i(\mathbf{r})$ describes the individual state i of the particle at $\mathbf{r}$. For example, in the case of a plane wave one has $\phi_i(\mathbf{r}) = \exp(i\mathbf{k}_i \cdot \mathbf{r})/\sqrt{\Omega}$ where $\mathbf{k}_i$ is the wave vector and Ω the system volume.

B.2.2 The Case of Fermions

In the case of fermions, one can demonstrate the general Hamiltonian starting from the Schrödinger equation can be written as Eq. (B.26) below (see demonstration in Ref. [88]), using the creation and annihilation operators b_f^+ and b_f defined by

$$b_f|\Psi\rangle = b_f|\cdots n_f \cdots\rangle = (-1)^{[f]} n_f |\cdots n_f - 1 \cdots\rangle \tag{B.16}$$

$$b_f^+|\Psi\rangle = b_f^+|\cdots n_f \cdots\rangle = (-1)^{[f]}(1 - n_f)|\cdots n_f + 1 \cdots\rangle \tag{B.17}$$

where $[f]$ is, by convention, the number of particles occupying the states on the left of the state f in the ket. It is noted that in some books the coefficients in front of the ket of (B.16)–(B.17) are given by $\sqrt{n_f}$ and $\sqrt{1 - n_f}$ instead of n_f and $(1 - n_f)$. However, one can verify that they are equivalent because n_f is 0 or 1. One has

$$\begin{aligned} b_f b_g|\cdots n_f \cdots n_g \cdots\rangle &= (-1)^{[g]} n_g b_f |\cdots n_f, \cdots n_g - 1, \cdots\rangle \\ &= (-1)^{[g]+[f]} n_g n_f |\cdots n_f - 1, \cdots n_g - 1, \cdots\rangle \\ b_g b_f|\cdots n_f \cdots n_g \cdots\rangle &= (-1)^{[f]} n_f b_g |\cdots n_f - 1 \cdots n_g, \cdots\rangle \\ &= (-1)^{[f]+[g]-1} n_f n_g |\cdots n_f - 1, \cdots n_g - 1, \cdots\rangle \end{aligned}$$

from which one has $b_f b_g = -b_g b_f$, or equivalently

$$\left[b_f, b_g\right]_+ \equiv b_f b_g + b_g b_f = 0 \tag{B.18}$$

In the same manner, one obtains for arbitrary f and g

$$\left[b_f^+, b_g^+\right]_+ = 0 \tag{B.19}$$

and

$$\left[b_f^+, b_g\right]_+ = 0 \text{ if } f \neq g \tag{B.20}$$

Now if $f = g$, one has

$$\begin{aligned} b_f^+ b_f|\cdots n_f \cdots\rangle &= (-1)^{[f]} n_f b_f |\cdots n_f - 1 \cdots\rangle \\ &= (-1)^{2[f]} n_f (1 - n_f + 1)|\cdots n_f \cdots\rangle \\ &= n_f(2 - n_f)|\cdots n_f \cdots >= n_f|\cdots n_f \cdots\rangle \end{aligned} \tag{B.21}$$

where in the last line, one has used $n_f(2-n_f)=n_f$ because

$$n_f = \left\{ \begin{array}{l} 0 \Rightarrow n_f(2-n_f)=0 \\ 1 \Rightarrow n_f(2-n_f)=1 \end{array} \right\} \Rightarrow n_f(2-n_f)=n_f \quad \text{(B.22)}$$

One calls $b_f^+ b_f$ operator "occupation number" because its eigenvalue when operating on the ket is n_f. Besides,

$$\begin{aligned} b_f b_f^+ |\cdots n_f \cdots\rangle &= (-1)^{[-f]+[f]}(n_f+1)(1-n_f)|\cdots n_f \cdots\rangle \\ &= (1+n_f)(1-n_f)|\cdots n_f \cdots\rangle \end{aligned} \quad \text{(B.23)}$$

$$\left. \begin{array}{l} \text{If } n_f = 0 \Rightarrow (1+n_f)(1-n_f)=1 \\ \text{If } n_f = 1 \Rightarrow (1+n_f)(1-n_f)=0 \end{array} \right\} \Rightarrow (1+n_f)(1-n_f) = 1-n_f$$

from which

$$b_f b_f^+ |\cdots n_f \cdots >= (1-n_f)|..n_f \cdots > \quad \text{(B.24)}$$

Comparing (B.24) to (B.21), one obtains $b_f b_f^+ = 1 - b_f^+ b_f$, namely $\left[b_f, b_f^+\right]_+ = 1$. Combining with (B.20), one can write

$$\left[b_g, b_f^+\right]_+ = \delta_{g,f} \quad \text{(B.25)}$$

Relations (B.18), (B.19) and (B.25) are called "fermion anticommutation relations."

Hamiltonian (B.1) in the case of fermions is written in the second quantization as (see demonstration in Ref. [88])

$$\hat{\mathcal{H}} = \sum_{i,k} \langle k|H|i\rangle b_k^+ b_i - \frac{1}{2}\sum_{i,j,k,l} \langle kl|V|ij\rangle b_k^+ b_l^+ b_i b_j \quad \text{(B.26)}$$

where

$$\langle k|H|i\rangle = \int d\mathbf{r}\phi_k^*(\mathbf{r})H(\mathbf{r})\phi_i(\mathbf{r}) \quad \text{(B.27)}$$

$$\langle kl|V|ij\rangle = \iint d\mathbf{r}d\mathbf{r}'\phi_k^*(\mathbf{r})\phi_l^*(\mathbf{r}')V(\mathbf{r},\mathbf{r}')\phi_i(\mathbf{r})\phi_j(\mathbf{r}') \quad \text{(B.28)}$$

Due to the anticommutation of the operators, one should respect the order of the operators as well as that of the arguments $\mathbf{r}$ and $\mathbf{r}'$ of ϕ functions in $\langle kl|V|ij\rangle$. A permutation of the operators should obey the anticommutation relations (B.18), (B.19) and (B.25).

In practice, one can use Hamiltonians (B.13) and (B.26) are starting points to study systems of bosons and fermions if one knows $H(\mathbf{r})$ and $V(\mathbf{r},\mathbf{r}')$. These forms of Hamiltonian are very useful in the study of systems of weakly interacting particles [186].

References

1. P. J. Ackerman, R. P. Trivedi, B. Senyuk, J. V. D. Lagemaat and I. I. Smalyukh, *Phys. Rev. E* **90**, 012505 (2014).
2. T. Adams, et al., *Phys. Rev. Lett.* **108**, 237204 (2012).
3. K. Akabli, H. T. Diep and S. Reynal, *J. Phys.: Condens. Matter* **19**, 356204 (2007).
4. K. Akabli and H. T. Diep, *J. Appl. Phys.* **103**, 07F307 (2008).
5. K. Akabli and H. T. Diep, *Phys. Rev. B* **77**, 165433 (2008).
6. K. Akabli, Y. Magnin, M. Oko, I. Harada and H. T. Diep, *Phys. Rev. B* **84**, 024428 (2011).
7. S. Alexander, J. S. Helman and I. Balberg, *Phys. Rev. B* **13**, 304 (1975).
8. J. W. Allen, G. Locovsky and J. C. Mikkelsen, Jr., *Solid State Commun.* **24**, 367 (1977).
9. R. M. C. de Almeida, N. Lemke and I. A. Campbell, *Brazilian J. Physics* **30**, 701 (2000); ibid. *J. Mag. Mag. Mater.* **226–230**, 1296 (2001).
10. D. J. Amit, *Field Theory, Renormalization Group and Critical Phenomena*, World Scientific, Singapore (1984).
11. P. W. Anderson, *Science* **235** (4793), 1196–1198 (1987).
12. S. A. Antonenko and A. I. Sokolov, *Phys. Rev. B* **49**, 15901 (1994).
13. S. S. Aplesnin, et al., *Phys. Solid State* **49**, Number 11, 2080–2085 (2007).
14. C. S. Arnold, D. P. Pappas and A. P. Popov, *Phys. Rev. Lett.* **83**, 3305 (1999).
15. N. W. Ashcroft and N. D. Mermin, *Solid State Physics*, Saunders College, Philadelphia (1976).
16. P. Azaria, H. T. Diep and H. Giacomini, *Phys. Rev. Lett.* **59**, 1629 (1987).
17. P. Azaria, H. T. Diep, H. Giacomini, *Europhys. Lett.* **9**, 755 (1989).
18. M. N. Baibich, et al., *Phys. Rev. Lett.* **61**, 2472 (1988).
19. P. Bak and M. H. Jensen, *J. Phys. C* **13**, L881 (1980).

20. S. Banerjee, J. Rowland, O. Erten and M. Randeria, *Phys. Rev. X* **4**, 031045 (2014).
21. M. N. Barber, in *Phase Transitions and Critical Phenomena*, vol. 8, ed. C. Domb and J. L. Lebowitz, Academic Press, London (1983).
22. A. Barthélémy, et al., *J. Magn. Magn. Mater.* **242–245**, 68 (2002).
23. See review on Oxide Spintronics by Manuel Bibes and Agnès Barthélémy, in a special issue of IEEE Transactions on electron devices on spintronics, *IEEE Trans. Electron. Devices* **54**, 1003 (2007).
24. A. Bauer and C. Pfleiderer, *Phys. Rev. B* **85**, 214418 (2012).
25. R. J. Baxter, *Exactly Solved Models in Statistical Mechanics*, Academic, New York (1982).
26. A. D. Beath and D. H. Ryan, *Phys. Rev. B* **73**, 174416 (2006).
27. B. Berge, H. T. Diep, A. Ghazali, P. Lallemand, *Phys. Rev. B* **34**, 3177 (1986).
28. E. F. Bertaut, *J. Phys. Radium* **22**, 321 (1961).
29. E. F. Bertaut, *J. Applied Phys.* **33**, 1138 (1962).
30. D. Blankschtein, M. Ma, A. Nihat-Berker, A. *Phys. Rev. B* **30**, 1362 (1984).
31. D. Blankschtein, M. Ma, A. Nihat-Berker, G. S. Grest and C. M. Soukoulis, *Phys. Rev. B* **29**, 5250 (1984).
32. A. Billoire and I. A. Campbell, *Phys. Rev. B* **84**, 054442 (2011).
33. K. Binder and D. W. Heermann, *Monte Carlo Simulation in Statistical Physics*, Springer-Verlag, Berlin (1992).
34. K. Binder, D. P. Landau and A. M. Ferrenberg, *Phys. Rev. E* **51**, 2823 (1995).
35. K. Binder, in *Phase Transitions and Critical Phenomena*, vol. 8, ed. C. Domb and J. L. Lebowitz, Academic Press, London (1983).
36. J. A. C. Bland and B. Heinrich (eds.), *Ultrathin Magnetic Structures*, vol. I and II, Springer-Verlag (1994).
37. A. N. Bogdanov and D. A. Yablonskii, *Sov. Phys. JETP* **68**, 101 (1989).
38. A. N. Bogdanov and D. A. Yablonskii, *Zh. Eksp. Teor. Fiz.* **95**, 178–182 (1989).
39. A. N. Bogdanov, U. K. Rößler and A. A. Shestakov, *Phys. Rev. E* **67**, 016602 (2003).
40. A. N. Bogdanov, U. K. Rössler and C. Pfleiderer, *Physica B* **359**, 1162–1164 (2005).
41. A. Bogdanov, A. Hubert, *J. Mag. Mag. Mater.* **138**, 255–269 (1994).

42. A. Bogdanov, U. Rößler, M. Wolf, K.-H. Müller, *Phys. Rev. B* **66**, 214410 (2002).
43. N. N. Bogolyubov and S. V. Tyablikov, *Doklady Akad. Nauk S. S. S. R.* **126**, 53 (1959) [translation: Soviet Phys.-Doklady **4**, 604 (1959)].
44. N. N. Bogoliubov, *Zhurnal Eksperimental'noi i Teoreticheskoi Fiziki* **34**, 58 (1958).
45. E. H. Boubcheur and H. T. Diep, *Phys. Rev. B* **58**, 5163 (1998).
46. E. H. Boucheur, R. Quartu, H. TT. Diep, O. Nagai, *Phys. Rev. B* **58**, 400 (1998).
47. S. T. Bramwell, M. J. P. Gingras and P. C. W. Holdsworth, in *Frustrated Spin Systems*, ed. H. T. Diep, 2nd ed., pp 383–474, and references therein; World Scientific: Singapore (2013).
48. S. Brooks, A. Gelman, S. L. Jones, X.-L. Meng, *Handbook of Markov Chain Monte Carlo*, CRC Press (2011).
49. G. Brown and T. C. Schulhess, *J. Appl. Phys.* **97**, 10E303 (2005).
50. A. Bunker, B. D. Gaulin and C. Kallin, *Phys. Rev.* **B 48**, 15861 (1993).
51. S. Buhrandt and L. Fritz, *Phys. Rev. B* **88**, 195137 (2013).
52. W. Cai, T. C. Lubensky, P. Nelson and T. Powers, *J. Phys. II France* **4**, 931 (1994).
53. T. W. Capehart and M. E. Fisher, *Phys. Rev. B* **13**, 5021 (1976).
54. J. Cardy, *Scaling and Renormalization in Statistical Physics*, Cambridge Lecture Notes in Physics, Cambridge University Press (2000).
55. M. Caselle and M. Hasenbusch, *Nucl.Phys. B* **470**, 435 (1996).
56. C. Cercignani, *The Boltzmann Equation and Its Applications*, Springer-Verlag, New York (1988).
57. A. Chacon, et al., *Phys. Rev. Lett.* **115** 267202 (2015).
58. P. M. Chaikin and T. C. Lubensky, *Principles of Condensed Matter Physics*, Cambridge University Press (1995).
59. S. Chandra, L. K. Malhotra, S. Dhara and A. C. Rastogi, *Phys. Rev. B* **54**, 13694 (1996).
60. J. Chen, et al., *Sci. Rep.* **7**, 7392 (2017).
61. S.-W. Cheong, M. Mostovoy, *Nat. Mat.* **6**, 13 (2007).
62. T. Chikyu and M. Suzuki, *Prog. Theor. Phys.* **78**, 1242 (1987).
63. M.-C. Chung, et al, *Eur. Phys. J. B* **18** 655 (2000).
64. F. Cinti, A. Cuccoli and A. Rettori, *Phys. Rev. B* **78**, 020402(R) (2008).
65. P. P. Craig, et al., *Phys. Rev. Lett.* 19, 1334 (1967).

66. M. Creutz, *Quarks, gluons and lattices*, (Cambridge University Press, Cambridge, 1983.
67. M. Debauche, H. T. Diep, H. Giacomini and P. Azaria, *Phys. Rev. B* **44**, 2369 (1991).
68. M. Debauche and H. T. Diep, *Phys. Rev. B* **46**, 8214 (1992).
69. P.-G. de Gennes and J. Friedel, *J. Phys. Chem. Solids* **4**, 71 (1958).
70. Youjin Deng and Henk W. J. Blöte, *Phys. Rev. E* **68**, 036125 (2003).
71. J. Dho, Y. N. Kim, Y. S. Hwang, J. C. Kim and N. H. Hur, *Appl. Phys. Lett.* **82**, 1434–1436 (2003).
72. H. W. Diehl, in *Phase Transitions and Critical Phenomena*, vol. 10, ed. C. Domb and J. L. Lebowitz, Academic Press, London (1986).
73. H. W. Diehl, *Int. J. Mod. Phys. B* **11**, 3503 (1997).
74. T. Dietl, *Review on Semiconductor Spintronics*,in Lectures Notes, vol. 712, Springer-Verlag, Berlin, pp. 1–46 (2007).
75. H. T. Diep, *Phys. Stat. Solidi (b)*, **103**, 809 (1981).
76. H. T. Diep, M. Debauche and H. Giacomini, *Phys. Rev. B* **43**, 8759(R) (1991).
77. H. T. Diep, M. Debauche and H. Giacomini,*J. Mag. Mag. Mater.* **104–107**, 184 (1992).
78. H. T. Diep, J. C. S. Levy and O. Nagai, *Phys. Stat. Solidi (b)* **93**, 351 (1979).
79. H. T. Diep, *Phys. Rev. B* **43**, 8509 (1991).
80. H. T. Diep, *Phys. Rev. B* **40**, 4818 (1989).
81. H. T. Diep, P. Lallemand and O. Nagai, *J. Phys. C* **18**, 1067 n(1985).
82. H. T. Diep and H. Kawamura, *Phys. Rev. B* **40**, 7019 (1989).
83. H. T. Diep, *Phys. Rev. B* **45**, 2863 (1992), and references therein. See section III. B for references on the phenomenon "order by disorder".
84. H. T. Diep (ed.), *Magnetic Systems with Competing Interactions*, World Scientific, Singapore (1994).
85. H. T. Diep (ed.), *Frustrated Spin Systems*, 2nd ed., World Scientific, Singapore (2013).
86. H. T. Diep and H. Giacomini, in *Frustrated Spin Systems*, ed. H. T. Diep, World Scientific, Singapore, pp. 1–58 (2013).
87. See, for example, H. T. Diep, *Theory of Magnetism: Application to Surface Physics*, World Scientific, Singapore (2014).
88. H. T. Diep, *Statistical Physics: Fundamentals and Application to Condensed Matter*, World Scientific, Singapore (2015).
89. H. T. Diep, *Phys. Rev. B* **91**, 014436 (2015).

90. H. T. Diep, *Phys. Rev. B* **39**, 397 (1989).
91. H. T. Diep, *Phys. Rev. B* **40**, 741 (1989).
92. H. T. Diep, S. El Hog and A. Bailly-Reyre, *AIP Adv.* **8**, 055707 (2018).
93. H. T. Diep, S. El Hog and H. Puszkarski, *AIP Adv.* **8**, 055706 (2018).
94. C. de Dominicis, H. Orland and F. Lainée, *J. Phys. Lett.* **46**, 463 (1985).
95. See the series *Phase Transitions and Critical Phenomena*, ed. C. Domb and J. L. Lebowitz, Academic Press, London.
96. S. Dong, S. Yunoki, X. Zhang, C. Sen, J.-M. Liu and E. Dagotto, *Phys. Rev. B* **82**, 035118 (2010).
97. S. Dong, X. Zhang, R. Yu, J.-M. Liu and E. Dagotto, *Phys. Rev. B* **84**, 155117 (2011).
98. J. Du, D. Li, Y. B. Li, N. K. Sun, J. Li and Z. D. Zhang, *Phys. Rev. B* **76**, 094401 (2007).
99. I. E. Dzyaloshinskii, *Sov. Phys. JETP* **5**, 1259 (1957).
100. C. Ederer and N. A. Spaldin, *Phys. Rev. B* **71**, 060401(R) (2005).
101. J. B. C. Efrem D'Sa, et al., *J. Mag. Mag. Mater.* **285**, 267 (2005).
102. S. El Hog and H. T.Diep, *J. Mag. Mag. Mater.* **400**, 276–281 (2016).
103. S. El Hog, A. Bailly-Reyre, H. T. Diep, *J. Mag. Mag. Mater.* **455**, 32–38 (2018).
104. S. El Hog, H. T. Diep and H. Puszkarski, *J. Phys. Condens. Matter* **29**, 305001 (2017); arXiv:1612.04147, hal-01415320.
105. S. El Hog, H. T. Diep, *J. Mag. Mag. Mater.* **429**, 102 (2017).
106. S. El Hog, F. Kato, H. Koibuchi and H. T. Diep, *J. Mag. Mag. Mater.* **498**, 166095(2020); arXiv:1910.01760.
107. K. Essafi, J. P. Kownacki and D. Mouhanna, *Phys. Rev. E* **89**, 042101(1–5) (2014).
108. F. H. L. Essler, H. Frahm, F. Göhmann, A. Kĺumper and V. E. Korepin, *The One-Dimensional Hubbard Model*, Cambridge University Press, London (2005).
109. M. Ezawa, *Phys. Rev. Lett.* **105**, 197202 (2010).
110. A. M. Ferrenberg and R. H. Swendsen, *Phys. Rev. Let.* **61**, 2635 (1988).
111. A. M. Ferrenberg and R. H. Swendsen, *Phys. Rev. Let.* **63**, 1195 (1989).
112. A. M. Ferrenberg and D. P. Landau, *Phys. Rev. B* **44**, 5081 (1991).
113. A. Fert, V. Cros and J. Sampaio, *Nat. Nanotechnol.* **8**, 152 (2013).
114. A. Fert and I. A. Campbell, *Phys. Rev. Lett.* **21**, 1190 (1968); I. A. Campbell, Phys. Rev. Lett. **24**, 269 (1970).

115. A. L. Fetter and J. D. Walecka, *Quantum Theory of Many-Particle Systems*, McGraw-Hill, New York (1971).

116. A. Finco, P.-J. Hsu, A. Kubetzka, K. von Bergmann and R. Wiesendanger, *Phys. Rev. B* **94**, 214402 (2016).

117. M. E. Fisher and J. S. Langer, *Phys. Rev. Lett.* **20**, 665 (1968).

118. M. E. Fisher and R. J. Burford, *Phys. Rev.* **156**, 583 (1967).

119. D. P. Foster, C. Gérard, I. Puha, *J. Phys. A: Math. Gen.* **34**, 5183 (2001).

120. D. P. Foster, C. Gérard, *Phys. Rev. B* **70**, 014411 (2004).

121. P. Fröbrich, P. J. Jensen and P. J. Kuntz, *Eur. Phys. J. B* **13**, 477 (2000), and references therein.

122. J. Fukuda and S. Žumer, *Nat. Comm.* **2**, 246 (2011).

123. A. Gaff and J. Hijmann, *Physica A* **80**, 149 (1975).

124. B. D. Gaulin and J. S. Gardner, in *Frustrated Spin Systems*, ed. H. T. Diep, 2nd ed., World Scientific, Singapore, pp. 474–507 (2013), and references therein.

125. A. Georges, G. Kotliar, W. Krauth and M. J. Rozenberg, *Rev. Mod. Phys,* **68**, 13 (1996).

126. D. A. Gilbert, et al., *Nat. Comm.*, published October 8, 2015, DOI:10.1038/ncomms9462.

127. G. Gompper and D. M. Kroll, In *Statistical Mechanics of Membranes and Surfaces*, ed. D. Nelson, T. Piran and S. Weinberg, 2nd ed., pp. 359–426, World Scientific (2004).

128. X. Granados, J. Fontcuberta, X. Obradors and J. B. Torrance, *Phys. Rev. B* **46**, 15683 (1992).

129. X. Granados, J. Fontcuberta, X. Obradors, Ll. Manosa and J. B. Torrance, *Phys. Rev. B* **48**, 11666 (1993).

130. E. Granato, J. M. Kosterlitz, J. Lee and M. P. Nightingale, *Phys. Rev. Lett.* **66**, 1090 (1991).

131. R. B. Griffiths and M. Kaufman, *Phys. Rev. B* **26**, 5022 (1982).

132. V. Grolier, J. Ferré, A. Maziewski, E. Stefanowicz and D. Renard, *J. Appl. Phys.* **73**, 5939 (1993).

133. C. Gros, D. Poilblanc, T. M. Rice and F. C. Zhang, *Physica C*, 153–155: 543–548 (1988).

134. P. Grunberg, R. Schreiber, Y. Pang, M. B. Brodsky and H. Sowers, *Phys. Rev. Lett.* **57**, 2442 (1986); G. Binasch, P. Grunberg, F. Saurenbach and W. Zinn, *Phys. Rev. B* **39**, 4828 (1989).

135. U. Güngördü, R. Nepal, O. A. Tretiakov, K. Belashchenko and A. A. Kovalev, *Phys. Rev. B* **93**, 064428 (2016).
136. M. C. Gutzwiller, *Phys. Rev. Lett.* **10**, 159 (1963); M. C. Gutzwiller, *Phys. Rev.* **137**, A1726 (1965).
137. M. V. Gvozdikova and M. E. Zhitomirsky, *JETP Lett.* **81**, 236 (2005).
138. C. Haas, *Phys. Rev.* **168**, 531 (1968).
139. I. Harada and K. Motizuki, *J. Phys. Soc. Jpn* **32**, 927 (1972).
140. A. Haug, *Theoretical Solid State Physics*, vol. II, p. 258, Pergamon Press (1972).
141. S. Hayami, S.-Z. Lin and C. D. Batista, *Phys. Rev. B* **93**, 184413 (2016).
142. X. He, Y. Q. Zhang and Z. D. Zhang, *J. Mater. Sci. Technol.* **27**, 64 (2011).
143. S. He, et al., *Nat. Mater.* **12**, 605 (2013).
144. M. Heide, G. Bihlmayer and S. Blügel, *Phys. Rev. B* **78**, 140403(R) (2008).
145. W. Helfrich, *Z. Naturforsch* **28c**, 693 (1973).
146. C. Henley, *J. Appl. Phys.* **61**, 3962 (1987); *Phys. Rev. Lett.* **62**, 2056 (1989).
147. C. Henley, *Phys. Rev. Lett.* **62**, 2056 (1989).
148. B. Hennion, W. Szuszkiewicz, E. Dynowska, E. Janik, T. Wojtowicz, *Phys. Rev. B* **66**, 224426 (2002).
149. A. Herpin, *Théorie du Magnétisme* (in French), Presse Universitaire de France (1968).
150. J. Heurich, J. König and A. H. MacDonald, *Phys. Rev. B* **68**, 064406 (2003).
151. D.-T. Hoang, Y. Magnin and H. T. Diep, *Mod. Phys. Lett. B* **25**, 937 (2011).
152. D.-T. Hoang and H. T. Diep, *Phys. Rev. E* **85**, 041107 (2012).
153. D.-T. Hoang, M. Kasperski, H. Puszkarski and and H. T. Diep, *J. Phys.: Condens. Matter* **25**, 056006 (2013).
154. P. C. Hohenberg and B. I. Halperin, *Rev. Mod. Phys.* **49**, 435 (1977).
155. T. Horiguchi, *Physica A* **146**, 613–621 (1987).
156. J. Hoshen and R. Kopelman, *Phys. Rev. B* **14**, 3438 (1974).
157. P.-J. Hsu, A. Finco, L. Schmidt, A. Kubetzka, K. von Bergmann and R. Wiesendanger, *Phys. Rev. Lett.* **116**, 017201 (2016).
158. Y. Huang, W. Kang, X. Zhang, Y. Zhou and W. Zhao, *Nanotechnology* **28**, 08LT02 (2017).
159. J. Hubbard, *Proc. Roy. Soc. London A* **276**, 238 (1963); J. Hubbard, *Proc. Roy. Soc. London A* **277**, 237 (1964).

160. Y. Imry and S.-K. Ma, *Phys. Rev. Lett.* **35**, 1399 (1975).
161. M. Inui, et al., *Phys. Rev. B* **37**, 5182 (1988).
162. M. Itakura, *J. Phys. Soc. Jpn* **72**, 74 (2003).
163. J. Iwasaki, M. Mochizuki and N. Nagaosa, *Nat. Commun.* **4**, 1463 (2013).
164. J. P Jamet, et al., *Phys. Rev.* **B** 57, 14320 (1998).
165. W. Jiang, et al., *Nat. Phys.* **13**, 162–169 (2017).
166. F. Jonietz, et al., *Science* **330**, 1648 (2010).
167. L. P. Kadanoff, et al., *Rev. Modern Phys.* **39**, 395 (1967).
168. H. Kadowaki, K. Ubukoshi, K. Hirakawa, J. L. Martinez and G. Shirane, *J. Phys. Soc. Jpn.* **56**, 4027 (1987).
169. A. Kallel, H. Boller and E. F. Bertaut, *J. Phys. Chem. Solids* **35**, 1139 (1974).
170. J. Kanamori, *Prog. Theor. Phys.* (Kyoto) **30**, 275 (1963).
171. W. Kang, Y. Huang, X. Zhang, Y. Zhou and W. Zhao, *Proc. IEEE*, August 2016, DOI: 10.1109/JPROC.2016.2591578.
172. K. Kanki, D. Loison and K.-D. Schotte, *J. Phys. Soc. Jpn.* **75**, 015001 (2006).
173. K. Kano and S. Naya, *Prog. Theor. Phys.* **10**, 158 (1953).
174. Y. Kantor and D. R. Nelson, *Phys. Rev. A* **36**, 4020 (1987).
175. T. A. Kaplan and N. Menyuk, *Phil. Mag.* **87**, 3711–3785 (2007), and references therein.
176. E. A. Karhu, et al., *Phys. Rev. B* **84**, 060404(R) (2011).
177. E. A. Karhu, et al., Chiral modulation and reorientation effects in MnSi thin films, *Phys. Rev. B* **85**, 094429 (2012).
178. T. Kasuya, *Prog. Theor. Phys.* **16**, 58 (1956).
179. M. Kataoka, *Phys. Rev. B* **63**, 134435 (2001).
180. H. Katsura, N. Nagaosa and A. V. Balatsky, *Phys. Rev. Lett.* **95**, 057205 (2005).
181. M. Kaufman, *Phys. Rev.* **36**, 3697 (1987).
182. M. P. Kawatra, J. A. Mydosh and J. I. Budnick, *Phys. Rev. B* **2**, 665 (1970).
183. J. Kim, et al., Voltage controlled propagating spin waves on a perpendicularly magnetized nanowire, arXiv: 1401.6910 (2014).
184. N. Kiselev, Ns. Kiselev, A. N. Bogdanov, R. Schäfer and U. K. Rößler, *J. Phys. D: Applied Phys.* **44**, 392001 (2011).
185. C. Kittel, *Introduction to Solid State Physics*, 8th ed., John Wiley & Sons, New York (2008).

186. C. Kittel, *Quantum Theory of Solids*, John Wiley & Sons, New York (1987).

187. M. V. Kompaniets, A. Kudlisa and A. I. Sokolov, *Nucl. Phys. B* **950**, 114874 (2020); arXiv:1911.01091v3 [cond-mat.stat-mech] 6 Dec 2019.

188. J. Kondo, *Prog. Theor. Phys.* **32**, 37 (1964).

189. L. Kong, L. Zhuang and S. Chu. *IEEE Trans Magn* **33**, 3019 (1997).

190. W. Koshibae and N. Nagaosa, *Sci. Rep.* **7**, 42645(1–10) (2017).

191. J. M. Kosterlitz and D. J. Thouless, *J. Phys. C: Solid State Phys.* **5**, L124–126 (1972); *J. Phys. C: Solid State Phys.* **6**, 1181 (1973).

192. J. M. Kosterlitz, *J. Phys. C* **7**, 1046 (1974).

193. L. Kramer, *Phys. Rev. B* **3**, 3822, (1971).

194. A. Kremen, S. Wissberg, N. Haham, E. Persky, Y. Frenkel and B. Kalisky, *Nano Lett.* **16**, 1626–1630 (2016).

195. H. Y. Kwon, S. P. Kang, Y. Z. Wu and C. Won, *J. Appl. Phys.* **113**, 133911 (2013).

196. H. Y. Kwon, K. M. Bu, Y. Z. Wu and C. Won, *J. Mag. Mag. Mater.* **324**, 2171 (2012).

197. P. Lallemand, H. T. Diep, A. Ghazali and G. Toulouse, *J. Phys. Lett.* (Paris) **46**, L1087 (1985).

198. D. P. Landau, *Phys. Rev. B* **13**, 2997 (1976).

199. D. P. Landau and K. Binder, *A Guide to Monte Carlo Simulations in Statistical Physics*, 3rd ed., Cambridge University Press, London (2009).

200. D. P. Landau and K. Binder, *Phys. Rev. B* **41**, 4786 (1990).

201. D. P. Landau and K. Binder, *Phys. Rev. B* **41**, 4633 (1990).

202. Review by Ph. Lecheminant, in *Frustrated Spin Systems*, ed. H. T. Diep, 2nd ed., World Scientific (2013).

203. S. Lee and K-C. Lee, *Phys. Rev. B* **49**, 15184 (1994).

204. J. Lee, J. M. Kosterlitz and E. Granato, *Phys. Rev. B* **43**, 11531 (1991).

205. S. Legvold, F. H. Spedding, F. Barson and J. F. Elliott, *Rev. Mod. Phys.* **25**, 129 (1953).

206. V. Leiner, D. Labergerie, R. Siebrecht, Ch. Sutter and H. Zabel, *Physica B* **283**, 167 (2000).

207. A. O. Leonov, I. E. Dragumov, U. K. Rößler and A. N. Bogdanov, *Phys. Rev. E* **90**, 042502 (2014).

208. A. I. Levatić, et al., *Sci. Rep.* **6**, 21347 (2016).

209. Y. B. Li, et al., *Phys. Rev. B* **72**, 193308 (2005).

210. S. Li, W. Kang, Y. Huang, X. Zhang, Y. Zhou and W. Zhao, *Nanotechnology* **28**, 31LT01 (2017).
211. E. H. Lieb and F. Y. Wu, *Phys. Rev. Lett.* **20**, 1445 (1967).
212. R. Liebmann, *Statistical Mechanics of Periodic Frustrated Ising Systems*, Lecture Notes in Physics, vol. 251 Springer-Verlag, Berlin (1986).
213. S.-Z. Lin, A. Saxena and C. D. Batista, *Phys. Rev. B* **91**, 224407 (2015).
214. D. Liu, et al., *Nat. Commun.* **3**, 931 (2012); arXiv:1202.5849.
215. C. L. Lu, et al., *Phys. Rev. B* **79**, 245105 (2009).
216. J. W. Lynn, *J. Appl. Phys.* **75**, 6806 (1994). See also the basics of neutron scattering and the flexibility of BT-7 triple-axis spectrometer in the "SPINS Experiment Handout" given at http://www.ncnr.nist.gov/instruments/bt7_new/index.html.
217. S.-K. Ma *Statistical Mechanics*, World Scientific, Singapore (1985).
218. Y. Magnin, K. Akabli, H. T. Diep and I. Harada, *Comput. Mater. Sci.* **49**, S204-S209 (2010).
219. Y. Magnin, K. Akabli and H. T. Diep, Spin resistivity in frustrated antiferromagnets, *Phys. Rev. B* **83**, 144406 (2011).
220. Y. Magnin, D.-T. Hoang and H. T. Diep, *Mod. Phys. Lett. B* **25**, 1029 (2011).
221. Y. Magnin and H. T. Diep, *Phys. Rev. B* **85**, 184413 (2012).
222. N. Majlis, *The Quantum Theory of Magnetism*, World Scientific, Singapore (2000).
223. A. Malakis, S. S. Martinos, I. A. Hadjiagapiou, N. G. Fytas and and P. Kalozoumis, *Phys. Rev. E* **72**, 066120 (2005).
224. S. V. Maleyev, *Phys. Rev. B* **73**, 174402 (2006).
225. A. Malinowski, V. L. Bezusyy, R. Minikayev, P. Dziawa, Y. Syryanyy and M. Sawicki, *Phys. Rev. B* **84**, 024409 (2011).
226. J. Martinez and M. Jalil, *New J. Phys.* **18**, 033008 (2016).
227. F. Matsukura, H. Ohno, A. Shen and Y. Sugawara, *Phys. Rev. B* **57**, R2037 (1998).
228. D. C. Mattis, *The Theory of Magnetism I: Statics and Dynamics*, 2nd ed., Springer-Verlag, Berlin (1988).
229. V. D. Mello, C. V. Chianca, Ana L. Danta and A. S. Carriç, *Phys. Rev. B* **67**, 012401 (2003).
230. E. Meloche, C. M. Pinciuc and M. L. Plumer, *Phys. Rev. B* **74**, 094424 (2006); E. Meloche, M. L. Plumer and C. M. Pinciuc, *Phys. Rev. B* **76**, 214402 (2007).
231. N. D. Mermin and H. Wagner, *Phys. Rev. Lett.* **17**, 1133 (1966).

232. P. Meyer, et al., *Phys. Rev. Lett.* **81**, 5656 (1998).
233. N. Metropolis, A. W. Rosenbluth, M. N. Rosenbluth, A. H. Teller and E. Teller, *J. Chem. Phys.* **21**, 1087 (1953).
234. M. Mézard, M. Parisi and M. Virasoro, *Spin Glass Theory and Beyond—An Introduction to the Replica Method and Its Applications*, World Scientific, Singapore (1986).
235. G. Misguich and C. Lhuillier, in *Frustrated Spin Systems*, ed. H. T. Diep, World Scientific, Singapore (2013).
236. S. R. Mobasser and T. R. Hart, *Proc. SPIE Conf. Ser.* **524**, 137 (1985).
237. J.-H. Moon, et al., *Phys. Rev. B* **88**, 184404 (2013).
238. A. Montorsi, *The Hubbard Model*, World Scientific, Singapore (1992).
239. T. Morita, *J. Phys. A* **19**, 1701 (1987).
240. T. Moriya, *Phys. Rev.* **120**, 91 (1960).
241. M. Mostovoy, *Phys. Rev. Lett.* **96**, 067601 (2006).
242. S. Mühlbauer, et al., *Science* **323**, 915 (2009).
243. W. Munzer, et al., *Phys. Rev. B* **81**, 041203(R) (2010).
244. O. Nagai, T. Horiguchi, S. Miyashita, in Frustrated Spin Systems, 2nd ed., H. T. Diep, ed., World Scientific, Singapore, chapter 4 (2013).
245. T. Nakajima, et al., *Phys. Rev. B* **83**, 220101(R) (2011).
246. D. Nelson, in *Statistical Mechanics of Membranes and Surfaces*, ed. D. Nelson, T. Piran and S. Weinberg, 2nd ed., pp. 1–17, World Scientific (2004).
247. M. E. J. Newman and G. T. Barkema, *Monte Carlo Methods in Statistical Physics*, Clarendon Press, Oxford (2002).
248. V. T. Ngo and H. T. Diep, *Phys. Rev. B* **75**, 035412 (2007), selected for *Vir. J. Nan. Sci. Tech.* **15**, 126 (2007).
249. V. T. Ngo and H. T. Diep, *J. Phys: Condens. Matter.* **19**, 386202 (2007).
250. V. T. Ngo and H. T. Diep, *J. Appl. Phys.* **103**, 07C712 (2007).
251. V. T. Ngo and H. T. Diep, *Phys. Rev. E* **78**, 031119 (2008).
252. See V. T. Ngo, H. V. Nguyen, H. T. Diep and V. L. Nguyen, *Phys. Rev. B.* **69**, 134429 (2004), and references on magnetic multilayers cited therein.
253. V. T. Ngo, D. T. Hoang and H. T. Diep, *J. Phys.: Cond. Matt.* **23**, 226002 (2011).
254. V. T. Ngo, D. Tien Hoang and H. T. Diep, *Phys. Rev. E* **82**, 041123 (2010).
255. V. T. Ngo, D. Tien Hoang and H. T. Diep, *Mod. Phys. Lett. B* **25**, 929–936 (2011).

256. V. T. Ngo, D. T. Hoang, H. T. Diep and I. A. *Mod. Phys. Lett. B* **28**, 1450067 (2014).

257. M. P. Nightingale, E. Granato and J. M. Kosterlitz, *Phys. Rev. B* **52**, 7402 (1995).

258. Y. Nii, A. Kikkawa, Y. Taguchi, Y. Tokura and Y. Iwasa, *Phys. Rev. Lett.* **113**, 267203 (2014).

259. Y. Nii, et al., *Nat. Comm.* **6**, 8539 (2015).

260. H. Noguchi, *J. Phys. Soc. Jpn.* **78**, 041007 (2009).

261. A. T. Ogielski, *Phys. Rev. B* **32**, 7384 (1985).

262. T. Oguchi, H. Nishimori and Y. Taguchi, *J. Phys. Soc. Jpn* **54**, 4494 (1985).

263. T. Okubo, S. Chung and H. Kawamura, *Phys. Rev. Lett.* **108**, 017206 (2012).

264. P. Olsson and S. Teitel, *Phys. Rev. B* **71**, 104423 (2005), and historical references cited therein.

265. L. Onsager, *Phys. Rev.* **65**, 117 (1944).

266. S. S. P. Parkin, M. Hayashi and L. Thomas, *Science* **320**, no. 5873, pp. 190–194 (2008).

267. P. Peczak and D. P. Landau, *J. Appl. Phys.* **67**, 5427 (1990).

268. A. Peles, B. W. Southern, B. Delamotte, D. Mouhanna and M. Tissier, *Phys. Rev. B* **69**, 220408 (2004).

269. A. E. Petrova, E. D. Bauer, V. Krasnorussky and S. M. Stishov, *Phys. Rev. B* **74**, 092401 (2006).

270. X. T. Pham Phu, V. T. Ngo and H. T. Diep, *Surf. Sci.* **603**, 109–116 (2009).

271. X. T. Pham Phu, V. T. Ngo and H. T. Diep, *Phys. Rev. E* **79**, 061106 (2009).

272. M. K. Phani, J. L. Lebowitz and M. H. Kalos, *Phys. Rev. B* **21**, 4027 (1980).

273. J. C. Phillips, *Rep. Prog. Phys.* **96**, 1193 (1996), and references therein.

274. C. Pinettes and H. T. Diep, *J. Appl. Phys.* **83**, 6317 (1998).

275. M. Plihal, D. L. Mills and J. Kirschner, *Phys. Rev. Lett.* **82**, 2579 (1999).

276. M. L. Plumer and M. B. Walker, *J. Phys. C* **14**, 4689 (1981).

277. M. L. Plumer, A. Mailhot, R. Ducharme, A. Caillé and H. T. Diep, *Phys. Rev. B* **47**, 14312 (1993).

278. T. L. Polgreen, *Phys. Rev. B* **29**, 1468 (1984).

279. A. M. Polyakov, *Nucl. Phys. B* **268**, 406 (1986).

280. J. Pommier, H. T. Diep, A. Ghazali and P. Lallemand, *J. Appl. Phys.* **63**, 3036 (1988).

281. H. Puszkarski, *Acta Phys. Pol. A* **38**, 217 (1970); *ibid.* A **38**, 899 (1970).

282. H. Puszkarski and P. E. Wigen, *Phys. Rev. Lett.* **35**, 1017 (1975).
283. A. Pyatakov, et al., *Europhys. Lett.* **93**, 17001 (2011).
284. A. Pyatakov, et al., *Ferroelectrics* **438**, 79–88 (2012).
285. See, for example, R. Quartu and H. T. Diep, *Phys. Rev. B* **55**, 2975 (1997).
286. R. Quartu and H. T. Diep, *J. Magn. Magn. Mater.* **182**, 38 (1998).
287. G. Ramirez-Santiago and J. V. Jose, *Phys. Rev. B* **49**, 9567 (1994).
288. M. Rasetti (ed.), *The Hubbard Model, Recent Results*, World Scientific, Singapore (1991).
289. E. Rastelli, L. Reatto and A. Tassi, *J. Phys. C* **18**, 353 (1985).
290. R. Ritz, et al., *Phys. Rev. B* **87**, 134424 (2013).
291. L. J. Rodrigues, V. D. Mello, D. H. A. L. Anselmo and M. S. Vasconcelos, *J. Mag. Mag. Mater.* **377**, 24 (2014).
292. S. Rohart and A. Thiaville, *Phys. Rev. B* **88**, 184422 (2013).
293. N. Romming, et al., *Science* **341** (6146), 636–639 (2013).
294. H. D. Rosales, D. C. Cabra and P. Pujol, Three-sublattice Skyrmions crystal in the antiferromagnetic triangular lattice, arXiv:1507.05109v1 (2015).
295. A. Rosch, *Proc. Natl. Acad. Sci.* **109**, 8793–8794 (2012).
296. U. K. Rößler, A. N. Bogdanov and C. Pfleiderer, *Nature* (London) **442**, 797 (2006).
297. P. Rujan, *J. Stat. Phys.* **49**, 139 (1987).
298. F. Rousseau, et al., *J. Vac. Sci Technol. B* **13**, 2787 (1995).
299. J. E. Sacco and F. Y. Wu, *J. Phys. A* **8**, 1780 (1975).
300. S. Sachdev and K. Park, *Ann. Phys.* **298**, 58 (2002).
301. S. Sachdev, in *Handbook of Magnetism and Advanced Magnetic Materials*, John Wiley & Sons, Ltd (2007).
302. J. Sampaio, V. Cros, S. Rohart, A. Thiaville and A. Fert, *Nat. Nanotechnol.* **8**, 839 (2013).
303. D. Sander, *Rep. Prog. Phys.* **62**, 809 (1999).
304. C. Santamaria and H. T. Diep, *J. Appl. Phys.* **81**, 5276 (1997).
305. C. Santamaria, R. Quartu and H. T. Diep, *J. Appl. Physics* **84**, 1953 (1998).
306. C. Santamaria and H. T. Diep, *J. Magn. Magn. Mater.* **212**, 23 (2000).
307. C. Santamaria and H. T. Diep, *J. Appl. Phys.* **91**, 6872 (2002).
308. T. S. Santos, S. J. May, J. L. Robertson and A. Bhattacharya, *Phys. Rev. B* **80**, 155114 (2009).
309. P. Schilbe, S. Siebentritt and K. H. Rieder, *Phys. Lett. A* **216**, 20 (1996).

310. A. Schilling, M. Cantoni, J. D. Guo and H. R. Ott, *Nature* **363**, 6424 (1993).
311. B. J. Schulz, K. Binder, M. Müller and D. P. Landau, *Phys. Rev. E* **67**, 067102 (2003).
312. K. Seeger, *Semiconductor Physics*, Springer-Verlag, Berlin (1982).
313. S. Seki, X. Z. Yu, S. Ishiwata and Y. Tokura, *Science* **336**, 198 (2012).
314. S. Seki, et al., *Phys. Rev. B* **96** 220404(R) (2017).
315. S. Seki, S. Ishiwata and Y. Tokura, *Phys. Rev. B* **86**, 060403(R) (2012).
316. I. A. Sergienko and E. Dagotto, *Phys. Rev. B* **73**, 094434 (2006).
317. I. F. Sharafullin, M. Kh. Kharrasov and H. T. Diep, *Phys. Rev. B* **99**, 214420 (2019).
318. I. F. Sharafullin and H. T. Diep, *Symmetry*, **12**, 26 (2020).
319. P. Sharma, *Science* **307** (5709), 531–533 (2005).
320. J. Shen, et al., *Phys. Rev.* **B** 56, 11134 (1997).
321. Y. Shi and J. Wang, *Phys. Rev. B* **97**, 224428 (2018).
322. K. Shibata, et al., *Nat. Nanotech.* **10**, 589 (2015).
323. Y. Shiota, et al., *Appl. Phys. Expr.* **4**, no. 4, article no. 43005 (2011).
324. F. C. Schwerer and L. J. Cuddy, *Phys. Rev.* **2**, 1575 (1970).
325. T. H. R. Skyrme, *Proc. Roy. Soc. A* **260** 127 (1961); *Nucl. Phys.* **31**, 556 (1962).
326. N. A. Spaldin, *Ferrimagnetism. Magnetic Materials: Fundamentals and Applications*, 2nd ed., Cambridge University Press, pp. 113–129 (2010).
327. J. M. Speight, *Phys. Rev D* **55**, 3830 (1997).
328. A. A. Stashkevich, et al., *Phys. Rev. B* **91**, 214409 (2015).
329. J. Stephenson, *J. Math. Phys.* **11**, 420 (1970); *Can. J. Phys.* **48**, 2118 (1970); *Phys. Rev. B* **1**, 4405, (1970).
330. S. M. Stishov, et al., *Phys. Rev. B* **76**, 052405 (2007).
331. L. Čanová, J. Strečka and M. Jaščur, *J. Phys.: Condens. Matter* **18**, 4967–4984 (2006).
332. J. Strečka, L. Čanová, M. Jaščur and M. Hagiwara, *Phys. Rev. B* **78**, 024427 (2008).
333. J. Strečka, O. Rojas and S. M. de Souza, *Phys. Lett. A* **376**,197–202 (2012).
334. D. F. Styer, *Phys. Rev. B* **32**, 393 (1985).
335. I. S. Suzuki and M. Suzuki, *Phys. Rev. B* **78**, 214404 (2008).
336. M. Suzuki, *J. Math. Phys.* **12**, 235 (1971).

337. R. H. Swendsen and J.-S. Wang, *Phys. Rev. Lett.* **58**, 86 (1987).
338. S. M. Sze and Kwok K. Ng, *Physics of Semiconductor Devices*, 3rd ed., John Wiley & Sons (2006).
339. W. Szuszkiewicz, E. Dynowska, B. Witkowska and B. Hennion, *Phys. Rev. B* **73**, 104403 (2006).
340. R. A. Tahir-Kheli and D. Ter Haar, *Phys. Rev.* **127**, 88 (1962).
341. Y. Takano and H. Koibuchi, *Phys. Rev. E*, **95**, 042411(1–11) (2017).
342. A. A. Thiele, *Phys. Rev. Lett.* **30**, 230–233 (1973).
343. M. Tissier, B. Delamotte and D. Mouhana, *Phys. Rev. Lett.* **84**, 5208 (2000).
344. G. Toulouse, *Commun. Phys.* **2**, 115 (1977).
345. See review by E. Y. Tsymbal and D. G. Pettifor, *Solid State Physics*, Academic Press, San Diego, vol. 56, pp. 113–237 (2001).
346. S. V. Tyablikov, *Ukr. Mat. Zh.* **11**, 289 (1959); S. V. Tyablikov, *Methods in the Quantum Theory of Magnetism*, Plenum Press, New York (1967).
347. V. Vaks, A. Larkin and Y. Ovchinnikov, *Sov. Phys. JEPT* **22**, 820 (1966).
348. J. Villain, *J. Phys.* C**10**, 1717 (1977).
349. J. Villain, R. Bideaux, J. P. Carton and R. Conte, *J. Phys.* (Paris) **41**, 1263 (1980).
350. J. Villain, *Phys. Chem. Solids* **11**, 303 (1959).
351. F. Wang and D. P. Landau, *Phys. Rev. Lett.* **86**, 2050 (2001); *Phys. Rev. E* **64**, 056101 (2001).
352. X. F. Wang, *Phys. Rev. Lett.* **102**, 117005 (2009).
353. W. Wang, *Phys. Rev. Lett.* **114**, 087203 (2015).
354. Q.-Y. Wang, et al., *Chin. Phys. Lett.* **29**, 037402 (2012); arXiv:1201.5694.
355. J. Wang, Y. Shi and M. Kamlah, *Phys. Rev. B.* **97**, 024429(1–7) (2018).
356. G. H. Wannier, *Phys. Rev.* **79**, 357 (1950); Erratum: Phys. Rev. B **7**, 5017 (E) (1973).
357. O. Wessely, B. Skubic and L. Nordstrom, Spin-transfer torque in helical spin-density waves, *Phys. Rev. B* **79**, 104433 (2009).
358. J. F. Wheater, *J. Phys. A Math. Gen.* **27**, 3323 (1994).
359. J. C. Wheatley, *Progress in Low Temperature Physics* **VI**, 77 (1970), ed. C. J. Gortor, North-Holland Publishing Co.
360. K. J. Wiese, in *Phase Transitions and Critical Phenomena 19*, ed. C. Domb and J. L. Lebowitz, Academic Press, pp. 253–498 (2000).
361. K. G. Wilson, *Phys. Rev. B* **4**, 3174 (1971).

362. K. G. Wilson and M. E. Fisher, *Phys. Rev. Lett.* **28**, 240 (1972).

363. K. G. Wilson, *Rev. Mod. Phys.* **47**, 773 (1975).

364. U. Wolff, *Phys. Rev. Lett.* **62**, 361 (1989); *Phys. Lett. B* **228**, 379 (1989).

365. J. Wosnitza, R. Deutschmann, H. von Löhneysen and R. K. Kremer, *J. Cond. Matter. Phys.* **6**, 8045 (1994).

366. F. Y. Wu, *Solid Stat. Comm.* **10**, 115–117 (1972).

367. A. L. Wysocki, K. D. Belashchenko, J. P. Velev and M. van Schilfgaarde, *J. Appl. Phys.* **101**, 09G506 (2007).

368. J. Xia, W. Siemons, G. Koster, M. R. Beasley and A. Kapitulnik, *Phys. Rev. B* **79**, R140407 (2009).

369. A. Yoshimori, *J. Phys. Soc. Jpn.* **14**, 807 (1959).

370. X. Z. Yu, et al., *Nat. Mater.* **10**, 106 (2011).

371. X. Z. Yu, *Nature* **465** (7300), 901 (2010).

372. X. Z. Yu, *PNAS* **109**, 8856 (2012).

373. Kh. Zakeri, *Phys. Rev. Lett.* **104**, 137203 (2010)

374. A. Zangwill, *Physics at Surfaces*, Cambridge University Press (1988).

375. G. Zarand, C. P. Moca and B. Janko, *Phys. Rev. Lett.* **94**, 247202 (2005).

376. Y. Q. Zhang, Z. D. Zhang and J. Aarts, *Phys. Rev. B* **79**, 224422 (2009).

377. X. Zhang, M. Ezawa and Y. Zhou, *Sci. Rep.* **5**, 9400 (2015).

378. X. Zhang, Y. Zhou, M. Ezawa, G. P. Zhao and W. Zhao, *Sci. Rep.* **5**, article no. 11369 (2015).

379. Y. Zhou and M. Ezawa, *Nat. Commun.* **5**, article no. 4652 (2014).

380. J. Zinn-Justin, *Quantum Field Theory and Critical Phenomena*, Oxford Unversity Press (2002).

381. V. I. Zverev, A. M. Tishin, A. S. Chernyshov, Ya Mudryk, K. A. Gschneidner Jr and V. K. Pecharsky, *J. Phys.: Condens. Matter* **26**, 066001 (2014).

382. V. I. Zverev, A. M. Tishin, Y. Min, Y. Mudryk, K. A. Gschneidner Jr and V. K. Pecharsky, *J. Phys.: Condens. Matter* **27**, 146002 (2015).

383. D. N. Zubarev, *Usp. Fiz. Nauk* **187**, 71 (1960)[translation: *Soviet Phys.-Uspekhi* **3**, 320 (1960)].

384. I. Žutić and S. Das Sarma, *Spintronics: Fundamentals and Applications*, Reviews of Modern Physics **76** (2), 323 (2004).

Index

For Product Safety Concerns and Information please contact our EU
representative GPSR@taylorandfrancis.com
Taylor & Francis Verlag GmbH, Kaufingerstraße 24, 80331 München, Germany

www.ingramcontent.com/pod-product-compliance
Ingram Content Group UK Ltd.
Pitfield, Milton Keynes, MK11 3LW, UK
UKHW022200190926
475702UK00003B/127

* 9 7 8 9 8 1 4 8 7 7 4 2 8 *